Control Engineering

El-Kébir Boukas

Stochastic Switching Systems

Analysis and Design

Birkhäuser
Boston • Basel • Berlin

El-Kébir Boukas
École Polytechnique de Montréal
Department of Mechanical Engineering
P.O. Box 6079, Station "centre-ville"
Montréal, QC H3C 3A7
Canada

AMS Subject Classification: 93E03, 93E15, 93E20

Library of Congress Cataloging-in-Publication Data
Boukas, El-Kébir.
 Stochastic switching systems : analysis and design / El-Kébir Boukas.
  p. cm. – (Control engineering)
  ISBN 0-8176-3782-6 (alk. paper)
   1. Stochastic systems. 2. Stability–Mathematical models. 3. Linear time invariant
systems. I. Title. II. Control engineering (Birkhäuser)

QA274.B668 2005
629.8'312–dc22   2005045213

ISBN-10 0-8176-3782-6   eISBN 0-8176-4452-0   Printed on acid-free paper.
ISBN-13 978-0-8176-3782-8

©2006 Birkhäuser Boston   *Birkhäuser*

Printed in the United States of America.   (IBT)

9 8 7 6 5 4 3 2 1

www.birkhauser.com

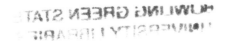

*This book is dedicated to
My wife Saida,
my daughters Imane and Ibtissama,
my son Anas,
and my mother for their love and support*

Contents

Preface

The success of any control system depends on the precision of the model (non-linear or linear) of the plant to be controlled. This model can be obtained using physical laws or identification techniques. Practicing control engineers can use models to synthesize the appropriate controller to guarantee the required performances.

For the nonlinear case, the techniques are few and in general hard to apply. However, if we linearize the nonlinear model to get a linear one in the state-space representation, for instance, we can find in the literature many techniques that can be used to get the controller that guarantees the desired performances. There are now many controllers that we can design for linear systems, such as the famous PID (proportional, integral, and derivative) controller, the $\mathscr{H}_2$ controller, the $\mathscr{H}_\infty$ controller, the state feedback controller, the output feedback controller, and the observer-based output feedback controller. The linear model we will design for our dynamical system will be locally valid and, to prevent performance degradations, uncertainties will be introduced to describe the neglected dynamics or any other phenomena, such as aging.

In the literature we can find different types of uncertainties, among them the norm bounded, the polytopic, and the linear fractional transformation. Nowadays, there are interesting results for the analysis and design of the class of linear systems with or without uncertainties. We are also able to control systems with some special nonlinearities, like saturation, using different types of controllers such as the state feedback controller and the output feedback controller. The last two decades we have brought new control design tools that can be used to design control systems that meet the required specifications.

In practical systems, the state vector is often not available for feedback for practical reasons such as, the nonavailability of the appropriate sensor to measure the components of the state vector or limitations in the budget. Therefore the design of an appropriate filter is required to estimate the state vector that can be used for control purposes. Many techniques can be used to estimate the state vector, including $\mathscr{H}_2$ filtering and $\mathscr{H}_\infty$ filtering.

In practice, some industrial systems, such as those with abrupt changes in their dynamics, can not be appropriately described by the famous linear time-invariant state-space representation. Such systems can be adequately described by the class of stochastic switching systems called piecewise deterministic systems or jump systems, which have two components in the state vector. The first component of this state vector takes values in $\mathbb{R}^n$ and evolves continuously in time, it represents the classical state vector generally used in modern control theory. The second takes values in a finite set and switches in a random manner between the finite number of states. This switching is represented by a continuous-time Markov process taking values in a finite space. The state vector of the class of piecewise deterministic systems is usually denoted by $(x(t), r(t))$. This class of systems has been successfully used to model different practical systems such as manufacturing systems, communications systems, aerospace systems, power systems, and economics systems.

This book gives up-to-date approaches for the analysis and design of control systems for the class of piecewise deterministic systems with or without uncertainties in the system matrices and/or in the transition probability rate matrix. This book can be used as a textbook for graduate-level engineering courses or as a reference for practicing control engineers and researchers in control engineering. Prerequisites to this book are elementary courses on mathematics, matrix theory, probability, optimization techniques, and control system theory.

We are deeply indebted to our colleagues P. Shi, V. Dragan, S. Al-Amer, A. Benzaouia, H. Liu and O. L. V. Costa for reading the manuscript, in full or in part, and making corrections and suggestions. We would also like to thank students J. Raouf and V. Remillard for their help in solving some of the examples in the book.

The draft of this book was completed in April 2004. We added new results that are related to the topics covered by this book as we became aware of them through journals and conference proceedings. However, because of the rapid developments of the subjects, it is possible that we inadvertently omitted some results and references. We apologize to any author or reader who feels that we have not given credit where it is due.

El-Kébir Boukas
Montréal, Canada
April 25th, 2005

1
Introduction

This chapter introduces the class of stochastic switching systems we discuss in this book by giving the motivation for studying it. After giving some practical systems, it also defines the problems we deal with. The contents of this book can be viewed as an extension of the class of linear time-invariant systems studied extensively in the last few decades. As will be shown by some examples, this class of systems is more general since it allows the modeling of systems with some abrupt changes in the state equation that cannot be described using the class of linear time-invariant systems. In this volume we concentrate mainly on the linear case, which has been extensively studied and reported in the literature. References [52, 12, 45, 51] and the references therein are particularly noted. But we would like to advise the reader that nonlinear models have also been introduced; we again refer the reader to [12, 45, 52] and the references therein.

1.1 Overview

Linear time-invariant systems have been and continue to be the engine of control theory development. They have been successfully used to model different industrial systems. Most running industrial plants are designed based on the theory of such a class of systems.

Systems with nonlinear behavior are generally linearized around an operating point; the theory of linear systems is then used for the analysis and design. Sometimes, when the nonlinearities are critical, it is preferable to use a nonlinear model for the analysis and design.

Nowadays there are interesting results on such a class of linear systems that can be used to analyze and design control systems. Among the problems that have been successfully solved are the stability problem, the stabilization problem, the filtering problem, and their robustness. Controllers such as the state feedback and the dynamic output feedback (or the special observer-based output control) are usually used in the stabilization problems. Various design approaches that use the algebraic Riccati equation (ARE), linear matrix inequalities (LMIs), among others, have been developed. For more details on these results, we refer the reader to [31, 41, 59, 66, 12] and the references therein.

In practice, some industrial systems cannot be represented by the class of linear time-invariant model since the behavior of the state equation of these systems is random with some special features. As examples we mention those with abrupt changes and breakdowns of components. Such classes of dynamical systems can be adequately described by the class of stochastic switching systems or the class of piecewise deterministic systems, which is the subject of this book.

If we restrict ourselves to the continuous-time version, this class of systems was introduced by Krasovskii and Lidskii [48]. These two authors built the formalism of this class of systems and studied the optimal control problem. In 1969, Sworder [60] studied the jump linear regulator. In 1971, Wonham [63] extended the formalism of the class of systems to include Gaussian noise in the state equation and studied the stability problem and the jump linear quadratic optimization problem. In 1990, Mariton summarized the established results, including his results and those of other researchers in his book [52]. In 1990, Ji and Chizeck [44] studied the controllability, observability, stability, and stabilizability problems. They also considered the jump linear regulator by developing the coupled set of Riccati equations. In 1993 de Souza and Fragoso [33] studied the $\mathscr{H}_\infty$ control problem. In 1995, Boukas [9] studied the robust stability of this class of systems. In all these contributions, the results are stated in the form of Riccati equations for the optimization problem or Lyapunov equations for the stability problem.

In the last decade, with the introduction of LMIs in control theory, we have seen the use of this technique for some results on the class of piecewise deterministic systems. Most of the problems like stability, stabilization, $\mathscr{H}_\infty$ control, and filtering. have been tackled and LMI results have been reported in the literature.

Among the authors who contributed to the stability problem and/or its robustness are Wonham [63], Ji and Chizeck [44], Feng et al. [40], Boukas [9], Dragan and Morozan [34, 35], Shi et al. [57], Benjelloun and Boukas [6], Boukas and Liu [14, 11, 13, 10], Boukas and Shi [16] , Boukas and Yang [20, 19], Costa and Boukas [25], Costa and Fragoso [28, 27], and Kats and Martynyuk [45]. For more details on the recent review of the literature on this topic, we refer the reader to Boukas and Liu [12], Kats and Martynyuk [45], Mahmoud and Shi [51], and the references therein. The existing results are either in the form of Lyapunov equations or LMIs. The stabilization problem has also attracted many researchers and interesting results have been reported in the literature: Ji and Chizeck [44], Benjelloun et al. [8], Boukas and Liu [13, 15, 11], Boukas et al. [18], Cao and Lam [22], Shi and Boukas [56], de Souza and Fragoso [33], Ait-Rami and El-Ghaoui [1], Bao et al. [5], Dragan and Morozan [34, 35], Ezzine and Karvaoglu [39], Costa et al. [26]. For more details, we refer the reader to

Boukas and Liu [12] and Mahmoud and Shi [51] and the references therein. Among the stabilization techniques that were studied are the state feedback stabilization, output feedback stabilization, $\mathscr{H}_\infty$ state feedback stabilization, and $\mathscr{H}_\infty$ output feedback stabilization. Among the authors who tackled the state feedback stabilization are Ji and Chizeck [44], Ait-Rami and El-Ghaoui [1], Bao et al. [5], Benjelloun et al. [8], Boukas and Liu [15, 11], Boukas et al. [18], Costa and Boukas [25], Dragan and Morozan [34, 35], and the references therein. For the $\mathscr{H}_\infty$ stabilization we quote the work of Aliyu and Boukas [2, 3], Benjelloun et al. [7], Boukas and Liu [10, 13, 14], Boukas and Shi [17], Cao and Lam [22, 23], Cao et al. [24], Costa and Marques [30], Dragan et al. [36], and the references therein. The filtering problem has been studied by Boukas and Liu [11], Costa and Guerra [29], Dufour and Bertrand [37, 38], Liu et al. [50], Shi et al. [58], Wang et al. [62], Xu et al. [65], and the references therein.

Manufacturing systems, power systems, communications systems, and aerospace systems are some applications in which this class of systems has been used successfully to model industrial plants. In manufacturing systems, for instance, piecewise deterministic systems were used to model production planning and/or maintenance planning. Olsder and Suri [54] were the first to use the formalism in manufacturing systems and studied the production planning with failure-prone machines. After 1980, the model was extended by many authors and other optimization problems were considered. Among the authors who contributed to the field are Gershwin and his coauthors, Zhang and his coauthors, and Boukas and his coauthors. The books of Gershwin [42] and Sethi and Zhang [55] and the references therein summarize most of the contributions in this area up to 1994. In this direction of research, most of the authors are interested by developing production and/or maintenance planning. Their methodology, used to develop the production and/or maintenance policies is, in general, dynamic programming and some computation tools.

1.2 State-Space Representation

Mathematically a dynamical system can be interpreted as an operator that maps the inputs to outputs. More specifically, if the system represents an industrial plant P that has as inputs $u(t)$ and $w(t)$ and outputs $y(t)$ and $z(t)$, the relationship between these inputs and outputs is given by the following equation:

$$\begin{bmatrix} y(t) \\ z(t) \end{bmatrix} = P \begin{bmatrix} u(t) \\ w(t) \end{bmatrix}. \tag{1.1}$$

The vectors $y(t)$ and $z(t)$ are referred to, respectively, as the measured output and the controlled output. More often, the measured output $y(t)$ is used to design a control law $u(.)$ that, maps this output to an action that will give to the closed loop system the desired behavior for the controlled output $z(t)$, despite the presence of exogenous input $w(t)$. Mathematically, this is represented by

$$u(t) = K(y(t)). \tag{1.2}$$

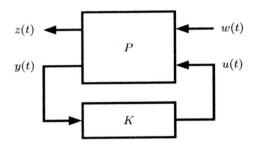

Fig. 1.1. Feedback system block diagram.

Control engineers often represent this operator when controlled in a closed loop by block diagram as illustrated in Figure 1.1. The inputs and the outputs are almost time varying and are linked by the following dynamics:

$$\begin{cases} \dot{x}(t) = f(x(t), u(t), w(t)), \\ y(t) = g(x(t), u(t)), \\ z(t) = h(x(t), u(t)), \end{cases} \tag{1.3}$$

where $x(t) \in \mathbb{R}^n$; $u(t) \in \mathbb{R}^m$; $y(t) \in \mathbb{R}^p$, $z(t) \in \mathbb{R}^q$, and $w(t) \in \mathbb{R}^s$ represent, respectively, the state vector, input vector, measured output vector, controlled output of the system at time t, and exogenous input that has to satisfy some conditions as it will be presented further, $f(.)$, $g(.)$, and $h(.)$ are given smooth vector-valued functions.

Remark 1. In (1.3) the functions $f(.)$, $g(.)$, and $h(.)$ are in general nonlinear in their arguments. The first equation is a differential equation that is referred to as the state equation and the second and the third are pure algebraic equations that represent, respectively, the output equations for $y(t)$ and $z(t)$.

This nonlinear model can always be linearized around the equilibrium point $(0, 0)$, which gives

$$\begin{cases} \dot{x}(t) = Ax(t) + Bu(t) + B_w w(t), \\ y(t) = C_y x(t) + D_y u(t), \\ z(t) = C_z x(t) + D_z u(t), \end{cases} \tag{1.4}$$

where A, B, B_w, C_y, D_y, C_z, and D_z are appropriate constant matrices with appropriate dimensions.

In general, this linearized model will never represent adequately the non-linear dynamical system. The following model is used to take care of the uncertainties that may represent the neglected dynamics, for instance, and the effect of external random disturbances:

$$
\begin{cases}
dx(t) = [A + \Delta A(t)]\, x(t)dt + [B + \Delta B(t)]\, u(t)dt \\
\qquad + B_w w dt + \mathbb{W}_1 x(t) d\omega(t), \\
y(t) = [C_y + \Delta C_y(t)]\, x(t) + [D_y + \Delta D_y(t)]\, u(t) + \mathbb{W}_2 w(t), \\
z(t) = [C_z + \Delta C_z(t)]\, x(t) + [D_z + \Delta D_z(t)]\, u(t),
\end{cases}
\tag{1.5}
$$

where the matrices A, B, B_w, C_y, D_y, C_z, and D_z keep the same meaning; as before; $\Delta A(t)$, $\Delta B(t)$, $\Delta C_y(t)$, $\Delta D_y(t)$, $\Delta C_z(t)$, and $\Delta D_z(t)$ represent, respectively, the uncertainties in the matrices A, B, C_y, D_y, C_z, and D_z; $\mathbb{W}_1$ and $\mathbb{W}_2$ are given matrices with appropriate dimensions; $\omega(t)$ and $w(t)$ are external disturbances that have some properties to be discussed later in this book.

Sometimes systems cannot be put in the previous form for physical reasons. These systems are referred to as singular systems. The state equation of such a class of systems is given by the following:

$$
\begin{cases}
Edx(t) = Ax(t)dt + Bu(t)dt + B_w w(t)dt \\
\qquad + \mathbb{W}_1 x(t) d\omega(t), \\
y(t) = C_y x(t) + D_y u(t) + \mathbb{W}_2 w(t), \\
z(t) = C_z x(t) + D_z u(t),
\end{cases}
\tag{1.6}
$$

where the matrices A, B, B_w, C_y, D_y, C_z, D_z, $\mathbb{W}_1$, and $\mathbb{W}_2$ keep the same meaning as before and E is singular matrix that has a rank equal to n_E, which is less than n (the dimension of the system).

The uncertain model is given in a similar way to the regular one as follows:

$$
\begin{cases}
Edx(t) = [A + \Delta A(t)]\, x(t)dt + [B + \Delta B(t)]\, u(t)dt \\
\qquad + B_w w(t)dt + \mathbb{W}_1 x(t) d\omega(t), \\
y(t) = [C_y + \Delta C_y(t)]\, x(t) + [D_y + \Delta D_y(t)]\, u(t) + \mathbb{W}_2 w(t), \\
z(t) = [C_z + \Delta C_z(t)]\, x(t) + [D_z + \Delta D_z(t)]\, u(t),
\end{cases}
\tag{1.7}
$$

where the different components keep the same meaning as before.

The models (1.4)–(1.7) have been extensively used to describe different type of systems. In the literature, we can find many references that deal with problems like stability, stabilizability, $\mathscr{H}_\infty$ control, filtering, and their robustness. For more information on these topics, we refer the reader to [31, 41, 59, 66, 12] and the references therein. Unfortunately these state equations cannot represent adequately some systems, such as those with abrupt changes. In the next section we will present a model that generalizes this one and that appropriately models the behavior of systems with breakdowns and abrupt changes in their dynamics.

1.3 Stochastic Switching Systems

Let us consider a simple system with the following dynamics:

$$\dot{x}(t) = a(t)x(t) + bu(t), x(0) = x_0, \tag{1.8}$$

where $x(t) \in \mathbb{R}$, $u(t) \in \mathbb{R}$, b is a given constant, and $a(t)$ is a Markov process that switches between two values a_1 and a_2 with the following transition rates matrix:

$$\Lambda = \begin{bmatrix} -p & p \\ q & -q \end{bmatrix},$$

where p and q are positive scalars.

The switches between the two modes are instantaneous and they occur randomly. Based on probability theory, we can find the steady-state probabilities that give how long the process $a(t)$ will spend in mode #1 and in mode #2, respectively. These two probabilities can be computed using the following relations:

$$\begin{bmatrix} \pi_1 & \pi_2 \end{bmatrix} \begin{bmatrix} -p & p \\ q & -q \end{bmatrix} = 0,$$
$$\pi_1 + \pi_2 = 1.$$

The resolution of these equations gives

$$\pi_1 = \frac{q}{p+q},$$
$$\pi_2 = \frac{p}{p+q}.$$

When time t evolves, the state equation of the system will switch in random between the following two dynamics:

$$\dot{x}(t) = a_1 x(t) + bu(t),$$
$$\dot{x}(t) = a_2 x(t) + bu(t).$$

This simple system belongs to the class of stochastic switching systems or piecewise deterministic systems. This class of systems is more general since it can be used to model practical systems with special features like breakdowns or abrupt changes in the parameters.

The question now is how to handle, for instance, the stability of such a system. Also, when the system with some appropriate scalars a_1, a_2, p, and q

is unstable, how can we design the appropriate controller that stochastically stabilizes the system? We can continue our list of problems until it is clear that the theory of LTI systems does not apply and some extensions are needed to handle the new problems raised.

Since the behavior of the system is stochastic, all the concepts should be stochastic. For the stability problem, the concept has been extended and two approaches are available. The first approach is due to Gihman and Skorohod [43]. The second one is due to Kushner [49] and is a direct extension of the Lyapunov approach that we will use extensively in the rest of this volume. Kushner's approach generalizes the Lyapunov approach to handle the stability of the class of systems we are dealing with here.

The class of piecewise deterministic systems is a switching class of systems that has two components in the state vector. The first component takes values in $\mathbb{R}^n$, evolves continuously in time, and represents the classical state vector that is usually used in the modern control theory. The second one takes values in a finite set and switches in a random manner between the finite number of states. This component is represented by a continuous-time Markov process. Usually the state vector of the class of piecewise deterministic systems is denoted by $(x(t), r(t))$. The evolution of this class of systems in time is comprised of two state equations, the switching and the continuous state equation described below.

- Switching: Let $\mathscr{S} = \{1, 2, \cdots, N\}$ be an index set. Let $\{r(t), t \geq 0\}$ be a continuous-time Markov process with right continuous trajectories taking values in $\mathscr{S}$ with the following stationary transition probabilities:

$$
P\left[r(t+h) = j | r(t) = i\right] = \begin{cases} \lambda_{ij} h + o(h), & i \neq j, \\ 1 + \lambda_{ii} h + o(h), & \text{otherwise,} \end{cases} \quad (1.9)
$$

where $h > 0$; $\lim_{h \to 0} \frac{o(h)}{h} = 0$; and $\lambda_{ij} \geq 0$ is the transition probability rate from the mode i to the mode j at time t and $\lambda_{ii} = -\sum_{\substack{j=1, \\ j \neq i}}^{N} \lambda_{ij}$.

- Continuous state equation:

$$
\begin{cases} dx(t) = A(r(t), t)x(t)dt + B(r(t), t)u(t)dt + B_w(r(t))w(t)dt \\ \qquad + \mathbb{W}_1(r(t))x(t)d\omega(t), x(0) = x_0, \\ y(t) = [C_y(r(t)) + \Delta C_y(r(t), t)] x(t) \\ \qquad + [D_y(r(t)) + \Delta D_y(r(t), t)] u(t) + \mathbb{W}_2(r(t))w(t), \\ z(t) = [C_z(r(t)) + \Delta C_z(r(t), t)] x(t) \\ \qquad + [D_z(r(t)) + \Delta D_z(r(t), t)] u(t), \end{cases} \quad (1.10)
$$

where $x(t) \in \mathbb{R}^n$ is the state vector at time t; $u(t) \in \mathbb{R}^p$ is the control at time t; $w(t) \in \mathbb{R}^m$ is an arbitrary external disturbance with norm-bounded energy or bounded average power; $\omega(t) \in \mathbb{R}$ is a standard Wiener process that is supposed to be independent of the Markov process $\{r(t), t \geq 0\}$, $B_w(r(t))$; and $\mathbb{W}_1(r(t))$ and $\mathbb{W}_2(r(t))$ are known matrices; $A(r(t), t)$ and $B(r(t), t)$ are, respectively, the state matrix and the control matrix that are assumed to contain uncertainties and their expressions for every $i \in \mathscr{S}$ are given by

$$A(i,t) = A(i) + D_A(i)F_A(i,t)E_A(i),$$
$$B(i,t) = B(i) + D_B(i)F_B(i,t)E_B(i),$$

with $A(i)$, $D_A(i)$, $E_A(i)$, $B(i)$, $D_B(i)$, and $E_B(i)$ are known matrices; and $F_A(i,t)$ and $F_B(i,t)$ are the uncertainty of the state matrix and the control matrix, respectively, that are assumed to satisfy the following:

$$\begin{cases} F_A^\top(i,t)F_A(i,t) \leq \mathbb{I}, \\ F_B^\top(i,t)F_B(i,t) \leq \mathbb{I}; \end{cases}$$

$C_y(i)$, $D_y(i)$, $C_z(i)$, and $D_z(i)$ are given matrices with appropriate dimensions. The matrices $\Delta C_y(i,t)$, $\Delta D_y(i,t)$, $\Delta C_z(i,t)$, and $\Delta D_z(i,t)$ are given by the following expression:

$$\begin{aligned} \Delta C_y(i,t) &= D_{C_y}(i)F_{C_y}(i,t)E_{C_y}(i), \\ \Delta D_y(i,t) &= D_{D_y}(i)F_{D_y}(i,t)E_{D_y}(i), \\ \Delta C_z(i,t) &= D_{C_z}(i)F_{C_z}(i,t)E_{C_z}(i), \\ \Delta D_z(i,t) &= D_{D_z}(i)F_{D_z}(i,t)E_{D_z}(i), \end{aligned}$$

where $D_{C_y}(i)$, $E_{C_y}(i)$, $D_{D_y}(i)$, $E_{D_y}(i)$, $D_{C_z}(i)$, $E_{C_z}(i)$, and $D_{D_z}(i)$, $E_{D_z}(i)$ are given matrices with appropriate dimensions and $F_{C_y}(i,t)$, $F_{D_y}(i,t)$, $F_{C_z}(i,t)$, and $F_{D_z}(i,t)$ are the uncertainties on the output matrices that are assumed to satisfy the following:

$$\begin{cases} F_{C_y}^\top(i,t)F_{C_y}(i,t) \leq \mathbb{I}, \\ F_{D_y}^\top(i,t)F_{D_y}(i,t) \leq \mathbb{I}, \\ F_{C_z}^\top(i,t)F_{C_z}(i,t) \leq \mathbb{I}, \\ F_{D_z}^\top(i,t)F_{D_z}(i,t) \leq \mathbb{I}. \end{cases}$$

An example of the evolution of the mode $r(t)$ and the state vector $x(t)$ in time t is illustrated in Figure 1.2 when $\mathscr{S} = \{1,2\}$ and $x(t) \in \mathbb{R}$.

Remark 2. It is well known that every sample path of the process $\{r(t), t \geq 0\}$ is a right-continuous step function (see Anderson [4]).

When the Wiener process is not acting on the state equation for all $t \geq 0$, the previous state equation becomes

$$\begin{cases} \dot{x}(t) = A(r(t),t)x(t) + B(r(t),t)u(t) + B_w(r(t))w(t), x(0) = x_0, \\ y(t) = C_y(r(t),t)x(t) + D_y(r(t),t)u(t), \\ z(t) = C_z(r(t),t)x(t) + D_z(r(t),t)u(t). \end{cases} \tag{1.11}$$

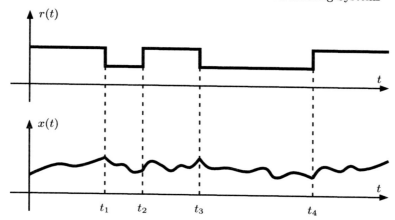

Fig. 1.2. Evolution of the mode $r(t)$ and the state $x(t)$ in time t.

To these dynamics we associate those of the nominal system that supposes all the uncertainties equal to zero, that is,

$$\begin{cases} \dot{x}(t) = A(r(t))x(t) + B(r(t))u(t) + B_w(r(t))w(t), x(0) = x_0, \\ y(t) = C_y(r(t))x(t) + D_y(r(t))u(t), \\ z(t) = C_z(r(t))x(t) + D_z(r(t))u(t). \end{cases} \qquad (1.12)$$

For the singular stochastic switching systems we have the following dynamics:

$$\begin{cases} Edx(t) = A(r(t),t)x(t)dt + B(r(t),t)u(t)dt + B_w(r(t))w(t)dt \\ \qquad + \mathbb{W}_1(r(t))x(t)d\omega(t), x(0) = x_0, \\ y(t) = C_y(r(t),t)x(t) + D_y(r(t),t)u(t) + \mathbb{W}_2(r(t))w(t), \\ z(t) = C_z(r(t),t)x(t) + D_z(r(t),t)u(t), \end{cases} \qquad (1.13)$$

where the matrices keep the same meaning as before and E is a singular matrix.

This book studies the class of piecewise deterministic systems (regular and singular). For the regular class of stochastic switching systems, we focus on the stochastic stability problem and the stabilization problem by using different types of controllers like the state feedback controller, the static output feedback, the dynamic output feedback controller (or the observer-based output feedback control), the $\mathscr{H}_\infty$ control problem, and the filtering problem. For the singular case an introduction to the subject is provided and only the stochastic stability and the stochastic stabilization using state feedback controllers are tackled.

In the next few pages we define each problem. Let us start with the stochastic stability problem. In theory there exist two different concepts of stability, which we refer to as internal stability and input-output stability. Therefore, a linear system is internally stable if the solution of the state vector corresponding to a zero input will converge to zero for any initial conditions. For the input-output stability, a linear system is said to be input-output stable if the state system remains bounded for all bounded inputs. If we denote by $x(t; x_0, r_0)$ the solution of the system (1.12) at time t, the system will be stochastically stable if the system state remains bounded for any initial conditions x_0 and r_0, respectively, of the state vector $x(t)$ and the mode $r(t)$. There are different concepts of stochastic stability and for more details on these concepts, we refer the reader to Chapter 2 as well as the References section.

The stochastic stabilization problem consists of designing a stabilizing controller that forces the system state to be stochastically stable and have the desired behavior. Different approaches of stochastic stabilization exist and can be used to attain the desired goal. Among these techniques are

- state feedback stabilization,
- output feedback stabilization.

These techniques will be discussed in Chapter 3 for nominal and uncertain systems.

Practical systems are always affected by external disturbances that may degrade the system performance. To overcome the negative effects of the external disturbances that are supposed to have finite energy or finite average power, the $\mathscr{H}_\infty$ technique was proposed. Contrary to optimal control, which handles the case of external disturbances that must satisfy some special assumptions, $\mathscr{H}_\infty$ control requires only that the external disturbance have finite energy or finite average power. $\mathscr{H}_\infty$ control is a way minimize the worst-case gain of the system. This optimization problem can be stated as a game optimization problem with two players; the designer, who is seeking a controller that minimizes the gain, and nature, which seeks an external disturbance that maximizes the gain.

The goal of $\mathscr{H}_\infty$ control is to seek a controller (state feedback, dynamic output feedback, static output feedback) that minimizes the $\mathscr{H}_\infty$ norm of the system closed-loop transfer matrix between the controlled output $z(t)$ and the external disturbance $w(t)$ that belongs to $\mathscr{L}_2[0, T]$, with $\Delta A(r(t), t) = \Delta B(r(t), t) \equiv 0$, that is,

$$\|G_{zw}\|_\infty = \sup_{\|w(t)\|_{2,[0,T]} \neq 0} \frac{\|z(t)\|_{2,[0,T]}}{\|w(t)\|_{2,[0,T]}}, \tag{1.14}$$

where $\|G_{zw}\|$ is the transfer matrix between the output $z(t)$ and the external disturbance $w(t)$.

The $\mathscr{H}_\infty$ control problem can be defined on either finite or infinite horizon $(T \to \infty)$. In the rest of this section, we develop the finite horizon case. To get the infinite horizon case, we let T go to infinity with the appropriate assumptions.

The $\mathcal{H}_\infty$-norm cost function (1.14) is not acceptable as an objective function since this cost depends on the controller; that is, the supremum makes this function independent of a particular disturbance input.

A quadratic objective function that yields tractable solutions of the differential game is referred to as a suboptimal solution to the $\mathcal{H}_\infty$ optimization control problem. It can be obtained by considering the following bound on the closed-loop $\mathcal{H}_\infty$ norm:

$$\|G_{zw}\|_\infty = \sup_{\|w(t)\|_{2,[0,T]}\neq 0} \frac{\|z(t)\|_{2,[0,T]}}{\|w(t)\|_{2,[0,T]}} < \gamma,$$

where γ is referred to as the performance bound.

This suboptimal controller must also satisfy the following bound:

$$\|G_{zw}\|_\infty^2 = \sup_{\|w(t)\|_{2,[0,T]}\neq 0} \frac{\|z(t)\|_{2,[0,T]}^2}{\|w(t)\|_{2,[0,T]}^2} < \gamma^2. \tag{1.15}$$

To make the supremum satisfy this inequality, the following should hold:

$$\frac{\|z(t)\|_{2,[0,T]}^2}{\|w(t)\|_{2,[0,T]}^2} \leq \gamma^2 - \varepsilon^2, \tag{1.16}$$

which gives

$$\|z(t)\|_{2,[0,T]}^2 - \gamma^2\|w(t)\|_{2,[0,T]}^2 \leq -\varepsilon^2\|w(t)\|_{2,[0,T]}^2. \tag{1.17}$$

Note that the satisfaction of this inequality for all disturbance inputs and some ε is equivalent to the bound on the closed-loop $\mathcal{H}_\infty$ norm (1.16). Therefore, the left-hand side of (1.17) can be used as an objective function of our $\mathcal{H}_\infty$ optimization control problem. The optimization problem we should solve is given by

$$\min_{u(.)} \mathbb{E}\left[\int_0^T \left[z^\top(t)z(t) - \gamma^2 w^\top(t)w(t)\right] dt\right],$$

subject to the autonomous state equation (1.12).

When the uncertainties are present in the dynamics, the robust $\mathcal{H}_\infty$ control consists of making the gain from the exogenous $w(t)$ to the controlled output $z(t)$, ($\mathcal{L}_2$-gain) less than or equal to $\gamma > 0$, that is,

$$\int_0^T \|z(t)\|^2 dt \leq \gamma^2 \int_0^T \|w(t)\|^2 dt,$$

for all $T > 0$ and for all admissible uncertainties. Note that T can be chosen to be infinite.

Mathematically the robust $\mathscr{H}_\infty$ control problem can be stated as follows. Given a positive γ, find a controller that robustly stabilizes the system and guarantees

$$\sup_{w(.)\in\mathscr{L}_2[0,\infty]} \frac{\|z_t\|_2^2}{\|w_t\|_2^2} \leq \gamma^2$$

for all admissible uncertainties. More details on this subject can be found in Chapter 4.

The state vector most often is not accessible for feedback for many practical and technological reasons, which limits the use of state feedback control. To overcome this, either we use output feedback control or get an estimate of the state vector using the filtering techniques and then use state feedback control. The filtering problem consists of determining an estimate $\hat{x}(t)$ of the state vector $x(t)$ or $\hat{z}(t)$ of the controlled output $z(t)$ using the measurement of the output $y(t)$. There are many techniques that can be used to accomplish such an estimation depending on the structure of the studied systems. Among these techniques are

- Kalman filtering,
- $\mathscr{H}_2$ filtering,
- $\mathscr{H}_\infty$ filtering.

The different filtering techniques will be covered in Chapter 5. In the rest of this section, we restrict ourselves to $\mathscr{H}_\infty$ filtering and determine how the filtering problem can be solved using $\mathscr{H}_\infty$ control theory. For this purpose, let the control $u(t) \equiv 0$ for all $t \geq 0$. Notice that this is not a restriction since if the control is not equal to zero, the way to handle this case is similar to the one we develop here. $\mathscr{H}_\infty$ filtering can be stated as follows. Given a nominal dynamical systems with exogenous input that can be deterministic but not known and measured output, design a filter to estimate an unmeasured output such that the mapping from the exogenous input to the estimation error is minimized or no larger than some prescribed level in terms of the $\mathscr{H}_\infty$ norm. Mathematically, the $\mathscr{H}_\infty$ filtering problem is stated as follows. Given $\gamma > 0$, find a filter such that

$$\sup_{w\in\mathscr{L}_2[0,\infty]} \frac{\|z(t) - \hat{z}(t)\|_2^2}{\|w(t)\|_2^2} < \gamma^2$$

holds for all $w(t) \in \mathscr{L}_2[0,\infty]$.

The filtering problem can be regarded as a special $\mathscr{H}_\infty$ control problem that keeps the $\mathscr{H}_\infty$ norm of the system transfer matrix between the estimation error and the exogenous disturbance less than a given positive constant γ.

The design of a linear time-invariant filter of order n for system (1.12) with the following form:

$$\begin{cases} \dot{x}_c(t) = K_A x_c(t) + K_B y(t), x_c(0) = 0, \\ \hat{z}(t) = K_C x_c(t), \end{cases} \tag{1.18}$$

is brought to an $\mathcal{H}_\infty$ control problem that can make the extended system $\{(x(t), e(t)), t \geq 0\}$ asymptotically stable when t goes to infinity and the estimation error $e(t) = z(t) - \hat{z}(t)$ satisfies the following condition:

$$\|e(t)\|_2 \leq \gamma \|w(t)\|_2. \tag{1.19}$$

When the uncertainties are acting on the system dynamics, the robust $\mathcal{H}_\infty$ filtering can be treated in the same way as the robust $\mathcal{H}_\infty$ control problem. It consists of making the extended system $\{(x(t), e(t)), t \geq 0\}$ asymptotically stable when t goes to infinity and the estimation error $e(t) = z(t) - \hat{z}(t)$ satisfies (1.19) for all admissible uncertainties.

1.4 Practical Examples

As a first example, let us consider a production system with failure-prone machines. For simplicity of presentation, let us assume that the production system consists of one machine and produces one part type. Let us also assume that the machine state is described by a continuous-time Markov process $\{r(t), t \geq 0\}$ with finite state space $\mathscr{S} = \{0, 1\}$. $r(t) = 0$ means that the machine is under repair and no part can be produced. $r(t) = 1$ means that the machine is operational and can produce parts. The switching between these two states is described by the following probability transitions:

$$Pr[r(t + \Delta t) = j | r(t) = i] = \begin{cases} \lambda_{ij} \Delta t + o(\Delta t), & \text{if } i \neq j, \\ 1 + \lambda_{ii} \Delta t + o(\Delta t), & \text{otherwise,} \end{cases}$$

with $\lambda_{ij} \geq 0$ when $i \neq j$, $\lambda_{ii} = -\sum_{j \neq i} \lambda_{ij}$, and $\lim_{\Delta t \to 0} \frac{o(\Delta t)}{\Delta t} = 0$.

Let $x(t)$ denote the stock level of the production system at time t. When $x(t) \geq 0$, it represents a surplus and when $x(t) < 0$ it represents the backlog. Let us assume that the produced parts deteriorate with time with a constant rate ρ. Let $u(t)$ and $d(t)$ represent, respectively, the production rate and the constant demand rate at time t.

The stock level is then described by the following dynamics:

$$\dot{x}(t) = -\rho x(t) + b(r(t))u(t) - d(t),$$

with

$$b(r(t)) = \begin{cases} 1, & \text{if } r(t) = 1, \\ 0, & \text{otherwise.} \end{cases}$$

When the stock level $x(t)$ is negative there is no deterioration of the stock.

The production rate $u(t)$ is assumed to satisfy the following constraints:

$$0 \leq u(t) \leq \bar{u},$$

with $\bar{u}$ a given positive constant. If we assume that the demand rate d is given, one of the problems that we can solve is determining the production rate $u(.)$ such that the stock level remains always close to zero.

The previous model can be extended to handle the more general cases of a production system producing p parts. The Markov process in this case represents the different state that the system production can occupy and that belongs to a finite set $\mathscr{S} = \{1, 2, \cdots, N\}$. The stock level $x(t)$ for this system is described by the following system of differential equations:

$$\dot{x}(t) = Ax(t) + B(r(t))u(t) - \mathbb{I}d(t),$$

where

$$\begin{aligned} x(t) &\in \mathbb{R}^p, u(t) \in \mathbb{R}^p, \\ A &= \text{diag}\left[-\rho_1, \cdots, -\rho_p\right], \\ B &= \text{diag}\left[b_1(r(t)), \cdots, b_p(r(t))\right], \\ \mathbb{I} &= \text{diag}\left[1, \cdots, 1\right], \end{aligned}$$

with ρ_i, $i = 1, \cdots, p$ is the deterioration rate of the part i and $b_j(r(t))$ is defined as previously. Similarly, if we fix the demand rate, how can we keep up production to ensure that the stock level $x(t)$ is always close to zero?

Our second practical example is borrowed from the aerospace industry. It consists of a VTOL (vertical take-off and landing) helicopter. The corresponding model presented here was developed by Narendra and Tripathi [53] and was used recently by de Farias et al. [32]. This system was also used by Kose et al. [47] in the deterministic framework. Let $x_1(t)$, $x_2(t)$, $x_3(t)$, and $x_4(t)$ denote, respectively, the horizontal velocity, vertical velocity, pitch rate, and pitch angle at time t. The evolution of the state vector with time can be described by the following system of differential equations:

$$\begin{cases} \dot{x}(t) = A(r(t))x(t) + B(r(t))u(t) + B_w w(t), \\ y(t) = C_y x(t) + D_y u(t), \\ z(t) = C_z x(t) + D_z w(t), \end{cases}$$

where $x(t)$, $y(t)$, and $z(t)$ represent, respectively, the state vector, measured output, and controller output; $w(t)$ is the external disturbance that is supposed to have finite energy. The different matrices in this model are assumed to be given by

$$A(r(t)) = \begin{bmatrix} -0.04 & 0.04 & 0.02 & -0.5 \\ 0.05 & -1.01 & 0.0 & -4.0 \\ 0.1 & a_{32}(r(t)) & -0.71 & a_{34}(r(t)) \\ 0.0 & 0.0 & 1.0 & 0 \end{bmatrix},$$

$$B(r(t)) = \begin{bmatrix} 0.44 & 0.18 \\ b_{21}(r(t)) & -7.60 \\ -5.52 & 4.49 \\ 0.0 & 0.0 \end{bmatrix}, \quad B_w = \begin{bmatrix} 0.1 & 0.0 \\ 0.0 & 0.1 \\ 0.1 & 0.0 \\ 0.0 & 0.1 \end{bmatrix},$$

$$B_z = \begin{bmatrix} 0.0\ 0.1 \\ 0.1\ 0.0 \end{bmatrix}, \qquad B_y = \begin{bmatrix} 0.0\ 0.1 \\ 0.1\ 0.0 \end{bmatrix}, \quad C_y = \begin{bmatrix} 1\ 0\ 1\ 0 \\ 0\ 1\ 0\ 1 \end{bmatrix},$$

$$D_y = \begin{bmatrix} 0\ 0.1\ 0\ 0 \\ 0\ 0\ 0.1\ 0 \end{bmatrix}, \quad C_z = \begin{bmatrix} 0\ 1\ 0\ 1 \\ 1\ 0\ 1\ 0 \end{bmatrix}, \quad D_z = \begin{bmatrix} 0.1\ 0.0 \\ 0.0\ 0.1 \end{bmatrix}.$$

The parameter $r(t)$ is modeled by a continuous-time Markov process taking values in a finite space $\mathscr{S} = \{1, 2, 3\}$ and describing the changes in the air speed, which we assume to have instantaneous switches. These modes correspond, respectively, to 135, 60, and 170 knots. The speed 135 corresponds to the nominal value. The switching between these modes is assumed to be described by the transition matrix that belongs to the polytope represented by

$$\Lambda_1 = \begin{bmatrix} -2.09 & 1.07 & 1.02 \\ 0.07 & -0.07 & 0.0 \\ 0.02 & 0.0 & -0.02 \end{bmatrix},$$

$$\Lambda_2 = \begin{bmatrix} -0.02 & 0.01 & 0.01 \\ 0.001 & -0.001 & 0.0 \\ 0.09 & 0.0 & -0.09 \end{bmatrix},$$

$$\Lambda_3 = \begin{bmatrix} -0.05 & 0.002 & 0.048 \\ 0.02 & -0.02 & 0.0 \\ 0.09 & 0.0 & -0.09 \end{bmatrix}.$$

The global transition matrix Λ is described by

$$\Lambda = \sum_{j=1}^{3} \alpha_j \Lambda_j,$$

with $\alpha_1 = 0.80$, $\alpha_2 = 0.1$, and $\alpha_3 = 0.1$.

The parameters $a_{32}(r(t))$, $a_{34}(r(t))$, and $b_{21}(r(t))$ are supposed to take the values of Table 1.1. This model will be used in different chapters to show the effectiveness of the developed results.

Airspeed (knots)	$a_{32}(r(t))$	$a_{34}(r(t))$	$b_{21}(r(t))$
135	0.37	1.42	3.55
60	0.07	0.12	1.0
170	0.51	2.52	5.11

Table 1.1. Parameters of the state and the input matrices.

As a third example, we consider an electrical circuit that can give us a stochastic singular system, illustrated by Figure 1.3. We assume that the position of the switch follows a continuous-time Markov process $\{r(t), t \geq 0\}$

with three states, $\mathscr{S} = \{1,2,3\}$. This Markov process is the consequence of a random request that may result from the choice of an operator. Let us assume that at time t, the Markov process $r(t)$ occupies the state 2, $r(t) = 2$. Letting the electrical current in the circuit be denoted by $i(t)$ and using the basic electrical circuits laws, we get

$$u(t) = v_R(t) + v_L(t) + v_C(t),$$
$$v_R(t) = Ri(t),$$
$$v_L(t) = L\frac{di(t)}{dt},$$
$$a(r(t))i(t) = \frac{dv_C(t)}{dt},$$

with

$$a(r(t)) = \begin{cases} \frac{1}{C_1} & \text{if } r(t) = 1, \\ \frac{1}{C_2} & \text{if } r(t) = 2, \\ \frac{1}{C_3} & \text{if } r(t) = 3. \end{cases}$$

These equations can be rewritten in matrix form as follows:

$$\begin{bmatrix} L & 0 & 0 & 0 \\ 0 & 0 & 1 & 0 \\ 0 & 0 & 0 & 0 \\ 0 & 0 & 0 & 0 \end{bmatrix} \begin{bmatrix} \frac{di(t)}{dt} \\ \frac{dv_L(t)}{dt} \\ \frac{dv_C(t)}{dt} \\ \frac{dv_R(t)}{dt} \end{bmatrix} = \begin{bmatrix} 0 & 1 & 0 & 0 \\ a(r(t)) & 0 & 0 & 0 \\ -R & 0 & 0 & 1 \\ 0 & 1 & 1 & 1 \end{bmatrix} \begin{bmatrix} i(t) \\ v_L(t) \\ v_C(t) \\ v_R(t) \end{bmatrix}$$
$$+ \begin{bmatrix} 0 \\ 0 \\ 0 \\ -1 \end{bmatrix} u(t).$$

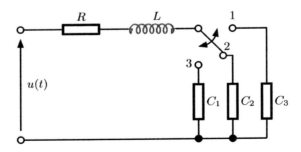

Fig. 1.3. Electrical circuit.

If we choose $x^\top(t) = [i(t), v_R(t) + v_L(t), v_R(t) + v_L(t) + v_C(t), v_R(t)]$, we get the following equivalent state representation:

$$\begin{bmatrix} L & 0 & 0 & 0 \\ L & 1 & -1 & 0 \\ 0 & -1 & 1 & 0 \\ L & 0 & 0 & 0 \end{bmatrix} \dot{x}(t) = \begin{bmatrix} 0 & 1 & 1 & -1 \\ -a(r(t)) & 1 & 1 & -1 \\ a(r(t)) - R & 0 & 0 & 1 \\ -R & 1 & 0 & 0 \end{bmatrix} x(t)$$

$$+ \begin{bmatrix} -1 \\ -1 \\ 0 \\ 0 \end{bmatrix} u(t),$$

which gives the following model:

$$E\dot{x}(t) = A(r(t))x(t) + Bu(t),$$

where

$$E = \begin{bmatrix} L & 0 & 0 & 0 \\ L & 1 & -1 & 0 \\ 0 & -1 & 1 & 0 \\ L & 0 & 0 & 0 \end{bmatrix}, \quad A(r(t)) = \begin{bmatrix} 0 & 1 & 1 & -1 \\ -a(r(t)) & 1 & 1 & -1 \\ a(r(t)) - R & 0 & 0 & 1 \\ -R & 1 & 0 & 0 \end{bmatrix},$$

$$B = \begin{bmatrix} -1 \\ -1 \\ 0 \\ 0 \end{bmatrix}.$$

This model gives the form of stochastic singular systems that we will treat in Chapter 6.

The three models in this section present the framework of the class of systems we will be considering in this volume. The second model will be used in the rest of this book to show the effectiveness of the proposed results in each chapter and the last example will be used in Chapter 6.

1.5 Organization of the Book

The rest of the book is organized as follows. Chapter 2 treats the stochastic stability problem and its robustness. Different concepts of stochastic stability are studied. The uncertainties are supposed to affect the system state matrix and/or the transition probability rate matrix. Most of the developed results are in the form of LMIs, which makes them solvable using the existing tools in the marketplace.

Chapter 3 deals with the stabilization problem and its robustness. Different types of controllers such as state feedback and output feedback, including the special case referred to as observer-based output control, are considered and design procedures for each controller are developed. LMI synthesis methods are developed for each controller. Many numerical examples are worked out to show the usefulness of the developed results.

In Chapter 4 we consider systems with external disturbances that can be arbitrary with the only restriction being bounded energy or bounded average power instead of Gaussian as it is the case for the linear quadratic Gaussian regulator. Different types of controllers that stochastically stabilize the class of systems that we are considering in this book, and assure the γ-disturbance rejection, are discussed. LMI design approaches are developed for this purpose.

Chapter 5 considers the filtering problem of the class of piecewise deterministic systems and its robustness. Kalman filtering and $\mathcal{H}_\infty$ filtering problems are treated and LMI conditions are developed to synthesize the gains of these filters.

In Chapter 6, the singular class of piecewise deterministic systems is introduced and some appropriate tools for the analysis and synthesis of this class of systems are developed. More specifically the stochastic stability and stochastic stabilization problems and their robustness are treated. LMI results are developed.

In all the chapters numerical examples are given to show the usefulness of each result. We include simple examples to allow the reader to follow the steps of the development.

1.6 Notation and Abbreviations

The notation used in this book is quite standard in control theory. The study of dynamical piecewise deterministic systems is mainly based on state-space representation; therefore, we will extensively use vectors and matrices in all our developments. Vectors and matrices are represented by lower and upper letters, respectively. The matrices in our volume will always depend on the system mode. When the mode occupies the state, $r(t) = i$, and any matrix $A(r(t))$ will be written as $A(i)$. The following table summarizes most of the symbols used in this book.

Symbol	Meaning
$\mathbb{R}^n$	Set of real n-dimensional vectors
$\mathbb{R}^{n \times m}$	Set of real $n \times m$ matrices
$\mathbb{N}$	Set of natural numbers
$\mathscr{S}$	Mode state space, $\mathscr{S} = \{1, 2, \cdots, N\}$, where N is a positive natural number
$x(t)$	State vector at time t, $x(t) \in \mathbb{R}^n$
$y(t)$	Measured output vector at time t, $y(t) \in \mathbb{R}^q$
$z(t)$	Controlled output vector at time t, $z(t) \in \mathbb{R}^p$
$u(t)$	Control input vector at time t, $u(t) \in \mathbb{R}^m$
$w(t)$	Exogenous input that has finite energy
$\omega(t)$	External disturbance vector at time t, $\omega(t) \in \mathbb{R}$ (Wiener process)

Symbol	Meaning
$\mathbb{I}$	Identity matrix (the size can be obtained from the context)
$\{r(t), t \geq 0\}$	Markov process describing the system mode at time t
$\mathbb{P}[.]$	Probability measure
$\Lambda\,[\lambda_{ij}]$	Transition probability rates matrix between the different modes of the system that belongs to $\mathscr{S}$
λ_{ij}	Jump rate from mode i to mode j, $(i, j \in \mathscr{S})$
$A(i)$	State matrix, $A \in \mathbb{R}^{n \times n}$
$\Delta A(i, t)$	Uncertainty on the matrix $A(i) \in \mathbb{R}^{n \times n}$
$B(i)$	Input matrix $B(i) \in \mathbb{R}^{n \times m}$
$F(i, t)$	Norm-bounded uncertainty
$\|.\|_2$	2-norm (vector, signal, or system)
$\|.\|_\infty$	∞-norm (vector, signal, or system)
$\mathbb{E}\,[.]$	Mathematical expectation
$\mathscr{L}_2\,[0, \infty[$	Set of integrable functions on $[0, \infty[$, $\mathscr{L}_2[0, \infty[\triangleq \{f(t) \mid \int_0^\infty f^\top(s)f(s)ds < \infty\}$
$\mathrm{diag}\,[A_1, \cdots, A_n]$	Real matrix with diagonal elements $A_1, \cdots, A_n$
$\mathrm{tr}\,(A)$	Trace of the square matrix A, $A \in \mathbb{R}^{n \times n}$
$A^\top$	Transpose of the real matrix A, $A \in \mathbb{R}^{n \times m}$
$\mathcal{S}_i(X)$	$\mathcal{S}_i(X) = \left[\sqrt{\lambda_{i1}}X_i, \sqrt{\lambda_{i2}}X_i, \cdots, \sqrt{\lambda_{ii-1}}X_i, \right.$ $\left. \sqrt{\lambda_{ii+1}}X_i, \cdots, \sqrt{\lambda_{iN}}X_i\right]$
$\mathcal{X}_i(X)$	$\mathcal{X}_i(X) = \mathrm{diag}\,[X_1, \cdots, X_{i-1}, X_{i+1}, \cdots, X_N]$
$\lambda_{max}(A)$	Maximum eigenvalue of the square matrix A
$\lambda_{min}(A)$	Minimum eigenvalue of the square matrix A
$V(x(t), i)$	Lyapunov candidate function that depends on the State vector $x(t)$ and the mode i
$\mathscr{L}V(.,.)$	Infinitesimal operator of the Lyapunov function
$P_i > 0$	Symmetric and positive-definite matrix
$P_i \geq 0$	Symmetric and semi-positive-definite matrix
$P_i < 0$	Symmetric and negative-definite matrix
$P_i \leq 0$	Symmetric and semi-negative-definite matrix
$Q > 0$	Set of symmetric and positive-definite matrix $(Q = (Q_1, \cdots, Q_N) > 0$, each component $Q_i > 0$, $i = 1, \cdots, N)$
$Q \geq 0$	Set of symmetric and semi-positive definite matrix $(Q = (Q_1, \cdots, Q_N) \geq 0$, each component $Q_i \geq 0$, $i = 1, \cdots, N)$
$Q < 0$	Set of symmetric and negative definite matrix $(Q = (Q_1, \cdots, Q_N) > 0$, each component $Q_i > 0$, $i = 1, \cdots, N)$

Symbol	Meaning
$Q \leq 0$	Set of symmetric and semi-negative definite matrix $(Q = (Q_1, \cdots, Q_N) \geq 0$, each component $Q_i \leq 0$, $i = 1, \cdots, N)$
$X > Y$	Means that $X - Y$ is symmetric and positive-definite matrix for symmetric and positive-definite matrices X and Y
$P^{-\top}(i)$	Inverse and the transpose of the nonsingular matrix $P(i)$
SS	Stochastically stable
MSQS	Mean square quadratically stable
LMI	Linear matrix inequality
GEVP	Generalized eigenvalue problem

2

Stability Problem

Consider a linear time-invariant system with the following dynamics:

$$\begin{cases} \dot{x}(t) = Ax(t), \\ x(0) = x_0, \end{cases}$$

where $x(t) \in \mathbb{R}^n$ is the state vector at time t, $x_0 \in \mathbb{R}^n$ is the initial state, and A is a constant known matrix with appropriate dimension. The stability of this class of systems has been extensively studied and many interesting results can be used to check the stability of a given system of this class. Lyapunov equations or equivalent LMI conditions are often used to check stability.

Since stability is the first requirement in any design specification, it continues to attract many researchers from the control and mathematics communities. For the class of systems we are considering in this book, the developed results for the class of linear time-invariant systems cannot be used directly and must be adapted to take care of the stochastic effect in the dynamic. For piecewise deterministic systems, the concept of stochastic stability is used since the state equations of these systems are stochastic. The stochastic stability problem is, in some sense, more complicated compared to the one of the deterministic dynamical systems. As we will see later in this chapter, the stability of the system in each mode does not imply the stochastic stability of the whole system. The reverse is also not true. Our goal in this chapter is to develop LMI-based stability conditions for the class of piecewise deterministic systems.

The rest of this chapter is organized as follows. In Section 2.1 the stochastic stability problem is stated and some useful definitions are given. Section 2.2 presents the results of stability. Both Lyapunov equations and LMI results are given. In Section 2.3 the robust stochastic stability is studied and LMI results are developed. Section 2.4 covers the stochastic stability and its robustness for the class of piecewise deterministic systems with Wiener process. Most of the results are illustrated by numerical examples to show the effectiveness of the results.

2.1 Problem Statement

Let us consider a dynamical system defined in a probability space $(\Omega, \mathcal{F}, \mathbb{P})$ and assume that its state equation is described by the following differential equation:

$$\begin{cases} \dot{x}(t) = A(r(t), t)x(t), \\ x(0) = x_0, \end{cases} \tag{2.1}$$

where $x(t) \in \mathbb{R}^n$ is the state vector; $x_0 \in \mathbb{R}^n$ is the initial state; $\{r(t), t \geq 0\}$ is a continuous-time Markov process taking values in a finite space $\mathscr{S} = \{1, 2, \cdots, N\}$ and describing the evolution of the mode at time t; $A(r(t), t) \in \mathbb{R}^{n \times n}$ is a matrix with an appropriate dimension that is supposed to have the following form for every $i \in \mathscr{S}$:

$$A(i, t) = A(i) + D_A(i)F_A(i, t)E_A(i), \tag{2.2}$$

with $A(i)$, $D_A(i)$, and $E_A(i)$ are real known matrices; and $F_A(i, t)$ is an unknown real matrix that satisfies the following:

$$F_A^\top(i, t)F_A(i, t) \leq \mathbb{I}. \tag{2.3}$$

The Markov process $\{r(t), t \geq 0\}$ with values in the finite set $\mathscr{S}$, describes the switching between the different modes and its evolution is governed by the following probability transitions:

$$\mathbb{P}\left[r(t+h) = j | r(t) = i\right]$$
$$= \begin{cases} \lambda_{ij}h + o(h), & \text{when } r(t) \text{ jumps from } i \text{ to } j, \\ 1 + \lambda_{ii}h + o(h), & \text{otherwise,} \end{cases} \tag{2.4}$$

where λ_{ij} is the transition rate from mode i to mode j with $\lambda_{ij} \geq 0$ when $i \neq j$ and $\lambda_{ii} = -\sum_{j=1, j \neq i}^N \lambda_{ij}$ and $o(h)$ is such that $\lim_{h \to 0} \frac{o(h)}{h} = 0$.

Notice that it is difficult to get the exact transition probability rate matrix, $\Lambda = [\lambda_{ij}]$, of the Markov process we are using to describe the switching between the different modes of the system. Therefore, uncertainties should sometimes be included in the model to correct the used dynamics that describe the switching. The analysis and the design approaches we will use should take care of these uncertainties. In the rest of this chapter we will assume sometimes that the matrix Λ belongs to a polytope, that is,

$$\Lambda = \sum_{k=1}^{\kappa} \alpha_k \Lambda_k, \tag{2.5}$$

where κ is a given positive integer; $0 \leq \alpha_k \leq 1$; and Λ_k is a known transition matrix and its expression is given by

$$\Lambda_k = \begin{bmatrix} \lambda_{11}^k & \cdots & \lambda_{1N}^k \\ \vdots & \ddots & \vdots \\ \lambda_{N1}^k & \cdots & \lambda_{NN}^k \end{bmatrix}, \tag{2.6}$$

where λ_{ij}^k keeps the same meaning as previous, with $\sum_{k=1}^{\kappa} \alpha_k = 1$.

Let $x(t; x_0, r_0)$ be the solution of system (2.1)–(2.5) at time t when the initial conditions are (x_0, r_0). In the rest of this chapter we will use $x(t)$ instead of $x(t; x_0, r_0)$.

Remark 3. The uncertainties that satisfy conditions (2.3) and (2.5) are referred to as admissible. The uncertainty term in (2.3) is supposed to depend on the system's mode $r(t)$ and on time t. The results developed in the rest of this chapter will remain valid even if the uncertainties are chosen to depend on the system's state $x(t)$, the mode $r(t)$, and time t.

Our aim in this chapter is to develop conditions that we can use to check if a dynamical system is stochastically stable and if it is robustly stochastically stable. It is preferable that these conditions be in the LMI formalism to allow the use of existing powerful tools such as Matlab LMI Toolbox or Scilab. Before giving results on stochastic stability and its robustness, let us define the different concepts of stability used in the rest of this book.

For system (2.1), when $F_A(i, t) \equiv 0$, for all modes and for $t \geq 0$, that is we drop the system's uncertainties, we have the following definitions.

Definition 1. *System (2.1), with $F_A(i, t) = 0$ for all modes and for all $t \geq 0$, is said to be*

1. *stochastically stable (SS) if there exists a finite positive constant $T(x_0, r_0)$ such that the following holds for any initial condition (x_0, r_0):*

$$\mathbb{E}\left[\int_0^\infty \|x(t)\|^2 dt | x_0, r_0\right] \leq T(x_0, r_0); \tag{2.7}$$

2. *mean square stable (MSS) if*

$$\lim_{t \to \infty} \mathbb{E}\|x(t)\|^2 = 0, \tag{2.8}$$

holds for any initial condition (x_0, r_0);

3. *mean exponentially stable (MES) if there exist positive constants α and β such that the following holds for any initial condition (x_0, r_0):*

$$\mathbb{E}\left[\|x(t)\|^2 | x_0, r_0\right] \leq \alpha \|x_0\| e^{-\beta t}. \tag{2.9}$$

Remark 4. From the previous definitions, we can see that MES implies MSS and SS.

When the system's uncertainties are not equal to zero, the concept of stochastic stability becomes robust stochastic stability and is defined for system (2.1) as follows.

Definition 2. *System (2.1) is said to be*

1. *robustly stochastically stable (RSS) if there exists a finite positive constant* $T(x_0, r_0)$ *such that condition (2.7) holds for any initial condition* (x_0, r_0) *and for all admissible uncertainties;*
2. *robust mean exponentially stable (RMES) if there exist positive constants* α *and* β *such that condition (2.9) holds for any initial condition* (x_0, r_0) *and for all admissible uncertainties.*

Remark 5. From these definitions, we can see that RMES implies RSS.

In the next section we give results on stochastic stability that can be used to check if a dynamical system of the class of piecewise deterministic systems is stochastically stable.

2.2 Stability

Let us now consider a dynamical system of the class of systems described by (2.1) and let the uncertainties be equal to zero for all the modes and for all $t \geq 0$.

Theorem 1. *Let* $Q = (Q(1), \cdots, Q(N)) > 0$ *be a given set of symmetric and positive-definite matrices. System (2.1) is stochastically stable if and only if there exists a set of symmetric and positive-definite matrices* $P = (P(1), \cdots, P(N)) > 0$ *such that the following holds for each* $i \in \mathscr{S}$:

$$A^{\top}(i)P(i) + P(i)A(i) + \sum_{j=1}^{N} \lambda_{ij} P(j) = -Q(i). \tag{2.10}$$

Furthermore, the stochastic stability implies the exponential mean square stability.

Proof: First notice that the joint process $\{(x(t), r(t)), t \geq 0\}$ is a time-homogeneous Markov process with the infinitesimal operator $\mathscr{L}$ acting on smooth functions $V(x(t), r(t))$.

Let us now start by the necessity proof. For this purpose, suppose that the system (2.1) is stochastically stable, which means that the following holds:

$$\mathbb{E}\left[\int_0^{\infty} \|x(t; x_0, r_0)\|^2 dt\right] < \infty,$$

for all $x_0 \in \mathbb{R}^n$. The last inequality implies that

$$\mathbb{E}\left[\int_0^\infty x^\top(t)Q(r(t))x(t)dt\,\Big|\,x_0,r_0\right] < \infty,$$

for any given symmetric and positive-definite matrix $Q(i) > 0, i \in \mathscr{S}$.
Let the function $\varPsi : \mathbb{R}^+ \times \mathbb{R}^+ \times \mathbb{R}^n \times \mathscr{S}$ be defined by

$$\varPsi(T,t,x(t),i) \triangleq \mathbb{E}\left[\int_t^T x^\top(s)Q(r(s))x(s)ds\,\Big|\,x(t)=x,r(t)=i\right]. \quad (2.11)$$

Using the time-homogeneous property, we have, with a slight abuse of notation, that

$$\varPsi(T,t,x,i) = \varPsi(T-t,x,i)$$

$$= \mathbb{E}\left[\int_0^{T-t} x^\top(s)Q(r(s))x(s)ds\,|\,x_0=x,r_0=i\right]$$

$$= x^\top(t)\mathbb{E}\left[\int_0^{T-t} \varPhi^\top(s)Q(r(s))\varPhi(s)ds\,|\,r_0=i\right]x(t)$$

$$\triangleq x^\top(t)P(T-t,i)x(t),$$

where $\varPhi(s)$ is the transition matrix.

Since the system is stochastically stable, $P(\cdot,i)$ is a monotonically increasing and positive-definite matrix-valued function and bounded from above. Thus

$$P(i) = \lim_{T\to\infty} P(T,i) \quad (2.12)$$

exists. Here, $P(i) > 0$ is also positive-definite. Let T be an arbitrarily fixed time. For any $T > s > t > 0$,

$$\frac{d}{dt}\mathbb{E}\left[\varPsi(T-t,x,i)\right| \text{ evaluated along the system trajectory}]$$

$$= \lim_{s\to t}\frac{1}{s-t}\left[\mathbb{E}\left(\varPsi(T-s,x(s),r(s))|x(t)=x,r(t)=i\right)-\varPsi(T-t,x,i)\right]$$

$$= \frac{\partial}{\partial t}\varPsi(T-t,x,i) + \mathscr{L}\varPsi(T-t,x,i)$$

$$= x^\top(t)\left[A^\top(i)P(T-t,i)+P(T-t,i)A(i)\right]x(t)$$

$$+ \sum_{j=1}^N \lambda_{ij}x^\top(t)P(T-t,j)x(t) + x^\top(t)\frac{\partial}{\partial t}P(T-t,x,i)x(t)$$

$$= x^\top(t)\left[\frac{\partial}{\partial t}P(T-t,i)+A^\top(i)P(T-t,i)+P(T-t,i)A(i)\right.$$

$$\left. + \sum_{j=1}^M \lambda_{ij}P(T-t,j)\right]x(t). \quad (2.13)$$

On the other hand, by (2.11) we have

$$
\begin{aligned}
&\mathbb{E}\left[\Psi(T-s,x(s),r(s))|x(t)=x,r(t)=i\right]-\Psi(T-t,x,i) \\
&= \left[\mathbb{E}(\Psi(T-s,x(s),r(s))-\Psi(T-t,x,i)|x(t)=x,r(t)=i)\right] \\
&= \mathbb{E}\left\{\mathbb{E}\left[\int_s^T x^\top(v)Q(r(v))x(v)dv|x(s),r(s)\right]\right. \\
&\qquad \left. -\mathbb{E}\left\{\int_t^T x^\top(v)Q(r(v))x(v)dv|x(t)=x,r(t)=i\right\}|x(t)=x,r(t)=i\right\} \\
&= \mathbb{E}\left\{\int_s^T x^\top(v)Q(r(v))x(v)dv-\int_t^T x^\top(v)Q(r(v))x(v)dv|x(t)=x,r(t)=i\right\} \\
&= -\mathbb{E}\left\{\int_0^{s-t} x^\top(u)Q(r(u))x(v)du|x_0=x,r_0=i\right\} \\
&= -x^\top\mathbb{E}\left\{\int_0^{s-t}\Phi^\top(u)Q(r(u))\Phi(u)du|r_0=i\right\}x. \qquad (2.14)
\end{aligned}
$$

However,

$$
\begin{aligned}
&\lim_{s\to t}\frac{1}{s-t}\mathbb{E}\left\{\int_0^{s-t}\Phi^\top(u)Q(r(u))\Phi(u)du|r_0=i\right\} \\
&= \lim_{s\to t}\mathbb{E}\left\{\frac{1}{s-t}\int_0^{s-t}Q(r(u))du|r(0)=i\right\} \\
&= \mathbb{E}\{Q(i)|r_0=i\}=Q(i)
\end{aligned}
$$

since $\Phi(0)=\mathbb{I}$.

This implies

$$
\frac{d}{dt}\mathbb{E}\Psi(T-t,x,i)=-x^\top Q(i)x. \qquad (2.15)
$$

Let T go to infinity in (2.13) and note that $\frac{\partial}{\partial t}P(T-t,j)$ tends to zero as T tends to infinity. Then (2.13), (2.14), and (2.15) give (2.10). This proves necessity.

For sufficiency, let us assume that $P(i)>0$, for $i\in\mathscr{S}$ solves (2.10). Define a stochastic Lyapunov function candidate $V(x(t),r(t))$ by the following expression:

$$
V(x(t),r(t))=x^\top(t)P(r(t))x(t), \qquad (2.16)
$$

If at time t, $x(t)=x$ and $r(t)=i$, for $i\in\mathscr{S}$, the infinitesimal operator acting on $V(.)$ and emanating from the point (x,i) at time t is given by:

$$
\mathscr{L}V(x(t),i)=\sum_{j=1}^N\lambda_{ij}V(x(t),j)+x^\top A^\top(i)\frac{\partial}{\partial x}V(x(t),i)
$$

$$= x^\top(t) \left[\sum_{j \in \mathscr{S}} \lambda_{ij} P(j) + A^\top(i) P(i) + P(i) A(i) \right] x(t)$$

$$= -x^\top(t) Q(i) x(t). \tag{2.17}$$

Thus,

$$\mathscr{L}V(x(t), i) \leq - \min_{i \in \mathscr{S}} \lambda_{min}[Q(i)] \, x^\top(t) x(t). \tag{2.18}$$

By Dynkin's formula, we obtain

$$\mathbb{E}[V(x(t), i)] - V(x_0, r_0) = \mathbb{E}\left[\int_0^t \mathscr{L}V(x(s), r(s)) ds \right]$$

$$\leq - \min_{i \in \mathscr{S}} \{\lambda_{min}(Q(i))\} \mathbb{E}\left[\int_0^t x^\top(s) x(s) ds | (x_0, r_0) \right],$$

implying, in turn,

$$\min_{i \in \mathscr{S}} \{\lambda_{min}(Q(i))\} \mathbb{E}\left[\int_0^t x^\top(s) x(s) ds | (x_0, r_0) \right]$$

$$\leq \mathbb{E}[V(x(0), r_0)] - \mathbb{E}[V(x(t), i)]$$

$$\leq \mathbb{E}[V(x(0), r_0)].$$

This yields that

$$\mathbb{E}\left[\int_0^t x^\top(s) x(s) ds | (x_0, r_0) \right] \leq \frac{\mathbb{E}[V(x(0), r_0)]}{\min_{i \in \mathscr{S}} \{\lambda_{min}(Q(i))\}}$$

holds for any $t > 0$. Letting t go to infinity implies that

$$\mathbb{E}\left[\int_0^\infty x^\top(s) x(s) ds | (x_0, r_0) \right]$$

is bounded by a constant $T(x_0, r_0)$ given by

$$T(x_0, r_0) = \frac{\mathbb{E}[V(x(0), r_0)]}{\min_{i \in \mathscr{S}} \{\lambda_{min}(Q(i))\}},$$

which implies that (2.1) is exponentially mean square stable and therefore stochastically stable. This completes the proof of sufficiency. The last statement of Theorem 1 is already proved in the proof of the sufficiency. This completes the proof of the theorem.   □

Note that when a set of symmetric and positive-definite matrices $\{P(i), i \in \mathscr{S}\}$ is given and we must solve for $Q(i), i \in \mathscr{S}$, via (2.10), the positive definiteness of $Q(i), i \in \mathscr{S}$ is only sufficient for the stability of the system (2.1), but not necessary. This can be illustrated by a simple example.

Example 1. Consider a system with one mode and the system parameters are given by

$$A = \begin{bmatrix} -1 & 4 \\ 0 & -1 \end{bmatrix}.$$

If we choose Q in (2.10) equal to the identity matrix, $\mathbb{I}$, then using (2.10) we obtain

$$P = \begin{bmatrix} -2 & 4 \\ 4 & -2 \end{bmatrix},$$

which is not positive-definite. However, this system is obviously stable since both poles of A are negative.

Note that if the system has only a single mode, then (2.10) reduces to the condition for deterministic stability. That is, the stochastic stability becomes one of the deterministic systems. However, for jump linear systems with multiple modes, stability in each mode is neither necessary nor sufficient for the stochastic stability of system (2.1).

To illustrate that the stochastic stability of system (2.1) does not imply that each model is deterministically stable, let us consider an example.

Example 2. Consider a system with two modes, that is, $\mathscr{S} = \{1, 2\}$. The system parameters are given by

- mode #1: $A(1) = 0.25$,
- mode #2: $A(2) = -1.5$.

Let the switching between the two modes be described by the following transition matrix:

$$\Lambda = \begin{bmatrix} -1 & 1 \\ 1 & -1 \end{bmatrix}.$$

This system is not stable in mode 1 because $A(1) > 0$. With this set of data and $Q(i) = 1$, (2.10) becomes

$$\begin{bmatrix} -0.5P(1) + P(2) = -1, \\ -4P(2) + P(1) = -1. \end{bmatrix}$$

Solving the above equations yields $P(1) = 5.0 > 0, P(2) = 1.5 > 0$. So this system is stochastically stable. This means that the stability of $A(i), i \in \mathscr{S}$ is not necessary for the system to be stochastically stable.

Theorem 1 provides a sufficient and necessary condition for system (2.1) to be stochastically stable, which is not easy to test. The following corollary gives a necessary condition for stochastic stability that is easier to test.

Corollary 1. *If system (2.1) is stochastically stable, then for each $i \in \mathscr{S}$, the matrix $A(i) + \frac{1}{2}\lambda_{ii}\mathbb{I}$ is stable; that is, all its eigenvalues have negative real parts.*

Proof: Since system (2.1) is stochastically stable, letting $Q(i) = \mathbb{I}$ in (2.10), we obtain that the set of coupled Lyapunov equations (2.10) has a set of symmetric and positive-definite solutions, denoted by $(P(1), \cdots, P(N)) > 0$, that is,

$$A^\top(i)P(i) + P(i)A(i) + \sum_{j \in \mathscr{S}} \lambda_{ij}P(j) = -\mathbb{I}, i \in \mathscr{S},$$

which can be rewritten as

$$\left[A^\top(i) + \frac{1}{2}\lambda_{ii}\mathbb{I}\right]P(i) + P(i)\left[A(i) + \frac{1}{2}\lambda_{ii}\mathbb{I}\right]$$
$$= -\left[\sum_{j \neq i} \lambda_{ij}P(j) + \mathbb{I}\right], i \in \mathscr{S}.$$

Since $\sum_{j \neq i} \lambda_{ij}P(j) + \mathbb{I}$ is symmetric and positive-definite, the last equation implies matrix $A(i) + \frac{1}{2}\lambda_{ii}\mathbb{I}$ is stable, that is, all its eigenvalues have negative real parts. This completes the proof of Corollary 1. $\qquad\square$

The above corollary gives a necessary condition for system (2.1) to be stochastically stable, but the following example shows it is not sufficient.

Example 3. Consider a two-mode system with the following data:

- mode #1: $A(1) = 1$,
- mode #2: $A(2) = 0.5$.

The switching between the two modes is described by the following transition matrix:

$$\Lambda = \begin{bmatrix} -3 & 3 \\ 2 & -2 \end{bmatrix}.$$

For this set of data, we obtain from (2.10)

$$\begin{bmatrix} 2P(1) - 3P(1) + 3P(2) = -1, \\ P(2) - 2P(2) + 2P(1) = -1. \end{bmatrix}$$

The solution of this system is $P(1) = -\frac{4}{5}, P(2) = -\frac{3}{5}$. $P(1)$ and $P(2)$ are not positive-definite. Thus, by Theorem 1, this system is not stable. But we have $A(1) + \frac{1}{2}\lambda_{11} = -\frac{1}{2}$ and $A(2) + \frac{1}{2}\lambda_{22} = -\frac{1}{2}$. That is, they are both stable. Therefore, we see that this necessary condition given by Corollary 1 is not sufficient for stochastic stability.

From the above theorem, we obtain the following one.

Theorem 2. *System (2.1) is stochastically stable if and only if there exists a set of symmetric and positive-definite matrices $P = (P(1), \cdots, P(N)) > 0$ such that the following coupled LMIs are feasible:*

$$A^\top(i)P(i) + P(i)A(i) + \sum_{j \neq i} \lambda_{ij} P(j) < 0, i \in \mathscr{S}. \tag{2.19}$$

Proof: First let us prove the necessity. Suppose that system (2.1) is stochastically stable. By letting $Q(i) = \mathbb{I}$ and using the theorem on stability, we have that

$$A^\top(i)P(i) + P(i)A(i) + \sum_{j \neq i} \lambda_{ij} P(j) = -\mathbb{I}, i \in \mathscr{S}$$

have a set of feasible solutions $P(i), i \in \mathscr{S}$ satisfying $P(i) > 0$. The last equation means that this set of matrices $P(i), i \in \mathscr{S}$, satisfies (2.19).

On the other hand, if (2.19) has a set of feasible solutions $P(i), i \in \mathscr{S}$, by defining $Q(i) \triangleq -[A^\top(i)P(i) + P(i)A(i) + \sum_{j \neq i} \lambda_{ij} P(i)]$, which is symmetric, positive-definite, and satisfies

$$A^\top(i)P(i) + P(i)A(i) + \sum_{j \neq i} \lambda_{ij} P(j) = -Q(i), i \in \mathscr{S},$$

we conclude that system (2.1) is stochastically stable.   □

Theorem 2 provides an LMI-based condition for the stability of system (2.1), which can be checked easily by using LMI Toolbox of Matlab or Scilab.

Example 4. To show the usefulness of Theorem 2, let us consider a system having dynamics (2.1) and the system parameters as follows: $\mathscr{S} = \{1, 2, 3\}$ and

$$\Lambda = \begin{bmatrix} -10 & 5 & 5 \\ 1 & -2 & 1 \\ 0.7 & 0.3 & -1 \end{bmatrix}, \quad A(1) = \begin{bmatrix} 2.5 & 0.3 & 0.8 \\ 0 & 3 & 0.2 \\ 0 & 0.5 & 2 \end{bmatrix}$$

$$A(2) = \begin{bmatrix} -2.5 & 1.2 & 0.3 \\ -0.5 & -5 & 1 \\ 0.25 & 1.2 & -5 \end{bmatrix}, \quad A(3) = \begin{bmatrix} -2 & 1.5 & -0.4 \\ 2.2 & -3 & 0.7 \\ 1.1 & 0.9 & -2 \end{bmatrix}.$$

With this set of data, solving (2.19) yields

$$P(1) = \begin{bmatrix} 35.8075 & 17.3278 & 7.7457 \\ 17.3278 & 35.8390 & 9.5721 \\ 7.7457 & 9.5721 & 19.1130 \end{bmatrix},$$

$$P(2) = \begin{bmatrix} 10.8900 & 3.3784 & 1.1620 \\ 3.3784 & 7.4710 & 2.0628 \\ 1.1620 & 2.0628 & 5.2354 \end{bmatrix},$$

$$P(3) = \begin{bmatrix} 20.6740 & 10.6676 & 1.7184 \\ 10.6676 & 12.4321 & 1.8633 \\ 1.7184 & 1.8633 & 9.1467 \end{bmatrix}.$$

Since the eigenvalues of $A(1)$ are $(2.5, 3.0916, 1.9084)$, mode 1 is not stable. However, direct computation gives

$$A^\top(i)P(i) + P(i)A(i) + \sum_{j \neq i} \lambda_{ij} P(j)$$

$$= \begin{bmatrix} -21.2175 & 6.8695 & 3.9122 \\ 6.8695 & -23.8719 & 4.6806 \\ 3.9122 & 4.6806 & -26.5458 \end{bmatrix},$$

with eigenvalues $(-30.4096, -28.3270, -12.8985)$. This means that this system is stochastically stable.

Next we proceed to consider the almost-sure stability. For this purpose, let us give the lemma.

Lemma 1. *Consider a matrix in the following companion form:*

$$A = \begin{bmatrix} 0 & 1 & 0 & \cdots & 0 \\ 0 & 0 & 1 & \cdots & 0 \\ \vdots & \vdots & \vdots & \ddots & \vdots \\ 0 & 0 & 0 & \vdots & 1 \\ a_1 & a_2 & a_3 & \cdots & a_n \end{bmatrix}, \tag{2.20}$$

with distinct-real eigenvalues $\lambda_1, \cdots, \lambda_n$ satisfying $|\lambda_i - \lambda_j| \geq 1$, $(i \neq j)$, then there exists a constant $M > 0$ and a positive integer k, both independent of $\lambda_1, \cdots, \lambda_n$, and a nonsingular matrix T such that

$$\|T\| \leq M \left(\max_{1 \leq i \leq n} |\lambda_i| \right)^k, \quad \|T^{-1}\| \leq M \left(\max_{1 \leq i \leq n} |\lambda_i| \right)^k, \tag{2.21}$$

and

$$T^{-1}AT = \text{diag}\{\lambda_1, \lambda_2, \cdots, \lambda_n\}.$$

Proof: Because A has distinct real eigenvalues, A can be put in diagonal form. The transformation matrix is given by

$$T = \begin{bmatrix} 1 & 1 & 1 & \cdots & 1 \\ \lambda_1 & \lambda_2 & \lambda_3 & \cdots & \lambda_n \\ \lambda_1^2 & \lambda_2^2 & \lambda_3^2 & \cdots & \lambda_n^2 \\ \vdots & \vdots & \vdots & \ddots & \vdots \\ \lambda_1^{n-1} & \lambda_2^{n-1} & \lambda_3^{n-1} & \cdots & \lambda_n^{n-1} \end{bmatrix},$$

and $T^{-1}AT = \text{diag}\{\lambda_1, \lambda_2, \cdots, \lambda_n\}$. To prove that T satisfies the required condition, we use the 1-norm. Recall that all matrix norms are equivalent over the real field. First notice that

$$T^{-1} = \frac{adj(T)}{det(T)} = \frac{adj(T)}{\prod_{1 \le i \le j \le n}(\lambda_i - \lambda_j)}.$$

With $|\lambda_i - \lambda_j| \ge 1$, $\|T^{-1}\| \le \|adj(T)\|$. All entries of T and $adj(T)$ are polynomials of $\lambda_1, \cdots, \lambda_n$, and there exists an $M > 0$ and a positive integer $k > 0$, both independent of $\lambda_1, \cdots, \lambda_n$, such that (2.21) holds. This completes the proof. □

Let us consider the following dynamics:

$$\begin{cases} \dot{x}(t) = A(r(t))x(t) + B(r(t))u(t), \\ x(0) = x_0, \end{cases} \tag{2.22}$$

where all the parameters of these dynamics keep the same meaning as before with $B(i)$ a known matrix for each $i \in \mathscr{S}$.

Let us now consider a state feedback controller of the form

$$u(t) = K(i)x(t), \text{when } r(t) = i,$$

and show the results of the following theorem.

Theorem 3. *Assume that $\{r(t), t \ge 0\}$ is a finite-state ergodic Markov chain with invariant measure π. If there exists an $i \in \mathscr{S}$ such that $\pi_i > 0$, then (2.22) is almost surely stabilizable. As a consequence, we conclude that individual mode controllability implies almost-sure stabilizability.*

Proof: We first prove the second statement, i.e., the individual mode controllability implies the almost-sure stabilizability. Without loss of generality, we only prove the single input case. For any $j \in \mathscr{S}$, the individual mode controllability assumption implies that $(A(j), B(j))$ is controllable. Therefore, there exists a nonsingular matrix $T_1(j)$ such that

$$T_1(j)A(j)T_1^{-1}(j) = \begin{bmatrix} 0 & 1 & 0 & \cdots & 0 \\ 0 & 0 & 1 & \cdots & 0 \\ \vdots & \vdots & \vdots & \ddots & \vdots \\ 0 & 0 & 0 & \vdots & 1 \\ a_1(j) & a_2(j) & a_3(j) & \cdots & a_n(j) \end{bmatrix} \triangleq A_1(j),$$

$$T_1(j)B(j) = \begin{bmatrix} 0 \\ 0 \\ \vdots \\ 0 \\ 1 \end{bmatrix} \triangleq B_1(j).$$

Let $\lambda_1, \cdots, \lambda_n$ be negative-real numbers satisfying $2n \geq |\lambda_i - \lambda_j| \geq 1 (i \neq j)$. Choose a matrix $K_1(j)$ such that

$$A_1(j) - B_1(j)K_1(j) \triangleq \bar{A}(j)$$

has eigenvalues $\lambda_1, \cdots, \lambda_n$ for any $j \in \mathscr{S}$. Now $\bar{A}(j)$ is in companion form and from Lemma 1 there exist $l > 0$, $M_1 > 0$, and nonsingular matrices $T_2(j)(j \in \mathscr{S})$ satisfying

$$\|T_2(j)\| \leq M_1 \left(\max_{1 \leq i \leq n} |\lambda_i| \right)^l,$$

$$\|T_2^{-1}(j)\| \leq M_1 \left(\max_{1 \leq i \leq n} |\lambda_i| \right)^l,$$

such that
$$T_2^{-1} \bar{A}(j) T_2(j) = \mathrm{diag}\{\lambda_1, \cdots, \lambda_n\} \triangleq D, j \in \mathscr{S}.$$

Choose the feedback control $u(t) = -K(r(t))x(t)$ where

$$K(j) = K_1(j)T_1(j), T(j) = T_1^{-1}T_2(j), j \in \mathscr{S}.$$

Then the closed-loop system becomes

$$\dot{x}(t) = T(j)DT^{-1}(j)x(t). \tag{2.23}$$

From the choice of $T_1(j)$ and $T_2(j)$, there exists an $M_2 > 0$ and $m > 0$, both independent of $\lambda_1, \cdots, \lambda_n$ and j, such that

$$\|T(j)\| \leq M_2 \left(\max_{1 \leq i \leq n} |\lambda_i| \right)^m,$$

$$\|T^{-1}(j)\| \leq M_2 \left(\max_{1 \leq i \leq n} |\lambda_i| \right)^m.$$

With $\lambda_j < 0$, let $\lambda = \max_{1 \leq i \leq n} \lambda_i$, then there exists an $M_3 > 0$, independent of $\lambda_1, \cdots, \lambda_n$ and j, such that

$$\|e^{Dt}\| \leq M_3 e^{\lambda t}, t \geq 0.$$

From the sojourn-time description of a finite-state Markov chain, (2.23) is almost surely stable if and only if the state transition matrix $\Phi(.,.)$ satisfies the following:

$$\Phi(t,0) = e^{\tilde{A}(r(k))(t-t_k)} e^{\tilde{A}(r(k-1))(\tau_{k-1})} \cdots e^{\tilde{A}(r_0)\tau_0} \to 0, a.s.(t \to \infty), (2.24)$$

where $\tilde{A}(j) = T(j)DT^{-1}(j), (j \in \mathscr{S})$. A simple computation yields

$$\|\Phi(t,0)\| = \|T(r(k))e^{D(t-t_k)}T^{-1}(r(k))T(r(k-1))e^{D\tau_{k-1}}T^{-1}(r(k-1))$$

$$\cdots T(r_0)e^{D\tau_0}T^{-1}(r_0)\|$$

$$\leq \|T(r(k))\|\|e^{D(t-t_k)}\|\|T^{-1}(r(k))\|\|T(r(k-1))\|\|e^{D\tau_{k-1}}\|$$

$$\cdots \|T(r_0)\|\|e^{D\tau_0}\|\|T^{-1}(r_0)\|$$

$$\leq \left[M_2\left(\max_{1\leq i\leq n}\lambda_i\right)^m\right]^{2(k+1)}M_3^{k+1}e^{\lambda(\tau_k+\tau_{k-1}+\cdots+\tau_0)}$$

$$\overset{\triangle}{=}\left[Me^{\lambda(\tau_k+\cdots+\tau_0)/(k+1)}\right]^{k+1}, \tag{2.25}$$

where $M = (M_2\max_{1\leq i\leq n}|\lambda_i|^l)^2 M_3$. Since $\{r(t), t \geq 0\}$ is a finite-state ergodic Markov chain and from the Law of Large Numbers, there exists a nonrandom constant $a > 0$, the average sojourn time, such that

$$\lim_{k\to\infty}\frac{\tau_k+\cdots+\tau_0}{k+1} = a, a.s.$$

Hence

$$\lim_{k\to\infty}Me^{\lambda(\tau_k+\cdots+\tau_0)/(k+1)} = Me^{\lambda a} \leq [M_2(|\lambda|+2n)^l]^2 M_3 e^{\lambda a} \to 0,$$
$$(\lambda \to -\infty).$$

Thus, we can choose $|\lambda|$ sufficiently large so that $Me^{\lambda a} < 1$.

From (2.25) we have

$$\lim_{t\to\infty}\Phi(t,0) = 0, a.s. \tag{2.26}$$

that is, (2.23) is almost surely stable. Therefore, (2.1) is almost surely stabilizable.

Next we prove the first statement. Without loss of generality, we assume that $(A(1), B(1))$ is controllable and $\pi_1 > 0$. We choose $K(2) = K(3) = \cdots = K(N) = 0$, and choose $K(1)$ and λ as in the first half of the proof. Then there exists an $M > 0$ and α, both independent of λ, such that

$$\|e^{A(i)t}\| \leq Me^{\alpha t}, (i \neq 1), \tag{2.27}$$

$$\|e^{(A(1)-B(1)K(1))t}\| \leq p(\lambda)e^{\lambda t}, \forall t \geq 0, \tag{2.28}$$

where $p(\lambda)$ is a polynomial with degree independent of λ. Let γ_k^1 denote the time occupied by state 1 during the time interval $(0, t_k)$ and γ_k^2 denote the time occupied by the states $2, 3, \cdots, N$ during the interval $(0, t_k)$. From the ergodicity of $\{r(t), t \geq 0\}$,

$$\lim_{k\to\infty}\frac{\gamma_k^1}{t_k} = \pi_1, \quad \lim_{k\to\infty}\frac{\gamma_k^2}{t+k} = 1 - \pi_1. \tag{2.29}$$

As in the first half of the proof, we obtain that

$$\|\Phi(t_k,0)\| \leq \left[(Mp(\lambda))^{(k+1)/t_k}e^{\lambda\gamma_k^1/t_k+\alpha\gamma_k^2/t_k}\right]^{t_k}, \tag{2.30}$$

and the term $[(Mp(\lambda))^{1/a}e^{\pi_1\lambda+(1-\pi_1)\alpha}]$ has the limit

$$(Mp(\lambda))^{1/a}e^{\pi_1\lambda+(1-\pi_1)\alpha} \to 0(\lambda \to -\infty).$$

Therefore, we can conclude that the system is almost surely stabilizable. This completes the proof of Theorem 3. □

The above discussion reveals that for jump linear systems, stochastic stability and mean square stability are equivalent. In fact, this conclusion holds only for the case that the mode process has finite states. The following example shows that when the system has an infinite number of modes, this conclusion is false.

Example 5. Consider a one-dimensional system with the system mode represented by a Poisson process $\theta(t)$ with parameter λ. Then the state equation system is described by

$$\dot{x}(t) = -b_{\theta(t)}x(t), t \ge 0, \tag{2.31}$$

where $x(t) \in \mathbb{R}$ and $b_i, i \ge 1$, are scalars defined by

$$b_i = \frac{\lambda}{2}\log\left(\frac{i+1}{i}\right), i \ge 1, \tag{2.32}$$

which means that b_i are positive and tends to zero when i increases (goes to infinity). In this setting, the trajectories of the state process are decreasing and connected solutions of (2.31) given by

$$x(t) = a_n e^{-b_{n+\theta_0-1}(t-\tau_{n-1})} \quad a.s. \ \tau_{n-1} \le t < \tau_n, \tag{2.33}$$

where $a_1 = x_0$ and $\tau_0 = 0 < \tau_1 < \tau_2 < \cdots$ is the sequence of jump times, and we appeal to the fact that once the Poisson process is known to be in state i, it can only arrive to state $i+1$ through one jump. A consequence of the continuity of (2.33) on every jump point is that

$$\lim_{t\to\tau_n} x(\tau_n) = a_n e^{-b_{n+\theta_0-1}(\tau_n-\tau_{n-1})} = a_{n+1}, n \ge 1, \tag{2.34}$$

from which calculation of a_2 and a_3 easily show us that

$$a_n = x_0 \exp\left\{-\sum_{i=1}^{n-1} b_{i+\theta_0-1}(\tau_i - \tau_{i-1}), \ n = 2, 3, \cdots, \right\}. \tag{2.35}$$

Furthermore, the sequence of jump times $\tau_0 = 0 < \tau_1 < \tau_2 < \cdots < \tau_n < \infty$ a.s. Now, since $-\lambda_{ii}$ are bounded from above, it follows that in any interval $[0, d], d < \infty$, almost all sample paths of the process $\{\theta(t), t \ge 0\}$ have only finitely many jump points τ_n. Consequently, with probability one, jump point sequences converging to some finite point $\bar{\tau}$ do not exist, so that

$$\lim_{n\to\infty} \tau_n = \infty, a.s. \tag{2.36}$$

and (2.33) indeed represents the trajectory of the state process $\{x(t), t \geq 0\}$.

We shall now show that the system is not stochastically stable for second-order random variable x_0, with $\mathbb{E}[x_0^2] \neq 0$ and deterministic $\theta_0 \in \mathscr{S}$.

From (2.36) and (2.34), we write that

$$\int_0^\infty x^2(t)dt = \sum_{n=1}^\infty \int_{\tau_{n-1}}^{\tau_n} x^2(t)dt \geq \sum_{n=1}^\infty a_{n+1}^2 (\tau_n - \tau_{n-1}) \quad a.s., \quad (2.37)$$

where we use the fact that (2.33) is a decreasing function.

Taking the mathematical expectation on both sides of (2.37) and using Fubini's Theorem, we obtain

$$\int_0^\infty \mathbb{E}[x^2(t)]dt \geq \sum_{n=1}^\infty \mathbb{E}\left[\left(x_0 \exp\left\{-\sum_{i=1}^n b_{i+\theta_0-1}(\tau_i - \tau_{i-1})\right\}\right)^2 (\tau_n - \tau_{n-1})\right]$$

$$= \sum_{n=1}^\infty \mathbb{E}\left[x_0^2 \exp\left\{-2\sum_{i=1}^{n-1} b_{i+\theta_0-1}(\tau_i - \tau_{i-1})\right\}\right]$$
$$\times \exp\{-2b_{n+\theta_0-1}(\tau_n - \tau_{n-1})\}(\tau_n - \tau_{n-1})$$

$$= \mathbb{E}[x_0^2]\left\{\sum_{n=1}^\infty \mathbb{E}\left[\exp\left\{-2\sum_{i=1}^{n-1} b_{i+\theta_0-1}(\tau_i - \tau_{i-1})\right\}\right]\right.$$
$$\left.\times\mathbb{E}[\exp\{-2b_{n+\theta_0-1}(\tau_n - \tau_{n-1}) + \log(\tau_n - \tau_{n-1})\}]\right\}$$

$$\geq \mathbb{E}[x_0^2]\left(\sum_{n=1}^\infty \left[-2\sum_{i=1}^{n-1} b_{i+\theta_0-1}\mathbb{E}[\tau_i - \tau_{i-1}]\right]\right.$$
$$\left.\times \exp\left\{-2b_{n+\theta_0-1}\mathbb{E}[\tau_n - \tau_{n-1}] + \mathbb{E}[\log(\tau_n - \tau_{n-1})]\right\}\right)$$

$$= E[x_0^2]\left(\sum_{n=1}^\infty \exp\left[-\frac{\lambda}{2}\sum_{i=1}^n \frac{\lambda}{2}\log\left(\frac{i+\theta_0}{i+\theta_0-1}\right)\right]\right)$$
$$\times \exp\{\mathbb{E}[\log(\tau_n - \tau_{n-1})]\}$$

$$= \mathbb{E}[x_0^2]\left(\sum_{n-1}^\infty \frac{\theta_0}{n+\theta_0} \exp\left\{\int_0^\infty \log(s)\lambda e^{-\lambda s}ds\right\}\right)$$

$$= \mathbb{E}[x_0^2]d_1 \sum_{n=1}^\infty \frac{\theta_0}{n+\theta_0} = \infty, \quad (2.38)$$

where $d_1 = \exp\left\{\int_0^\infty \log(s)\lambda e^{\lambda s}ds\right\}$ is finite and positive and does not depend on n. This means that the system under study is not stochastically stable.

Next we proceed to show that the system is mean square stable. For this purpose, we write

$$\mathbb{E}\left[x^2(\tau_n)|x_0, \theta_0\right] = \mathbb{E}\left[a_{n+1}^2|x_0, \theta_0\right]$$

$$= x_0^2 \left[\prod_{i=1}^{n} \exp\{-2b_{i+\theta_0-1}(\tau_i - \tau_{i-1})\} | \theta_0 \right]$$

$$= x_0^2 \prod_{i=1}^{n} \mathbb{E}[\exp\{-2b_{i+\theta_0-1}\}(\tau_i - \tau_{i-1}) | \theta_0]$$

$$= x_0^2 \prod_{i=1}^{n} \left\{ \frac{-\exp(-2(b_{i+\theta_0-1} + \lambda)s)}{1 + \frac{2b_{i+\theta_0-1}}{\lambda}} \Big|_0^\infty \right\}$$

$$= x_0^2 \left(\prod_{i=1}^{n} \left(1 + \log \left(\frac{i+\theta_0}{i+\theta_0 - 1} \right) \right) \right)^{-1}.$$

Now,

$$\sum_{i=1}^{n} \log \left(\frac{i+\theta_0}{i+\theta_0 - 1} \right) = \log(n + \theta_0) - \log(\theta_0) \to \infty (n \to \infty).$$

Therefore, we have

$$\mathbb{E}[x^2(\tau_n)|x_0, \theta_0] \to 0 (n \to \infty).$$

Since almost all trajectories of $\{x\}$ are decreasing, we have from (2.36) that

$$\lim_{t \to \infty} x^2(t) = \lim_{\tau_n \to \infty} x^2(\tau_n) = \lim_{n \to \infty} x^2(\tau_n) \quad a.s.$$

Appealing to the Lebesgue monotone convergence theorem, we have, for any joint distribution of (x_0, θ_0), that

$$\lim_{t \to \infty} \mathbb{E}\left[x^2(t)\right] = \mathbb{E}\left[\lim_{t \to \infty} x^2(t)\right] = \lim_{n \to \infty} \mathbb{E}\left[x^2(\tau_n)\right] = \lim_{n \to \infty} \mathbb{E}\left[\mathbb{E}\left[x^2(\tau_n)\right]\right]$$

$$= \mathbb{E}\left[\lim_{n \to \infty} \mathbb{E}(x^2(\tau_n))\right] = 0.$$

This proves that the system under study is mean square stable.

The next theorem states another result using a LMI framework for stochastic stability. It gives an LMI condition that we should satisfy to guarantee the stochastic stability of our class of systems.

Theorem 4. *System (2.1) with all the uncertainties equal to zero is stochastically stable if and only if there exists a set of symmetric and positive-definite matrices $P = (P(1), \cdots, P(N)) > 0$ such that the following holds for every $i \in \mathscr{S}$:*

$$A^\top(i)P(i) + P(i)A(i) + \sum_{j=1}^{N} \lambda_{ij} P(j) < 0. \tag{2.39}$$

Proof: The proof of necessity is similar to the one of Theorem 2 and the detail is omitted. To prove the sufficiency of this theorem, let us consider the following Lyapunov function:

$$V(x(t), r(t)) = x^\top(t)P(r(t))x(t),$$

where $P(i) > 0$ is a symmetric and positive-definite matrix for every $i \in \mathscr{S}$.

Let $\mathscr{L}$ denote the infinitesimal operator of the Markov process $\{(x(t), r(t)),\ t \geq 0\}$. If at time t, $x(t) = x$ and $r(t) = i$ for $i \in \mathscr{S}$, the expression of the infinitesimal operator acting on $V(.)$ and emanating from the point (x, i) at time t is given by (see Appendix A)

$$
\begin{aligned}
\mathscr{L}V(x(t), i) &= \lim_{h \to 0} \frac{\mathbb{E}\left[V(x(t+h), r(t+h))|x(t) = x, r(t) = i\right] - V(x(t), i)}{h} \\
&= \dot{x}^\top(t)P(i)x(t) + x^\top(t)P(i)\dot{x}(t) + \sum_{j=1}^{N} \lambda_{ij} x^\top(t)P(j)x(t) \\
&= x^\top(t)A^\top(i)P(i)x(t) + x^\top(t)P(i)A(i)x(t) \\
&\quad + \sum_{j=1}^{N} \lambda_{ij} x^\top(t)P(j)x(t) \\
&= x^\top(t)\left[A^\top(i)P(i) + P(i)A(i) + \sum_{j=1}^{N} \lambda_{ij}P(j)\right]x(t) \\
&= x^\top(t)\Lambda_n(i)x(t).
\end{aligned}
$$

Using condition (2.39) we get

$$\mathscr{L}V(x(t), i) \leq -\min_{i \in \mathscr{S}}\{\lambda_{min}(-\Lambda_n(i))\}x^\top(t)x(t). \tag{2.40}$$

Combining this again with Dynkin's formula (see Appendix A) yields

$$
\begin{aligned}
\mathbb{E}\left[V(x(t), i)\right] - \mathbb{E}\left[V(x(0), r_0)\right] &= \mathbb{E}\left[\int_0^t \mathscr{L}V(x(s), r(s))ds|(x_0, r_0)\right] \\
&\leq -\min_{i \in \mathscr{S}}\{\lambda_{min}(-\Lambda_n(i))\}\mathbb{E}\left[\int_0^t x^\top(s)x(s)ds|(x_0, r_0)\right],
\end{aligned}
$$

implying, in turn,

$$
\begin{aligned}
\min_{i \in \mathscr{S}}\{\lambda_{min}(-\Lambda_n(i))\}\mathbb{E}&\left[\int_0^t x^\top(s)x(s)ds|(x_0, r_0)\right] \\
&\leq \mathbb{E}\left[V(x(0), r_0)\right] - \mathbb{E}\left[V(x(t), i)\right] \\
&\leq \mathbb{E}\left[V(x(0), r_0)\right].
\end{aligned}
$$

This yields that

$$\mathbb{E}\left[\int_0^t x^\top(s)x(s)ds|(x_0,r_0)\right] \le \frac{\mathbb{E}\left[V(x(0),r_0)\right]}{\min_{i\in\mathscr{S}}\{\lambda_{min}(-\Lambda_n(i))\}}$$

holds for any $t > 0$. Letting t go to infinity implies that

$$\mathbb{E}\left[\int_0^\infty x^\top(s)x(s)ds|(x_0,r_0)\right]$$

is bounded by a constant $T(x_0,r_0)$ given by

$$T(x_0,r_0) = \frac{\mathbb{E}\left[V(x(0),r_0)\right]}{\min_{i\in\mathscr{S}}\{\lambda_{min}(-\Lambda_n(i))\}},$$

which ends the proof of Theorem 4.   □

Example 6. Let us consider a numerical example to show the validity of the results of the previous theorem. For this purpose, let us assume that the dynamical system considered in this example has two modes and its state space belongs to $\mathbb{R}^2$. Its state equation is supposed to be described by (2.1) with all the uncertainties equal to zero and the different remaining matrices as follows:

$$\Lambda = \begin{bmatrix} -3 & 3 \\ 2 & -2 \end{bmatrix}, \quad A(1) = \begin{bmatrix} -2.1 & 0.1 \\ 0.1 & 1.1 \end{bmatrix}, \quad A(2) = \begin{bmatrix} -1.9 & -0.1 \\ -0.1 & 0.9 \end{bmatrix}.$$

First notice that the two modes are unstable in the deterministic sense since the eigenvalues in mode 1 and mode 2 are, respectively, $(-2.1031, 1.1031)$ and $(-1.9036, 0.9036)$.

Solving the set of coupled LMIs (2.39), we conclude that the system is infeasible. Therefore the system is not stochastically stable.

Example 7. In this second example, we show that the stability of one mode or both is not a necessary condition for the stochastic stability of the whole system. For this purpose, let us consider a dynamical system that has two modes and the switching between these two modes is described by the following transition probability rate matrix:

$$\Lambda = \begin{bmatrix} -4 & 4 \\ 7 & -7 \end{bmatrix}.$$

The state matrices $A(r(t))$ in each mode are given by

$$A(1) = \begin{bmatrix} -2 & 0.2 \\ 2 & -5 \end{bmatrix}, \quad A(2) = \begin{bmatrix} 1 & 0.2 \\ 0 & -5 \end{bmatrix}.$$

Notice that mode 1 is stable while mode 2 is unstable in the deterministic sense since the eigenvalues for these modes are, respectively, $(-1.8721, -5.1279)$ and $(1.0, -5.0)$.

Solving now the set of coupled LMIs (2.39), we get

$$P(1) = \begin{bmatrix} 0.9673 & 0.0653 \\ 0.0653 & 0.1389 \end{bmatrix}, \quad P(2) = \begin{bmatrix} 1.5775 & 0.0689 \\ 0.0689 & 0.1378 \end{bmatrix}.$$

As can be seen, the two matrices are symmetric and positive-definite, therefore based on the result of the previous theorem, the studied system is stochastically stable.

Example 8. In this example let us consider a system with two modes. The state equation of this system in each mode is described by (2.1) with the following matrices:

$$A(1) = \begin{bmatrix} -1.0 & 0.1 \\ 0.0 & -2.0 \end{bmatrix}, \quad A(2) = \begin{bmatrix} -0.1 & 0.0 \\ 0.0 & 0.2 \end{bmatrix}.$$

The transition probability rate matrix between these modes is given by

$$\Lambda = \begin{bmatrix} -1 & 1 \\ 5 & -5 \end{bmatrix}.$$

Notice that mode 1 is stable and mode 2 is unstable in the deterministic sense since the eigenvalues for these modes are, respectively, $(-1.0, -2.0)$ and $(-0.1, 0.2)$.

For the steady-state probabilities transition (see Appendix A), we have

$$\begin{bmatrix} \pi_1 & \pi_2 \end{bmatrix} \begin{bmatrix} -1.0 & 1.0 \\ 5.0 & -5.0 \end{bmatrix} = 0,$$
$$\pi_1 + \pi_2 = 1,$$

which gives $\pi_1 = \frac{5}{6}$ and $\pi_2 = \frac{1}{6}$. This means that the system spends more time in the stable mode, which therefore affects the stability of the global system.

Solving now the set of coupled LMIs (2.39), we get

$$P(1) = \begin{bmatrix} 39.6196 & 1.2015 \\ 1.2015 & 21.8963 \end{bmatrix}, \quad P(2) = \begin{bmatrix} 49.7876 & 1.1642 \\ 1.1642 & 37.8007 \end{bmatrix}.$$

As can be seen, the two matrices are symmetric and positive-definite, therefore based on the results of the previous theorem, the studied system is stochastically stable.

Example 9. In this example we want to show that the stochastic stability of the whole dynamical system does not imply the stability of each mode. For this purpose let us consider a dynamical system of the class we are studying with the following data:

$$A(1) = \begin{bmatrix} 0.01 & 0.0 \\ 0.0 & -0.2 \end{bmatrix}, \quad A(2) = \begin{bmatrix} -0.1 & 0.0 \\ 0.0 & 0.2 \end{bmatrix}.$$

The transition probability rate matrix between these modes is given by

$$\Lambda = \begin{bmatrix} -4 & 4 \\ 5 & -5 \end{bmatrix}.$$

Notice that the two modes are unstable in the deterministic sense since the eigenvalues for these modes are, respectively, $(0.01, -0.2)$ and $(-0.1, 0.2)$.
For the steady-state probabilities we have

$$\begin{bmatrix} \pi_1 & \pi_2 \end{bmatrix} \begin{bmatrix} -4.0 & 4.0 \\ 5.0 & -5.0 \end{bmatrix} = 0,$$

$$\pi_1 + \pi_2 = 1,$$

which gives $\pi_1 = \frac{5}{9}$ and $\pi_2 = \frac{4}{9}$. This means that the system spends almost the same time at the two modes.
Solving now the set of coupled LMIs (2.39), we get

$$P(1) = \begin{bmatrix} 34.4895 & 0.0 \\ 0.0 & 22.4475 \end{bmatrix}, \quad P(2) = \begin{bmatrix} 33.7245 & 0.0 \\ 0.0 & 24.5555 \end{bmatrix}.$$

As can be seen, the two matrices are symmetric and positive-definite; therefore based on the results of the previous theorem, the system is stochastically stable.
For this system, if we change the transition probability rate matrix to

$$\Lambda = \begin{bmatrix} -2 & 2 \\ 5 & -5 \end{bmatrix},$$

which gives $\pi_1 = \frac{5}{7}$ and $\pi_2 = \frac{2}{7}$, the system spends more time in the first mode.
Solving now the LMI (2.39), we get

$$P(1) = \begin{bmatrix} 0.7032 & 0.0 \\ 0.0 & 0.8306 \end{bmatrix}, \quad P(2) = \begin{bmatrix} 0.6948 & 0.0 \\ 0.0 & 0.9464 \end{bmatrix}.$$

As can be seen the two matrices are symmetric and positive-definite; therefore based on the result of the previous theorem, the studied system is stochastically stable. Based on these two cases we conclude that the stability in the deterministic sense for each mode is not a necessary condition for the stochastic stability of the global system.

Example 10. In this example we show that stability in each mode is not necessary for the stochastic stability of the system. For this purpose, let us consider the following data:

- mode #1:

$$A(1) = \begin{bmatrix} 0.25 & -1 \\ 0 & -2 \end{bmatrix},$$

- mode #2:

$$A(2) = \begin{bmatrix} -2 & -1 \\ 0 & 0.25 \end{bmatrix}.$$

Let the switching between the two modes be described by the following transition rate matrix:

$$\Lambda = \begin{bmatrix} -p & p \\ q & -q \end{bmatrix}.$$

For the steady-state probabilities we have

$$\begin{bmatrix} \pi_1 & \pi_2 \end{bmatrix} \begin{bmatrix} -p & p \\ q & -q \end{bmatrix} = 0,$$
$$\pi_1 + \pi_2 = 1,$$

which gives $\pi_1 = \frac{q}{p+q}$ and $\pi_2 = \frac{p}{p+q}$. If we let $p = q$, this means that the system spends almost the same time at the two modes. Solving now the set of coupled LMIs (2.39) with $p = q = 2$, we get

$$P(1) = \begin{bmatrix} 0.9988 & -0.3388 \\ -0.3388 & 0.9447 \end{bmatrix}, \quad P(2) = \begin{bmatrix} 0.4782 & -0.2452 \\ -0.2452 & 2.0443 \end{bmatrix},$$

which are both symmetric and positive-definite matrices. This implies that the system is stochastically stable even if the two modes are unstable in the deterministic sense.

Now if we solve the set of coupled LMIs (2.39) with $p = q = 0.02$, we find that the system is stochastically unstable.

Example 11. In this example we show that instability in each mode is not necessary for the stochastic stability of the system. For this purpose, let us consider the following data:

- mode #1:

$$A(1) = \begin{bmatrix} -1 & 5 \\ 0 & -2 \end{bmatrix},$$

- mode #2:

$$A(2) = \begin{bmatrix} -2 & 0 \\ 5 & -1 \end{bmatrix}.$$

Let the switching between the two modes be described by the following transition rate matrix:

$$\Lambda = \begin{bmatrix} -p & p \\ q & -q \end{bmatrix}.$$

For the steady-state probabilities we have

$$\begin{bmatrix} \pi_1 & \pi_2 \end{bmatrix} \begin{bmatrix} -p & p \\ q & -q \end{bmatrix} = 0,$$
$$\pi_1 + \pi_2 = 1,$$

which gives $\pi_1 = \frac{q}{p+q}$ and $\pi_2 = \frac{p}{p+q}$. Solving now the set of coupled LMIs (2.39) with $p = q = 0.02$, we get

$$P(1) = \begin{bmatrix} 0.4677 & 0.6361 \\ 0.6361 & 1.9124 \end{bmatrix}, \quad P(2) = \begin{bmatrix} 1.9124 & 0.6361 \\ 0.6361 & 0.4677 \end{bmatrix},$$

which are both symmetric and positive-definite matrices. This implies that the system is stochastically stable even if the two modes are stable in the deterministic sense.

If we solve the set of the coupled LMIs (2.39) with $p = q = 2$, we find that the system is stochastically unstable.

Theorem 5. *The following statements are equivalent:*

1. *System (2.1) is MSS*
2. *The N-coupled LMIs*

$$A^\top(i)P(i) + P(i)A(i) + \sum_{j=1}^{N} \lambda_{ij} P(j) < 0 \qquad (2.41)$$

are feasible for some symmetric and positive-definite matrices $P = (P(1), \cdots, P(N)) > 0$.
3. *For any given set of symmetric and positive-definite matrices $Q = (Q(1), \cdots, Q(N)) > 0$, there exists a unique set of matrices $P = (P(1), \cdots, P(N)) > 0$ satisfying the following:*

$$A^\top(i)P(i) + P(i)A(i) + \sum_{j=1}^{N} \lambda_{ij} P(j) = -Q(i). \qquad (2.42)$$

For well-known reasons (for instance that parameters change due to wearing), it is always desirable that the systems should have a prescribed degree of stability α. The question we are facing now is if a stochastically stable dynamical system of our class has the degree of stability α (α is a given positive constant). To answer this question, we should develop the type of conditions we have to use in this case. For this purpose, let us make the following variable change:

$$\tilde{x}(t) = e^{\alpha t} x(t).$$

Differentiating this relation with respect to time gives

$$\dot{\tilde{x}}(t) = \alpha e^{\alpha t} x(t) + e^{\alpha t} \dot{x}(t) = [A(r(t), t) + \alpha \mathbb{I}] e^{\alpha t} x(t) = [A(r(t), t) + \alpha \mathbb{I}] \tilde{x}(t)$$
$$= A_\alpha(r(t), t) \tilde{x}(t), \tag{2.43}$$

where $A_\alpha(r(t), t) = A(r(t), t) + \alpha \mathbb{I}$.

Letting the uncertainties be equal to zero and applying the stability results of Theorem 4 to system (2.43) when the uncertainties are fixed to zero, the following must hold for every $i \in \mathscr{S}$ to guarantee its stability:

$$A_\alpha^\top(i) P(i) + P(i) A_\alpha(i) + \sum_{j=1}^N \lambda_{ij} P(j) < 0,$$

which gives

$$[A(i) + \alpha \mathbb{I}]^\top P(i) + P(i) [A(i) + \alpha \mathbb{I}] + \sum_{j=1}^N \lambda_{ij} P(j) < 0.$$

The following theorem gives the stochastic stability results for systems with a prescribed degree of stability α.

Corollary 2. *Let α be a given positive constant. System (2.1) with all the uncertainties equal to zero is stochastically stable and has a degree of stochastic stability equal to α if and only if there exists a set of symmetric and positive-definite matrices $P = (P(1), \cdots, P(N)) > 0$ such that the following holds for every $i \in \mathscr{S}$:*

$$[A(i) + \alpha \mathbb{I}]^\top P(i) + P(i) [A(i) + \alpha \mathbb{I}] + \sum_{j=1}^N \lambda_{ij} P(j) < 0. \tag{2.44}$$

From the practical point of view, the determination of the maximum degree of stability is of great importance. This maximum can be obtained in our case by solving the following optimization problem:

$$\begin{cases} \max\limits_{\substack{\alpha \geq 0, \\ P=(P(1),\cdots,P(N))>0}} \alpha, \\ \text{s.t.:} \\ [A(i) + \alpha \mathbb{I}]^\top P(i) + P(i) [A(i) + \alpha \mathbb{I}] + \sum_{j=1}^N \lambda_{ij} P(j) < 0. \end{cases}$$

Notice that the constraints of this optimization problem are nonlinear in the decision variables α and $P(i)$, $i = 1, 2, \cdots, N$ that we can cast into quasi-convex problem with respect to these parameters, and therefore we can use the generalized eigenvalue minimization problem (GEVP) to solve it. For this purpose notice that the previous constraints can be rewritten as follows:

$$A^\top(i) P(i) + P(i) A(i) + 2\alpha P(i) + \sum_{j=1}^N \lambda_{ij} P(j) < 0,$$

which gives

$$2P(i) < -\frac{1}{\alpha} \left[A^\top(i)P(i) + P(i)A(i) + \sum_{j=1}^{N} \lambda_{ij} P(j) \right].$$

Letting $\gamma = \frac{1}{\alpha}$ the previous optimization becomes

$$\text{Pn:} \begin{cases} \min\limits_{\substack{\gamma > 0, \\ P=(P(1),\cdots,P(N))>0}} \gamma, \\ \text{s.t.:} \\ 2P(i) < -\gamma \left[A^\top(i)P(i) + P(i)A(i) + \sum_{j=1}^{N} \lambda_{ij} P(j) \right]. \end{cases} \tag{2.45}$$

If we denote by $\bar{\alpha}$ the solution of this optimization problem, the result tells us that the closed-loop state equation will have decaying behavior with a rate equal to $\bar{\alpha}$.

Remark 6. Notice that the system should be stable, that is:

$$A^\top(i)P(i) + P(i)A(i) + \sum_{j=1}^{N} \lambda_{ij} P(j) < 0,$$

and this requires adding an extra condition when solving the GEVP.

Example 12. To show the usefulness of the results of the previous corollary, let us consider a system with two modes and the following data:

- mode #1:

$$A(1) = \begin{bmatrix} 0.2 & 0.1 \\ 0.0 & -1.0 \end{bmatrix},$$

- mode #2:

$$A(2) = \begin{bmatrix} -1.0 & 0.1 \\ 0.2 & -2.0 \end{bmatrix}.$$

The switching of the mode is described by the following transition rates matrix:

$$\Lambda = \begin{bmatrix} -5.0 & 5.0 \\ 4.0 & -4.0 \end{bmatrix}.$$

Solving the previous GEVP we get

$$P(1) = 10^{-4} \cdot \begin{bmatrix} 0.4028 & 0.0356 \\ 0.0356 & 0.0034 \end{bmatrix}, \quad P(2) = 10^{-4} \cdot \begin{bmatrix} 0.3095 & 0.0279 \\ 0.0279 & 0.0027 \end{bmatrix},$$

which are both symmetric and positive-definite matrices. The corresponding degree of stability is $\alpha = 0.3791$.

All the results developed in this section are valid only for nominal dynamical systems of the class of piecewise deterministic systems. When functioning, there is no guarantee that the real system will remain stable even if the conditions used to check the stability are satisfied, since the uncertainties that may be caused by different phenomena are not taken into account in the analysis. To avoid this, it is necessary to establish conditions that take care of these uncertainties. The next section will deal with such problems. It will consider the stability problem with uncertainties of the state matrix. Section 2.4 will treat the class of systems with Markovian jumps and Wiener process and chain up by considering uncertainties on both the state matrix and the transition probability rate matrix.

2.3 Robust Stability

Let us now return to the class of dynamical systems governed by (2.1) and assume this time that the uncertainties in the state matrix are not equal to zero. Our goal is to establish conditions that allow us to check if the considered system is robustly stochastically stable.

The following theorem gives sufficient conditions in the LMI formalism that can be used to reach our goal.

Theorem 6. *System (2.1) is robustly stochastically stable if there exist a set of symmetric and positive-definite matrices $P = (P(1), \cdots, P(N)) > 0$ and a set of positive scalars $\varepsilon_A = (\varepsilon_A(1), \cdots, \varepsilon_A(N)) > 0$ such that the following holds for every $i \in \mathscr{S}$ and for all admissible uncertainties:*

$$\begin{bmatrix} J_u(i) & P(i)D_A(i) \\ D_A^\top(i)P(i) & -\varepsilon_A(i)\mathbb{I} \end{bmatrix} < 0, \tag{2.46}$$

where $J_u(i) = A^\top(i)P(i) + P(i)A(i) + \sum_{j=1}^{N} \lambda_{ij} P(j) + \varepsilon_A(i)E_A^\top(i)E_A(i)$.

Proof: To prove this theorem, let us consider a Lyapunov candidate function with the following expression:

$$V(x(t), i) = x^\top(t)P(i)x(t), \tag{2.47}$$

where $P(i)$ is a symmetric and positive-definite matrix for every $i \in \mathscr{S}$, solution of (2.46).

Let $\mathscr{L}$ be the infinitesimal operator of the Markov process $\{(x(t), r(t)), t \geq 0\}$. If at time t, $x(t) = x$ and $r(t) = i$ for $i \in \mathscr{S}$, the infinitesimal operator acting on $V(.)$ and emanating from the point (x, i) at time t is given by

$$\mathscr{L}V(x(t), i) = \lim_{h \to 0} \frac{\mathbb{E}\left[V(x(t+h), r(t+h))|x(t) = x(t), r(t) = i\right] - V(x(t), i)}{h}$$

$$= \dot{x}^\top(t)P(i)x(t) + x^\top(t)P(i)\dot{x}(t) + \sum_{j=1}^{N} \lambda_{ij} x^\top(t)P(j)x(t)$$

$$= x^\top(t)A^\top(i,t)P(i)x(t) + x^\top(t)P(i)A(i,t)x(t)$$
$$+ \sum_{j=1}^{N}\lambda_{ij}x^\top(t)P(j)x(t).$$

Using the structure of the uncertainties we get

$$\mathscr{L}V(x(t),i) = x^\top(t)\left[A(i) + D_A(i)F_A(i,t)E_A(i)\right]^\top P(i)x(t)$$
$$+ x^\top(t)P(i)\left[A(i) + D_A(i)F_A(i,t)E_A(i)\right]x(t)$$
$$+ \sum_{j=1}^{N}\lambda_{ij}x^\top(t)P(j)x(t)$$
$$= x^\top(t)\left[A^\top(i)P(i) + P(i)A(i)\right]x(t)$$
$$+ 2x^\top(t)P(i)D_A(i)F_A(i,t)E_A(i)x(t)$$
$$+ x^\top(t)\sum_{j=1}^{N}\lambda_{ij}P(j)x(t). \tag{2.48}$$

Notice that using Lemma 7 in Appendix A we have

$$2x^\top(t)P(i)D_A(i)F_A(i,t)E_A(i)x(t) \le \varepsilon_A^{-1}(i)x^\top(t)P(i)D_A(i)D_A^\top(i)P(i)x(t)$$
$$+ \varepsilon_A(i)x^\top(t)E_A^\top(i)E_A(i)x(t).$$

Using this relation, (2.48) becomes

$$\mathscr{L}V(x(t),i) \le x^\top(t)\left[A^\top(i)P(i) + P(i)A(i) + \sum_{j=1}^{N}\lambda_{ij}P(j)\right]x(t)$$
$$+ \varepsilon_A^{-1}(i)x^\top(t)P(i)D_A(i)D_A^\top(i)P(i)x(t)$$
$$+ \varepsilon_A(i)x^\top(t)E_A^\top(i)E_A(i)x(t)$$
$$= x^\top(t)\Lambda_u(i)x(t),$$

with

$$\Lambda_u(i) = A^\top(i)P(i) + P(i)A(i) + \sum_{j=1}^{N}\lambda_{ij}P(j)$$
$$+ \varepsilon_A^{-1}(i)P(i)D_A(i)D_A^\top(i)P(i) + \varepsilon_A(i)E_A^\top(i)E_A(i).$$

Using (2.46) and the Schur complement we conclude that $\Lambda_u(i) < 0$ for every $i \in \mathscr{S}$. Based on this we get in turn that

$$\mathscr{L}V(x(t),i) \le -\min_{i\in\mathscr{S}}\{\lambda_{min}(-\Lambda_u(i))\}x^\top(t)x(t).$$

The rest of the proof is similar to the one of Theorem 4 and the details are omitted. This ends the proof of Theorem 6.   □

Remark 7. The previous proof can be replaced by the following one. In fact, from the results of Theorem 6, the uncertain system will be stable if the following LMI holds for each $i \in \mathscr{S}$ and for all admissible uncertainties:

$$A^\top(i,t)P(i) + P(i)A(i,t) + \sum_{j=1}^{N} \lambda_{ij}P(j) < 0.$$

Using the expression of the matrix $A(i,t)$, this inequality is equivalent to the following one:

$$A^\top(i)P(i) + P(i)A(i) + P(i)D_A(i)F_A(i,t)E_A(i)$$
$$+E_A^\top(i)F_A^\top(i,t)D_A^\top(i)P(i) + \sum_{j=1}^{N} \lambda_{ij}P(j) < 0.$$

Based on Lemma 7 in Appendix A we get

$$P(i)D_A(i)F_A(i,t)E_A(i) + E_A^\top(i)F_A^\top(i,t)D_A^\top(i)P(i)$$
$$\leq \varepsilon_A^{-1}(i)P(i)D_A(i)D_A^\top(i)P(i) + \varepsilon_A(i)E_A^\top(i)E_A(i).$$

Using this, the previous inequality will hold if the following holds for every $i \in \mathscr{S}$:

$$A^\top(i)P(i) + P(i)A(i) + \sum_{j=1}^{N} \lambda_{ij}P(j) + \varepsilon_A(i)E_A^\top(i)E_A(i)$$
$$+\varepsilon_A^{-1}(i)P(i)D_A(i)D_A^\top(i)P(i) < 0.$$

After using the Schur complement we get the following:

$$\left[\begin{array}{cc} \begin{bmatrix} A^\top(i)P(i) \\ +P(i)A(i) \\ +\varepsilon_A(i)E_A^\top(i)E_A(i) \\ +\sum_{j=1}^{N} \lambda_{ij}P(j) \end{bmatrix} & P(i)D_A(i) \\ \\ D_A^\top(i)P(i) & -\varepsilon_A(i)\mathbb{I} \end{array} \right] < 0.$$

This ends the proof. $\square$

Remark 8. The condition we gave in Theorem 6 is sufficient, which means that if we are not able to find a set of symmetric and positive-definite matrices $P = (P(1), \cdots, P(N)) > 0$ that satisfies the condition (2.46), this does not imply that the dynamical system is not robustly stochastically stable.

Example 13. In this example we show how the results on robust stability given in the previous theorem can be used. For this purpose let us consider a dynamical system with two modes. The different matrices are given by

- mode #1:

$$A(1) = \begin{bmatrix} 0.2 & 0.1 \\ 0.0 & -1.0 \end{bmatrix}, \quad D_A(1) = \begin{bmatrix} 0.1 \\ 0.2 \end{bmatrix}, \quad E_A(1) = \begin{bmatrix} 0.2 & 0.1 \end{bmatrix},$$

- mode #2:

$$A(2) = \begin{bmatrix} -1.0 & 0.1 \\ 0.2 & -2.0 \end{bmatrix}, \quad D_A(2) = \begin{bmatrix} 0.13 \\ 0.1 \end{bmatrix}, \quad E_A(2) = \begin{bmatrix} 0.1 & 0.2 \end{bmatrix}.$$

The transition probability rate matrix is given by

$$\Lambda = \begin{bmatrix} -4 & 4 \\ 1 & -1 \end{bmatrix}.$$

Letting $\varepsilon_A(1) = \varepsilon_A(2) = 0.5$ and solving the set of coupled LMIs (2.46), we get

$$P(1) = \begin{bmatrix} 1.8947 & 0.0972 \\ 0.0972 & 0.7245 \end{bmatrix}, \quad P(2) = \begin{bmatrix} 1.2669 & 0.0761 \\ 0.0761 & 0.5849 \end{bmatrix}.$$

These two matrices are symmetric and positive-definite and therefore, following the previous theorem, the considered system is robustly stochastically stable for all the admissible uncertainties.

As we did for the nominal systems, let us see how we can determine if a given system belonging to the class of systems we are considering has a prescribed degree of stability α. To answer this question, we will proceed as we did for the previous section. By applying the results on robust stability of Theorem 6 for system (2.1), we get for a given positive constant α that this system will be robustly stochastically stable if there exist a set of symmetric and positive-definite matrices $P = (P(1), \cdots, P(N)) > 0$ and a set of positive scalars $\varepsilon_A = (\varepsilon_A(1), \cdots, \varepsilon_A(N)) > 0$ such that the following holds for each $i \in \mathscr{S}$ and for all admissible uncertainties:

$$\begin{bmatrix} J_u^\alpha(i) & P(i)D_A(i) \\ D_A^\top(i)P(i) & -\varepsilon_A(i)\mathbb{I} \end{bmatrix} < 0, \tag{2.49}$$

where $J_u^\alpha(i) = [A(i) + \alpha\mathbb{I}]^\top P(i) + P(i)[A(i) + \alpha\mathbb{I}] + \sum_{j=1}^N \lambda_{ij} P(j) + \varepsilon_A(i) E_A^\top(i) E_A(i)$.

The following theorem gives the results on robust stochastic stability for a dynamical system of the class of piecewise deterministic systems we are considering to have a prescribed degree of stability equal to α.

Corollary 3. *Let α be a given positive constant. If there exists a set of symmetric and positive-definite matrices $P = (P(1), \cdots, P(N)) > 0$ and a set of positive scalars $\varepsilon_A = (\varepsilon_A(1), \cdots, \varepsilon_A(N)) > 0$ such that (2.49) holds for every $i \in \mathscr{S}$ and for all admissible uncertainties, then system (2.1) is robustly stochastically stable.*

In the previous section we computed the maximum prescribed degree of stability for the nominal dynamical system of our class of systems by solving the optimization problem (2.45). In a similar way we can solve the following optimization problem to get the maximum prescribed degree for robust stability:

$$
\text{Pu:} \begin{cases} \min\limits_{\substack{\gamma>0, \\ \varepsilon_A=(\varepsilon_A(1),\cdots,\varepsilon_A(N))>0, \\ P=(P(1),\cdots,P(N))>0,}} \gamma, \\[2em] \text{s.t.:} \\[0.5em] \begin{bmatrix} 2P(i) & 0 \\ 0 & 0 \end{bmatrix} < -\gamma \begin{bmatrix} \tilde{J}_u^{\alpha}(i) & P(i)D_A(i) \\ D_A^{\top}(i)P(i) & -\varepsilon_A(i)\mathbb{I} \end{bmatrix}, \end{cases}
$$

where $\tilde{J}_u^{\alpha}(i) = A^{\top}(i)P(i) + P(i)A(i) + \sum_{j=1}^{N} \lambda_{ij}P(j) + \varepsilon_A(i)E_A^{\top}(i)E_A(i)$.

Example 14. To show the usefulness of the results of the previous corollary, let us consider the system with two modes of the previous example with the same data. Letting $\varepsilon_A(1) = \varepsilon_A(2) = 0.5$ and solving the previous generalized eigenvalue problem optimization problem we get

$$
P(1) = \begin{bmatrix} 0.7666 & 0.0359 \\ 0.0359 & 0.3271 \end{bmatrix}, \quad P(2) = \begin{bmatrix} 0.4475 & 0.0223 \\ 0.0223 & 0.2474 \end{bmatrix},
$$

which are both symmetric and positive-definite matrices. The corresponding degree of stability is $\alpha = 0.6055$.

In the next section we will consider the case when our class of system is perturbed by a Wiener process and see how this will affect the stability and robust stability conditions developed earlier.

2.4 Stability of Systems with Wiener Process

In this section we consider the case when an external Wiener process is acting on the class of systems we are studying and see the effect on the stability and robust stability conditions developed earlier. In the rest of this section, we treat the case of systems without uncertainties followed by the case with norm-bounded uncertainties in the state matrix and polytopic uncertainties on the transition probability rate matrix.

When the external Wiener process is acting on our class of systems the state equation becomes

$$
\begin{cases} dx(t) = A(r(t), t)x(t)dt + \mathbb{W}(r(t))x(t)d\omega(t), \\ x(0) = x_0, \end{cases} \tag{2.50}
$$

where $x(t) \in \mathbb{R}^n$ is the state vector at time t; $\omega(t) \in \mathbb{R}$ is a standard Wiener process that is assumed to be independent of the Markov process $\{r(t), t \geq 0\}$;

$A(r(t), t)$ is a matrix with appropriate dimension and its expression is given by

$$A(r(t), t) = A(r(t)) + D_A(r(t))F_A(r(t), t)E_A(r(t)),$$

with $A(r(t))$; $D_A(r(t))$, $E_A(r(t))$ are known matrices with appropriate dimensions and $F_A(r(t), t)$ satisfies (2.3); and the matrix $\mathbb{W}(i)$ is supposed to be given for each mode $i \in \mathscr{S}$.

Let us now assume that all the uncertainties are equal to zero in the previous state equation and see how we can establish the conditions that can be used to check the stochastic stability of this class of systems with Wiener process.

The result on stochastic stability is given by the following theorem.

Theorem 7. *System (2.50) is stochastically stable if and only if there exists a set of symmetric and positive-definite matrices $P = (P(1), \cdots, P(N)) > 0$ such that the following LMI holds for every $i \in \mathscr{S}$:*

$$A^\top(i)P(i) + P(i)A(i) + \mathbb{W}^\top(i)P(i)\mathbb{W}(i) + \sum_{j=1}^N \lambda_{ij}P(j) < 0. \qquad (2.51)$$

Proof: Let us start by the proof of necessity. For this purpose, let us assume that the system is stochastically stable and denote by $\Phi(.)$ the transition matrix corresponding to

$$d\Phi(t) = A(r(t))\Phi(t) + \mathbb{W}(r(t))\Phi(t)d\omega(t), \Phi(0) = \mathbb{I}.$$

Since the system is stable, this implies that for any symmetric and positive-definite matrix $Q(i) > 0, i \in \mathscr{S}$ the following holds:

$$\mathbb{E}\left[\int_0^\infty x^\top(s)Q(r(s))x(s)ds \Big| x(0) = x, r(0) = r_0\right] < \infty.$$

Let the function $\Psi : \mathbb{R}^+ \times \mathbb{R}^+ \times \mathbb{R}^n \times \mathscr{S} \to \mathbb{R}^n$ be defined by

$$\Psi(T, t, x, i) = \mathbb{E}\left[\int_t^T x^\top(s)Q(r(s))x(s)ds \Big| x(t) = x, r(t) = i\right].$$

As we did previously, by the time-homogeneous property, we get

$$\begin{aligned}
\Psi(T, t, x, i) &= \Psi(T - t, 0, x, i) \\
&= \mathbb{E}\left[\int_0^{T-t} x^\top(s)Q(r(s))x(s)ds \Big| x(0) = x, r(0) = i\right] \\
&= x^\top \mathbb{E}\left[\int_t^T \Phi^\top(s)Q(r(s))\Phi(s)ds \Big| r(0) = i\right] x
\end{aligned}$$

$$= x^\top P(T-t,i)x$$
$$= \Theta(T-t,x,i).$$

From the other side, since the system $P(.,i)$ is monotonically increasing, positive-definite matrix-valued function and bounded from above, thus

$$P(i) = \lim_{T\to\infty} P(T,i)$$

exists. Hence $P(i)$ is also positive-definite.

Let T be arbitrary fixed time. For any $T > s > t > 0$, we have

$$\frac{d}{dt}\mathbb{E}\left[\Theta(T-t,x,i)\middle| \text{ evaluated along the system trajectory}\right]$$

$$= \lim_{s\to t}\frac{1}{s-t}\left[\mathbb{E}\left(\Theta(T-s,x(s),r(s))|x(t)=x,r(t)=i\right) - \Theta(T-t,x,i)\right]$$

$$= \frac{\partial}{\partial t}\Theta(T-t,x,i) + \mathscr{L}\Theta(T-t,x,i)$$

$$= x^\top(t)\frac{\partial}{\partial t}P(T-t,x,i)x(t) + x^\top(t)\left[A^\top(i)P(T-t,i) + P(T-t,i)A(i)\right]x(t)$$

$$+ \sum_{j=1}^{N}\lambda_{ij}x^\top(t)P(T-t,j)x(t)$$

$$= x^\top(t)\left[A^\top(i)P(T-t,i) + P(T-t,i)A(i)\right.$$

$$\left. + \sum_{j=1}^{N}\lambda_{ij}P(T-t,j)\right]x(t). \tag{2.52}$$

On the other hand, by (2.11) we have

$$\mathbb{E}\left[\Theta(T-s,x(s),r(s))|x(t)=x,r(t)=i\right] - \Theta(T-t,x,i)$$

$$= \left[\mathbb{E}(\Theta(T-s,x(s),r(s)) - \Theta(T-t,x,i)|x(t)=x,r(t)=i)\right]$$

$$= \mathbb{E}\left\{\mathbb{E}\left[\int_s^T x^\top(v)Q(r(v))x(v)dv|x(s),r(s)\right]\right.$$

$$\left. -\mathbb{E}\left\{\int_t^T x^\top(v)Q(r(v))x(v)dv|x(t)=x,r(t)=i\right\}|x(t)=x,r(t)=i\right\}$$

$$= \mathbb{E}\left\{\int_s^T x^\top(v)Q(r(v))x(v)dv - \int_t^T x^\top(v)Q(r(v))x(v)dv|x(t)=x,r(t)=i\right\}$$

$$= -\mathbb{E}\left\{\int_0^{s-t} x^\top(v)Q(r(v))x(v)dv|x_0=x,r_0=i\right\}$$

$$= -x^\top(t)\mathbb{E}\left\{\int_0^{s-t}\Phi^\top(u)\Phi(u)du|r_0=i\right\}x(t). \tag{2.53}$$

However,

$$
\lim_{s \to t} \frac{1}{s-t} \mathbb{E} \left\{ \int_0^{s-t} \Phi^\top(u) Q(r(u)) \Phi(u) du \big| r_0 = i \right\}
$$
$$
= \lim_{s \to t} \mathbb{E} \left\{ \frac{1}{s-t} \int_0^{s-t} \Phi^\top(u) Q(r(u)) \Phi(u) du \big| r(0) = i \right\}
$$
$$
= \mathbb{E}\{Q(i) | r_0 = i\} = Q(i),
$$

which implies

$$
\frac{d}{dt} \mathbb{E}\left[\Theta(T-t, x, i) \right] = -x^\top(t) Q(i) x(t). \tag{2.54}
$$

Let T go to infinity in (2.52) and note that $\frac{\partial}{\partial t} P(T-t, i)$ tends to zero as T tends to infinity. Then (2.52), (2.53), and (2.54) give the desired result. This proves necessity.

To prove the sufficiency, let $P(i) > 0$, $i \in \mathscr{S}$ be a symmetric and positive-definite matrix, that represents a solution of the LMI (2.51). Then a Lyapunov candidate function can be given by the following expression:

$$
V(x(t), i) = x^\top(t) P(i) x(t), \text{ when } r(t) = i.
$$

Using the results of Appendix A, the infinitesimal operator of the Markov process $\{(x(t), r(t)), t \geq 0\}$, acting on $V(.)$ and emanating from the point (x, i) at time t, when at time t, $x(t) = x$ and $r(t) = i$ for $i \in \mathscr{S}$, is given by

$$
\mathscr{L}V(x(t), i) = [A(i)x(t)]^\top V_x(x(t), i) + \sum_{j=1}^N \lambda_{ij} V(x(t), j)
$$
$$
+ \frac{1}{2} \text{tr} \left[x^\top(t) \mathbb{W}^\top(i) V_{xx}(x(t), i) \mathbb{W}(i) x(t) \right].
$$

Notice that $V_x(x(t), i) = 2P(i)x(t)$ and $V_{xx}(x(t), i) = 2P(i)$, which implies

$$
\mathscr{L}V(x(t), i) = 2x^\top(t) A^\top(i) P(i) x(t) + \sum_{j=1}^N \lambda_{ij} x^\top(t) P(j) x(t)
$$
$$
+ x^\top(t) \mathbb{W}^\top(i) P(i) \mathbb{W}(i) x(t)
$$
$$
= x^\top(t) \left[A^\top(i) P(i) + P(i) A(i) \right] x(t)
$$
$$
+ x^\top(t) \left[\mathbb{W}^\top(i) P(i) \mathbb{W}(i) + \sum_{j=1}^N \lambda_{ij} P(j) \right] x(t)
$$
$$
= x^\top(t) \Gamma(i) x(t),
$$

with $\Gamma(i) = A^\top(i) P(i) + P(i) A(i) + \mathbb{W}^\top(i) P(i) \mathbb{W}(i) + \sum_{j=1}^N \lambda_{ij} P(j)$.

Using condition (2.51) we get

$$\mathscr{L}V(x(t),i) \leq -\min_{i \in \mathscr{S}}\{\lambda_{min}(-\Gamma(i))\}x^\top(t)x(t),$$

The rest of the proof is similar to the one of Theorem 4 and the details are omitted. This ends the proof of Theorem 7. □

Example 15. To illustrate the usefulness of the results of this theorem let us consider a two-mode dynamical system of the class studied in this section. For simplicity of presentation, let the state vector $x(t)$ belong to $\mathbb{R}^2$ and the state matrices and the disturbance noise matrices be given by

- mode #1:

$$A(1) = \begin{bmatrix} 0.01 & 0.0 \\ 0.0 & -0.2 \end{bmatrix}, \quad \mathbb{W}(1) = \begin{bmatrix} 0.2 & 0.0 \\ 0.0 & 0.2 \end{bmatrix},$$

- mode #2:

$$A(2) = \begin{bmatrix} -0.1 & 0.0 \\ 0.0 & 0.2 \end{bmatrix}, \quad \mathbb{W}(2) = \begin{bmatrix} 0.1 & 0.0 \\ 0.0 & 0.1 \end{bmatrix}.$$

The transition probability rate matrix between the two modes is given by

$$\Lambda = \begin{bmatrix} -2.0 & 2.0 \\ 5.0 & -5.0 \end{bmatrix}.$$

Solving the set of coupled LMIs (2.51), we get the following matrices:

$$P(1) = \begin{bmatrix} 645.4258 & 0.0 \\ 0.0 & 635.4497 \end{bmatrix}, \quad P(2) = \begin{bmatrix} 621.7449 & 0.0 \\ 0.0 & 715.8490 \end{bmatrix}.$$

These two matrices are symmetric and positive-definite and, based on the previous theorem, the system is stochastically stable.

Let us now return to the initial state equation with uncertainties and Wiener process and establish the stochastic stability conditions.

The following theorem summarizes the results in this case.

Theorem 8. *System (2.50) is robustly stochastically stable if there exist a set of symmetric and positive-definite matrices $P = (P(1), \cdots, P(N)) > 0$ and a set of positive scalars $\varepsilon_A = (\varepsilon_A(1), \cdots, \varepsilon_A(N)) > 0$ such that the following LMI holds for each $i \in \mathscr{S}$ and for all admissible uncertainties:*

$$\begin{bmatrix} J_w(i) & P(i)D_A(i) \\ D_A^\top(i)P(i) & -\varepsilon_A(i)\mathbb{I} \end{bmatrix} < 0, \tag{2.55}$$

with $J_w(i) = A^\top(i)P(i) + P(i)A(i) + \mathbb{W}^\top(i)P(i)\mathbb{W}(i) + \varepsilon_A(i)E_A^\top(i)E_A(i) + \sum_{j=1}^N \lambda_{ij}P(j)$.

Proof: Let $P(i) > 0$, $i \in \mathscr{S}$, be a symmetric and positive-definite matrix that represents a solution of the LMI (2.55). Then a Lyapunov function candidate can be given by the following expression:

$$V(x(t), i) = x^\top(t) P(i) x(t), \text{ when } r(t) = i.$$

Using the results of Appendix A, the infinitesimal operator of the Markov process $\{(x(t), r(t)), t \geq 0\}$ acting on $V(.)$ and emanating from the point (x, i) at time t, where $x(t) = x$ and $r(t) = i$ for $i \in \mathscr{S}$, is given by

$$\mathscr{L}V(x(t), i) = [A(i,t)x(t)]^\top V_x(x(t)) + \sum_{j=1}^{N} \lambda_{ij} V(x(t), j)$$

$$+ \frac{1}{2} \text{tr} \left[x^\top(t) \mathbb{W}^\top(i) V_{xx}(x(t), i) \mathbb{W}(i) x(t) \right].$$

Using the following expressions for $V_x(x(t), i)$ and $V_{xx}(x(t), i)$ given, respectively, by $V_x(x(t), i) = 2P(i)x(t)$ and $V_{xx}(x(t), i) = 2P(i)$, we obtain

$$\mathscr{L}V(x(t), i) = 2x^\top(t) A^\top(i,t) P(i) x(t) + \sum_{j=1}^{N} \lambda_{ij} x^\top(t) P(j) x(t)$$

$$+ x^\top(t) \mathbb{W}^\top(i) P(i) \mathbb{W}(i) x(t)$$

$$= x^\top(t) \Big[A^\top(i) P(i)$$

$$+ P(i)A(i) + E_A^\top(i) F_A^\top(i,t) D_A^\top(i) P(i)$$

$$+ P(i) D_A(i) F_A(i,t) E_A(i)$$

$$+ \mathbb{W}^\top(i) P(i) \mathbb{W}(i) + \sum_{j=1}^{N} \lambda_{ij} P(j) \Big] x(t).$$

Based on the results of Lemma 7 in Appendix A, we have

$$2x^\top(t) P(i) D_A(i) F_A(i,t) E_A(i) x(t)$$

$$\leq \varepsilon_A^{-1}(i) x^\top(t) P(i) D_A(i) D^\top(i) P(i) x(t)$$

$$+ \varepsilon_A(i) x^\top(t) E_A^\top(i) E_A(i) x(t).$$

Considering this, the previous relation becomes

$$\mathscr{L}V(x(t), i) \leq x^\top(t) \Big[A^\top(i) P(i) + P(i)A(i)$$

$$+ \varepsilon_A^{-1}(i) P(i) D_A(i) D_A^\top(i) P(i)$$

$$+ \varepsilon_A(i) E_A^\top(i) E_A(i) + \mathbb{W}^\top(i) P(i) \mathbb{W}(i)$$

$$+ \sum_{j=1}^{N} \lambda_{ij} P(j) \Big] x(t)$$

$$\leq x^\top(t)\Gamma_u(i)x(t), \tag{2.56}$$

with

$$\Gamma_u(i) = A^\top(i)P(i) + P(i)A(i) + \varepsilon_A^{-1}(i)P(i)D_A(i)D_A^\top(i)P(i)$$

$$+\varepsilon_A(i)E_A^\top(i)E_A(i) + \mathbb{W}^\top(i)P(i)\mathbb{W}(i) + \sum_{j=1}^{N}\lambda_{ij}P(j).$$

Using condition (2.55) and the Schur complement, we conclude that $\Gamma_u(i) < 0$ and therefore we get

$$\mathcal{L}V(x(t), i) \leq -\min_{i\in\mathscr{S}}\{\lambda_{min}(-\Gamma_u(i))\}x^\top(t)x(t). \tag{2.57}$$

The rest of the proof is similar to the one of Theorem 4 and the details are omitted. This ends the proof of Theorem 8.   □

Remark 9. Notice that $\mathbb{W}(i)P(i)\mathbb{W}(i) = \mathbb{W}(i)P(i)P^{-1}(i)P(i)\mathbb{W}(i)$, which gives another form for the LMI of the previous theorem, that is:

$$\begin{bmatrix} \tilde{J}_w(i) & P(i)D_A(i) & \mathbb{W}^\top(i)P(i) \\ D_A^\top(i)P(i) & -\varepsilon_A(i)\mathbb{I} & 0 \\ P(i)\mathbb{W}(i) & 0 & -P(i) \end{bmatrix} < 0,$$

with $\tilde{J}_w(i) = A^\top(i)P(i) + P(i)A(i) + \varepsilon_A(i)E_A^\top(i)E_A(i) + \sum_{j=1}^{N}\lambda_{ij}P(j)$.

Example 16. To show the results of Theorem 8, let us consider an uncertain system with two modes. The switching between these two modes is described by the following transition probability rate matrix:

$$\Lambda = \begin{bmatrix} -4.0 & 4.0 \\ 1.0 & -1.0 \end{bmatrix}.$$

The state matrices $A(i), i = 1, 2$; noise matrices, $\mathbb{W}(i), i = 1, 2$; and the uncertainties matrices are given by

- mode #1:

$$A(1) = \begin{bmatrix} 0.2 & 0.1 \\ 0 & -1.0 \end{bmatrix}, \quad D_A(1) = \begin{bmatrix} 0.1 \\ 0.2 \end{bmatrix}, \quad E_A(1) = \begin{bmatrix} 0.2 & 0.1 \end{bmatrix}$$

$$W(1) = \begin{bmatrix} 0.1 & 0.0 \\ 0.0 & 0.1 \end{bmatrix},$$

- mode #2:

$$A(2) = \begin{bmatrix} -1.0 & 0.1 \\ 0.2 & -2.0 \end{bmatrix}, \quad D_A(2) = \begin{bmatrix} 0.13 \\ 0.1 \end{bmatrix}, \quad E_A(2) = \begin{bmatrix} 0.1 & 0.2 \end{bmatrix}$$

$$W(2) = \begin{bmatrix} 0.2 & 0.0 \\ 0.0 & 0.2 \end{bmatrix}.$$

Solving the set of coupled LMIs (2.55), we get the following matrices:

$$P(1) = \begin{bmatrix} 1.9207 & 0.0988 \\ 0.0988 & 0.7316 \end{bmatrix}, \quad P(2) = \begin{bmatrix} 1.2875 & 0.0773 \\ 0.0773 & 0.5925 \end{bmatrix}.$$

The eigenvalues of these two matrices are all positive, which implies that they are symmetric and positive-definite. The conditions of the previous theorem are satisfied and consequently the system under study in this example is robustly stochastically stable.

In most of the previous results we have assumed that the jump rate was free of uncertainties, which is not real in practice since it is always difficult to get the exact transition probability matrix. In the rest of this chapter we will try to take care of the uncertainties that may affect the jump rates and establish equivalent results to the previous ones. The uncertainties we will consider for the transition probability rate matrix are polytopic and were presented earlier in this chapter.

Let us now assume that we have uncertainties on the state matrix and on the transition probability rate matrix that satisfy the previous given conditions. In this case, the system will be stable if the following holds for all admissible uncertainties:

$$A^\top(i,t)P(i) + P(i)A(i,t) + \sum_{k=1}^\kappa \sum_{j=1}^N \alpha_k \lambda_{ij}^k P(j) < 0,$$

which gives

$$A^\top(i,t)P(i) + P(i)A(i,t) + \sum_{j=1}^N \mu_{ij} P(j) < 0,$$

with $\mu_{ij} = \sum_{k=1}^\kappa \alpha_k \lambda_{ij}^k$.

Moreover, if we take care of the expression of $A(i,t)$, we get in a similar way that

$$\begin{bmatrix} J_{uu}(i) & P(i)D_A(i) \\ D_A^\top(i)P(i) & -\varepsilon_A(i)\mathbb{I} \end{bmatrix} < 0,$$

where $J_{uu} = A^\top(i)P(i) + P(i)A(i) + \sum_{j=1}^N \mu_{ij} P(j) + \varepsilon_A(i)E_A^\top(i)E_A(i)$.

The following theorem gives the results that can be used to check if a given system with uncertainties on the state matrix and on the jump rates is robustly stochastically stable.

Theorem 9. *System (2.1) is robustly stochastically stable if there exist a set of symmetric and positive-definite matrices $P = (P(1), \cdots, P(N)) > 0$ and a set of positive scalars $\varepsilon_A = (\varepsilon_A(1), \cdots, \varepsilon_A(N)) > 0$ such that the following holds for every $i \in \mathscr{S}$ and for all admissible uncertainties:*

$$\begin{bmatrix} J_{uu}(i) & P(i)D_A(i) \\ D_A^\top(i)P(i) & -\varepsilon_A(i)\mathbb{I} \end{bmatrix} < 0,$$

where $J_{uu} = A^\top(i)P(i) + P(i)A(i) + \sum_{j=1}^{N} \mu_{ij}P(j) + \varepsilon_A(i)E_A^\top(i)E_A(i)$.

Example 17. To show the validity of the results of the previous theorem, let us consider the two-mode system of the previous example with the same data and $\mathbb{W}(1) = \mathbb{W}(2) = 0$.

The switching between the two modes is described by the following transition rate matrices:

$$\Lambda_1 = \begin{bmatrix} -4.0 & 4.0 \\ 1.0 & -1.0 \end{bmatrix}, \quad \alpha_1 = 0.6,$$

$$\Lambda2 = \begin{bmatrix} -3.0 & 3.0 \\ 1.0 & -1.0 \end{bmatrix}, \quad \alpha_2 = 0.4,$$

which gives

$$\mu = \begin{bmatrix} -3.6 & 3.6 \\ 1.0 & -1.0 \end{bmatrix}.$$

Letting $\varepsilon_A(1) = \varepsilon_A(2) = 0.5$ and solving the set of coupled LMIs gives

$$P(1) = \begin{bmatrix} 1.9941 & 0.1012 \\ 0.1012 & 0.7441 \end{bmatrix}, \quad P(2) = \begin{bmatrix} 1.2945 & 0.0773 \\ 0.0773 & 0.5929 \end{bmatrix},$$

which are both symmetric and positive-definite matrices. This implies that the system is stochastically stable.

We can establish similar results when a Wiener process disturbance is acting on the system. The corresponding result is summarized in the following theorem, which can be proved following the steps used previously.

Theorem 10. *System (2.50) is robustly stochastically stable if there exist a set of symmetric and positive-definite matrices $P = (P(1), \cdots, P(N)) > 0$ and a set of positive scalars $\varepsilon_A = (\varepsilon_A(1), \cdots, \varepsilon_A(N)) > 0$ such the following LMI holds for every $i \in \mathscr{S}$ and for all admissible uncertainties:*

$$\begin{bmatrix} J_w(i) & P(i)D_A(i) \\ D_A^\top(i)P(i) & -\varepsilon_A(i)\mathbb{I} \end{bmatrix} < 0, \tag{2.58}$$

with

$$J_w(i) = A^\top(i)P(i) + P(i)A(i) + \mathbb{W}^\top(i)P(i)\mathbb{W}(i)$$

$$+ \varepsilon_A(i)E_A^\top(i)E_A(i) + \sum_{j=1}^{N} \mu_{ij}P(j).$$

Example 18. To show the validity of the results of the previous theorem, let us consider the two-mode system of Example 16 with the same data. The switching between the two modes is described by the following transition rate matrices:

$$\Lambda_1 = \begin{bmatrix} -4.0 & 4.0 \\ 1.0 & -1.0 \end{bmatrix}, \quad \alpha_1 = 0.6,$$

$$\Lambda2 = \begin{bmatrix} -3.0 & 3.0 \\ 1.0 & -1.0 \end{bmatrix}, \quad \alpha_2 = 0.4,$$

which gives

$$\mu = \begin{bmatrix} -3.6 & 3.6 \\ 1.0 & -1.0 \end{bmatrix}.$$

Letting $\varepsilon_A(1) = \varepsilon_A(2) = 0.5$ and solving the LMI gives:

$$P(1) = \begin{bmatrix} 1.9941 & 0.1012 \\ 0.1012 & 0.7441 \end{bmatrix}, \quad P(2) = \begin{bmatrix} 1.2945 & 0.0773 \\ 0.0773 & 0.5929 \end{bmatrix},$$

which are both symmetric and positive-definite matrices. This implies that the system is stochastically stable.

Previously we proposed a way to compute the maximum prescribed degree of stability for the nominal and uncertain dynamical system of our class of systems by solving the optimization problem (2.45). In a similar manner we can solve the following optimization to get the maximum prescribed degree for robust stability when the system is perturbed by external Wiener process with polytopic uncertainties of the transition probability rate matrix:

$$\text{Pw:} \begin{cases} \min\limits_{\substack{\gamma>0, \\ \varepsilon_A=(\varepsilon_A(1),\cdots,\varepsilon_A(N))>0, \\ P=(P(1),\cdots,P(N))>0,}} \gamma, \\ \text{s.t.:} \\ \begin{bmatrix} 2P(i) & 0 \\ 0 & 0 \end{bmatrix} < -\gamma \begin{bmatrix} \tilde{J}_w^\alpha(i) & P(i)D_A(i) \\ D_A^\top(i)P(i) & -\varepsilon_A(i)\mathbb{I} \end{bmatrix}, \end{cases}$$

where $\tilde{J}_w^\alpha(i) = A^\top(i)P(i) + P(i)A(i) + \mathbb{W}(i)P(i)\mathbb{W}(i)\sum_{j=1}^N \mu_{ij}P(j) + \varepsilon_A(i)E_A^\top(i)E_A(i)$.

Example 19. For this example we take the same data as the previous one and compute the minimum degree of stability the system can have. Solving the optimization (2.59) with these date gives

$$P(1) = \begin{bmatrix} 0.8401 & 0.0009 \\ 0.0009 & 0.6666 \end{bmatrix}, \quad P(2) = \begin{bmatrix} 0.4722 & 0.0073 \\ 0.0073 & 0.4207 \end{bmatrix},$$

which are both symmetric and positive-definite matrices. This implies that the system is stochastically stable. The minimum degree of stability is $\alpha = 0.5548$.

2.5 Notes

In this chapter we covered the stability problem and its robustness. We started by studying the stochastic stability of the nominal system of the class of piece-wise deterministic systems and established LMI conditions for this purpose. Then we considered the case of uncertain systems with uncertainties on the state matrix and/or on the transition probability rates matrix. Stochastic stability and robust stochastic stability LMI conditions have been developed for uncertain systems. The case of Wiener process external disturbance was also considered and stochastic stability and robust stochastic stability LMI conditions were developed. The material presented in this chapter is based on many references, among them Ji and Chizeck [44], Boukas [9], Dragan and Morozan [35], and personal work.

3

Stabilization Problem

One of the most popular control problems, the stabilization problem consists of determining a control law that forces the closed-loop state equation of a given system to guarantee the desired design performances. This problem has and continues to attract many researchers from the control community and many techniques can be used to solve the stabilization problem for dynamical systems. From the practical point of view when designing any control system, the stabilization problem is the most important in the design phase since it will give the desired performances to the designed control system.

The concepts of stochastic stability and its robustness for the class of piecewise deterministic systems were presented in the previous chapter. Most of the developed results are LMI-based conditions that can be used easily to check if a dynamical system of the class we are considering is stochastically stable and robustly stochastically stable.

In practice some systems are unstable or their performances are not acceptable. To stabilize or improve the performances of such systems, we examine the design of an appropriate controller. Once combined with the system this controller should stabilize the closed loop and at the same time guarantee the required performances.

In the literature, we can find different techniques of stabilization that can be divided into two groups. The first group gathers all the techniques that assume the complete access to the state vector and the other group is composed of techniques that are based on partial state vector observation.

For the class of systems under consideration, the following techniques can be used:

- state feedback stabilization,
- output feedback stabilization.

This chapter will focus on these two techniques and develop LMI-based procedures to design the corresponding gains. The rest of this chapter is organized as follows. In Section 3.1, the stabilization problem is stated and some useful definitions are given. Section 3.2 treats the state feedback stabilization for nominal and uncertain classes of piecewise deterministic systems. Section 3.3 covers the stabilization with the static output feedback controller. In Section 3.4, output feedback is covered. Section 3.5 deals with observer-output feedback stabilization. Section 3.6 develops the design of the state feedback controller with constant gain. All the developed results are in LMI framework, which makes the resolution of the stabilization problem easier. Many numerical examples are provided to show the usefulness of the developed results.

3.1 Problem Statement

Let us consider a dynamical system defined in a probability space $(\Omega, \mathcal{F}, \mathbb{P})$ and assume that its state equation is described by the following differential equations:

$$\begin{cases} \dot{x}(t) = A(r(t), t)x(t) + B(r(t), t)u(t), \\ x(0) = x_0, \end{cases} \tag{3.1}$$

where $x(t) \in \mathbb{R}^n$ is the state vector; $x_0 \in \mathbb{R}^n$ is the initial state; $u(t) \in \mathbb{R}^m$ is the control input; $\{r(t), t \geq 0\}$ is the continuous-time Markov process taking values in a finite space $\mathscr{S} = \{1, 2, \cdots, N\}$ and describes the evolution of the mode at time t; $A(r(t), t) \in \mathbb{R}^{n \times n}$ and $B(r(t), t) \in \mathbb{R}^{n \times m}$ are matrices with the following forms:

$$A(r(t), t) = A(r(t)) + D_A(r(t))F_A(r(t), t)E_A(r(t)),$$
$$B(r(t), t) = B(r(t)) + D_B(r(t))F_B(r(t), t)E_B(r(t)),$$

with $A(r(t))$, $D_A(r(t))$, $E_A(r(t))$, $B(r(t))$, $D_B(r(t))$, and $E_B(r(t))$ being real known matrices with appropriate dimensions; and $F_A(r(t), t)$ and $F_B(r(t), t)$ are unknown real matrices that satisfy the following for every $i \in \mathscr{S}$:

$$\begin{cases} F_A^\top(i, t)F_A(i, t) \leq \mathbb{I}, \\ F_B^\top(i, t)F_B(i, t) \leq \mathbb{I}. \end{cases} \tag{3.2}$$

The Markov process $\{r(t), t \geq 0\}$ beside taking values in the finite set $\mathscr{S}$ describing the switching between the different modes is governed by the following probability transitions:

$$\mathbb{P}\left[r(t+h) = j | r(t) = i\right]$$
$$= \begin{cases} \lambda_{ij}h + o(h) & \text{when } r(t) \text{ jumps from } i \text{ to } j, \\ 1 + \lambda_{ii}h + o(h) & \text{otherwise,} \end{cases} \tag{3.3}$$

where λ_{ij} is the transition rate from mode i to mode j with $\lambda_{ij} \geq 0$ when $i \neq j$ and $\lambda_{ii} = -\sum_{j=1, j \neq i}^{N} \lambda_{ij}$ and $o(h)$ are such that $\lim_{h \to 0} \frac{o(h)}{h} = 0$.

As in Chapter 2, we assume that the matrix Λ belongs to a polytope, that is,

$$\Lambda = \sum_{k=1}^{\kappa} \alpha_k \Lambda_k, \tag{3.4}$$

with κ a positive given integer, $0 \leq \alpha_k \leq 1$, and Λ_k a known transition matrix. Its expression is given by

$$\Lambda_k = \begin{bmatrix} \lambda_{11}^k & \cdots & \lambda_{1N}^k \\ \vdots & \ddots & \vdots \\ \lambda_{N1}^k & \cdots & \lambda_{NN}^k \end{bmatrix}, \tag{3.5}$$

where λ_{ij}^k keeps the same meaning as before, with $\sum_{k=1}^{\kappa} \alpha_k = 1$.

Remark 10. The uncertainties that satisfy the conditions (3.2) and (3.4) are referred to as admissible. The uncertainty term in (3.2) is supposed to depend on the system's mode $r(t)$ and on time t. The results developed in the rest of this chapter will remain valid even if the uncertainties are chosen to depend on the system's state $x(t)$, the mode $r(t)$, and time t.

The problem we face in this chapter is designing a control law $u(\cdot)$ that maps the state vector $x(t)$ and/or the output vector $y(t)$ and the mode at time t $r(t)$ into a control action that will give the system the required performances. Different structures for the control law can be used. Among these are the

- state feedback controller,
- output feedback controller,
- observer-based output feedback controller.

When we have complete access to the system mode $r(t)$, the choice among these techniques will depend on the problem. For instance, if the state vector is completely available for feedback, the use of state feedback stabilization is more appropriate. But in case of partial observation of the state vector, two solutions are possible. The first one requires an estimation of the state vector and then the use of the state feedback controller. The second one suggests the use of output feedback controller or observer-based output feedback control. In this case, the output is used to compute the control law.

In the rest of this chapter, we show how we can compute the gains for the mentioned controllers. Both the stabilization and the robust stabilization problems are treated. The assumption of the complete access to the system mode $r(t)$ at each time t will be made when necessary.

Definition 3. *System (3.1) with $F_A(r(t),t) = F_B(r(t),t) = 0$ for all modes and for $t \geq 0$ is said to be stabilizable in the SS (MES, MSQS) sense if there exists a controller such that the closed-loop system is SS (MES, MSQS) for every initial condition (x_0, r_0).*

When the uncertainties are not equal to zero, the previous definition is replaced by the following one.

Definition 4. *System (3.1) is said to be robustly stabilizable in the stochastic sense if there exists a controller such that the closed-loop system is stochastically stable for every initial condition (x_0, r_0) and for all admissible uncertainties.*

The type of controllers considered in this chapter are presented in the next sections and the design procedures that can be used to compute the controllers gains are also developed.

3.2 State Feedback Stabilization

Let us consider a dynamical system and assume that we have complete access to the state vector $x(t)$ and to the mode $r(t)$ at each time t. The controller we are planning to design is described by the following structure:

$$u(t) = K(r(t))x(t), \tag{3.6}$$

where $K(r(t))$ is a gain with appropriate dimension to be determined for each mode $r(t) \in \mathscr{S}$.

Remark 11. Notice that the gain of the controller (3.6) is mode dependent, which requires the knowledge of the mode $r(t)$ at time t to choose the appropriate gain among the set of gains $K = (K(1), K(2), \cdots, K(N))$. In this case, we determine N gains. Thus when the mode switches from mode i, which uses the gain $K(i)$, to mode j the controller gain must be switched instantaneously to the gain $K(j)$ to guarantee the desired performances.

The block diagram of the closed-loop system under the state feedback controller is represented by Figure 3.1.

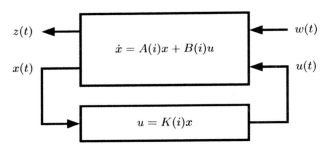

Fig. 3.1. State feedback stabilization block diagram (nominal system).

Let us consider the stabilization of the nominal system of the class under study. Letting $F_A(r(t), t) = F_B(r(t), t) = 0$ in the state equation (3.1) and replacing the control $u(t)$ by its expression given in (3.6), the closed-loop state equation becomes

$$\begin{cases} \dot{x}(t) = [A(r(t)) + B(r(t))K(r(t))] x(t), \\ x(0) = x_0. \end{cases} \tag{3.7}$$

Using the stability results of Chapter 2, the closed-loop state equation is stochastically stable if and only if the following is satisfied for every $i \in \mathscr{S}$:

$$[A(i) + B(i)K(i)]^\top P(i) + P(i) [A(i) + B(i)K(i)]$$
$$+ \sum_{j=1}^N \lambda_{ij} P(j) < 0. \tag{3.8}$$

This condition is nonlinear in $P(i)$ and $K(i)$. To transform it into an LMI form, let $X(i) = P^{-1}(i)$. Pre- and post-multiply (3.8) by $X(i)$ to get

$$X(i) [A(i) + B(i)K(i)]^\top + [A(i) + B(i)K(i)] X(i)$$
$$+ X(i) \left[\sum_{j=1}^N \lambda_{ij} X^{-1}(j) \right] X(i) < 0,$$

which gives in turn

$$X(i)A^\top(i) + X(i)K^\top(i)B^\top(i) + A(i)X(i) + B(i)K(i)X(i)$$
$$+ X(i) \left[\sum_{j=1}^N \lambda_{ij} X^{-1}(j) \right] X(i) < 0.$$

Letting $\mathcal{S}_i(X)$ and $\mathcal{X}_i(X)$ be defined as follows:

$$\begin{cases} \mathcal{S}_i(X) = \left[\sqrt{\lambda_{i1}} X(i), \cdots, \sqrt{\lambda_{ii-1}} X(i), \sqrt{\lambda_{ii+1}} X(i), \\ \qquad \cdots, \sqrt{\lambda_{iN}} X(i) \right] \\ \mathcal{X}_i(X) = \text{diag} \left[X(1), \cdots, X(i-1), X(i+1), \cdots, X(N) \right], \end{cases}$$

the term $X(i) \left[\sum_{j=1}^N \lambda_{ij} X^{-1}(j) \right] X(i)$ can be rewritten as follows:

$$X(i) \left[\sum_{j=1}^N \lambda_{ij} X^{-1}(j) \right] X(i) = \lambda_{ii} X(i) + \mathcal{S}_i(X)\mathcal{X}_i^{-1}(X)\mathcal{S}_i^\top(X).$$

Letting $Y(i) = K(i)X(i)$ and using the expressions $\mathcal{S}_i(X)$ and $\mathcal{X}_i(X)$ we get

$$X(i)A^{\top}(i) + Y^{\top}(i)B^{\top}(i) + A(i)X(i) + B(i)Y(i)$$
$$+\lambda_{ii}X(i) + \mathcal{S}_i(X)\mathcal{X}_i^{-1}(X)\mathcal{S}_i^{\top}(X) < 0,$$

which after using the Schur complement gives the following set of coupled LMIs:

$$\begin{bmatrix} J_n(i) & \mathcal{S}_i(X) \\ \mathcal{S}_i^{\top}(X) & -\mathcal{X}_i(X) \end{bmatrix} < 0, \tag{3.9}$$

where $J_n(i) = X(i)A^{\top}(i) + Y^{\top}(i)B^{\top}(i) + A(i)X(i) + B(i)Y(i) + \lambda_{ii}X(i)$.

The following theorem summarizes the results of this development.

Theorem 11. *If there exists a set of symmetric and positive-definite matrices $X = (X(1), \cdots, X(N)) > 0$ and a set of matrices $Y = (Y(1), \cdots, Y(N))$ satisfying the following set of coupled LMIs (3.9) for each $i \in \mathscr{S}$, then the controller (3.6), with $K(i) = Y(i)X^{-1}(i)$, stabilizes system (3.1) in the stochastic sense when the uncertainties are equal to zero.*

Example 20. In this example we show how the results of this theorem can be used to synthesize a state feedback controller that stochastically stabilizes a dynamical system. For this purpose, let us consider a system with two modes and the following data:

- mode #1:

$$A(1) = \begin{bmatrix} 1.0 & -0.5 \\ 0.1 & 1.0 \end{bmatrix}, \quad B(1) = \begin{bmatrix} 1.0 & 0.0 \\ 0.0 & 1.0 \end{bmatrix},$$

- mode #2:

$$A(2) = \begin{bmatrix} -0.2 & -0.5 \\ 0.5 & -0.25 \end{bmatrix}, \quad B(2) = \begin{bmatrix} 1.0 & 0.0 \\ 0.0 & 1.0 \end{bmatrix}.$$

The transition probability rates matrix between the two modes is given by

$$\Lambda = \begin{bmatrix} -2.0 & 2.0 \\ 3.0 & -3.0 \end{bmatrix}.$$

First notice that the first mode is instable in the deterministic sense and the global system is instable. We can stabilize this system using the results of the previous theorem. Solving the LMIs (3.9), we get

$$X(1) = \begin{bmatrix} 0.3408 & 0.0 \\ 0.0 & 0.3408 \end{bmatrix}, \quad X(2) = \begin{bmatrix} 0.2556 & 0.0 \\ 0.0 & 0.2556 \end{bmatrix},$$

$$Y(1) = \begin{bmatrix} -0.5112 & 0.0059 \\ 0.1304 & -0.5112 \end{bmatrix}, \quad Y(2) = \begin{bmatrix} -0.0767 & 0.0000 \\ 0.0 & -0.0639 \end{bmatrix},$$

which gives the following gains:

$$K(1) = \begin{bmatrix} -1.5000 & 0.0174 \\ 0.3826 & -1.5000 \end{bmatrix}, \quad K(2) = \begin{bmatrix} -0.3000 & 0.0000 \\ 0.0000 & -0.2500 \end{bmatrix}.$$

The conditions of the previous theorem are satisfied and therefore the closed-loop system under the state feedback with the computed gains is stochastically stable.

In fact, in a similar way, we can also stochastically stabilize the dynamical system of the class we are considering with a prescribed degree α by using the following theorem.

Theorem 12. *Let α be a positive constant. If there exist a set of symmetric and positive-definite matrices $X = (X(1), \cdots, X(N)) > 0$ and a set of matrices $Y = (Y(1), \cdots, Y(N))$ satisfying the following set of coupled LMIs for each $i \in \mathscr{S}$:*

$$\begin{bmatrix} J_n^\alpha(i) & \mathcal{S}_i(X) \\ \mathcal{S}_i^\top(X) & -\mathcal{X}_i(X) \end{bmatrix} < 0, \tag{3.10}$$

where $J_n^\alpha(i) = X(i)[A(i) + \alpha \mathbb{I}]^\top + Y^\top(i)B^\top(i) + [A(i) + \alpha \mathbb{I}]X(i) + B(i)Y(i) + \lambda_{ii}X(i)$, then the controller (3.6), with $K(i) = Y(i)X^{-1}(i)$, stabilizes system (3.1) in the stochastic sense when the uncertainties are equal to zero.

Proof: The proof of this theorem follows the same lines of Theorem 11, replacing the matrix $A(i)$ by $A(i) + \alpha \mathbb{I}$ in the previous state equation and the details are omitted. $\qquad \square$

We can also use the results of the following nonlinear optimization to determine a stabilizing controller that guarantees a prescribed degree of stability equal to $\alpha = \gamma^{-1}$:

$$\text{Pns:} \begin{cases} \min\limits_{\substack{\gamma > 0, \\ X = (X(1), \cdots, X(N)) > 0, \\ Y = (Y(1), \cdots, Y(N)),}} \gamma, \\ \text{s.t.:} \\ \begin{bmatrix} 2X(i) & 0 \\ 0 & 0 \end{bmatrix} < -\gamma \begin{bmatrix} \tilde{J}_u^\alpha(i) & \mathcal{S}_i(X) \\ \mathcal{S}_i^\top(X) & -\mathcal{X}_i(X) \end{bmatrix}, \end{cases}$$

where $\tilde{J}_n^\alpha(i) = X(i)A^\top(i) + Y^\top(i)B^\top(i) + A(i)X(i) + B(i)Y(i) + \lambda_{ii}X(i)$.

If the optimization problem Pns has a solution $\bar{\alpha}$, then the controller that stabilizes the dynamical system (3.1) and guarantees the prescribed degree of stability $\bar{\alpha}$ is given by

$$u(t) = Y(i)X^{-1}(i)x(t), i = 1, 2, \cdots, N.$$

Example 21. In this example, let us consider the same data as in Example 20 and see how we can determine the maximum degree of stochastic stability. To solve the previous optimization problem, we added some extra constraints on $X(i)$ and $Y(i)$, $i = 1, 2$. We have imposed these variables to satisfy

$$X(i) > \mathbb{I},$$
$$Y(i) < 10\mathbb{I},$$

for $i = 1, 2$.

With these constraints the solution of the previous optimization problem gives

$$X(1) = \begin{bmatrix} 1.0058 & 0.0056 \\ 0.0056 & 1.0056 \end{bmatrix}, \qquad X(2) = \begin{bmatrix} 1.0852 & -0.0321 \\ -0.0321 & 1.0892 \end{bmatrix},$$

$$Y(1) = \begin{bmatrix} -9.9425 & 3.0890 \\ -2.9741 & -9.9414 \end{bmatrix}, \qquad Y(2) = \begin{bmatrix} -9.7965 & 0.0891 \\ 0.2575 & -9.7740 \end{bmatrix}.$$

The corresponding gains are

$$K(1) = \begin{bmatrix} -9.9026 & 3.1270 \\ -2.9020 & -9.8699 \end{bmatrix}, \qquad K(2) = \begin{bmatrix} -9.0327 & -0.1840 \\ -0.0278 & -8.9746 \end{bmatrix}.$$

The maximal degree of stability is $\bar{\alpha} = 8.9048$.

Let us now assume that the previous state equation is changed to

$$\begin{cases} dx(t) = A(r(t), t)x(t)dt + B(r(t), t)u(t)dt + \mathbb{W}(r(t))x(t)d\omega(t), \\ x(0) = x_0, \end{cases} \qquad (3.11)$$

where the different components in this equation keep the same meaning, while $\mathbb{W}(r(t))$ is a given matrix with appropriate dimension and $w(t) \in \mathbb{R}$ is a Wiener process acting on the system that we assume to be independent of the Markov process $\{r(t), t \geq 0\}$ as described before.

Let all the uncertainties in the state equation (3.11) be zero and see how we can design a state feedback controller in the form (3.6). Plugging the expression of this controller in (3.11) gives

$$\begin{cases} dx(t) = \bar{A}(r(t))x(t)dt + \mathbb{W}(r(t))x(t)d\omega(t), \\ x(0) = x_0, \end{cases} \qquad (3.12)$$

where $\bar{A}(r(t)) = A(r(t)) + B(r(t))K(r(t))$, and $K(r(t))$ is the controller gain to be determined for each $r(t) \in \mathscr{S}$.

Using the stability results we established in Chapter 2, we should have the following set of coupled LMIs verified for every $i \in \mathscr{S}$ to guarantee that the closed loop is stochastically stable:

$$\bar{A}^\top(i)P(i) + P(i)\bar{A}(i) + \mathbb{W}^\top(i)P(i)\mathbb{W}(i) + \sum_{j=1}^N \lambda_{ij}P(j) < 0.$$

Using the expression of $\bar{A}(i)$ we get

$$A^\top(i)P(i) + P(i)A(i) + K^\top(i)B^\top(i)P(i) + P(i)B(i)K(i)$$

$$+\mathbb{W}^\top(i)P(i)\mathbb{W}(i) + \sum_{j=1}^{N} \lambda_{ij}P(j) < 0, i = 1, 2, \cdots, N.$$

This set of matrix inequalities is nonlinear in the decision variables $P(i)$ and $K(i)$. To transform it into an LMI, let $X(i) = P^{-1}(i)$. Pre- and post-multiply the previous set of matrix inequalities by $X(i)$ to get

$$X(i)A^\top(i) + A(i)X(i) + X(i)K^\top(i)B^\top(i) + B(i)K(i)X(i)$$

$$+X(i)\mathbb{W}^\top(i)X^{-1}(i)\mathbb{W}(i)X(i) + \sum_{j=1}^{N} \lambda_{ij}X(i)X^{-1}(j)X(i) < 0.$$

Letting $Y(i) = K(i)X(i)$ and using the expression of

$$\sum_{j=1}^{N} \lambda_{ij}X(i)X^{-1}(j)X(i),$$

as it was established before, and the Schur complement we get

$$\begin{bmatrix} J_w(i) & X(i)\mathbb{W}^\top(i) & \mathcal{S}_i(X) \\ \mathbb{W}(i)X(i) & -X(i) & 0 \\ \mathcal{S}_i^\top(X) & 0 & -\mathcal{X}_i(X) \end{bmatrix} < 0, \tag{3.13}$$

where

$$J_w(i) = X(i)A^\top(i) + A(i)X(i) + Y^\top(i)B^\top(i) + B(i)Y(i) + \lambda_{ii}X(i).$$

The following theorem summarizes the results of the stabilization problem.

Theorem 13. *If there exist a set of symmetric and positive-definite matrices* $X = (X(1), \cdots, X(N)) > 0$ *and a set of matrices* $Y = (Y(1), \cdots, Y(N))$ *satisfying the following set of coupled LMIs (3.13) for each* $i \in \mathscr{S}$, *then the controller (3.6), with* $K(i) = Y(i)X^{-1}(i)$, *stabilizes system (3.11) in the stochastic sense when the uncertainties are equal to zero.*

Example 22. To show the usefulness of this theorem, let us consider the two-mode dynamical system considered in Example 20 with the following extra data:

- mode #1:

$$\mathbb{W}(1) = \begin{bmatrix} 0.1 & 0.0 \\ 0.0 & 0.1 \end{bmatrix},$$

- mode #2:

$$W(2) = \begin{bmatrix} 0.2 & 0.0 \\ 0.0 & 0.2 \end{bmatrix}.$$

Solving the LMI (3.13), we get

$$X(1) = \begin{bmatrix} 18.2135 & -0.0000 \\ -0.0000 & 18.2135 \end{bmatrix}, \qquad X(2) = \begin{bmatrix} 16.1415 & 0.0000 \\ 0.0000 & 16.1415 \end{bmatrix},$$

$$Y(1) = \begin{bmatrix} -34.9977 & -265.2702 \\ 272.5556 & -34.9977 \end{bmatrix}, \qquad Y(2) = \begin{bmatrix} -8.3439 & -5.1312 \\ 5.1312 & -7.5368 \end{bmatrix}.$$

The corresponding gains are given by

$$K(1) = \begin{bmatrix} -1.9215 & -14.5644 \\ 14.9644 & -1.9215 \end{bmatrix}, \qquad K(2) = \begin{bmatrix} -0.5169 & -0.3179 \\ 0.3179 & -0.4669 \end{bmatrix}.$$

Based on the results of the previous theorem, we conclude that the system of this example is stochastically stable under the state feedback controller with the computed gains.

Let us return to the initial problem and see how we can design a controller that robustly stabilizes the uncertain class of piecewise deterministic systems when the uncertainties are acting on the state matrix only. For this purpose, plugging the controller (3.6) in the state equation (3.1), we get

$$\begin{cases} \dot{x}(t) = [A(r(t), t) + B(r(t), t)K(r(t))] \, x(t), \\ x(0) = x_0. \end{cases}$$

Using the robust stability results of Chapter 2, the closed-loop state equation is then stochastically stable if there exists a set of symmetric and positive-definite matrices $P = (P(1), \cdots, P(N)) > 0$ such that the following is satisfied for all the admissible uncertainties and for every $i \in \mathscr{S}$:

$$[A(i, t) + B(i, t)K(i)]^\top P(i) + P(i) [A(i, t) + B(i, t)K(i)]$$

$$+ \sum_{j=1}^N \lambda_{ij} P(j) < 0. \qquad (3.14)$$

This condition is nonlinear in $P(i)$ and $K(i)$ and depends on the uncertainties. To transform it into an LMI form that does not depend on uncertainties, let $X(i) = P^{-1}(i)$. Pre- and post-multiply (3.14) by $X(i)$ to get

$$X(i) [A(i, t) + B(i, t)K(i)]^\top + [A(i, t) + B(i, t)K(i)] X(i)$$

$$+ X(i) \left[\sum_{j=1}^N \lambda_{ij} X^{-1}(j) \right] X(i) < 0,$$

which gives in turn, after using the expressions of the uncertainties on the matrices A and B:

$$X(i)A^\top(i) + X(i)K^\top(i)B^\top(i) + A(i)X(i) + B(i)K(i)X(i)$$
$$+D_A(i)F_A(i,t)E_A(i)X(i) + X(i)E_A^\top(i)F_A^\top(i,t)D_A^\top(i)$$
$$+D_B(i)F_B(i,t)E_B(i)K(i)X(i)$$
$$+X(i)K^\top(i)E_B^\top(i)F_B^\top(i,t)D_B^\top(i)$$
$$+X(i)\left[\sum_{j=1}^N \lambda_{ij}X^{-1}(j)\right]X(i) < 0.$$

Using Lemma 7 in Appendix A, we have

$$D_A(i)F_A(i,t)E_A(i)X(i) + X(i)E_A^\top(i)F_A^\top(i,t)D_A^\top(i)$$
$$\leq \varepsilon_A(i)D_A(i)D_A^\top(i) + \varepsilon_A^{-1}(i)X(i)E_A^\top(i)E_A(i)X(i)$$
$$D_B(i)F_B(i,t)E_B(i)K(i)X(i) + X(i)K^\top(i)E_B^\top(i)F_B^\top(i,t)D_B^\top(i)$$
$$\leq \varepsilon_B(i)D_B(i)D_B^\top(i) + \varepsilon_B^{-1}(i)X(i)K^\top(i)E_B^\top(i)E_B(i)K(i)X(i).$$

Using the same definitions for $\mathcal{S}_i(X)$ and $\mathcal{X}_i(X)$ as before and after letting $Y(i) = K(i)X(i)$, the closed-loop system will be robustly stochastically stable if the following holds:

$$X(i)A^\top(i) + Y^\top(i)B^\top(i) + A(i)X(i) + B(i)Y(i) + \lambda_{ii}X(i)$$
$$+\varepsilon_A(i)D_A(i)D_A^\top(i) + \varepsilon_B(i)D_B(i)D_B^\top(i)$$
$$+\varepsilon_A^{-1}(i)X(i)E_A^\top(i)E_A(i)X(i)$$
$$+\varepsilon_B^{-1}(i)Y^\top(i)E_B^\top(i)E_B(i)Y(i)$$
$$+\mathcal{S}_i(X)\mathcal{X}_i^{-1}(X)\mathcal{S}_i^\top(X) < 0,$$

which gives in turn the following set of coupled LMIs after using the Schur complement:

$$\begin{bmatrix} J_u(i) & X(i)E_A^\top(i) & Y^\top(i)E_B^\top(i) & \mathcal{S}_i(X) \\ E_A(i)X(i) & -\varepsilon_A(i)\mathbb{I} & 0 & 0 \\ E_B(i)Y(i) & 0 & -\varepsilon_B(i)\mathbb{I} & 0 \\ \mathcal{S}_i^\top(X) & 0 & 0 & -\mathcal{X}_i(X) \end{bmatrix} < 0, \qquad (3.15)$$

where

$$J_u(i) = X(i)A^\top(i) + Y^\top(i)B^\top(i) + A(i)X(i) + B(i)Y(i)$$
$$+\lambda_{ii}X(i) + \varepsilon_A(i)D_A(i)D_A^\top(i) + \varepsilon_B(i)D_B(i)D_B^\top(i).$$

The following theorem summarizes the results on the design of a robust stabilizing state feedback controller.

Theorem 14. *If there exist a set of symmetric and positive-definite matrices* $X = (X(1), \cdots, X(N)) > 0$, *a set of matrices* $Y = (Y(1), \cdots, Y(N))$, *and a set of positive scalars* $\varepsilon_A = (\varepsilon_A(1), \cdots, \varepsilon_A(N)) > 0$ *and* $\varepsilon_B = (\varepsilon_B(1), \cdots, \varepsilon_B(N)) > 0$ *such that the following set of coupled LMIs (3.15) holds for each* $i \in \mathscr{S}$ *and all admissible uncertainties, then the controller (3.6), with* $K(i) = Y(i)X^{-1}(i)$, *robustly stabilizes system (3.1) in the stochastic sense.*

Example 23. In this numerical example we show how to design a robust stabilizing state feedback controller using the results of this theorem. For this purpose let us consider a system with two modes and the following data:

- mode #1:

$$A(1) = \begin{bmatrix} 1.0 & -0.5 \\ 0.1 & 1.0 \end{bmatrix}, \quad D_A(1) = \begin{bmatrix} 0.1 \\ 0.2 \end{bmatrix}, \quad E_A(1) = \begin{bmatrix} 0.2 & 0.1 \end{bmatrix},$$

$$B(1) = \begin{bmatrix} 1.0 & 0.0 \\ 0.0 & 1.0 \end{bmatrix}, \quad D_B(1) = \begin{bmatrix} 0.1 \\ 0.2 \end{bmatrix}, \quad E_B(1) = \begin{bmatrix} 0.2 & 0.1 \end{bmatrix},$$

- mode #2:

$$A(2) = \begin{bmatrix} -0.2 & 0.5 \\ 0.0 & -0.25 \end{bmatrix}, \quad D_A(2) = \begin{bmatrix} 0.13 \\ 0.1 \end{bmatrix}, \quad E_A(2) = \begin{bmatrix} 0.1 & 0.2 \end{bmatrix},$$

$$B(2) = \begin{bmatrix} 1.0 & 0.0 \\ 0.0 & 1.0 \end{bmatrix}, \quad D_B(2) = \begin{bmatrix} 0.13 \\ 0.1 \end{bmatrix}, \quad E_B(2) = \begin{bmatrix} 0.1 & 0.2 \end{bmatrix}.$$

The transition probability rate matrix between the two modes is given by

$$\Lambda = \begin{bmatrix} -2.0 & 2.0 \\ 3.0 & -3.0 \end{bmatrix}.$$

Letting $\varepsilon_A(1) = \varepsilon_A(2) = 0.5$ and $\varepsilon_B(1) = \varepsilon_B(2) = 0.1$ and solving the set of coupled LMIs (3.15), we get

$$X(1) = \begin{bmatrix} 0.3140 & -0.0046 \\ -0.0046 & 0.3297 \end{bmatrix}, \quad X(2) = \begin{bmatrix} 0.2828 & -0.0016 \\ -0.0016 & 0.2867 \end{bmatrix},$$

$$Y(1) = \begin{bmatrix} -0.6668 & 0.0482 \\ 0.1075 & -0.7130 \end{bmatrix}, \quad Y(2) = \begin{bmatrix} -0.2181 & -0.1268 \\ -0.0160 & -0.1860 \end{bmatrix},$$

which gives the following gains:

$$K(1) = \begin{bmatrix} -2.1219 & 0.1169 \\ 0.3110 & -2.1586 \end{bmatrix}, \quad K(2) = \begin{bmatrix} -0.7737 & -0.4466 \\ -0.0604 & -0.6489 \end{bmatrix}.$$

Using the results of the previous theorem, we conclude that the system of this example is stochastically stable under the state feedback controller with the computed gains.

As we did for the stabilizing controller that prescribes the desired degree of stability α, similar results for the robustness can be obtained by the following theorem.

Theorem 15. *Let α be a given positive constant. If there exist a set of symmetric and positive-definite matrices $X = (X(1), \cdots, X(N)) > 0$, a set of matrices $Y = (Y(1), \cdots, Y(N))$, and a set of positive scalars $\varepsilon_A = (\varepsilon_A(1), \cdots, \varepsilon_A(N))$ and $\varepsilon_B = (\varepsilon_B(1), \cdots, \varepsilon_B(N))$ such that the following set of coupled LMIs holds for each $i \in \mathscr{S}$ and all admissible uncertainties:*

$$\begin{bmatrix} J_u^\alpha(i) & X(i)E_A^\top(i) & Y^\top(i)E_B^\top(i) & \mathcal{S}_i(X) \\ E_A(i)X(i) & -\varepsilon_A(i)\mathbb{I} & 0 & 0 \\ E_B(i)Y(i) & 0 & -\varepsilon_B(i)\mathbb{I} & 0 \\ \mathcal{S}_i^\top(X) & 0 & 0 & -\mathcal{X}_i(X) \end{bmatrix} < 0, \qquad (3.16)$$

where

$$J_u^\alpha(i) = X(i)\left[A(i) + \alpha\mathbb{I}\right]^\top + Y^\top(i)B^\top(i) + \left[A(i) + \alpha\mathbb{I}\right]X(i)$$
$$+ B(i)Y(i) + \lambda_{ii}X(i) + \varepsilon_A(i)D_A(i)D_A^\top(i) + \varepsilon_B(i)D_B(i)D_B^\top(i),$$

then the controller (3.6), with $K(i) = Y(i)X^{-1}(i)$, robustly stabilizes system (3.1) in the stochastic sense.

Proof: The proof of this theorem is straightforward and follows the same lines as before and the details are omitted.   □

We can also use the results of the following optimization problem to determine a stabilizing controller that guarantees a prescribed degree of stability equal to $\alpha = \gamma^{-1}$:

Pus: $\begin{cases} \min\limits_{\substack{\gamma > 0, \\ \varepsilon_A = (\varepsilon_A(1), \cdots, \varepsilon_A(N)) > 0, \\ \varepsilon_B = (\varepsilon_B(1), \cdots, \varepsilon_B(N)) > 0, \\ X = (X(1), \cdots, X(N)) > 0, \\ Y = (Y(1), \cdots, Y(N)),}} \gamma, \\[2em] \text{s.t.:} \\[1em] \begin{bmatrix} 2X(i) & 0 & 0 & 0 \\ 0 & 0 & 0 & 0 \\ 0 & 0 & 0 & 0 \\ 0 & 0 & 0 & 0 \end{bmatrix} \\[3em] < -\gamma \begin{bmatrix} \tilde{J}_u^\alpha(i) & X(i)E_A^\top(i) & Y^\top(i)E_B^\top(i) & \mathcal{S}_i(X) \\ E_A(i)X(i) & -\varepsilon_A(i)\mathbb{I} & 0 & 0 \\ E_B(i)Y(i) & 0 & -\varepsilon_B(i)\mathbb{I} & 0 \\ \mathcal{S}_i^\top(X) & 0 & 0 & -\mathcal{X}_i(X) \end{bmatrix}, \end{cases}$

where

$$\tilde{J}_u^\alpha(i) = X(i)A^\top(i) + Y^\top(i)B^\top(i) + A(i)X(i) + B(i)Y(i)$$

$$+\lambda_{ii}X(i) + \varepsilon_A(i)D_A(i)D_A^\top(i) + \varepsilon_B(i)D_B(i)D_B^\top(i).$$

If the optimization problem Pus has a solution denoted by $\bar{\alpha}$, then the controller that robustly stabilizes the dynamical system (3.1) and guarantees the prescribed degree of stability $\bar{\alpha}$, is given by

$$u(t) = Y(i)X^{-1}(i)x(t), i = 1, 2, \cdots, N.$$

Example 24. To show the usefulness of the results of this theorem, let us consider the system of the previous example and see how we can determine the maximum degree of stochastic stability. Solving the optimization problem with the constraints $X(i) > \mathbb{I}$ and $Y(i) < 10\mathbb{I}$ for $i = 1, 2$, we get

$$X(1) = \begin{bmatrix} 1.6357 & -1.4260 \\ -1.4260 & 4.1989 \end{bmatrix}, \quad X(2) = \begin{bmatrix} 3.8032 & -1.6532 \\ -1.6532 & 3.0954 \end{bmatrix},$$

$$Y(1) = \begin{bmatrix} -3.4676 & 3.3010 \\ 2.8947 & -8.5208 \end{bmatrix}, \quad Y(2) = \begin{bmatrix} -9.7095 & -1.3889 \\ 4.4376 & -1.5969 \end{bmatrix},$$

which give the following gains:

$$K(1) = \begin{bmatrix} -2.0379 & 0.0941 \\ 0.0008 & -2.0290 \end{bmatrix}, \quad K(2) = \begin{bmatrix} -3.5789 & -2.3602 \\ 1.2275 & 0.1397 \end{bmatrix}.$$

The corresponding $\bar{\alpha}$ is equal to 0.5998.

In all the results we developed previously we assumed that the controller gain is known and no uncertainty is attached. Practically, this is not true since we will always have errors in implementing the computed gains. As an example, we can mention the case when the gains are realized electronically using operational amplifiers and passive electronic components. Keel and Bhattacharyya [46] have shown that the controller may be very sensitive to the errors in the controller parameters even if the design takes care of the system uncertainties. To overcome this, the variations parameter as well as the system uncertainties should be included in the controller design phase. The goal becomes then how to design a controller that is nonfragile in the sense that the closed-loop system tolerates a certain change in the controller parameters as well as the system uncertainties that may affect the different matrices.

In practice the implementation of the controller is quite different from the expression (3.6), and an extra term should be added to take care of the errors that may be caused by reasons different than those mentioned previously. The controller gain becomes then

$$K(i, t) = K(i) + \Delta K(i, t), \tag{3.17}$$

with $\Delta K(i, t) = \rho(i)F_K(i, t)K(i)$, where $\rho(i)$ is an uncertain real parameter indicating the measure of nonfragility against controller gain variations and

$F_K(i,t)$ is the uncertainty that will be supposed to satisfy the following for every $i \in \mathscr{S}$:

$$F_K^\top(i,t)F_K(i,t) \leq \mathbb{I}. \tag{3.18}$$

Let us now assume that the uncertainties of the control matrix $B(i)$ are equal to zero and see how to synthesize the gain for the state feedback controller with the following form for every $i \in \mathscr{S}$:

$$K(i) = \varrho(i)B^\top(i)P(i), \tag{3.19}$$

where $\varrho(i)$ is a real number and $P(i) > 0$ is symmetric and positive-definite matrix for every $i \in \mathscr{S}$.

Plugging the controller in the state equation (3.1) with $F_B(i,t)$ fixed to zero for all $t > 0$, we get the following closed-loop dynamics:

$$
\begin{aligned}
\dot{x}(t) = {} & [A(i,t) + B(i)K(i,t)]\, x(t) \\
= {} & \big[A(i) + D_A(i)F_A(i,t)E_A(i) + B(i)\big[\varrho(i)B^\top(i)P(i) \\
& + \rho(i)F_K(i,t)\varrho(i)B^\top(i)P(i)\big]\big]\, x(t).
\end{aligned} \tag{3.20}
$$

The following theorem gives a design procedure that may be used to synthesize a controller in the form (3.17) that robustly stabilizes the system with nonfragility $\rho(i)$.

Theorem 16. *If there exist a set of symmetric and positive-definite matrices $X = (X(1), \cdots, X(N)) > 0$ and a set of positive scalars $\varepsilon_A = (\varepsilon_A(1), \cdots, \varepsilon_A(N))$, $\mu = (\mu(1), \cdots, \mu(N))$, $\nu = (\nu(1), \cdots, \nu(N))$ and a set of scalars $\varrho = (\varrho(1), \cdots, \varrho(N))$ satisfying the following set of coupled LMIs for every $i \in S$ and for all admissible uncertainties:*

$$
\begin{bmatrix}
J(i) & X(i)E_A^\top(i) & \varrho(i)B(i) & S_i(X) \\
E_A(i)X(i) & -\varepsilon_A(i)\mathbb{I} & 0 & 0 \\
\varrho(i)B^\top(i) & 0 & -\mu(i)\mathbb{I} & 0 \\
S_i^\top(X) & 0 & 0 & -\mathcal{X}_i(X)
\end{bmatrix} < 0, \tag{3.21}
$$

where

$$
\begin{aligned}
J(i) = {} & X(i)A^\top(i) + A(i)X(i) + \varepsilon_A(i)D_A(i)D_A^\top(i) \\
& + 2\varrho(i)B(i)B^\top(i) + \lambda_{ii}X(i) + \nu(i)B(i)B^\top(i), \\
\mu(i) = {} & \frac{\varepsilon_K(i)}{\rho(i)}, \\
\nu(i) = {} & \varepsilon_K(i)\rho(i),
\end{aligned}
$$

then the closed-loop system is robustly stochastically stable with nonfragility $\rho(i)$ under the controller (3.6) with the gain $K(i) = \varrho(i)B^\top(i)X^{-1}(i)$.

Proof: Let $r(t) = i$ and $P(r(t)) > 0$ be a symmetric and positive-definite matrix and consider the following candidate Lyapunov function given as follows:

$$V(x(t), i) = x^\top(t)P(i)x(t),$$

where $P(i)$ is the solution of (3.21).

The infinitesimal generator of the Markov process $\{(x(t), r(t)), t \geq 0\}$ acting on $V(.)$ and emanating from the point (x, i) at time t, where $x(t) = x$ and $r(t) = i$ for $i \in \mathscr{S}$, is given by

$$
\begin{aligned}
\mathscr{L}V(x(t), i) = {} & \lim_{h \to 0} \frac{\mathbb{E}\left[V(x_{t+h}, r(t+h)) - V(x(t), i) | x(t), i\right]}{h} \\
= {} & \dot{x}^\top(t)P(i)x(t) + x^\top(t)P(i)\dot{x}(t) + \sum_{j=1}^{N} \lambda_{ij}x^\top(t)P(j)x(t) \\
= {} & \left[[A(i,t) + B(i)K(i,t)]\, x(t)\right]^\top P(i)x(t) + x^\top(t)P(i)\,[A(i,t) \\
& + B(i)K(i,t)]\, x(t) + \sum_{j=1}^{N} \lambda_{ij}x^\top(t)P(j)x(t) \\
= {} & x^\top(t)A^\top(i)P(i)x(t) + x^\top(t)P(i)A(i)x(t) \\
& + 2x^\top(t)P(i)D_A(i)F_A(i,t)E_A(i)x(t) \\
& 2\varrho(i)x^\top(t)P(i)B(i)B^\top(i)P(i)x(t) + \sum_{j=1}^{N} \lambda_{ij}x^\top(t)P(j)x(t) \\
& + 2\rho(i)\varrho(i)x^\top(t)P(i)B(i)F_K(i,t)B^\top(i)P(i)x(t).
\end{aligned}
$$

Using Lemma 7 in Appendix A, we get

$$
\begin{aligned}
2x^\top(t)P(i)D_A(i)F_A(i,t)E_A(i)x(t) &\leq \varepsilon_A(i)x^\top(t)P(i)D_A(i)D_A^\top(i)P(i)x(t) \\
&\quad + \varepsilon_A^{-1}(i)x^\top(t)E_A^\top(i)E_A(i)x(t), \\
2\rho(i)\varrho(i)x^\top(t)P(i)B(i)F_K(i,t)B^\top(i)P(i)x(t) \\
\leq \varepsilon_K^{-1}(i)\rho(i)\varrho^2(i)x^\top(t)P(i)B(i)B^\top(i)P(i)x(t) \\
\quad + \varepsilon_K(i)\rho(i)x^\top(t)P(i)B(i)F_K^\top(i,t)F_K(i,t)B^\top(i)P(i)x(t) \\
\leq \varepsilon_K^{-1}(i)\rho(i)\varrho^2(i)x^\top(t)P(i)B(i)B^\top(i)P(i)x(t) \\
\quad + \varepsilon_K(i)\rho(i)x^\top(t)P(i)B(i)B^\top(i)P(i)x(t).
\end{aligned}
$$

Based on this, the previous expression of $\mathscr{L}V(x(t), i)$ becomes

$$
\begin{aligned}
\mathscr{L}V(x(t), i) \leq {} & x^\top(t)A^\top(i)P(i)x(t) + x^\top(t)P(i)A(i)x(t) \\
& + \varepsilon_A(i)x^\top(t)P(i)D_A(i)D_A^\top(i)P(i)x(t) \\
& + \varepsilon_A^{-1}(i)x^\top(t)E_A^\top(i)E_A(i)x(t) + \sum_{j=1}^{N} \lambda_{ij}x^\top(t)P(j)x(t)
\end{aligned}
$$

$$+\varepsilon_K^{-1}(i)\rho(i)\varrho^2(i)x^\top(t)P(i)B(i)B^\top(i)P(i)x(t)$$
$$+\varepsilon_K(i)\rho(i)x^\top(t)P(i)B(i)B^\top(i)P(i)x(t)$$
$$+2\varrho(i)x^\top(t)P(i)B(i)B^\top(i)P(i)x(t)$$
$$\leq x^\top(t)\Gamma(i)x(t),$$

with

$$\Gamma(i) = A^\top(i)P(i) + P(i)A(i) + \varepsilon_A(i)P(i)D_A(i)D_A^\top(i)P(i)$$
$$+\varepsilon_A^{-1}(i)E_A^\top(i)E_A(i) + \sum_{j=1}^{N}\lambda_{ij}P(j)$$
$$+\varepsilon_K^{-1}(i)\rho(i)\varrho^2(i)P(i)B(i)B^\top(i)P(i)$$
$$+\varepsilon_K(i)\rho(i)P(i)B(i)B^\top(i)P(i)$$
$$+2\varrho(i)P(i)B(i)B^\top(i)P(i).$$

The expression $\Gamma(i)$ is nonlinear in the design parameters $\varrho(i)$ and $P(i)$ for every $i \in \mathscr{S}$. To cast it into an LMI, let us put $X(i) = P^{-1}(i)$ for each $i \in S$. Let us pre- and post-multiply $\Gamma(i)$ by $X(i)$ to get

$$X(i)\Gamma(i)X(i) = X(i)A^\top(i) + A(i)X(i) + \varepsilon_A(i)D_A(i)D_A^\top(i)$$
$$+\varepsilon_A^{-1}(i)X(i)E_A^\top(i)E_A(i)X(i) + \sum_{j=1}^{N}\lambda_{ij}X(i)X^{-1}(j)X(i)$$
$$+\varepsilon_K^{-1}(i)\rho(i)\varrho^2(i)B(i)B^\top(i) + \varepsilon_K(i)\rho(i)B(i)B^\top(i)$$
$$+2\varrho(i)B(i)B^\top(i). \qquad (3.22)$$

Letting $\mathcal{S}_i(X)$ and $\mathcal{X}_i(X)$ be defined as before, and using the Schur complement, we get

$$\begin{bmatrix} J(i) & X(i)E_A^\top(i) & \varrho(i)B(i) & \mathcal{S}_i(X) \\ E_A(i)X(i) & -\varepsilon_A(i)\mathbb{I} & 0 & 0 \\ \varrho(i)B^\top(i) & 0 & -\frac{\varepsilon_K(i)}{\rho(i)}\mathbb{I} & 0 \\ \mathcal{S}_i^\top(X) & 0 & 0 & -\mathcal{X}_i(X) \end{bmatrix} < 0,$$

where

$$J(i) = X(i)A^\top(i) + A(i)X(i) + \lambda_{ii}X(i) + \varepsilon_A(i)D_A(i)D_A^\top(i)$$
$$+2\varrho(i)B(i)B^\top(i) + \varepsilon_K(i)\rho(i)B(i)B^\top(i),$$

which is symmetric and negative-definite by hypothesis and therefore we conclude that $\Gamma(i) < 0$, which implies

$$\mathscr{L}V(x(t),i) \leq -\min_{i\in\mathscr{S}}\{\lambda_{min}(-\Gamma(i))\}x^\top(t)x(t).$$

The rest of the proof is similar to what has been done previously and the details are omitted. This proves that the closed-loop system is stable under the chosen controller.   □

Example 25. In this example we show the usefulness of the proposed results. For this purpose let us consider a system with two modes and two components in the state vector. Let the data in each mode be given by

- mode #1:

$$A(1) = \begin{bmatrix} 1.0 & -0.5 \\ 0.1 & 1.0 \end{bmatrix}, \qquad B(1) = \begin{bmatrix} 1.0 & 0.0 \\ 0.0 & 1.0 \end{bmatrix},$$

$$D_A(1) = \begin{bmatrix} 0.1 \\ 0.2 \end{bmatrix}, \qquad E_A(1) = \begin{bmatrix} 0.2 & 0.1 \end{bmatrix},$$

- mode #2:

$$A(2) = \begin{bmatrix} -0.2 & 0.5 \\ 0.0 & -0.25 \end{bmatrix}, \qquad B(2) = \begin{bmatrix} 1.0 & 0.0 \\ 0.0 & 1.0 \end{bmatrix},$$

$$D_A(2) = \begin{bmatrix} 0.13 \\ 0.1 \end{bmatrix}, \qquad E_A(2) = \begin{bmatrix} 0.1 & 0.2 \end{bmatrix}.$$

Let the transition probability rate matrix between these two modes be given by

$$\Lambda = \begin{bmatrix} -2.0 & 2.0 \\ 3.0 & -3.0 \end{bmatrix}.$$

Letting $\varepsilon_A(1) = \varepsilon_A(2) = 0.5$, $\varepsilon_K(1) = \varepsilon_K(2) = 0.1$, and $\rho(1) = 0.5$, $\rho(2) = 0.6$ and solving the LMI (3.21), we get

$$X(1) = \begin{bmatrix} 0.0623 & 0.002 \\ 0.002 & 0.056 \end{bmatrix}, \quad X(2) = \begin{bmatrix} 0.0749 & -0.0038 \\ -0.0038 & 0.0695 \end{bmatrix},$$

$$\varrho(1) = -0.1930, \qquad \varrho(2) = -0.1592,$$

which gives the following controller gains:

$$K(1) = \begin{bmatrix} -3.1004 & 0.1113 \\ 0.1113 & -3.4498 \end{bmatrix}, \quad K(2) = \begin{bmatrix} -2.1297 & -0.1153 \\ -0.1153 & -2.2966 \end{bmatrix}.$$

Let us return to the system with Wiener process disturbance as described by the state equation (3.11). Assume that the uncertainties are not equal to zero. How can we synthesize a state feedback controller that robustly stabilizes the closed-loop system? Plugging the controller (3.6) in (3.11) we get

$$dx(t) = \bar{A}(r(t), t)x(t)dt + \mathbb{W}(r(t))x(t)d\omega(t), \tag{3.23}$$

where $\bar{A}(r(t), t) = A(r(t), t) + B(r(t), t)K(r(t))$.

Using the results of Chapter 2 on the stability condition of piecewise deterministic systems with Wiener process, to guarantee the stability of (3.23) we should satisfy the following for every $i \in \mathscr{S}$:

$$\bar{A}^{\top}(i,t)P(i) + P(i)\bar{A}(i,t) + \mathbb{W}^{\top}(i)P(i)\mathbb{W}(i) + \sum_{j=1}^{N}\lambda_{ij}P(j) < 0.$$

Using the expression of $\bar{A}(i,t)$ we obtain

$$A^{\top}(i,t)P(i) + P(i)A(i,t) + K^{\top}(i)B^{\top}(i,t)P(i) + P(i)B(i,t)K(i)$$
$$+\mathbb{W}^{\top}(i)P(i)\mathbb{W}(i) + \sum_{j=1}^{N}\lambda_{ij}P(j) < 0.$$

This inequality is nonlinear in the decision variables $P(i)$ and $K(i)$ that need to be transformed to an equivalent LMI. For this purpose let $X(i) = P^{-1}(i)$. Pre- and post-multiplying this inequality by $X(i)$ gives

$$X(i)A^{\top}(i,t) + A(i,t)X(i) + X(i)K^{\top}(i)B^{\top}(i,t)$$
$$+B(i,t)K(i)X(i) + X(i)\mathbb{W}^{\top}(i)X^{-1}(i)\mathbb{W}(i)X(i)$$
$$+\sum_{j=1}^{N}\lambda_{ij}X(i)X^{-1}(j)X(i) < 0.$$

Using the expression of the uncertainties this inequality becomes

$$X(i)A^{\top}(i) + A(i)X(i) + X(i)K^{\top}(i)B^{\top}(i) + B(i)K(i)X(i)$$
$$+D_A(i)F_A(i,t)E_A(i)X(i) + X(i)E_A^{\top}(i)F_A^{\top}(i,t)D_A(i)$$
$$+D_B(i)F_B(i,t)E_B(i)K(i)X(i) + X(i)K^{\top}(i)E_B^{\top}(i)F_B^{\top}(i,t)D_B^{\top}(i)$$
$$+X(i)\mathbb{W}^{\top}(i)X^{-1}(i)\mathbb{W}(i)X(i) + \sum_{j=1}^{N}\lambda_{ij}X(i)X^{-1}(j)X(i) < 0.$$

Notice that from Lemma 7 in Appendix A we have:

$$D_A(i)F_A(i,t)E_A(i)X(i) + X(i)E_A^{\top}(i)F_A^{\top}(i,t)D_A(i)$$
$$\le \varepsilon_A(i)D_A(i)D_A^{\top}(i) + \varepsilon_A^{-1}(i)X(i)E_A^{\top}(i)E_A(i)X(i)$$
$$D_B(i)F_B(i,t)E_B(i)K(i)X(i) + X(i)K^{\top}(i)E_B^{\top}(i)F_B^{\top}(i,t)D_B^{\top}(i)$$
$$\le \varepsilon_B(i)D_B(i)D_B^{\top}(i) + \varepsilon_B^{-1}(i)X(i)K^{\top}(i)E_B^{\top}(i)E_B(i)K(i)X(i).$$

Letting $Y(i) = K(i)X(i)$ and using these inequalities and the Schur complement, the previous inequality will be satisfied if the following holds:

$$\begin{bmatrix} \tilde{J}_w(i) & X(i)E_A^{\top}(i) & Y^{\top}(i)E_B^{\top}(i) \\ E_A(i)X(i) & -\varepsilon_A(i)\mathbb{I} & 0 \\ E_B(i)Y(i) & 0 & -\varepsilon_B(i)\mathbb{I} \\ \mathbb{W}(i)X(i) & 0 & 0 \\ \mathcal{S}_i^{\top}(X) & 0 & 0 \end{bmatrix}$$

$$\begin{bmatrix} X(i)\mathbb{W}^{\top}(i) & \mathcal{S}_i(X) \\ 0 & 0 \\ 0 & 0 \\ -X(i) & 0 \\ 0 & -\mathcal{X}_i(X) \end{bmatrix} < 0, \tag{3.24}$$

where

$$\begin{aligned} \tilde{J}_w(i) = \ & X(i)A^{\top}(i) + A(i)X(i) + Y^{\top}(i)B^{\top}(i) + B(i)Y(i) \\ & + \lambda_{ii}X(i) + \varepsilon_A(i)D_A(i)D_A^{\top}(i) + \varepsilon_B(i)D_B(i)D_B^{\top}(i). \end{aligned}$$

The following theorem summarizes the results on the design of a state feedback controller that robustly stabilizes the class of systems we are studying when an external disturbance of Wiener process type is acting on the dynamics.

Theorem 17. *If there exist a set of symmetric and positive-definite matrices $X = (X(1), \cdots, X(N)) > 0$, a set of matrices $Y = (Y(1), \cdots, Y(N))$, and a set of positive scalars $\varepsilon_A = (\varepsilon_A(1), \cdots, \varepsilon_A(N)) > 0$ and $\varepsilon_B = (\varepsilon_B(1), \cdots, \varepsilon_B(N)) > 0$ such that the following set of coupled LMIs (3.24) holds for each $i \in \mathscr{S}$ and all admissible uncertainties, then the controller (3.6), with $K(i) = Y(i)X^{-1}(i)$, robustly stabilizes system (3.11) in the stochastic sense.*

Example 26. To illustrate the usefulness of the results of this theorem let us consider a two-mode system with the following data:

- mode #1:

$$A(1) = \begin{bmatrix} 1.0 & -0.5 \\ 0.1 & 1.0 \end{bmatrix}, \qquad B(1) = \begin{bmatrix} 1.0 & 0.0 \\ 0.0 & 1.0 \end{bmatrix},$$

$$D_A(1) = \begin{bmatrix} 0.1 \\ 0.2 \end{bmatrix}, \qquad D_B(1) = \begin{bmatrix} 0.1 \\ 0.2 \end{bmatrix},$$

$$E_A(1) = \begin{bmatrix} 0.2 & 0.1 \end{bmatrix}, \qquad E_B(1) = \begin{bmatrix} 0.2 & 0.1 \end{bmatrix},$$

$$\mathbb{W}(1) = \begin{bmatrix} 0.1 & 0.0 \\ 0.0 & 0.1 \end{bmatrix},$$

- mode #2:

$$A(2) = \begin{bmatrix} -0.2 & 0.5 \\ 0.0 & -0.25 \end{bmatrix}, \qquad B(2) = \begin{bmatrix} 1.0 & 0.0 \\ 0.0 & 1.0 \end{bmatrix},$$

$$D_A(2) = \begin{bmatrix} 0.13 \\ 0.1 \end{bmatrix}, \qquad D_B(2) = \begin{bmatrix} 0.13 \\ 0.1 \end{bmatrix},$$

$$E_A(2) = \begin{bmatrix} 0.1 & 0.2 \end{bmatrix}, \qquad E_B(2) = \begin{bmatrix} 0.1 & 0.2 \end{bmatrix}$$

$$\mathbb{W}(2) = \begin{bmatrix} 0.2 & 0.0 \\ 0.0 & 0.2 \end{bmatrix}.$$

The chosen positive scalars $\varepsilon_A(i)$ and $\varepsilon_B(i)$, $i = 1, 2$, are given by

$$\varepsilon_A(1) = \varepsilon_A(2) = 0.5,$$
$$\varepsilon_B(1) = \varepsilon_B(2) = 0.1.$$

The switching between the different modes is described by the following transition matrix:

$$\Lambda = \begin{bmatrix} -2.0 & 2.0 \\ 3.0 & -3.0 \end{bmatrix}.$$

Solving the LMI (3.24), we get

$$X(1) = \begin{bmatrix} 0.3240 & -0.0033 \\ -0.0033 & 0.3351 \end{bmatrix}, \quad X(2) = \begin{bmatrix} 0.2943 & -0.0009 \\ -0.0009 & 0.2964 \end{bmatrix},$$
$$Y(1) = \begin{bmatrix} -0.6019 & 0.0423 \\ 0.1048 & -0.6362 \end{bmatrix}, \quad Y(2) = \begin{bmatrix} -0.1425 & -0.1320 \\ -0.0200 & -0.1165 \end{bmatrix}.$$

The corresponding controller gains are given by

$$K(1) = \begin{bmatrix} -1.8568 & 0.1080 \\ 0.3042 & -1.8953 \end{bmatrix}, \quad K(2) = \begin{bmatrix} -0.4856 & -0.4469 \\ -0.0692 & -0.3933 \end{bmatrix}.$$

Based on the results of this theorem, the system of this example is stochastically stable under the state feedback controller with these gains.

As we did previously, let us try to synthesize a nonfragile controller for this class of systems when the uncertainties on the control matrix $B(r(t))$ are all equal to zero. Plugging the controller with the same expression as before for the design of a nonfragile controller in the dynamics, we get the following closed-loop dynamics:

$$\begin{aligned} dx(t) &= [A(r(t), t) + B(r(t))K(r(t), t)]\, x(t)dt + \mathbb{W}(r(t))x(t)d\omega(t) \\ &= [A(r(t)) + D_A(r(t))F_A(r(t), t)E_A(r(t)) \\ &\quad + B(r(t)) [\varrho(r(t))B^\top(r(t))P(r(t)) \\ &\quad + \rho(r(t))F_K(r(t), t)\varrho(r(t))B^\top(r(t))P(r(t)))]]\, x(t)dt \\ &\quad + \mathbb{W}(r(t))x(t)d\omega(t). \end{aligned}$$

Based on Theorem 8, the closed-loop system will be stable if the following holds for every $i \in \mathscr{S}$:

$$\begin{aligned} & [A(i, t) + B(i) [K(i) + \rho(i)F_K(i, t)K(i)]]^\top P(i) \\ & + P(i) [A(i, t) + B(i) [K(i) + \rho(i)F_K(i, t)K(i)]] \\ & + \mathbb{W}^\top(i)P(i)\mathbb{W}(i) + \sum_{j=1}^{N} \lambda_{ij}P(j) < 0. \end{aligned}$$

Using the fact that $K(i) = \varrho(i)B^\top(i)P(i)$ and Lemma 7, we get

$$2x^\top(t)P(i)D_A(i)F_A(i,t)E_A(i)x(t) \leq \varepsilon_A(i)x^\top(t)P(i)D_A(i)D_A^\top(i)P(i)x(t)$$
$$+\varepsilon_A^{-1}(i)x^\top(t)(i)E_A^\top(i)E_A(i)x(t),$$
$$2\rho(i)\varrho(i)x^\top(t)P(i)B(i)F_K(i,t)B^\top(i)P(i)x(t)$$
$$\leq \varepsilon_K^{-1}(i)\rho(i)\varrho^2(i)x^\top(t)P(i)B(i)B^\top(i)P(i)x(t)$$
$$+\varepsilon_K(i)\rho(i)x^\top(t)P(i)B(i)F_K^\top(i,t)F_K(i,t)B^\top(i)P(i)x(t)$$
$$\leq \varepsilon_K^{-1}(i)\rho(i)\varrho^2(i)x^\top(t)P(i)B(i)B^\top(i)P(i)x(t)$$
$$+\varepsilon_K(i)\rho(i)x^\top(t)P(i)B(i)B^\top(i)P(i)x(t).$$

Based on this, the left-hand side of the previous inequality becomes

$$
\begin{aligned}
\Gamma(i) =\ & A^\top(i)P(i) + P(i)A(i) + \varepsilon_A(i)P(i)D_A(i)D_A^\top(i)P(i)\\
&+\varepsilon_A^{-1}(i)E_A^\top(i)E_A(i) + \sum_{j=1}^{N}\lambda_{ij}P(j)\\
&+\varepsilon_K^{-1}(i)\rho(i)\varrho^2(i)P(i)B(i)B^\top(i)P(i)\\
&+\varepsilon_K(i)\rho(i)P(i)B(i)B^\top(i)P(i)\\
&+2\varrho(i)P(i)B(i)B^\top(i)P(i) + \mathbb{W}^\top(i)P(i)\mathbb{W}(i). \quad (3.25)
\end{aligned}
$$

This inequality is nonlinear in the design parameters $\varrho(i)$ and $P(i)$ for every $i \in \mathscr{S}$. To cast it into an LMI, let us put $X(i) = P^{-1}(i)$ for each $i \in \mathscr{S}$. Pre- and post-multiplying (3.25) by $X(i)$ gives

$$
\begin{aligned}
X(i)\Gamma(i)X(i) =\ & X(i)A^\top(i) + A(i)X(i) + \varepsilon_A(i)D_A(i)D_A^\top(i)\\
&+\varepsilon_A^{-1}(i)X(i)E_A^\top(i)E_A(i)X(i) + \sum_{j=1}^{N}\lambda_{ij}X(i)X^{-1}(j)X(i)\\
&+\varepsilon_K^{-1}(i)\rho(i)\varrho^2(i)B(i)B^\top(i) + \varepsilon_K(i)\rho(i)B(i)B^\top(i)\\
&+2\varrho(i)B(i)B^\top(i) + X(i)\mathbb{W}^\top(i)X^{-1}(i)\mathbb{W}(i)X(i). \quad (3.26)
\end{aligned}
$$

Defining $\mathcal{S}_i(X)$ and $\mathcal{X}_i(X)$ as before and using the Schur complement, we get

$$
\begin{bmatrix}
J(i) & X(i)E_A^\top(i) & \varrho(i)B(i) & X(i)\mathbb{W}^\top(i) & \mathcal{S}_i(X)\\
E_A(i)X(i) & -\varepsilon_A(i)\mathbb{I} & 0 & 0 & 0\\
\varrho(i)B^\top(i) & 0 & -\mu(i)\mathbb{I} & 0 & 0\\
\mathbb{W}(i)X(i) & 0 & 0 & -X(i) & 0\\
\mathcal{S}_i^\top(X) & 0 & 0 & 0 & -\mathcal{X}_i(X)
\end{bmatrix} < 0,
$$

where

$$
\begin{aligned}
J(i) =\ & X(i)A^\top(i) + A(i)X(i) + \varepsilon_A(i)D_A(i)D_A^\top(i) + \lambda_{ii}X(i)\\
&+2\varrho(i)B(i)B^\top(i) + \varepsilon_K(i)\rho(i)B(i)B^\top(i).
\end{aligned}
$$

The following theorem gives a result in the LMI framework that can be used to design a nonfragile robust controller for the class of systems we are considering.

Theorem 18. *If there exist a set of symmetric and positive-definite matrices $X = (X(1), \cdots , X(N)) > 0$, a set of positive scalars $\varepsilon_A = (\varepsilon_A(1), \cdots , \varepsilon_A(N))$, $\mu = (\mu(1), \cdots , \mu(N))$, $\nu = (\nu(1), \cdots , \nu(N))$, and a set of scalars $\varrho = (\varrho(1), \cdots , \varrho(N))$ satisfying the following set of coupled LMIs for every $i \in \mathcal{S}$ and for all admissible uncertainties:*

$$\begin{bmatrix} J(i) & X(i)E_A^\top(i) & \varrho(i)B(i) & X(i)\mathbb{W}^\top(i) & \mathcal{S}_i(X) \\ E_A(i)X(i) & -\varepsilon_A(i)\mathbb{I} & 0 & 0 & 0 \\ \varrho(i)B^\top(i) & 0 & -\mu(i)\mathbb{I} & 0 & 0 \\ \mathbb{W}(i)X(i) & 0 & 0 & -X(i) & 0 \\ \mathcal{S}_i^\top(X) & 0 & 0 & 0 & -\mathcal{X}_i(X) \end{bmatrix} < 0, \quad (3.27)$$

where

$$\begin{aligned} J(i) = {}& X(i)A^\top(i) + A(i)X(i) + \varepsilon_A(i)D_A(i)D_A^\top(i) + \lambda_{ii}X(i) \\ & + 2\varrho(i)B(i)B^\top(i) + \nu(i)B(i)B^\top(i), \\ \mu(i) = {}& \frac{\varepsilon_K(i)}{\rho(i)}, \\ \nu(i) = {}& \varepsilon_K(i)\rho(i), \end{aligned}$$

then the closed-loop system is robustly stochastically stable with nonfragility $\rho(i)$ under the controller (3.6) with the gain $K(i) = \varrho(i)B^\top(i)X^{-1}(i), i = 1, 2, \cdots , N$.

Example 27. In this example, we show the usefulness of the proposed results in this section. For this purpose let us consider the two-mode system of Example 26 with $\Delta B(i) = 0, i = 1, 2$ and

$$\mathbb{W}(1) = \begin{bmatrix} 0.2 & 0.0 \\ 0.0 & 0.2 \end{bmatrix}, \quad \mathbb{W}(2) = \begin{bmatrix} 0.1 & 0.0 \\ 0.0 & 0.1 \end{bmatrix}.$$

Letting $\varepsilon_A(1) = \varepsilon_A(2) = 0.5$, $\varepsilon_K(1) = \varepsilon_K(2) = 0.1$, and $\rho(1) = 0.5$, $\rho(2) = 0.6$ and solving the LMI (3.27), we get

$$X(1) = \begin{bmatrix} 0.0648 & 0.0028 \\ 0.0028 & 0.0582 \end{bmatrix}, \quad X(2) = \begin{bmatrix} 0.0807 & -0.0036 \\ -0.0036 & 0.0751 \end{bmatrix},$$
$$\nu(1) = -0.1803, \qquad\qquad \nu(2) = -0.1431,$$

which gives the following controller gains:

$$K(1) = \begin{bmatrix} -2.7883 & 0.1341 \\ 0.1341 & -3.1053 \end{bmatrix}, \quad K(2) = \begin{bmatrix} -1.7766 & -0.0849 \\ -0.0849 & -1.9081 \end{bmatrix}.$$

Based on the results of this theorem, the system of this example is stochastically stable under the state feedback controller with these gains.

The results established in this section do not consider the uncertainties on the jump rates. In the rest of this section, we discuss these uncertainties and synthesize a controller that can robustly stabilize the class of systems we are considering whether the Wiener process external disturbance is acting or not.

Let us consider the case when the system is not perturbed by the external Wiener process. In this case, the closed-loop system will be stable when the two uncertainties are present if the following holds for every $i \in \mathscr{S}$:

$$[A(i,t) + B(i,t)K(i)]^\top P(i) + P(i)[A(i,t) + B(i,t)K(i)]$$
$$+ \sum_{k=1}^{\kappa}\sum_{j=1}^{N} \alpha_k \lambda_{ij}^k P(j) < 0,$$

which gives

$$A^\top(i,t)P(i) + P(i)A(i,t) + K^\top(i)B^\top(i,t)P(i)$$
$$+ P(i)B(i,t)K(i) + \sum_{j=1}^{N} \mu_{ij}P(j) < 0,$$

with $\mu_{ij} = \sum_{k=1}^{\kappa} \alpha_k \lambda_{ij}^k$.

Using the expression of the uncertainties of the state and control matrices and Lemma 7 in Appendix A, after letting $X(i) = P^{-1}(i)$ and pre- and post-multiplying the previous expression by $X(i)$, we get

$$X(i)A^\top(i) + A(i)X(i) + X(i)K^\top(i)B^\top(i)$$
$$+ B(i)K(i)X(i) + \varepsilon_A(i)D_A(i)D_A^\top(i)$$
$$+ \varepsilon_B(i)D_B(i)D_B^\top(i) + \varepsilon_A^{-1}(i)X(i)E_A^\top(i)E_A(i)X(i)$$
$$+ \varepsilon_B^{-1}(i)X(i)K^\top(i)E_B^\top(i)E_B(i)K(i)X(i)$$
$$+ \sum_{j=1}^{N} \mu_{ij}X(i)X^{-1}(j)X(i) < 0.$$

Letting $Y(i) = K(i)X(i)$, and $\mathcal{S}_i(X)$ and $\mathcal{X}_i(X)$ be defined as before by replacing λ_{ij} by $\mu_{ij}, j = 1, \cdots, i-1, i+1, \cdots, N$, we get

$$X(i)A^\top(i) + A(i)X(i) + Y^\top(i)B^\top(i)$$
$$+ B(i)Y(i) + \varepsilon_A(i)D_A(i)D_A^\top(i)$$
$$+ \varepsilon_B(i)D_B(i)D_B^\top(i) + \varepsilon_A^{-1}(i)X(i)E_A^\top(i)E_A(i)X(i)$$
$$+ \varepsilon_B^{-1}(i)Y^\top(i)E_B^\top(i)E_B(i)Y(i)$$
$$+ \mathcal{S}_i(X)\mathcal{X}_i^{-1}(X)\mathcal{S}_i^\top(X) + \mu_{ii}X(i) < 0,$$

which gives in turn after using the Schur complement:

$$\begin{bmatrix} J_{u1}(i) & X(i)E_A^\top(i) & Y^\top(i)E_B^\top(i) & \mathcal{S}_i(X) \\ E_A(i)X(i) & -\varepsilon_A(i)\mathbb{I} & 0 & 0 \\ E_B(i)Y(i) & 0 & -\varepsilon_B(i)\mathbb{I} & 0 \\ \mathcal{S}_i^\top(X) & 0 & 0 & -\mathcal{X}_i(X) \end{bmatrix} < 0, \qquad (3.28)$$

where $J_{u1}(i) = X(i)A^\top(i) + A(i)X(i) + Y^\top(i)B^\top(i) + B(i)Y(i) + \mu_{ii}X(i) + \varepsilon_A(i)D_A(i)D_A^\top(i) + \varepsilon_B(i)D_B(i)D_B^\top(i)$.

The following theorem gives results that can be used to check if a given system with uncertainties on the state matrix and on the jump rates is robustly stochastically stable.

Theorem 19. *If there exist a set of symmetric and positive-definite matrices $X = (X(1), \cdots, X(N)) > 0$ and a set of matrices $Y = (Y(1), \cdots, Y(N))$ and a set of positive scalars $\varepsilon_A = (\varepsilon_A(1), \cdots, \varepsilon_A(N)) > 0$ and $\varepsilon_B = (\varepsilon_B(1), \cdots, \varepsilon_B(N)) > 0$ such that the following set of coupled LMIs (3.28) holds for every $i \in \mathscr{S}$ and for all admissible uncertainties, then the controller (3.6), with $K(i) = Y(i)X^{-1}(i)$, robustly stabilizes system (3.1) in the stochastic sense.*

Example 28. To illustrate the results of this theorem let us consider the two-mode system of Example 26 with $\mathbb{W}(i) = 0, i = 1, 2$.

The switching between the different modes is governed by the following transition matrices:

$$\Lambda_1 = \begin{bmatrix} -2.0 & 2.0 \\ 3.0 & -3.0 \end{bmatrix}, \quad \alpha_1 = 0.6,$$

$$\Lambda 2 = \begin{bmatrix} -1.5 & 1.5 \\ 3.50 & -3.5 \end{bmatrix}, \quad \alpha_2 = 0.4.$$

The corresponding μ is given by

$$\mu = \begin{bmatrix} -1.80 & 1.80 \\ 3.20 & -3.20 \end{bmatrix}.$$

The positive scalars $\varepsilon_A(i)$ and $\varepsilon_B(i)$, $i = 1, 2$ are chosen as follows:

$$\varepsilon_A(1) = \varepsilon_A(2) = 0.5,$$
$$\varepsilon_B(1) = \varepsilon_B(2) = 0.1.$$

Solving the LMI (3.28) we get

$$X(1) = \begin{bmatrix} 0.3262 & -0.0054 \\ -0.0054 & 0.3442 \end{bmatrix}, \quad X(2) = \begin{bmatrix} 0.2789 & -0.0017 \\ -0.0017 & 0.2836 \end{bmatrix},$$

$$Y(1) = \begin{bmatrix} -0.7014 & 0.0513 \\ 0.1133 & -0.7512 \end{bmatrix}, \quad Y(2) = \begin{bmatrix} -0.1995 & -0.1270 \\ -0.0160 & -0.1671 \end{bmatrix}.$$

The corresponding controller gains are given by

$$K(1) = \begin{bmatrix} -2.1484 & 0.1155 \\ 0.3116 & -2.1779 \end{bmatrix}, \quad K(2) = \begin{bmatrix} -0.7183 & -0.4524 \\ -0.0610 & -0.5897 \end{bmatrix}.$$

Based on the results of this theorem, the system of this example is stochastically stable under the state feedback controller with these gains.

When the Wiener process external disturbance is acting on the class of systems we are considering, similar results can be obtained following the same steps presented previously. Omitting the details, these results can be summarized by the following theorem.

Theorem 20. *If there exists a set of symmetric and positive-definite matrices* $X = (X(1), \cdots, X(N)) > 0$ *and a set of matrices* $Y = (Y(1), \cdots, Y(N))$ *and a set of positive scalars* $\varepsilon_A = (\varepsilon_A(1), \cdots, \varepsilon_A(N)) > 0$ *and* $\varepsilon_B = (\varepsilon_B(1), \cdots, \varepsilon_B(N)) > 0$ *such that the following set of coupled LMIs holds for every* $i \in \mathscr{S}$ *and for all admissible uncertainties:*

$$
\begin{bmatrix}
J_{w1}(i) & X(i)E_A^\top(i) & Y^\top(i)E_B^\top(i) \\
E_A(i)X(i) & -\varepsilon_A(i)\mathbb{I} & 0 \\
E_B(i)Y(i) & 0 & -\varepsilon_B(i)\mathbb{I} \\
\mathbb{W}(i)X(i) & 0 & 0 \\
\mathcal{S}_i^\top(X) & 0 & 0
\end{bmatrix}
$$

$$
\left.
\begin{matrix}
X(i)\mathbb{W}^\top(i) & \mathcal{S}_i(X) \\
0 & 0 \\
0 & 0 \\
-X(i) & 0 \\
0 & -\mathcal{X}_i(X)
\end{matrix}
\right] < 0, \qquad (3.29)
$$

where $J_{w1}(i) = X(i)A^\top(i) + A(i)X(i) + Y^\top(i)B^\top(i) + B(i)Y(i) + \mu_{ii}X(i) + \varepsilon_A(i)D_A(i)D_A^\top(i) + \varepsilon_B(i)D_B(i)D_B^\top(i)$, *then the controller (3.6), with* $K(i) = Y(i)X^{-1}(i)$, *robustly stabilizes system (3.11) in the stochastic sense.*

Example 29. To illustrate the results developed in this theorem let us consider the two-mode system of Example 26.

Let the positive scalars $\varepsilon_A(i)$ and $\varepsilon_B(i)$, $i = 1, 2$ be fixed as follows:

$$
\varepsilon_A(1) = \varepsilon_A(2) = 0.50,
$$
$$
\varepsilon_B(1) = \varepsilon_B(2) = 0.10.
$$

The switching between the different modes is described by the following transition matrices with the appropriate weights:

$$
\Lambda_1 = \begin{bmatrix} -2.0 & 2.0 \\ 3.0 & -3.0 \end{bmatrix}, \qquad \alpha_1 = 0.6,
$$

$$
\Lambda2 = \begin{bmatrix} -1.50 & 1.50 \\ 3.50 & -3.50 \end{bmatrix}, \qquad \alpha_2 = 0.4.
$$

The corresponding μ is given by

$$
\mu = \begin{bmatrix} -1.80 & 1.80 \\ 3.20 & -3.20 \end{bmatrix}.
$$

Solving the LMI (3.29), we get

$$X(1) = \begin{bmatrix} 0.3134 & -0.0033 \\ -0.0033 & 0.3242 \end{bmatrix}, \quad X(2) = \begin{bmatrix} 0.2692 & -0.0009 \\ -0.0009 & 0.2717 \end{bmatrix},$$

$$Y(1) = \begin{bmatrix} -0.5738 & 0.0399 \\ 0.1004 & -0.6052 \end{bmatrix}, \quad Y(2) = \begin{bmatrix} -0.0948 & -0.1226 \\ -0.0193 & -0.0720 \end{bmatrix}.$$

The corresponding controller gains are given by

$$K(1) = \begin{bmatrix} -1.8300 & 0.1042 \\ 0.3006 & -1.8638 \end{bmatrix}, \quad K(2) = \begin{bmatrix} -0.3536 & -0.4524 \\ -0.0725 & -0.2654 \end{bmatrix}.$$

Based on the results of this theorem, the system of this example is stochastically stable under the state feedback controller with these gains.

For the nonfragile controller, we can establish the following results.

Theorem 21. *If there exist a set of symmetric and positive-definite matrices* $X = (X(1), \cdots, X(N)) > 0$, *and a set of positive scalars* $\varepsilon_A = (\varepsilon_A(1), \cdots, \varepsilon_A(N))$, $\mu = (\mu(1), \cdots, \mu(N))$, $\nu = (\nu(1), \cdots, \nu(N))$, *and a set of scalars* $\varrho = (\varrho(1), \cdots, \varrho(N))$ *satisfying the following set of coupled LMIs for every* $i \in S$ *and for all admissible uncertainties:*

$$\begin{bmatrix} J_f(i) & X(i)E_A^\top(i) & \varrho(i)B(i) & X(i)\mathbb{W}^\top(i) \\ E_A(i)X(i) & -\varepsilon_A(i)\mathbb{I} & 0 & 0 \\ \varrho(i)B^\top(i) & 0 & -\mu(i)\mathbb{I} & 0 \\ \mathbb{W}(i)X(i) & 0 & 0 & -X(i) \\ S_i^\top(X) & 0 & 0 & 0 \end{bmatrix}$$

$$\left. \begin{matrix} S_i(X) \\ 0 \\ 0 \\ 0 \\ -\mathcal{X}_i(X) \end{matrix} \right] < 0, \tag{3.30}$$

where

$$\begin{aligned} J_f(i) = {}& X(i)A^\top(i) + A(i)X(i) + \varepsilon_A(i)D_A(i)D_A^\top(i) + \mu_{ii}X(i) \\ & + 2\varrho(i)B(i)B^\top(i) + \nu(i)B(i)B^\top(i), \\ \mu(i) = {}& \frac{\varepsilon_K(i)}{\rho(i)}, \\ \nu(i) = {}& \varepsilon_K(i)\rho(i), \end{aligned}$$

then closed-loop system is robustly stochastically stable with nonfragility $\rho(i)$ *under the controller (3.6) with the gain* $K(i) = \varrho(i)B^\top(i)X^{-1}(i), i = 1, 2, \cdots, N.$

Example 30. To illustrate the results of this theorem, let us consider the two-mode system of Example 26 with $\Delta B(i) = 0, i = 1, 2.$

The positive scalars $\varepsilon_A(i)$ and $\varepsilon_B(i)$, $i = 1, 2$ are fixed as follows:

$$\varepsilon_A(1) = \varepsilon_A(2) = 0.5,$$
$$\varepsilon_K(1) = \varepsilon_K(2) = 0.1,$$
$$\rho(1) = \rho(2) = 0.6.$$

The switching between the two modes is described by the following transition rate matrices with the appropriate weights:

$$\Lambda(1) = \begin{bmatrix} -2.0 & 2.0 \\ 3.0 & -3.0 \end{bmatrix}, \qquad \alpha_1 = 0.6,$$

$$\Lambda(2) = \begin{bmatrix} -1.50 & 1.50 \\ 3.50 & -3.50 \end{bmatrix}, \qquad \alpha2 = 0.4.$$

The corresponding μ is given by

$$\mu = \begin{bmatrix} -1.80 & 1.80 \\ 3.20 & -3.20 \end{bmatrix}.$$

Solving the LMI (3.30), we get

$$X(1) = \begin{bmatrix} 0.0687 & 0.0030 \\ 0.0030 & 0.0612 \end{bmatrix}, \quad X(2) = \begin{bmatrix} 0.0838 & -0.0037 \\ -0.0037 & 0.0771 \end{bmatrix},$$
$$\nu(1) = -0.1918, \qquad\qquad \nu(2) = -0.1587.$$

The corresponding controller gains are given by

$$K(1) = \begin{bmatrix} -2.7972 & 0.1351 \\ 0.1351 & -3.1431 \end{bmatrix}, \quad K(2) = \begin{bmatrix} -1.8977 & -0.0917 \\ -0.0917 & -2.0629 \end{bmatrix}.$$

Based on the results of this theorem, the system of this example is stochastically stable under the state feedback controller with these gains.

This section covered the state feedback stabilization problem. By assuming complete access to the mode and to the state vector at time t, we developed many design approaches for different classes of systems. In the next section, we will relax the assumption on the availability of the state vector $x(t)$ and try to design an output feedback controller that ensures the same performance.

3.3 Static Output Feedback Stabilization

In the previous section we assumed complete access to the state vector, which may not always be valid for technological or cost reasons. This assumption limits the use of the developed results. In this section we concentrate on the design of a stabilizing static output feedback controller that uses the system output

$$y(t) = C(r(t))x(t)$$

to compute the control with the form

$$u(t) = F(r(t))y(t) = F(r(t))C(r(t))x(t), \tag{3.31}$$

with $F(r(t))$ the gain to be determined for each $r(t) \in \mathscr{S}$.

Remark 12. In this section we assume that the output matrix $C(r(t))$ has no uncertainties.

Let us now concentrate on the design of the static output feedback controller (3.31). Plugging the controller expression in the system dynamics (3.1) gives

$$\begin{aligned} \dot{x}(t) &= [A(i) + B(i)F(i)C(i)]\, x(t), \\ &= A_{cl}(i)x(i), \end{aligned}$$

with $A_{cl}(i) = A(i) + B(i)F(i)C(i)$.

Based on Theorem 4, the closed-loop system is stochastically stable if there exists a set of symmetric and positive-definite matrices $P = (P(1), \cdots, P(N)) > 0$ such that the following holds:

$$P(i)A_{cl}(i) + A_{cl}^{\top}(i)P(i) + \sum_{j=1}^{N} \lambda_{ij} P(j) < 0,$$

which gives

$$P(i)A(i) + A^{\top}(i)P(i) + P(i)B(i)F(i)C(i) + [P(i)B(i)F(i)C(i)]^{\top}$$
$$+ \sum_{j=1}^{N} \lambda_{ij} P(j) < 0.$$

This matrix inequality is nonlinear in the design parameters $P(i)$ and $F(i)$. To put it into the LMI form, let $X(i) = P^{-1}(i)$. Pre- and post-multiplying this inequality by $X(i)$ gives

$$A(i)X(i) + X(i)A^{\top}(i) + B(i)F(i)C(i)X(i) + X(i)C^{\top}(i)F^{\top}(i)B^{\top}(i)$$
$$+ \sum_{j=1}^{N} \lambda_{ij} X(i)X^{-1}(j)X(i) < 0.$$

Notice that

$$\sum_{j=1}^{N} \lambda_{ij} X(i)X^{-1}(j)X(i) = \lambda_{ii}X(i) + \mathcal{S}_i(X)\mathcal{X}_i^{-1}(X)\mathcal{S}_i^{\top}(X),$$

with $\mathcal{S}_i(X)$ and $\mathcal{X}_i(X)$ keeping the same definition as before.

If we let $F(i) = Z(i)Y^{-1}(i)$ and $Y(i)C(i) = C(i)X(i)$ hold for every $i \in \mathscr{S}$ for some appropriate matrices that we have to determine, we get

$$A(i)X(i) + X(i)A^\top(i) + B(i)Z(i)C(i) + C^\top(i)Z^\top(i)B^\top(i)$$
$$+\lambda_{ii}X(i) + \mathcal{S}_i(X)\mathcal{X}_i^{-1}(X)\mathcal{S}_i^\top(X) < 0.$$

Finally, using the Schur complement gives

$$\begin{bmatrix} J(i) & \mathcal{S}_i(X) \\ \mathcal{S}_i^\top(X) & -\mathcal{X}_i(X) \end{bmatrix} < 0,$$

with $J(i) = A(i)X(i) + X(i)A^\top(i) + B(i)Z(i)C(i) + C^\top(i)Z^\top(i)B^\top(i) + \lambda_{ii}X(i)$.

The following theorem summarizes the results of this development.

Theorem 22. *If there exist sets of symmetric and positive-definite matrices $X = (X(1), \cdots, X(N)) > 0$, and $Y = (Y(1), \cdots, Y(N)) > 0$ and a set of matrices $Z = (Z(1), \cdots, Z(N))$, such that the following holds for each $i \in \mathscr{S}$:*

$$\begin{cases} Y(i)C(i) = C(i)X(i), \\ \begin{bmatrix} J(i) & \mathcal{S}_i(X) \\ \mathcal{S}_i^\top(X) & -\mathcal{X}_i(X) \end{bmatrix} < 0, \end{cases} \tag{3.32}$$

where

$$J(i) = A(i)X(i) + X(i)A^\top(i) + B(i)Z(i)C(i) + C^\top(i)Z^\top(i)B^\top(i)$$
$$+\lambda_{ii}X(i),$$

then system (3.1) is stochastically stable and the controller gain is given by $F(i) = Z(i)Y^{-1}(i)$, $i \in \mathscr{S}$.

Let us now consider the effect of the uncertainties. Based on the results of Chapter 2, (3.1) is robust stochastically stable if there exist a set of symmetric and positive-definite matrices $P = (P(1), \cdots, P(N)) > 0$ and a set of positive scalars $\varepsilon_A = (\varepsilon_A(1), \cdots, \varepsilon_A(N))$, such that the following coupled LMIs hold for every $i \in \mathscr{S}$:

$$\begin{bmatrix} J_u(i) & P(i)D_A(i) \\ D_A^\top(i)P(i) & -\varepsilon_A(i)\mathbb{I} \end{bmatrix} < 0, \tag{3.33}$$

with $J_u(i) = P(i)A(i) + A^\top(i)P(i) + \sum_{j=1}^N \lambda_{ij}P(j) + \varepsilon_A(i)E_A^\top(i)E_A(i)$.

Following the same steps as for the nominal system, we can establish the following result for an uncertain system. This result allows the design of a static output feedback that robustly stochastically stabilizes the class of systems we are considering in this book.

Corollary 4. *If there exist sets of symmetric and positive-definite matrices $X = (X(1), \cdots, X(N)) > 0$, and $Y = (Y(1), \cdots, Y(N)) > 0$ and a set of matrices $Z = (Z(1), \cdots, Z(N))$ and sets of positive scalars $\varepsilon_A = (\varepsilon_A(1), \cdots, \varepsilon_A(N))$ and $\varepsilon_B = (\varepsilon_B(1), \cdots, \varepsilon_B(N))$, such that the following holds for each $i \in \mathscr{S}$:*

$$
\begin{cases}
Y(i)C(i) = C(i)X(i), \\
\begin{bmatrix}
J_X(i) & X(i)E_A^\top(i) & C^\top(i)Z^\top(i)E_B^\top(i) & S_i(X) \\
E_A(i)X(i) & -\varepsilon_A(i)\mathbb{I} & 0 & 0 \\
E_B^\top(i)Z(i)C(i) & 0 & -\varepsilon_B(i)\mathbb{I} & 0 \\
S_i^\top(X) & 0 & 0 & -\mathcal{X}_i(X)
\end{bmatrix} < 0,
\end{cases} \tag{3.34}
$$

where

$$
\begin{aligned}
J_X(i) = \; & A(i)X(i) + X(i)A^\top(i) + B(i)Z(i)C(i) + C^\top(i)Z^\top(i)B^\top(i) \\
& + \lambda_{ii}(i)X(i) + \varepsilon_A(i)D_A(i)D_A^\top(i) + \varepsilon_B(i)D_B(i)D_B^\top(i),
\end{aligned}
$$

then (3.1) is robustly stochastically stable and the controller gain is given by $F(i) = Z(i)Y^{-1}(i)$, $i \in \mathscr{S}$.

Example 31. To show the validity of our results, let us consider a two-mode system with the following data:

- mode #1:

$$
A(1) = \begin{bmatrix} 0.0 & -1.0 & 1.0 \\ -1.0 & 3.0 & 0.0 \\ 0.0 & 0.0 & 0.0 \end{bmatrix}, \quad B(1) = \begin{bmatrix} 0.0 & 0.2 \\ 1.0 & 0.0 \\ -0.1 & 1.0 \end{bmatrix},
$$

$$
C(1) = \begin{bmatrix} 1.0 & 0.0 & 1.0 \\ 0.3 & 1.0 & 0.0 \end{bmatrix},
$$

- mode #2:

$$
A(2) = \begin{bmatrix} 0.0 & 1.5 & 1.5 \\ -1.0 & -3.0 & 0.0 \\ 0.0 & 0.0 & 0.0 \end{bmatrix}, \quad B(2) = \begin{bmatrix} 0.0 & -0.2 \\ 1.2 & 0.0 \\ 0.1 & 1.2 \end{bmatrix},
$$

$$
C(2) = \begin{bmatrix} 1.0 & 0.0 & 1.0 \\ 0.1 & 1.0 & 0.0 \end{bmatrix}.
$$

The switching between the two modes is described by

$$
\Lambda = \begin{bmatrix} -2.0 & 2.0 \\ 1.0 & -1.0 \end{bmatrix}.
$$

Solving the set of LMIs (3.34) gives

$$X(1) = \begin{bmatrix} 1.1035 & -0.0776 & -0.2586 \\ -0.0776 & 1.0839 & 0.0129 \\ -0.2586 & 0.0129 & 1.1229 \end{bmatrix}, Y(1) = \begin{bmatrix} 0.8643 & -0.0647 \\ -0.0647 & 1.0606 \end{bmatrix},$$

$$Z(1) = \begin{bmatrix} 4.5134 & -4.2276 \\ -0.2412 & -5.0995 \end{bmatrix}, \qquad X(2) = \begin{bmatrix} 1.4311 & -0.0637 & -0.6375 \\ -0.0637 & 1.2542 & 0.0183 \\ -0.6375 & 0.0183 & 1.4356 \end{bmatrix},$$

$$Y(2) = \begin{bmatrix} 0.7982 & -0.0454 \\ -0.0454 & 1.2478 \end{bmatrix}, \qquad Z(2) = \begin{bmatrix} -0.6999 & 2.0917 \\ -1.1875 & 0.0496 \end{bmatrix}.$$

This gives the following gains:

$$F(1) = \begin{bmatrix} 4.9461 & -3.6842 \\ -0.6420 & -4.8472 \end{bmatrix}, \quad F(2) = \begin{bmatrix} -0.7831 & 1.6478 \\ -1.4886 & -0.0145 \end{bmatrix}.$$

Let us now consider the case of the state equation (3.11) and see how we can design a stabilizing static output feedback controller. Plugging the controller (3.31) in the system dynamics (3.11) gives

$$\begin{aligned} dx(t) &= [A(i) + B(i)F(i)C(i)]\, x(t)dt + \mathbb{W}(i)x(t)d\omega(t) \\ &= A_{cl}(i)x(t)dt + \mathbb{W}(i)x(t)d\omega(t), \end{aligned}$$

with $A_{cl}(i) = A(i) + B(i)F(i)C(i)$.

Based on Theorem 7, the closed-loop system is stochastically stable if there exists a set of symmetric and positive-definite matrices such that the following holds:

$$P(i)A_{cl}(i) + A_{cl}^{\top}(i)P(i) + \mathbb{W}^{\top}(i)P(i)\mathbb{W}(i) + \sum_{j=1}^{N} \lambda_{ij}P(j) < 0,$$

which gives

$$P(i)A(i) + A^{\top}(i)P(i) + P(i)B(i)F(i)C(i) + [P(i)B(i)F(i)C(i)]^{\top}$$
$$+\mathbb{W}^{\top}(i)P(i)\mathbb{W}(i) + \sum_{j=1}^{N} \lambda_{ij}P(j) < 0.$$

This matrix inequality is nonlinear in the design parameters $P(i)$ and $F(i)$. To put it into the LMI form, let $X(i) = P^{-1}(i)$. Pre- and post-multiplying this inequality by $X(i)$ gives

$$A(i)X(i) + X(i)A^{\top}(i) + B(i)F(i)C(i)X(i)$$
$$+X(i)C^{\top}(i)F^{\top}(i)B^{\top}(i) + X(i)\mathbb{W}^{\top}(i)X^{-1}(i)\mathbb{W}(i)X(i)$$
$$+ \sum_{j=1}^{N} \lambda_{ij}X(i)X^{-1}(j)X(i) < 0.$$

Notice that

$$\sum_{j=1}^{N} \lambda_{ij} X(i) X^{-1}(j) X(i) = \lambda_{ii} X(i) + \mathcal{S}_i(X) \mathcal{X}_i^{-1}(X) \mathcal{S}_i^{\top}(X),$$

with $\mathcal{S}_i(X)$ and $\mathcal{X}_i(X)$ keeping the same definitions as before.

Now if we let $F(i) = Z(i)Y^{-1}(i)$ and $Y(i)C(i) = C(i)X(i)$ hold for every $i \in \mathscr{S}$ for some appropriate matrices that we have to determine, we get

$$A(i)X(i) + X(i)A^{\top}(i) + B(i)Z(i)C(i) + C^{\top}(i)Z^{\top}(i)B^{\top}(i)$$
$$+ X(i)\mathbb{W}^{\top}(i)X^{-1}(i)\mathbb{W}(i)X(i) + \lambda_{ii}X(i) + \mathcal{S}_i(X)\mathcal{X}_i^{-1}(X)\mathcal{S}_i^{\top}(X) < 0.$$

Finally, using the Schur complement gives

$$\begin{bmatrix} J(i) & X(i)\mathbb{W}^{\top}(i) & \mathcal{S}_i(X) \\ \mathbb{W}(i)X(i) & -X(i) & 0 \\ \mathcal{S}_i^{\top}(X) & 0 & -\mathcal{X}_i(X) \end{bmatrix} < 0,$$

with $J(i) = A(i)X(i) + X(i)A^{\top}(i) + B(i)Z(i)C(i) + C^{\top}(i)Z^{\top}(i)B^{\top}(i) + \lambda_{ii}X(i)$.

The following theorem summarizes the results of this development.

Theorem 23. *If there exist sets of symmetric and positive-definite matrices $X = (X(1), \cdots, X(N)) > 0$, and $Y = (Y(1), \cdots, Y(N)) > 0$ and a set of matrices $Z = (Z(1), \cdots, Z(N))$, such that the following holds for each $i \in \mathscr{S}$:*

$$\begin{cases} \begin{bmatrix} J(i) & X(i)\mathbb{W}^{\top}(i) & \mathcal{S}_i(X) \\ \mathbb{W}(i)X(i) & -X(i) & 0 \\ \mathcal{S}_i^{\top}(X) & 0 & -\mathcal{X}_i(X) \end{bmatrix} < 0, \\ Y(i)C(i) = C(i)X(i), \end{cases} \tag{3.35}$$

where

$$\begin{aligned} J(i) = & A(i)X(i) + X(i)A^{\top}(i) + B(i)Z(i)C(i) + C^{\top}(i)Z^{\top}(i)B^{\top}(i) \\ & + \lambda_{ii}X(i), \end{aligned}$$

then (3.11) is stochastically stable and the controller gain is given by $F(i) = Z(i)Y^{-1}(i)$, $i \in \mathscr{S}$.

For the uncertain system, based on Chapter 2, (3.11) is robustly stochastically stable if there exist a set of symmetric and positive-definite matrices $P = (P(1), \cdots, P(N)) > 0$ and a set of positive scalars $\varepsilon_A = (\varepsilon_A(1), \cdots, \varepsilon_A(N))$, such that the following coupled LMIs hold for every $i \in \mathscr{S}$:

$$\begin{bmatrix} J_u(i) & P(i)D_A(i) \\ D_A^{\top}(i)P(i) & -\varepsilon_A(i)\mathbb{I} \end{bmatrix} < 0, \tag{3.36}$$

with $J_u(i) = P(i)A(i) + A^{\top}(i)P(i) + \sum_{j=1}^{N} \lambda_{ij}P(j) + \varepsilon_A(i)E_A^{\top}(i)E_A(i) + \mathbb{W}^{\top}(i)P(i)\mathbb{W}(i)$.

Following the same steps as for the nominal system, we can establish the following result for uncertain system using Lemma 7 and the Schur complement Lemma 4. This result allows the design of a static output feedback that robustly stochastically stabilizes the class of systems we are considering in this book.

Corollary 5. *If there exist sets of symmetric and positive-definite matrices* $X = (X(1), \cdots, X(N)) > 0$, *and* $Y = (Y(1), \cdots, Y(N)) > 0$ *and a set of matrices* $Z = (Z(1), \cdots, Z(N))$, *and sets of positive scalars* $\varepsilon_A = (\varepsilon_A(1), \cdots, \varepsilon_A(N))$, *and* $\varepsilon_B = (\varepsilon_B(1), \cdots, \varepsilon_B(N))$, *such that the following holds for each* $i \in \mathscr{S}$:

$$
\left\{
\begin{array}{l}
Y(i)C(i) = C(i)X(i), \\
\begin{bmatrix}
J_X(i) & X(i)E_A^\top(i) & C^\top(i)Z^\top(i)E_B^\top(i) \\
E_A(i)X(i) & -\varepsilon_A(i)\mathbb{I} & 0 \\
E_B^\top(i)Z(i)C(i) & 0 & -\varepsilon_B(i)\mathbb{I} \\
\mathbb{W}(i)X(i) & 0 & 0 \\
\mathcal{S}_i^\top(X) & 0 & 0
\end{array}
\right.
$$

$$
\left.
\begin{array}{cc}
X(i)\mathbb{W}^\top(i)E^\top & \mathcal{S}_i(X) \\
0 & 0 \\
0 & 0 \\
-X(i) & 0 \\
0 & -\mathcal{X}_i(X)
\end{bmatrix} < 0,
\end{array}
\right.
\tag{3.37}
$$

where

$$
\begin{aligned}
J_X(i) =\ & A(i)X(i) + X(i)A^\top(i) + B(i)Z(i)C(i) + C^\top(i)Z^\top(i)B^\top(i) \\
& + \lambda_{ii}X(i) + \varepsilon_A(i)D_A(i)D_A^\top(i) + \varepsilon_B(i)D_B(i)D_B^\top(i),
\end{aligned}
$$

then (3.11) is robustly stochastically stable and the controller gain is given by $F(i) = Z(i)Y^{-1}(i)$, $i \in \mathscr{S}$.

Example 32. To show the validity of our results, let us consider a two-mode system of Example 31 with

$$
\mathbb{W}(1) = \begin{bmatrix} 1.0 & 0.0 & 0.0 \\ 0.0 & 1.0 & 0.0 \\ 0.0 & 0.0 & 1.0 \end{bmatrix}, \quad
\mathbb{W}(2) = \begin{bmatrix} 1.0 & 0.0 & 0.0 \\ 0.0 & 1.0 & 0.0 \\ 0.0 & 0.0 & 1.0 \end{bmatrix}.
$$

Solving LMI (3.35) gives

$$
X(1) = \begin{bmatrix} 1.2691 & -0.0653 & -0.6528 \\ -0.0653 & 1.0389 & 0.0237 \\ -0.6528 & 0.0237 & 1.2733 \end{bmatrix}, Y(1) = \begin{bmatrix} 0.6205 & -0.0416 \\ -0.0416 & 1.0324 \end{bmatrix},
$$

$$Z(1) = \begin{bmatrix} 3.8368 & -4.8147 \\ -0.7412 & -4.9618 \end{bmatrix}, \qquad X(2) = \begin{bmatrix} 1.5880 & -0.0984 & -0.9841 \\ -0.0984 & 1.2118 & 0.0386 \\ -0.9841 & 0.0386 & 1.5940 \end{bmatrix},$$

$$Y(2) = \begin{bmatrix} 0.6099 & -0.0598 \\ -0.0598 & 1.2020 \end{bmatrix}, \qquad Z(2) = \begin{bmatrix} -0.5203 & 1.2902 \\ -1.2420 & -0.2225 \end{bmatrix}.$$

This gives the following gains:

$$F(1) = \begin{bmatrix} 5.8870 & -4.4263 \\ -1.5211 & -4.8674 \end{bmatrix}, \qquad F(2) = \begin{bmatrix} -0.7515 & 1.0360 \\ -2.0645 & -0.2878 \end{bmatrix}.$$

3.4 Output Feedback Stabilization

Previously we covered the design of a state feedback controller that assumes the availability of the state vector $x(t)$ at each time t. But in reality this strong assumption is not always true. In fact, some of the state variables are not measurable by their construction or the lack of appropriate sensors to give information. The state feedback control that we developed is impossible to use since the state vector is not available for feedback. Alternatively we can use the output feedback controller that uses the system's measurement to compute the control law. Notice that we can estimate the state vector and continue to use the state feedback controller. This method will be covered in Chapter 5.

Let us now focus on output feedback stabilization to see how we can design the controller that stochastically stabilizes the nominal system and robustly stochastically stabilizes the uncertain system of the class we are considering.

The structure of the controller we use in this section is given by the following expression:

$$\begin{cases} \dot{x}_c(t) = K_A(r(t))x_c(t) + K_B(r(t))y(t), \, x_c(0) = 0, \\ u(t) = K_C(r(t))x_c(t), \end{cases} \tag{3.38}$$

where $x_c(t) \in \mathbb{R}^n$ is the controller state; $y(t) \in \mathbb{R}^p$ is the system's measurement; and $K_A(r(t))$, $K_B(r(t))$, and $K_C(r(t))$ are the design gains to be determined.

If we add the system's measurement to the previous state equation we get

$$\begin{cases} \dot{x}(t) = A(r(t), t)x(t) + B(r(t), t)u(t), \, x(0) = x_0, \\ y(t) = C_y(r(t), t)x(t), \end{cases} \tag{3.39}$$

where the matrices $A(r(t), t)$ and $B(r(t), t)$ keep the same meaning as before and the matrix $C_y(r(t), t)$ is defined as follows:

$$C_y(r(t), t) = C_y(r(t)) + D_{C_y}(r(t))F_{C_y}(r(t), t)E_{C_y}(r(t)),$$

where $C_y(r(t))$, $D_{C_y}(r(t))$, and $E_{C_y}(r(t))$ are real known matrices and F_{C_y} $(r(t), t)$ is an unknown real matrix that satisfies the following for every $i \in \mathscr{S}$:

$$F_{C_y}^\top(i, t) F_{C_y}(i, t) \le \mathbb{I}.$$

The block diagram of the closed-loop system (nominal system) under the output feedback controller is represented by Figure 3.2.

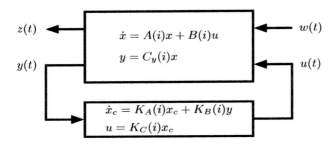

Fig. 3.2. Output feedback stabilization block diagram (nominal system).

As we did previously for the state feedback controller, let us see how we can design the output controller for a nominal system. Combining the system dynamics (3.39), with all the uncertainties equal to zero, and the controller dynamics (3.38), we get the following extended dynamics:

$$\begin{cases} \dot{\eta}(t) = \begin{bmatrix} A(i) & B(i)K_C(i) \\ K_B(i)C_y(i) & K_A(i) \end{bmatrix} \eta(t), \\ \eta(0) = \begin{bmatrix} x_0 \\ 0 \end{bmatrix} \end{cases} \tag{3.40}$$

where $\eta(t) = \begin{bmatrix} x(t) \\ x_c(t) \end{bmatrix}$.

Let $\tilde{A}(i)$ be defined as follows:

$$\tilde{A}(i) = \begin{bmatrix} A(i) & B(i)K_C(i) \\ K_B(i)C_y(i) & K_A(i) \end{bmatrix}.$$

Based on the results of Chapter 2, the closed-loop dynamics of the extended system will be stable if there exists a set of symmetric and positive-definite matrices $P = (P(1), \cdots, P(N)) > 0$ such that the following holds for each $i \in \mathscr{S}$:

$$\tilde{A}^\top(i)P(i) + P(i)\tilde{A}(i) + \sum_{j=1}^N \lambda_{ij} P(j) < 0. \tag{3.41}$$

Let $P(i)$, $i \in \mathscr{S}$ be defined by

$$P(i) = \begin{bmatrix} P_1(i) & P_2(i) \\ P_2^\top(i) & P_3(i) \end{bmatrix},$$

where $P_1(i) > 0$, $P_3(i) > 0$ are symmetric and positive-definite matrices. Let us define the following matrices:

$$W(i) = \left[P_1(i) - P_2(i)P_3^{-1}(i)P_2^\top(i) \right]^{-1},$$

$$U(i) = \begin{bmatrix} W(i) & \mathbb{I} \\ W(i) & 0 \end{bmatrix},$$

$$V(i) = \begin{bmatrix} \mathbb{I} & 0 \\ 0 & -P_3^{-1}(i)P_2^\top(i) \end{bmatrix}.$$

Based on these definitions we conclude that

$$V(i)U(i) = \begin{bmatrix} W(i) & \mathbb{I} \\ -P_3^{-1}(i)P_2^\top(i)W(i) & 0 \end{bmatrix}.$$

Pre- and post-multiply the left-hand side of (3.41) by $U^\top(i)V^\top(i)$ and $V(i)U(i)$ to get

$$U^\top(i)V^\top(i)\tilde{A}^\top(i)P(i)V(i)U(i) + U^\top(i)V^\top(i)P(i)\tilde{A}(i)V(i)U(i)$$

$$+ \sum_{j=1}^{N} \lambda_{ij} U^\top(i)V^\top(i)P(j)V(i)U(i).$$

Let us now compute

$$U^\top(i)V^\top(i)P(i)\tilde{A}(i)V(i)U(i) \text{ and } U^\top(i)V^\top(i)P(j)V(i)U(i),$$

in function of the system matrices. In fact, for the first term we have

$$U^\top(i)V^\top(i)P(i)\tilde{A}(i)V(i)U(i)$$
$$= \begin{bmatrix} W^\top(i) & -W^\top(i)P_2(i)P_3^{-1}(i) \\ \mathbb{I} & 0 \end{bmatrix} \begin{bmatrix} P_1(i) & P_2(i) \\ P_2^\top(i) & P_3(i) \end{bmatrix}$$
$$\times \begin{bmatrix} A(i) & B(i)K_C(i) \\ K_B(i)C_y(i) & K_A(i) \end{bmatrix} \begin{bmatrix} W(i) & \mathbb{I} \\ -P_3^{-1}(i)P_2^\top(i)W(i) & 0 \end{bmatrix}$$
$$= \begin{bmatrix} \mathscr{L}_1(i) & \mathscr{L}_2(i) \\ \mathscr{L}_3(i) & \mathscr{L}_4(i) \end{bmatrix}.$$

Performing the multiplication we get

$$\mathscr{L}_1(i) = W^\top(i)P_1(i)A(i)W(i) + W^\top(i)P_2(i)K_B(i)C_y(i)W(i)$$
$$- W^\top(i)P_2(i)P_3^{-1}(i)P_2^\top(i)A(i)W(i)$$
$$- W^\top(i)P_2(i)K_B(i)C_y(i)W(i)$$

$$-W^\top(i)P_1(i)B(i)K_C(i)P_3^{-1}(i)P_2^\top(i)W(i)$$
$$-W^\top(i)P_2(i)K_A(i)P_3^{-1}(i)P_2^\top(i)W(i)$$
$$+W^\top(i)P_2(i)P_3^{-1}(i)P_2^\top(i)B(i)K_C(i)P_3^{-1}(i)P_2^\top(i)W(i)$$
$$+W^\top(i)P_2(i)K_A(i)P_3^{-1}(i)P_2^\top(i)W(i),$$

$$\mathscr{Z}_2(i) = W^\top(i)P_1(i)A(i) - W^\top(i)P_2(i)P_3^{-1}(i)P_2^\top(i)A(i)$$
$$-W^\top(i)P_2(i)K_B(i)C_y(i) + W^\top(i)P_2(i)K_B(i)C_y(i),$$

$$\mathscr{Z}_3(i) = P_1(i)A(i)W(i)$$
$$+P_2(i)K_B(i)C_y(i)W(i) - P_1(i)B(i)K_C(i)P_3^{-1}(i)P_2^\top(i)W(i)$$
$$-P_2(i)K_A(i)P_3^{-1}(i)P_2^\top(i)W(i),$$

$$\mathscr{Z}_4(i) = P_1(i)A(i) + P_2(i)K_B(i)C_y(i).$$

Using some basic algebraic manipulations and the fact that

$$W(i)\left[P_1(i) - P_2(i)P_3^{-1}(i)P_2^\top(i)\right] = \mathbb{I},$$

the previous elements become

$$\mathscr{Z}_1(i) = W^\top(i)\left[P_1(i) - P_2(i)P_3^{-1}(i)P_2^\top(i)\right]A(i)W(i)$$
$$-W^\top(i)\left[P_1(i) - P_2(i)P_3^{-1}(i)P_2^\top(i)\right]B(i)K_C(i)P_3^{-1}(i)P_2^\top(i)W(i)$$
$$= A(i)W(i) - B(i)K_C(i)P_3^{-1}(i)P_2^\top(i)W(i),$$

$$\mathscr{Z}_2(i) = W^\top(i)\left[P_1(i) - P_2(i)P_3^{-1}(i)P_2^\top(i)\right]A(i) = A(i),$$

$$\mathscr{Z}_3(i) = P_1(i)A(i)W(i) + P_2(i)K_B(i)C_y(i)W(i)$$
$$-P_1(i)B(i)K_C(i)P_3^{-1}(i)P_2^\top(i)W(i)$$
$$-P_2(i)K_A(i)P_3^{-1}(i)P_2^\top(i)W(i),$$

$$\mathscr{Z}_4(i) = P_1(i)A(i) + P_2(i)K_B(i)C_y(i).$$

Using all these computations, we get

$$U^\top(i)V^\top(i)P(i)\tilde{A}(i)V(i)U(i)$$

$$= \begin{bmatrix} \begin{bmatrix} A(i)W(i) \\ -B(i)K_C(i)P_3^{-1}(i)P_2^\top(i)W(i) \end{bmatrix} & \begin{bmatrix} A(i) \\ P_1(i)A(i) \\ +P_2(i)K_B(i)C_y(i) \end{bmatrix} \\ \begin{bmatrix} P_1(i)A(i)W(i) \\ +P_2(i)K_B(i)C_y(i)W(i) \\ -P_1(i)B(i)K_C(i)P_3^{-1}(i)P_2^\top(i)W(i) \\ -P_2(i)K_A(i)P_3^{-1}(i)P_2^\top(i)W(i) \end{bmatrix} & \end{bmatrix}.$$

Using the fact that $U^\top(i)V^\top(i)\tilde{A}^\top(i)P(i)V(i)U(i)$ is the transpose of $U^\top(i)V^\top(i)P(i)\tilde{A}(i)V(i)U(i)$ we get

$$U^\top(i)V^\top(i)\tilde{A}^\top(i)P(i)V(i)U(i)$$

$$= \begin{bmatrix} \begin{bmatrix} W^\top(i)A^\top(i) \\ -W^\top(i)P_2(i)P_3^{-1}(i)K_C^\top(i)B^\top(i) \end{bmatrix} & \\ A^\top(i) & \\ \begin{bmatrix} W^\top(i)A^\top(i)P_1(i) \\ +W^\top(i)C_y^\top(i)K_B^\top(i)P_2^\top(i) \\ -W^\top(i)P_2(i)P_3^{-1}(i)K_C^\top(i)B^\top(i)P_1(i) \\ -W^\top(i)P_2(i)P_3^{-1}(i)K_A^\top(i)P_2^\top(i) \end{bmatrix} & \\ \begin{bmatrix} A^\top(i)P_1(i) \\ +C_y^\top(i)K_B^\top(i)P_2^\top(i) \end{bmatrix} & \end{bmatrix}.$$

For the term $U^\top(i)V^\top(i)P(j)V(i)U(i)$, we have

$$\begin{bmatrix} W^\top(i) & -W^\top(i)P_2(i)P_3^{-1}(i) \\ \mathbb{I} & 0 \end{bmatrix} \begin{bmatrix} P_1(j) & P_2(j) \\ P_2^\top(j) & P_3(j) \end{bmatrix}$$

$$\times \begin{bmatrix} W(i) & \mathbb{I} \\ -P_3^{-1}(i)P_2^\top(i)W(i) & 0 \end{bmatrix}$$

$$= \begin{bmatrix} \begin{bmatrix} W^\top(i)P_1(j)W(i) \\ -W^\top(i)P_2(i)P_3^{-1}(i)P_2^\top(j)W(i) \\ -W^\top(i)P_2(j)P_3^{-1}(i)P_2^\top(i)W(i) \\ +W^\top(i)P_2(i)P_3^{-1}(i)P_3(j)P_3^{-1}(i)P_2^\top(i)W(i) \end{bmatrix} & \\ P_1(j)W(i) - P_2(j)P_3^{-1}(i)P_2^\top(i)W(i) & \\ \begin{bmatrix} W^\top(i)P_1(j) \\ -W^\top(i)P_2(i)P_3^{-1}(i)P_2^\top(j) \end{bmatrix} & \\ P_1(j) & \end{bmatrix},$$

which can be rewritten as follows using the fact that $W^{-1}(j) = P_1(j) - P_2(j)P_3^{-1}(j)P_2^\top(j)$:

$$\left[\begin{bmatrix} W^\top(i)W^{-1}(j)W(i) \\ +W^\top(i)\left[P_2(i)P_3^{-1}(i)P_3(j) - P_2^\top(j)\right] \\ \times P_3^{-1}(j)\left[P_2(i)P_3^{-1}(i)P_3(j) - P_2^\top(j)\right]^\top W(i) \\ \left[P_1(j) - P_2(j)P_3^{-1}(i)P_2^\top(i)\right]W(i) \end{bmatrix} \quad \star \\ \qquad\qquad P_1(j) \right].$$

Using all the previous algebraic manipulations, the stochastic stability condition for the closed-loop system becomes

$$\begin{bmatrix} \widehat{\mathcal{M}_1(i)} & \mathcal{M}_2(i) \\ \mathcal{M}_2^\top(i) & \mathcal{M}_3(i) \end{bmatrix} < 0,$$

with

$$\begin{aligned}
\widehat{\mathcal{M}_1(i)} &= \mathcal{M}_1(i) + \sum_{j=1}^N \lambda_{ij} W^\top(i)\left[P_2(i)P_3^{-1}(i)P_3(j) - P_2^\top(j)\right] \\
&\quad \times P_3^{-1}(j)\left[P_2(i)P_3^{-1}(i)P_3(j) - P_2^\top(j)\right]^\top W(i), \\
\mathcal{M}_1(i) &= A(i)W(i) + W^\top(i)A^\top(i) - B(i)K_C(i)P_3^{-1}(i)P_2^\top(i)W(i) \\
&\quad - W^\top(i)P_2(i)P_3^{-1}(i)K_C^\top(i)B^\top(i) \\
&\quad + \sum_{j=1}^N \lambda_{ij}W^\top(i)W^{-1}(j)W(i), \\
\mathcal{M}_2(i) &= A(i) + W^\top(i)A^\top(i)P_1(i) + W^\top(i)C_y^\top(i)K_B^\top(i)P_2^\top(i) \\
&\quad - W^\top(i)P_2(i)P_3^{-1}(i)K_C^\top(i)B^\top(i)P_1(i) \\
&\quad - W^\top(i)P_2(i)P_3^{-1}(i)K_A^\top(i)P_2^\top(i) \\
&\quad + \sum_{j=1}^N \lambda_{ij}W^\top(i)\left[P_1(j) - P_2(j)P_3^{-1}(i)P_2^\top(i)\right]^\top, \\
\mathcal{M}_3(i) &= P_1(i)A(i) + P_2(i)K_B(i)C_y(i) + A^\top(i)P_1(i) \\
&\quad + C_y^\top(i)K_B^\top(i)P_2^\top(i) + \sum_{j=1}^N \lambda_{ij}P_1(j).
\end{aligned}$$

Since

$$\sum_{j=1}^N \lambda_{ij}W^\top(i)\left[P_2(i)P_3^{-1}(i)P_3(j) - P_2^\top(j)\right]P_3^{-1}(j)$$

$$\times \left[P_2(i)P_3^{-1}(i)P_3(j) - P_2^\top(j)\right]^\top W(i) \geq 0,$$

we get the following equivalent condition:

$$\begin{bmatrix} \mathcal{M}_1(i) & \mathcal{M}_2(i) \\ \mathcal{M}_2^\top(i) & \mathcal{M}_3(i) \end{bmatrix} < 0.$$

Letting

$$P(i) = \begin{bmatrix} X(i) & Y^{-1}(i) - X(i) \\ Y^{-1}(i) - X(i) & X(i) - Y^{-1}(i) \end{bmatrix},$$

that is,

$$
\begin{aligned}
P_1(i) &= X(i), \\
P_2(i) &= Y^{-1}(i) - X(i), \\
P_3(i) &= X(i) - Y^{-1}(i),
\end{aligned}
$$

implies $W(i) = \left[P_1(i) - P_2(i)P_3^{-1}(i)P_2^{\top}(i) \right]^{-1} = Y(i)$ and $P_3^{-1}(i)P_2^{\top}(i) = -\mathbb{I}$.
If we define $\mathcal{K}_B(i)$ and $\mathcal{K}_C(i)$ by

$$
\begin{aligned}
\mathcal{K}_B(i) &= P_2(i)K_B(i) = \left[Y^{-1}(i) - X(i) \right] K_B(i), \\
\mathcal{K}_C(i) &= -K_C(i)P_3^{-1}(i)P_2^{\top}(i)W(i) = K_C(i)Y(i),
\end{aligned}
$$

and use all the previous algebraic manipulations, we get

$$\begin{bmatrix} \mathcal{M}_1(i) & \mathcal{M}_2(i) \\ \mathcal{M}_2^{\top}(i) & \mathcal{M}_3(i) \end{bmatrix} < 0,$$

with

$$
\begin{aligned}
\mathcal{M}_1(i) =\ & A(i)Y(i) + Y^{\top}(i)A^{\top}(i) + B(i)\mathcal{K}_C(i) \\
& + \mathcal{K}_C^{\top}(i)B^{\top}(i) + \sum_{j=1}^{N} \lambda_{ij} Y^{\top}(i)Y^{-1}(j)Y(i), \\
\mathcal{M}_2(i) =\ & A(i) + Y^{\top}(i)A^{\top}(i)X(i) + Y^{\top}(i)C_y^{\top}(i)\mathcal{K}_B^{\top}(i) \\
& + \mathcal{K}_C^{\top}(i)B^{\top}(i)X(i) \\
& + Y^{\top}(i)K_A^{\top}(i) \left[Y^{-1}(i) - X(i) \right]^{\top} \\
& + \sum_{j=1}^{N} \lambda_{ij} Y^{\top}(i)Y^{-1}(j), \\
\mathcal{M}_3(i) =\ & X(i)A(i) + \mathcal{K}_B(i)C_y(i) + A^{\top}(i)X(i) \\
& + C_y^{\top}(i)\mathcal{K}_B^{\top}(i) + \sum_{j=1}^{N} \lambda_{ij} X(j).
\end{aligned}
$$

Using the expression of the controller given by

$$
\begin{cases}
K_A(i) = \left[X(i) - Y^{-1}(i) \right]^{-1} \left[A^{\top}(i) + X(i)A(i)Y(i) \right. \\
\qquad\quad + X(i)B(i)\mathcal{K}_C(i) + \mathcal{K}_B(i)C_y(i)Y(i) \\
\qquad\quad \left. + \sum_{j=1}^{N} \lambda_{ij} Y^{-1}(j)Y(i) \right] Y^{-1}(i), \\
K_B(i) = \left[Y^{-1}(i) - X(i) \right]^{-1} \mathcal{K}_B(i), \\
K_C(i) = \mathcal{K}_C(i)Y^{-1}(i),
\end{cases}
$$

we have $\mathcal{M}_2(i) = 0$. This implies that the stability condition is equivalent to the following conditions:

$$\mathcal{M}_1(i) < 0,$$
$$\mathcal{M}_3(i) < 0,$$

which gives

$$A(i)Y(i) + Y^\top(i)A^\top(i) + B(i)\mathcal{K}_C(i) + \mathcal{K}_C^\top(i)B^\top(i)$$
$$+ \sum_{j=1}^N \lambda_{ij} Y^\top(i)Y^{-1}(j)Y(i) < 0,$$
$$X(i)A(i) + \mathcal{K}_B(i)C_y(i) + A^\top(i)X(i) + C_y^\top(i)\mathcal{K}_B^\top(i)$$
$$+ \sum_{j=1}^N \lambda_{ij} X(j) < 0.$$

Notice that

$$\sum_{j=1}^N \lambda_{ij} Y^\top(i)Y^{-1}(j)Y(i) = \lambda_{ii}Y(i) + \mathcal{S}_i(Y)\mathcal{Y}_i^{-1}(Y)\mathcal{S}_i^\top(Y),$$

with

$$\mathcal{S}_i(Y) = \left[\sqrt{\lambda_{i1}}Y(i), \cdots, \sqrt{\lambda_{ii-1}}Y(i), \sqrt{\lambda_{ii+1}}Y(i), \right.$$
$$\left. \cdots, \sqrt{\lambda_{iN}}Y(i) \right],$$
$$\mathcal{Y}_i(Y) = \mathrm{diag}\left[Y(1), \cdots, Y(i-1), Y(i+1), \cdots, Y(N) \right].$$

Using this, the previous stability conditions become

$$\left[\begin{array}{cc} \left[\begin{array}{c} A(i)Y(i) + Y^\top(i)A^\top(i) + B(i)\mathcal{K}_C(i) \\ +\mathcal{K}_C^\top(i)B^\top(i) + \lambda_{ii}Y(i) \end{array} \right] & \mathcal{S}_i(Y) \\ \mathcal{S}_i^\top(Y) & -\mathcal{Y}_i(Y) \end{array} \right] < 0,$$
$$X(i)A(i) + \mathcal{K}_B(i)C_y(i) + A^\top(i)X(i) + C_y^\top(i)\mathcal{K}_B^\top(i)$$
$$+ \sum_{j=1}^N \lambda_{ij} X(j) < 0.$$

Finally, notice that

$$U^\top(i)V^\top(i)P(i)V(i)U(i) = \left[\begin{array}{cc} Y(i) & \mathbb{I} \\ \mathbb{I} & X(i) \end{array} \right] > 0.$$

The results of the previous algebraic manipulations are summarized by the following theorem.

Theorem 24. *Nominal system (3.39) is stochastically stable if and only if for every* $i \in \mathscr{S}$, *the following set of coupled LMIs is feasible for some symmetric and positive-definite matrices* $X = (X(1), \cdots, X(N)) > 0$ *and* $Y = (Y(1), \cdots, Y(N)) > 0$, *and matrices* $\mathcal{K}_B = (\mathcal{K}_B(1), \cdots, \mathcal{K}_B(N))$ *and* $\mathcal{K}_C = (\mathcal{K}_C(1), \cdots, \mathcal{K}_C(N))$:

$$
\left[\begin{array}{cc} \begin{bmatrix} A(i)Y(i) + Y^\top(i)A^\top(i) \\ +B(i)\mathcal{K}_C(i) \\ +\mathcal{K}_C^\top(i)B^\top(i) + \lambda_{ii}Y(i) \end{bmatrix} & \mathcal{S}_i(Y) \\ \mathcal{S}_i^\top(Y) & -\mathcal{Y}_i(Y) \end{array} \right] < 0, \qquad (3.42)
$$

$$
X(i)A(i) + \mathcal{K}_B(i)C_y(i) + A^\top(i)X(i) + C_y^\top(i)\mathcal{K}_B^\top(i)
$$

$$
+ \sum_{j=1}^{N} \lambda_{ij}X(j) < 0, \qquad (3.43)
$$

$$
\begin{bmatrix} Y(i) & \mathbb{I} \\ \mathbb{I} & X(i) \end{bmatrix} > 0, \qquad (3.44)
$$

with

$$
\mathcal{S}_i(Y) = \left[\sqrt{\lambda_{i1}}Y(i), \cdots, \sqrt{\lambda_{ii-1}}Y(i), \sqrt{\lambda_{ii+1}}Y(i), \cdots, \sqrt{\lambda_{iN}}Y(i) \right],
$$

$$
\mathcal{Y}_i(Y) = \text{diag}\left[Y(1), \cdots, Y(i-1), Y(i+1), \cdots, Y(N) \right].
$$

Furthermore the dynamic output-feedback controller is given by

$$
\begin{cases} K_A(i) = \left[X(i) - Y^{-1}(i) \right]^{-1} \left[A^\top(i) + X(i)A(i)Y(i) \right. \\ \qquad\quad +X(i)B(i)\mathcal{K}_C(i) + \mathcal{K}_B(i)C_y(i)Y(i) \\ \qquad\quad \left. + \sum_{j=1}^{N} \lambda_{ij}Y^{-1}(j)Y(i) \right] Y^{-1}(i), \\ K_B(i) = \left[Y^{-1}(i) - X(i) \right]^{-1} \mathcal{K}_B(i), \\ K_C(i) = \mathcal{K}_C(i)Y^{-1}(i). \end{cases} \qquad (3.45)
$$

Example 33. In this example let us consider a system with two modes and the following data:

- mode #1:

$$
A(1) = \begin{bmatrix} 1.0 & -0.5 \\ 0.1 & 1.0 \end{bmatrix}, \quad B(1) = \begin{bmatrix} 1.0 & 0.0 \\ 0.0 & 1.0 \end{bmatrix}, \quad C(1) = \begin{bmatrix} 1.0 & 0.0 \\ 0.0 & 1.0 \end{bmatrix},
$$

- mode #2:

$$
A(2) = \begin{bmatrix} -0.2 & -0.5 \\ 0.5 & -0.25 \end{bmatrix}, \quad B(2) = \begin{bmatrix} 1.0 & 0.0 \\ 0.0 & 1.0 \end{bmatrix}, \quad C(2) = \begin{bmatrix} 1.0 & 0.0 \\ 0.0 & 1.0 \end{bmatrix}.
$$

Let us assume that the switching between the two modes is described by the following transition matrix:

$$\Lambda = \begin{bmatrix} -2.0 & 2.0 \\ 3.0 & -3.0 \end{bmatrix}.$$

Solving the LMIs (3.42)–(3.44), we get

$$X(1) = \begin{bmatrix} 155.3339 & 0.0000 \\ 0.0000 & 155.3339 \end{bmatrix}, \qquad X(2) = \begin{bmatrix} 156.7142 & -0.0000 \\ -0.0000 & 156.7142 \end{bmatrix},$$

$$Y(1) = \begin{bmatrix} 85.5704 & -0.0000 \\ -0.0000 & 85.5704 \end{bmatrix}, \qquad Y(2) = \begin{bmatrix} 76.2129 & 0.0000 \\ 0.0000 & 76.2129 \end{bmatrix},$$

$$\mathcal{K}_B(1) = \begin{bmatrix} -240.2401 & 78.9301 \\ -16.7965 & -240.2401 \end{bmatrix}, \qquad \mathcal{K}_B(2) = \begin{bmatrix} -50.1127 & -5.2440 \\ 5.2440 & -42.2770 \end{bmatrix},$$

$$\mathcal{K}_C(1) = 10^3 \cdot \begin{bmatrix} -0.1707 & 2.5137 \\ -2.4795 & -0.1707 \end{bmatrix}, \qquad \mathcal{K}_C(2) = \begin{bmatrix} -45.2177 & -5.8319 \\ 5.8319 & -41.4071 \end{bmatrix}.$$

The corresponding controller gains are given by

$$K_A(1) = \begin{bmatrix} -2.5413 & 29.3867 \\ -28.9867 & -2.5413 \end{bmatrix}, \qquad K_B(1) = \begin{bmatrix} 1.5467 & -0.5082 \\ 0.1081 & 1.5467 \end{bmatrix},$$

$$K_C(1) = \begin{bmatrix} -1.9946 & 29.3764 \\ -28.9764 & -1.9946 \end{bmatrix}, \qquad K_A(2) = \begin{bmatrix} -1.1132 & -0.6100 \\ 0.6100 & -1.0632 \end{bmatrix},$$

$$K_B(2) = \begin{bmatrix} 0.3198 & 0.0335 \\ -0.0335 & 0.2698 \end{bmatrix}, \qquad K_C(2) = \begin{bmatrix} -0.5933 & -0.0765 \\ 0.0765 & -0.5433 \end{bmatrix}.$$

Based on the results of this theorem, the system of this example is stochastically stable under the output feedback controller with the computed gains.

Let us now consider that the uncertainties are acting on the dynamics and focus on the design of the output feedback controller with the form given in (3.38). Using the controller dynamics and system dynamics we get the following:

$$\dot{\eta}(t) = \begin{bmatrix} A(i) & B(i)K_C(i) \\ K_B(i)C_y(i) & K_A(i) \end{bmatrix} \eta(t)$$

$$+ \begin{bmatrix} D_A(i)F_A(i,t)E_A(i) & \begin{bmatrix} D_B(i)F_B(i,t) \\ \times E_B(i)K_C(i) \end{bmatrix} \\ K_B(i)D_{C_y}(i)F_{C_y}(i,t)E_{C_y}(i) & 0 \end{bmatrix} \eta(t),$$

with

$$\eta(t) = \begin{bmatrix} x(t) \\ x_c(t) \end{bmatrix}.$$

This dynamics can be rewritten as follows:

$$\dot{\eta}(t) = \left[\tilde{A}(i) + \Delta\tilde{A}(i,t)\right]\eta(t),$$

with

$$\tilde{A}(i) = \begin{bmatrix} A(i) & B(i)K_C(i) \\ K_B(i)C_y(i) & K_A(i) \end{bmatrix},$$
$$\Delta\tilde{A}(i,t) = \Delta\tilde{A}_A(i,t) + \Delta\tilde{B}_B(i,t) + \Delta\tilde{C}_{C_y}(i,t),$$

where

$$\Delta\tilde{A}_A(i,t) = \begin{bmatrix} D_A(i)F_A(i,t)E_A(i) & 0 \\ 0 & 0 \end{bmatrix},$$
$$\Delta\tilde{B}_B(i,t) = \begin{bmatrix} 0 & D_B(i)F_B(i,t)E_B(i)K_C(i) \\ 0 & 0 \end{bmatrix},$$
$$\Delta\tilde{C}_{C_y}(i,t) = \begin{bmatrix} 0 & 0 \\ K_B(i)D_{C_y}(i)F_{C_y}(i,t)E_{C_y}(i) & 0 \end{bmatrix}.$$

Notice that

$$\Delta\tilde{A}_A(i,t) = \begin{bmatrix} D_A(i) & 0 \\ 0 & 0 \end{bmatrix}\begin{bmatrix} F_A(i,t) & 0 \\ 0 & 0 \end{bmatrix}\begin{bmatrix} E_A(i) & 0 \\ 0 & 0 \end{bmatrix}$$
$$= \tilde{D}_A(i)\tilde{F}_A(i,t)\tilde{E}_A(i),$$
$$\Delta\tilde{B}_B(i,t) = \begin{bmatrix} 0 & D_B(i) \\ 0 & 0 \end{bmatrix}\begin{bmatrix} 0 & 0 \\ 0 & F_B(i,t) \end{bmatrix}\begin{bmatrix} 0 & 0 \\ 0 & E_B(i)K_C(i) \end{bmatrix}$$
$$= \tilde{D}_B(i)\tilde{F}_B(i,t)\tilde{E}_B(i),$$
$$\Delta\tilde{C}_{C_y}(i,t) = \begin{bmatrix} 0 & 0 \\ 0 & K_B(i)D_{C_y}(i) \end{bmatrix}\begin{bmatrix} 0 & 0 \\ 0 & F_{C_y}(i,t) \end{bmatrix}\begin{bmatrix} 0 & 0 \\ E_{C_y}(i) & 0 \end{bmatrix}$$
$$= \tilde{D}_{C_y}(i)\tilde{F}_{C_y}(i,t)\tilde{E}_{C_y}(i).$$

Let us now study the stability of the extended dynamics. Using the results of Chapter 2, the dynamics are stable if there exists a set of symmetric and positive-definite matrices $P = (P(1), \cdots, P(N)) > 0$ such that the following holds for each $i \in \mathscr{S}$:

$$\tilde{A}^\top(i,t)P(i) + P(i)\tilde{A}(i,t) + \sum_{j=1}^{N}\lambda_{ij}P(j) < 0.$$

Using the expression of $\tilde{A}(i,t)$, we get

$$\tilde{A}^\top(i)P(i) + P(i)\tilde{A}(i) + P(i)\Delta\tilde{A}_A(i,t) + \Delta\tilde{A}_A^\top(i,t)P(i)$$
$$+P(i)\Delta\tilde{B}_B(i,t) + \Delta\tilde{B}_B^\top(i,t)P(i) + P(i)\Delta\tilde{C}_{C_y}(i,t)$$

$$+\Delta\tilde{C}_{C_y}^{\top}(i,t)P(i) + \sum_{j=1}^{N} \lambda_{ij}P(j) < 0.$$

Based on the Lemma 7 in Appendix A, we have

$$P(i)\Delta\tilde{A}_A(i,t) + \Delta\tilde{A}_A^{\top}(i,t)P(i)$$
$$\leq \tilde{\varepsilon}_A^{-1}(i)P(i)\tilde{D}_A(i)\tilde{D}_A^{\top}(i)P(i) + \tilde{\varepsilon}_A(i)\tilde{E}_A^{\top}(i)\tilde{E}_A(i),$$
$$P(i)\Delta\tilde{B}_B(i,t) + \Delta\tilde{B}_B^{\top}(i,t)P(i)$$
$$\leq \tilde{\varepsilon}_B^{-1}(i)P(i)\tilde{D}_B(i)\tilde{D}_B^{\top}(i)P(i) + \tilde{\varepsilon}_B(i)\tilde{E}_B^{\top}(i)\tilde{E}_B(i),$$
$$P(i)\Delta\tilde{C}_{C_y}(i,t) + \Delta\tilde{C}_{C_y}^{\top}(i,t)P(i)$$
$$\leq \tilde{\varepsilon}_{C_y}^{-1}(i)P(i)\tilde{D}_{C_y}(i)\tilde{D}_{C_y}^{\top}(i)P(i) + \tilde{\varepsilon}_{C_y}(i)\tilde{E}_{C_y}^{\top}(i)\tilde{E}_{C_y}(i).$$

Using this and the Schur complement, the previous stability condition will be satisfied if the following holds:

$$\begin{bmatrix} \tilde{J}(i) & P(i)\tilde{D}_A(i) & P(i)\tilde{D}_B(i) & P(i)\tilde{D}_{C_y}(i) \\ \tilde{D}_A^{\top}(i)P(i) & -\tilde{\varepsilon}_A(i)\mathbb{I} & 0 & 0 \\ \tilde{D}_B^{\top}(i)P(i) & 0 & -\tilde{\varepsilon}_B(i)\mathbb{I} & 0 \\ \tilde{D}_{C_y}^{\top}(i)P(i) & 0 & 0 & -\tilde{\varepsilon}_{C_y}(i)\mathbb{I} \end{bmatrix} < 0,$$

with

$$\tilde{J}(i) = \tilde{A}^{\top}(i)P(i) + P(i)\tilde{A}(i) + \sum_{j=1}^{N} \lambda_{ij}P(j) + \tilde{\varepsilon}_A(i)\tilde{E}_A^{\top}(i)\tilde{E}_A(i)$$
$$+\tilde{\varepsilon}_B(i)\tilde{E}_B^{\top}(i)\tilde{E}_B(i) + \tilde{\varepsilon}_B(i)\tilde{E}_{C_y}^{\top}(i)\tilde{E}_{C_y}(i).$$

Again using the Schur complement, we get

$$J(i) + P(i)\left[\tilde{D}_A(i)\ \tilde{D}_B(i)\ \tilde{D}_{C_y}(i)\right]\Upsilon^{-1}(i)\begin{bmatrix}\tilde{D}_A^{\top}(i) \\ \tilde{D}_B^{\top}(i) \\ \tilde{D}_{C_y}^{\top}(i)\end{bmatrix}P(i)$$
$$+\left[\tilde{E}_A^{\top}(i)\ \tilde{E}_B^{\top}(i)\ \tilde{E}_{C_y}^{\top}(i)\right]\Upsilon(i)\begin{bmatrix}\tilde{E}_A(i) \\ \tilde{E}_B(i) \\ \tilde{E}_{C_y}(i)\end{bmatrix} < 0,$$
$$\Upsilon(i) = \begin{bmatrix} \tilde{\varepsilon}_A(i)\mathbb{I} & 0 & 0 \\ 0 & \tilde{\varepsilon}_B(i)\mathbb{I} & 0 \\ 0 & 0 & \tilde{\varepsilon}_{C_y}(i)\mathbb{I} \end{bmatrix},$$

with $J(i) = \tilde{A}^{\top}(i)P(i) + P(i)\tilde{A}(i) + \sum_{j=1}^{N} \lambda_{ij}P(j)$.

This set of coupled matrix inequalities that guarantees robust stochastic stability is nonlinear in $P(i)$ and the controller gains, $K_A(i)$, $K_B(i)$, and $K_C(i)$. To cast it into an LMI form let us pre- and post-multiply this inequality by $U^{\top}(i)V^{\top}(i)$ and $V(i)U(i)$, respectively, as we did previously. Before multiplying, notice that $U^\top(i)V^\top(i)J(i)V(i)U(i)$ has already been computed and we do not need to recompute it. For the two other terms we have

$$U^\top(i)V^\top(i) \left[\tilde{E}_A^\top(i) \ \tilde{E}_B^\top(i) \ \tilde{E}_{C_y}^\top(i) \right] \Upsilon(i)$$

$$\times \begin{bmatrix} \tilde{E}_A(i) \\ \tilde{E}_B(i) \\ \tilde{E}_{C_y}(i) \end{bmatrix} V(i)U(i)$$

$$= U^\top(i)V^\top(i) \begin{bmatrix} \tilde{\varepsilon}_A(i)\tilde{E}_A^\top(i)\tilde{E}_A(i) \\ +\tilde{\varepsilon}_B(i)\tilde{E}_B^\top(i)\tilde{E}_B(i) \\ +\tilde{\varepsilon}_{C_y}(i)\tilde{E}_{C_y}^\top(i)\tilde{E}_{c_y}(i) \end{bmatrix} V(i)U(i)$$

$$= \tilde{\varepsilon}_A(i) \begin{bmatrix} W^\top(i)E_A^\top(i)E_A(i)W(i) & W^\top(i)E_A^\top(i)E_A(i) \\ E_A^\top(i)E_A(i)W(i) & E_A^\top(i)E_A(i) \end{bmatrix}$$

$$+\tilde{\varepsilon}_B(i) \begin{bmatrix} \begin{bmatrix} W^\top(i)P_2(i)P_3^{-1}(i)K_C^\top(i)E_B^\top(i) \\ \times E_B(i)K_C(i)P_3^{-1}(i)P_2^\top(i)W(i) \end{bmatrix} & 0 \\ 0 & 0 \end{bmatrix}$$

$$+\tilde{\varepsilon}_{C_y} \begin{bmatrix} W^\top(i)E_{C_y}^\top(i)E_{C_y}(i)W(i) & W^\top(i)E_{C_y}^\top(i)E_{C_y}(i) \\ E_{C_y}^\top(i)E_{C_y}(i)W(i) & E_{C_y}^\top(i)E_{C_y}(i) \end{bmatrix}$$

and

$$U^\top(i)V^\top(i)P(i) \left[\tilde{D}_A(i) \ \tilde{D}_B(i) \ \tilde{D}_{C_y}(i) \right] \Upsilon^{-1}(i)$$

$$\times \begin{bmatrix} \tilde{D}_A^\top(i) \\ \tilde{D}_B^\top(i) \\ \tilde{D}_{C_y}^\top(i) \end{bmatrix} P(i)V(i)U(i)$$

$$= U^\top(i)V^\top(i) \begin{bmatrix} \tilde{\varepsilon}_A^{-1}(i)P(i)\tilde{D}_A(i)\tilde{D}_A^\top(i)P(i) \\ \tilde{\varepsilon}_B^{-1}(i)P(i)\tilde{D}_B(i)\tilde{D}_B^\top(i)P(i) \\ \tilde{\varepsilon}_{C_y}^{-1}(i)P(i)\tilde{D}_{C_y}(i)\tilde{D}_{C_y}^\top(i)P(i) \end{bmatrix} V(i)U(i).$$

To compute the expression of this term, notice that for the first term we have

$$U^\top(i)V^\top(i) \left[P(i)\tilde{D}_A(i)\tilde{D}_A^\top(i)P(i) \right] V(i)U(i)$$

$$= U^\top(i)V^\top(i) \begin{bmatrix} P_1(i)D_A(i)D_A^\top(i)P_1(i) \\ P_2^\top(i)D_A(i)D_A^\top(i)P_1(i) \end{bmatrix}$$

$$\begin{matrix} P_1(i)D_A(i)D_A^\top(i)P_2(i) \\ P_2^\top(i)D_A(i)D_A^\top(i)P_2(i) \end{matrix} \Bigg] V(i)U(i)$$

$$= \begin{bmatrix} W^\top(i) & -W^\top(i)P_2(i)P_3^{-1}(i) \\ \mathbb{I} & 0 \end{bmatrix}$$

$$\begin{bmatrix} P_1(i)D_A(i)D_A^\top(i)P_1(i) & P_1(i)D_A(i)D_A^\top(i)P_2(i) \\ P_2^\top(i)D_A(i)D_A^\top(i)P_1(i) & P_2^\top(i)D_A(i)D_A^\top(i)P_2(i) \end{bmatrix}$$

$$\begin{bmatrix} W(i) & \mathbb{I} \\ -P_3^{-1}(i)P_2^\top(i)W(i) & 0 \end{bmatrix}$$

$$= \begin{bmatrix} \begin{bmatrix} W^\top(i)P_1(i)D_A(i)D_A^\top(i) \\ \times \left[P_1(i) - P_2(i)P_3^{-1}(i)P_2^\top(i)\right]W(i) \\ -W^\top(i)P_2(i)P_3^{-1}(i)P_2^\top(i)D_A(i)D_A^\top(i) \\ \times \left[P_1(i) - P_2(i)P_3^{-1}(i)P_2^\top(i)\right]W(i) \end{bmatrix} \\ P_1(i)D_A(i)D_A^\top(i)\left[P_1(i) - P_2(i)P_3^{-1}(i)P_2^\top(i)\right]W(i) \end{bmatrix}$$

$$\begin{matrix} W^\top(i)\left[P_1(i) - P_2(i)P_3^{-1}(i)P_2^\top(i)\right]D_A(i)D_A^\top(i)P_1(i) \\ P_1(i)D_A(i)D_A^\top(i)P_1(i) \end{matrix}\Bigg].$$

Using the fact that $\left[P_1(i) - P_2(i)P_3^{-1}(i)P_2^\top(i)\right]W(i) = \mathbb{I}$, we get

$$U^\top(i)V^\top(i)\left[P(i)\tilde{D}_A(i)\tilde{D}_A^\top(i)P(i)\right]V(i)U(i)$$

$$= \begin{bmatrix} D_A(i)D_A^\top(i) & D_A(i)D_A^\top(i)P_1(i) \\ P_1(i)D_A(i)D_A^\top(i) & P_1(i)D_A(i)D_A^\top(i)P_1(i) \end{bmatrix}.$$

For the second term, we have

$$U^\top(i)V^\top(i)\begin{bmatrix} P_1(i)D_B(i)D_B^\top(i)P_1(i) \\ P_2^\top(i)D_B(i)D_B^\top(i)P_1(i) \end{bmatrix}$$

$$\begin{matrix} P_1(i)D_B(i)D_B^\top(i)P_2(i) \\ P_2^\top(i)D_B(i)D_B^\top(i)P_2(i) \end{matrix}\Bigg] V(i)U(i)$$

$$= \begin{bmatrix} W^\top(i) & -W^\top(i)P_2(i)P_3^{-1}(i) \\ \mathbb{I} & 0 \end{bmatrix}$$

$$\begin{bmatrix} P_1(i)D_B(i)D_B^\top(i)P_1(i) & P_1(i)D_B(i)D_B^\top(i)P_2(i) \\ P_2^\top(i)D_B(i)D_B^\top(i)P_1(i) & P_2^\top(i)D_B(i)D_B^\top(i)P_2(i) \end{bmatrix}$$

$$\begin{bmatrix} W(i) & \mathbb{I} \\ -P_3^{-1}(i)P_2^\top(i)W(i) & 0 \end{bmatrix}$$

$$= \begin{bmatrix} \begin{bmatrix} W^\top(i)P_1(i)D_B(i)D_B^\top(i) \\ \times \left[P_1(i) - P_2(i)P_3^{-1}(i)P_2^\top(i)\right]W(i) \\ -W^\top(i)P_2(i)P_3^{-1}(i)P_2^\top(i)D_B(i)D_B^\top(i) \\ \times \left[P_1(i) - P_2(i)P_3^{-1}(i)P_2^\top(i)\right]W(i) \end{bmatrix} \\ P_1(i)D_B(i)D_B^\top(i)\left[P_1(i) - P_2(i)P_3^{-1}(i)P_2^\top(i)\right]W(i) \end{bmatrix}$$

$$\begin{matrix} W^\top(i)\left[P_1(i) - P_2(i)P_3^{-1}(i)P_2^\top(i)\right]D_B(i)D_B^\top(i)P_1(i) \\ P_1(i)D_B(i)D_B^\top(i)P_1(i) \end{matrix}\Bigg].$$

$$= \begin{bmatrix} D_B(i)D_B^\top(i) & D_B(i)D_B^\top(i)P_1(i) \\ P_1(i)D_B(i)D_B^\top(i) & P_1(i)D_B(i)D_B^\top(i)P_1(i) \end{bmatrix}.$$

For the third term, we have

$$U^\top(i)V^\top(i)$$

$$\times \begin{bmatrix} P_2(i)K_B(i)D_{C_y}(i)D_{C_y}^\top(i)K_B^\top(i)P_2^\top(i) \\ P_3(i)K_B(i)D_{C_y}(i)D_{C_y}^\top(i)K_B^\top(i)P_2^\top(i) \end{bmatrix}$$

$$\left.\begin{array}{c} P_2(i)K_B(i)D_{C_y}(i)D_{C_y}^\top(i)K_B^\top(i)P_3(i) \\ P_3(i)K_B(i)D_{C_y}(i)D_{C_y}^\top(i)K_B^\top(i)P_3(i) \end{array}\right]$$

$$\times V(i)U(i)$$

$$= \left[\begin{array}{cc} W^\top(i) & -W^\top(i)P_2(i)P_3^{-1}(i) \\ \mathbb{I} & 0 \end{array}\right]$$

$$\left[\begin{array}{c} P_2(i)K_B(i)D_{C_y}(i)D_{C_y}^\top(i)K_B^\top(i)P_2^\top(i) \\ P_3(i)K_B(i)D_{C_y}(i)D_{C_y}^\top(i)K_B^\top(i)P_2^\top(i) \end{array}\right.$$

$$\left.\begin{array}{c} P_2(i)K_B(i)D_{C_y}(i)D_{C_y}^\top(i)K_B^\top(i)P_3(i) \\ P_3(i)K_B(i)D_{C_y}(i)D_{C_y}^\top(i)K_B^\top(i)P_3(i) \end{array}\right]$$

$$\left[\begin{array}{cc} W(i) & \mathbb{I} \\ -P_3^{-1}(i)P_2^\top(i)W(i) & 0 \end{array}\right]$$

$$= \left[\begin{array}{cc} 0 & 0 \\ 0 & P_2(i)K_B(i)D_{C_y}(i)D_{C_y}^\top(i)K_B^\top(i)P_2^\top(i)W(i) \end{array}\right].$$

Taking into account these computations, we have

$$U^\top(i)V^\top(i)\left[\begin{array}{c} \tilde\varepsilon_A^{-1}(i)P(i)\tilde D_A(i)\tilde D_A^\top(i)P(i) \\ \tilde\varepsilon_B^{-1}(i)P(i)\tilde D_B(i)\tilde D_B^\top(i)P(i) \\ \tilde\varepsilon_{C_y}^{-1}(i)P(i)\tilde D_{C_y}(i)\tilde D_{C_y}^\top(i)P(i) \end{array}\right]V(i)U(i)$$

$$= \tilde\varepsilon_A^{-1}(i)\left[\begin{array}{cc} D_A(i)D_A^\top(i) & D_A(i)D_A^\top(i)P_1(i) \\ P_1(i)D_A(i)D_A^\top(i) & P_1(i)D_A(i)D_A^\top(i)P_1(i) \end{array}\right]$$

$$+\tilde\varepsilon_B^{-1}(i)\left[\begin{array}{cc} D_B(i)D_B^\top(i) & D_B(i)D_B^\top(i)P_1(i) \\ P_1(i)D_B(i)D_B^\top(i) & P_1(i)D_B(i)D_B^\top(i)P_1(i) \end{array}\right]$$

$$+\tilde\varepsilon_{C_y}^{-1}(i)\left[\begin{array}{cc} 0 & 0 \\ 0 & P_2(i)K_B(i)D_{C_y}(i)D_{C_y}^\top(i)K_B^\top(i)P_2^\top(i)W(i) \end{array}\right].$$

Using all these transformations, the previous stochastic stability condition for the extended state equation becomes

$$\left[\begin{array}{cc} \widehat{M_1(i)} & \mathcal{M}_2(i) \\ \mathcal{M}_2^\top(i) & \mathcal{M}_3(i) \end{array}\right]$$

$$+\tilde\varepsilon_A^{-1}(i)\left[\begin{array}{cc} D_A(i)D_A^\top(i) & D_A(i)D_A^\top(i)P_1(i) \\ P_1(i)D_A(i)D_A^\top(i) & P_1(i)D_A(i)D_A^\top(i)P_1(i) \end{array}\right]$$

$$+\tilde\varepsilon_B^{-1}(i)\left[\begin{array}{cc} D_B(i)D_B^\top(i) & D_B(i)D_B^\top(i)P_1(i) \\ P_1(i)D_B(i)D_B^\top(i) & P_1(i)D_B(i)D_B^\top(i)P_1(i) \end{array}\right]$$

$$+\tilde\varepsilon_{C_y}^{-1}(i)\left[\begin{array}{cc} 0 & 0 \\ 0 & P_2(i)K_B(i)D_{C_y}(i)D_{C_y}^\top(i)K_B^\top(i)P_2^\top(i)W(i) \end{array}\right]$$

$$+\tilde\varepsilon_A(i)\left[\begin{array}{cc} W^\top(i)E_A^\top(i)E_A(i)W(i) & W^\top(i)E_A^\top(i)E_A(i) \\ E_A^\top(i)E_A(i)W(i) & E_A^\top(i)E_A(i) \end{array}\right]$$

$$+\tilde{\varepsilon}_B(i)\left[\begin{array}{cc}\left[\begin{array}{c}W^\top(i)P_2(i)P_3^{-1}(i)K_C^\top(i)E_B^\top(i)\\ \times E_B(i)K_C(i)P_3^{-1}(i)P_2^\top(i)W(i)\end{array}\right] & 0\\ 0 & 0\end{array}\right]$$

$$+\tilde{\varepsilon}_{C_y}(i)\left[\begin{array}{cc}W^\top(i)E_{C_y}^\top(i)E_{C_y}(i)W(i) & W^\top(i)E_{C_y}^\top(i)E_{C_y}(i)\\ E_{C_y}^\top(i)E_{C_y}(i)W(i) & E_{C_y}^\top(i)E_{C_y}(i)\end{array}\right]<0,$$

which can be rewritten as follows:

$$\left[\begin{array}{cc}\widehat{H_1(i)} & H_2(i)\\ H_2^\top(i) & H_3(i)\end{array}\right]<0,$$

with

$$\begin{aligned}\widehat{H_1(i)}=&\ \widehat{\mathcal{M}_1(i)}+\tilde{\varepsilon}_A^{-1}(i)D_A(i)D_A^\top(i)+\tilde{\varepsilon}_B^{-1}(i)D_B(i)D_B^\top(i)\\ &+\tilde{\varepsilon}_A W^\top(i)E_A^\top(i)E_A(i)W(i)\\ &+\tilde{\varepsilon}_B W^\top(i)P_2(i)P_3^{-1}(i)K_C^\top(i)E_B^\top(i)\\ &\times E_B(i)K_C(i)P_3^{-1}(i)P_2^\top(i)W(i)\\ &+\tilde{\varepsilon}_{C_y}(i)W^\top(i)E_{C_y}^\top(i)E_{C_y}(i)W(i),\\ H_2(i)=&\ \mathcal{M}_2(i)+\tilde{\varepsilon}_A^{-1}(i)D_A(i)D_A^\top(i)P_1(i)\\ &+\tilde{\varepsilon}_B^{-1}(i)D_B(i)D_B^\top(i)P_1(i)\\ &+\tilde{\varepsilon}_A(i)W^\top(i)E_A^\top(i)E_A(i)\\ &+\tilde{\varepsilon}_{C_y}(i)W^\top(i)E_{C_y}^\top(i)E_{C_y}(i),\\ H_3(i)=&\ \mathcal{M}_3(i)+\tilde{\varepsilon}_A^{-1}(i)P_1(i)D_A(i)D_A^\top(i)P_1(i)\\ &+\tilde{\varepsilon}_B^{-1}(i)P_1(i)D_B(i)D_B^\top(i)P_1(i)\\ &+\tilde{\varepsilon}_{C_y}^{-1}(i)P_2(i)K_B(i)D_{C_y}(i)D_{C_y}^\top(i)K_B(i)P_2(i)W(i)\\ &+\tilde{\varepsilon}_A(i)E_A^\top(i)E_A(i)+\tilde{\varepsilon}_{C_y}(i)E_{C_y}^\top(i)E_{C_y}(i).\end{aligned}$$

If we define $\mathcal{K}_B(i)$ and $\mathcal{K}_C(i)$ as we did previously, that is,

$$\begin{aligned}P(i)=&\ \left[\begin{array}{cc}X(i) & Y^{-1}(i)-X(i)\\ Y^{-1}(i)-X(i) & X(i)-Y^{-1}(i)\end{array}\right],\\ \mathcal{K}_B(i)=&\ P_2(i)K_B(i)=\left(Y^{-1}(i)-X(i)\right)K_B(i),\\ \mathcal{K}_C(i)=&\ -K_C(i)P_3^{-1}(i)P_2^\top(i)W(i)=K_C(i)Y(i),\end{aligned}$$

and we follow the same steps after choosing the controller gains (putting $H_2(i)=0$) as follows:

$$\begin{cases} K_A(i) = \left[X(i) - Y^{-1}(i)\right]^{-1}\left[A^\top(i) + X(i)A(i)Y(i)\right. \\ \qquad\quad + X(i)B(i)\mathcal{K}_C(i) + \mathcal{K}_B(i)C_y(i)Y(i) \\ \qquad\quad + \sum_{j=1}^N \lambda_{ij}Y^{-1}(j)Y(i) \\ \qquad\quad + \tilde\varepsilon_A^{-1}(i)X(i)D_A(i)D_A^\top(i) + \tilde\varepsilon_B^{-1}(i)X(i)D_B(i)D_B^\top(i) \\ \qquad\quad + \tilde\varepsilon_A(i)E_A^\top(i)E_A(i)Y(i) \\ \qquad\quad \left. + \tilde\varepsilon_{C_y}(i)E_{C_y}^\top(i)E_{C_y}(i)Y(i)\right]Y^{-1}(i), \\ K_B(i) = \left[Y^{-1}(i) - X(i)\right]^{-1}\mathcal{K}_B(i), \\ K_C(i) = \mathcal{K}_C(i)Y^{-1}(i), \end{cases}$$

we get

$$\mathcal{H}_1(i) < 0,$$
$$\mathcal{H}_3(i) < 0,$$

with

$$\begin{aligned} \mathcal{H}_1(i) =\ & A(i)Y(i) + Y^\top(i)A^\top(i) + B(i)\mathcal{K}_C(i) + \mathcal{K}_C^\top(i)B^\top(i) \\ & + \sum_{j=1}^N \lambda_{ij}Y(i)Y^{-1}(j)Y(i) + \tilde\varepsilon_A^{-1}(i)D_A(i)D_A^\top(i) \\ & + \tilde\varepsilon_B^{-1}(i)D_B(i)D_B^\top(i) + \tilde\varepsilon_A Y^\top(i)E_A^\top(i)E_A(i)Y(i) \\ & + \tilde\varepsilon_B\mathcal{K}_C^\top(i)E_B^\top(i)E_B(i)\mathcal{K}_C(i) \\ & + \tilde\varepsilon_{C_y}(i)Y^\top(i)E_{C_y}^\top(i)E_{C_y}(i)Y(i), \\ \mathcal{H}_3(i) =\ & X(i)A(i) + A^\top(i)X(i) + \mathcal{K}_B(i)C_y(i) \\ & + \mathcal{K}_B^\top(i)C_y^\top(i) + \sum_{j=1}^N \lambda_{ij}X(j) \\ & + \tilde\varepsilon_A^{-1}(i)X(i)D_A(i)D_A^\top(i)X(i) \\ & + \tilde\varepsilon_B^{-1}(i)X(i)D_B(i)D_B^\top(i)X(i) \\ & + \tilde\varepsilon_{C_y}^{-1}(i)\mathcal{K}_B(i)D_{C_y}(i)D_{C_y}^\top(i)\mathcal{K}_B^\top(i) \\ & + \tilde\varepsilon_A(i)E_A^\top(i)E_A(i) \\ & + \tilde\varepsilon_{C_y}(i)E_{C_y}^\top(i)E_{C_y}(i). \end{aligned}$$

Using the fact that

$$\sum_{j=1}^N \lambda_{ij}Y(i)Y^{-1}(j)Y(i) = \lambda_{ii}Y(i) + \mathcal{S}_i(Y)\mathcal{Y}_i^{-1}(Y)\mathcal{S}_i^\top(Y),$$

where $\mathcal{S}_i(Y)$ and $\mathcal{Y}_i(Y)$ keep the same definitions as before, these two matrix inequalities are equivalent to the following LMIs:

$$
\begin{bmatrix}
J_{\mathcal{H}_1}(i) & Y^\top(i)E_A^\top(i) & \mathcal{K}_C^\top(i)E_B^\top(i) \\
E_A(i)Y(i) & -\tilde{\varepsilon}_A^{-1}(i)\mathbb{I} & 0 \\
E_B(i)\mathcal{K}_C(i) & 0 & -\tilde{\varepsilon}_B^{-1}(i)\mathbb{I} \\
E_{C_y}(i)Y(i) & 0 & 0 \\
\mathcal{S}_i^\top(Y) & 0 & 0
\end{bmatrix}
$$

$$
\left.\begin{matrix}
Y^\top(i)E_{C_y}^\top(i) & \mathcal{S}_i(Y) \\
0 & 0 \\
0 & 0 \\
\tilde{\varepsilon}_{C_y}^{-1}(i)\mathbb{I} & 0 \\
0 & -\mathcal{Y}_i(Y)
\end{matrix}\right] < 0,
$$

$$
\begin{bmatrix}
J_{\mathcal{H}_2}(i) & X(i)D_A(i) & X(i)D_B(i) & \mathcal{K}_B(i)D_{C_y}(i) \\
D_A^\top(i)X(i) & -\tilde{\varepsilon}_A(i)\mathbb{I} & 0 & 0 \\
D_B^\top(i)X(i) & 0 & -\tilde{\varepsilon}_B(i)\mathbb{I} & 0 \\
D_{C_y}^\top(i)\mathcal{K}_B^\top(i) & 0 & 0 & -\tilde{\varepsilon}_{C_y}(i)\mathbb{I}
\end{bmatrix} < 0,
$$

with

$$
\begin{aligned}
J_{\mathcal{H}_1}(i) =\ & A(i)Y(i) + Y^\top(i)A^\top(i) + B(i)\mathcal{K}_C(i) + \mathcal{K}_C^\top(i)B^\top(i) \\
& + \lambda_{ii}Y(i) + \tilde{\varepsilon}_A^{-1}(i)D_A(i)D_A^\top(i) + \tilde{\varepsilon}_B^{-1}(i)D_B(i)D_B^\top(i), \\
J_{\mathcal{H}_2}(i) =\ & X(i)A(i) + A^\top(i)X(i) + \mathcal{K}_B(i)C_y(i) + \mathcal{K}_B^\top(i)C_y^\top(i) \\
& + \sum_{j=1}^{N} \lambda_{ij}X(j) + \tilde{\varepsilon}_A(i)E_A^\top(i)E_A(i) + \tilde{\varepsilon}_{C_y}(i)E_{C_y}^\top(i)E_{C_y}(i).
\end{aligned}
$$

The results of all these developments are summarized by the following theorem.

Theorem 25. *Let* $\tilde{\varepsilon}_A = (\tilde{\varepsilon}_A(1), \cdots, \tilde{\varepsilon}_A(N))$, $\tilde{\varepsilon}_B = (\tilde{\varepsilon}_B(1), \cdots, \tilde{\varepsilon}_B(N))$, *and* $\tilde{\varepsilon}_{C_y} = (\tilde{\varepsilon}_{C_y}(1), \cdots, \tilde{\varepsilon}_{C_y}(N))$ *be given sets of positive scalars. System (3.51) is stochastically stable if and only if for every* $i \in \mathscr{S}$, *the following LMIs are feasible for some symmetric and positive-definite matrices* $X = (X(1), \cdots, X(N)) > 0$, *and* $Y = (Y(1), \cdots, Y(N)) > 0$, *and matrices* $\mathcal{K}_B = (\mathcal{K}_B(1), \cdots, \mathcal{K}_B(N))$, *and* $\mathcal{K}_C = (\mathcal{K}_C(1), \cdots, \mathcal{K}_C(N))$:

$$
\begin{bmatrix}
J_{\mathcal{H}_1}(i) & Y^\top(i)E_A^\top(i) & \mathcal{K}_C^\top(i)E_B^\top(i) \\
E_A(i)Y(i) & -\tilde{\varepsilon}_A^{-1}(i)\mathbb{I} & 0 \\
E_B(i)\mathcal{K}_C(i) & 0 & -\tilde{\varepsilon}_B^{-1}(i)\mathbb{I} \\
E_{C_y}(i)Y(i) & 0 & 0 \\
\mathcal{S}_i^\top(Y) & 0 & 0
\end{bmatrix}
$$

$$
\left.\begin{matrix}
Y^\top(i)E_{C_y}^\top(i) & \mathcal{S}_i(Y) \\
0 & 0 \\
0 & 0 \\
\tilde{\varepsilon}_{C_y}^{-1}(i)\mathbb{I} & 0 \\
0 & -\mathcal{Y}_i(Y)
\end{matrix}\right] < 0, \tag{3.46}
$$

$$\begin{bmatrix} J_{\mathcal{H}_2}(i) & X(i)D_A(i) & X(i)D_B(i) \\ D_A^\top(i)X(i) & -\tilde{\varepsilon}_A(i)\mathbb{I} & 0 \\ D_B^\top(i)X(i) & 0 & -\tilde{\varepsilon}_B(i)\mathbb{I} \\ D_{C_y}^\top(i)\mathcal{K}_B^\top(i) & 0 & 0 \end{bmatrix}$$

$$\begin{matrix} \mathcal{K}_B(i)D_{C_y}(i) \\ 0 \\ 0 \\ -\tilde{\varepsilon}_{C_y}(i)\mathbb{I} \end{matrix} \Bigg] < 0, \tag{3.47}$$

$$\begin{bmatrix} Y(i) & \mathbb{I} \\ \mathbb{I} & X(i) \end{bmatrix} > 0, \tag{3.48}$$

with

$$\mathcal{S}_i(Y) = \left[\sqrt{\lambda_{i1}}Y(i), \cdots, \sqrt{\lambda_{ii-1}}Y(i), \sqrt{\lambda_{ii+1}}Y(i), \cdots, \sqrt{\lambda_{iN}}Y(i) \right],$$
$$\mathcal{Y}_i(Y) = \mathrm{diag}\left[Y(1), \cdots, Y(i-1), Y(i+1), \cdots, Y(N) \right].$$

Furthermore, the dynamic output-feedback controller is given by

$$\begin{cases} K_A(i) = \left[X(i) - Y^{-1}(i) \right]^{-1} \left[A^\top(i) + X(i)A(i)Y(i) \right. \\ \qquad + X(i)B(i)\mathcal{K}_C(i) + \mathcal{K}_B(i)C_y(i)Y(i) \\ \qquad + \sum_{j=1}^{N} \lambda_{ij}Y^{-1}(j)Y(i) \\ \qquad + \tilde{\varepsilon}_A^{-1}(i)X(i)D_A(i)D_A^\top(i) \\ \qquad + \tilde{\varepsilon}_B^{-1}(i)X(i)D_B(i)D_B^\top(i) \\ \qquad + \tilde{\varepsilon}_A(i)E_A^\top(i)E_A(i)Y(i) \\ \qquad \left. + \tilde{\varepsilon}_{C_y}(i)E_{C_y}^\top(i)E_{C_y}(i)Y(i) \right] Y^{-1}(i), \\ K_B(i) = \left[Y^{-1}(i) - X(i) \right]^{-1} \mathcal{K}_B(i), \\ K_C(i) = \mathcal{K}_C(i)Y^{-1}(i). \end{cases} \tag{3.49}$$

Example 34. Let us consider a system with two modes and see how we can illustrate the developed results in this theorem. Let us assume that the corresponding data are as follows:

- mode #1:

$$A(1) = \begin{bmatrix} 1.00 & -0.50 \\ 0.10 & 1.00 \end{bmatrix}, \quad B(1) = \begin{bmatrix} 1.0 & 0.0 \\ 0.0 & 1.0 \end{bmatrix}, \quad C(1) = \begin{bmatrix} 1.0 & 0.0 \\ 0.0 & 1.0 \end{bmatrix},$$

$$D_A(1) = \begin{bmatrix} 0.10 \\ 0.20 \end{bmatrix}, \quad D_B(1) = \begin{bmatrix} 0.10 \\ 0.20 \end{bmatrix}, \quad D_C(1) = \begin{bmatrix} 0.10 \\ 0.20 \end{bmatrix},$$

$$E_A(1) = \begin{bmatrix} 0.20 & 0.10 \end{bmatrix}, \quad E_B(1) = \begin{bmatrix} 0.20 & 0.10 \end{bmatrix}, \quad E_C(1) = \begin{bmatrix} 0.20 & 0.10 \end{bmatrix},$$

- mode #2:

$$A(2) = \begin{bmatrix} -0.20 & 0.50 \\ 0.0 & -0.25 \end{bmatrix}, \quad B(2) = \begin{bmatrix} 1.0 & 0.0 \\ 0.0 & 1.0 \end{bmatrix}, \quad C(2) = \begin{bmatrix} 1.0 & 0.0 \\ 0.0 & 1.0 \end{bmatrix},$$

$$D_A(2) = \begin{bmatrix} 0.13 \\ 0.10 \end{bmatrix}, \qquad D_B(2) = \begin{bmatrix} 0.13 \\ 0.10 \end{bmatrix}, \qquad D_C(2) = \begin{bmatrix} 0.13 \\ 0.10 \end{bmatrix},$$

$$E_A(2) = \begin{bmatrix} 0.10 \ 0.20 \end{bmatrix}, \qquad E_B(2) = \begin{bmatrix} 0.10 \ 0.20 \end{bmatrix}, \ E_C(2) = \begin{bmatrix} 0.10 \ 0.20 \end{bmatrix}.$$

The switching between the two modes is described by the following transition matrix:

$$\Lambda = \begin{bmatrix} -2.0 & 2.0 \\ 3.0 & -3.0 \end{bmatrix}.$$

The positive scalars $\varepsilon_A(i)$, $\varepsilon_B(i)$, and $\varepsilon_{C_y}(i)$, $i = 1, 2$ are chosen as follows:

$$\varepsilon_A(1) = \varepsilon_A(2) = 0.50,$$
$$\varepsilon_B(1) = \varepsilon_B(2) = 0.10,$$
$$\varepsilon_C(1) = \varepsilon_C(2) = 0.10.$$

Solving the LMIs (3.46)–(3.48), we get

$$X(1) = \begin{bmatrix} 2.8679 & -1.1001 \\ -1.1001 & 1.0058 \end{bmatrix}, \qquad X(2) = \begin{bmatrix} 11.8504 & -5.9146 \\ -5.9146 & 3.4179 \end{bmatrix},$$

$$Y(1) = \begin{bmatrix} 65.8510 & -37.4644 \\ -37.4644 & 116.3295 \end{bmatrix}, \qquad Y(2) = \begin{bmatrix} 86.0951 & -29.9875 \\ -29.9875 & 81.6347 \end{bmatrix},$$

$$\mathcal{K}_B(1) = \begin{bmatrix} -120.3571 & 59.6013 \\ 61.2617 & -31.7377 \end{bmatrix}, \qquad \mathcal{K}_B(2) = \begin{bmatrix} -44.3132 & 55.2044 \\ 58.9262 & -79.4786 \end{bmatrix},$$

$$\mathcal{K}_C(1) = \begin{bmatrix} -198.1368 & 69.5958 \\ 90.6906 & -292.8053 \end{bmatrix}, \qquad \mathcal{K}_C(2) = \begin{bmatrix} -141.1144 & -59.3008 \\ 7.5585 & -96.6843 \end{bmatrix}.$$

The corresponding controller gains are given by

$$K_A(1) = \begin{bmatrix} -34.5409 & 14.0604 \\ 25.7147 & -16.7391 \end{bmatrix}, \qquad K_B(1) = \begin{bmatrix} 32.2672 & -15.0181 \\ -25.6925 & 15.1979 \end{bmatrix},$$

$$K_C(1) = \begin{bmatrix} -3.2671 & -0.4539 \\ -0.0671 & -2.5386 \end{bmatrix}, \qquad K_A(2) = \begin{bmatrix} 35.3248 & -54.8428 \\ 82.4890 & -118.5219 \end{bmatrix}.$$

$$K_B(2) = \begin{bmatrix} -37.6957 & 53.8016 \\ -82.8657 & 116.9123 \end{bmatrix}, \qquad K_C(2) = \begin{bmatrix} -2.1697 & -1.5234 \\ -0.3724 & -1.3211 \end{bmatrix}.$$

Based on the results of this theorem, the system of this example is stochastically stable under the output feedback controller with the computed gains.

Let us now study the effect of external Wiener process disturbance of the results we developed. For this purpose consider a continuous-time linear piecewise deterministic system defined in a probability space $(\Omega, \mathcal{F}, \mathbb{P})$ with the following dynamics:

$$\begin{cases} dx(t) = A(r(t))x(t)dt + B(r(t))u(t)dt + \mathbb{W}(r(t))x(t)d\omega(t), \\ y(t) = C_y(r(t))x(t), x(0) = x_0. \end{cases} \quad (3.50)$$

The problem consists of designing an output feedback controller that stabilizes the closed loop of the considered class of system. The structure of this output feedback controller is given by (3.38).

Our goal in this section is to synthesize the gains of the output feedback controller given by (3.38). Plugging the controller into the dynamics we get the following closed-loop dynamics:

$$\begin{aligned} d\eta(t) &= \begin{bmatrix} A(i) & B(i)K_C(i) \\ K_B(i)C_y(i) & K_A(i) \end{bmatrix} \eta(t)dt + \begin{bmatrix} \mathbb{W}(i) & 0 \\ 0 & 0 \end{bmatrix} \eta(t)d\omega(t) \\ &= \tilde{A}(i)\eta(t)dt + \tilde{\mathbb{W}}(i)\eta(t)d\omega(t), \end{aligned} \quad (3.51)$$

where $\tilde{A}(i)$ and $\tilde{\mathbb{W}}(i)$ are defined by

$$\eta(t) = \begin{bmatrix} x(t) \\ x_c(t) \end{bmatrix},$$

$$\tilde{A}(i) = \begin{bmatrix} A(i) & B(i)K_C(i) \\ K_B(i)C_y(i) & K_A(i) \end{bmatrix},$$

$$\tilde{\mathbb{W}}(i) = \begin{bmatrix} \mathbb{W}(i) & 0 \\ 0 & 0 \end{bmatrix}.$$

Let us now return to the initial problem and see how to design the output feedback controller with the form given by (3.38).

Based on Theorem 8, the closed-loop system will be stable if the following holds for every $i \in \mathscr{S}$:

$$\tilde{A}^\top(i)P(i) + P(i)\tilde{A}(i) + \tilde{\mathbb{W}}^\top(i)P(i)\tilde{\mathbb{W}}(i) + \sum_{j=1}^{N} \lambda_{ij}P(j) < 0. \quad (3.52)$$

This inequality is nonlinear in the design parameters $K_A(i)$, $K_B(i)$, $K_C(i)$, and $P(i)$ for every $i \in \mathscr{S}$. To cast it into an LMI, let us define $P(i)$, $W(i)$, $U(i)$, and $V(i)$ as done previously, and pre- and post-multiply (3.52) by $U^\top(i)V^\top(i)$ and $V(i)U(i)$, respectively, to get

$$\begin{aligned} U^\top(i)V^\top(i)\tilde{A}^\top(i)P(i)V(i)U(i) + U^\top(i)V^\top(i)P(i)\tilde{A}(i)V(i)U(i) \\ + U^\top(i)V^\top(i)\tilde{\mathbb{W}}^\top(i)P(i)\tilde{\mathbb{W}}(i)V(i)U(i) \\ + \sum_{j=1}^{N} \lambda_{ij}U^\top(i)V^\top(i)P(j)V(i)U(i) < 0. \quad (3.53) \end{aligned}$$

Let us now compute the different terms in this inequality. Using standard algebraic manipulations as done previously we get

$$U^\top(i)V^\top(i)P(i)\tilde{A}(i)V(i)U(i)$$

$$= \begin{bmatrix} \begin{bmatrix} A(i)W(i) \\ -B(i)K_C(i)P_3^{-1}(i)P_2^\top(i)W(i) \end{bmatrix} \\ \begin{bmatrix} P_1(i)A(i)W(i) \\ +P_2(i)K_B(i)C_y(i)W(i) \\ -P_1(i)B(i)K_C(i)P_3^{-1}(i)P_2^\top(i)W(i) \\ -P_2(i)K_A(i)P_3^{-1}(i)P_2^\top(i)W(i) \end{bmatrix} \\ \begin{bmatrix} A(i) \\ \begin{bmatrix} P_1(i)A(i) \\ +P_2(i)K_B(i)C_y(i) \end{bmatrix} \end{bmatrix} \end{bmatrix}.$$

Using the fact that

$$U^\top(i)V^\top(i)\tilde{A}^\top(i)P(i)V(i)U(i)$$

is the transpose of

$$U^\top(i)V^\top(i)P(i)\tilde{A}(i)V(i)U(i),$$

we get

$$U^\top(i)V^\top(i)\tilde{A}^\top(i)P(i)V(i)U(i)$$

$$= \begin{bmatrix} \begin{bmatrix} W^\top(i)A^\top(i) \\ -W^\top(i)P_2(i)P_3^{-1}(i)K_C^\top(i)B^\top(i) \end{bmatrix} \\ A^\top(i) \\ \begin{bmatrix} W^\top(i)A^\top(i)P_1(i) \\ +W^\top(i)C_y^\top(i)K_B^\top(i)P_2^\top(i) \\ -W^\top(i)P_2(i)P_3^{-1}(i)K_C^\top(i)B^\top(i)P_1(i) \\ -W^\top(i)P_2(i)P_3^{-1}(i)K_A^\top(i)P_2^\top(i) \end{bmatrix} \\ \begin{bmatrix} A^\top(i)P_1(i) \\ +C_y^\top(i)K_B^\top(i)P_2^\top(i) \end{bmatrix} \end{bmatrix}.$$

For the term $U^\top(i)V^\top(i)P(j)V(i)U(i)$, we have

$$\begin{bmatrix} W^\top(i) & -W^\top(i)P_2(i)P_3^{-1}(i) \\ \mathbb{I} & 0 \end{bmatrix} \begin{bmatrix} P_1(j) & P_2(j) \\ P_2^\top(j) & P_3(j) \end{bmatrix}$$

$$\times \begin{bmatrix} W(i) & \mathbb{I} \\ -P_3^{-1}(i)P_2^\top(i)W(i) & 0 \end{bmatrix}$$

$$= \begin{bmatrix} \begin{bmatrix} W^\top(i)P_1(j)W(i) \\ -W^\top(i)P_2(i)P_3^{-1}(i)P_2^\top(j)W(i) \\ -W^\top(i)P_2(j)P_3^{-1}(i)P_2^\top(i)W(i) \\ +W^\top(i)P_2(i)P_3^{-1}(i)P_3(j)P_3^{-1}(i)P_2^\top(i)W(i) \end{bmatrix} \\ P_1(j)W(i) - P_2(j)P_3^{-1}(i)P_2^\top(i)W(i) \\ \begin{bmatrix} W^\top(i)P_1(j) \\ -W^\top(i)P_2(i)P_3^{-1}(i)P_2^\top(j) \end{bmatrix} \\ P_1(j) \end{bmatrix},$$

which can be rewritten as follows:

$$
\left[
\begin{array}{cc}
\begin{array}{l}
W^{\top}(i)W^{-1}(j)W(i) \\
+W^{\top}(i)\left[P_2(i)P_3^{-1}(i)P_3(j) - P_2^{\top}(j)\right] \\
\times P_3^{-1}(j)\left[P_2(i)P_3^{-1}(i)P_3(j) - P_2^{\top}(j)\right]^{\top}W(i)
\end{array} & \star \\
\left[P_1^{\top}(j) - P_2(j)P_3^{-1}(i)P_2^{\top}(i)\right]W(i) & P_1(j)
\end{array}
\right].
$$

For the term $U^{\top}(i)V^{\top}(i)\tilde{\mathbb{W}}^{\top}(i)P(i)\tilde{\mathbb{W}}(i)V(i)U(i)$, notice that

$$
\tilde{\mathbb{W}}^{\top}(i)P(i)\tilde{\mathbb{W}}(i) = \begin{bmatrix} \mathbb{W}^{\top}(i)P_1(i)\mathbb{W}(i) & 0 \\ 0 & 0 \end{bmatrix},
$$

which implies

$$
\begin{aligned}
& U^{\top}(i)V^{\top}(i)\tilde{\mathbb{W}}^{\top}(i)P(i)\tilde{\mathbb{W}}(i)V(i)U(i) \\
&= \begin{bmatrix} W^{\top}(i)\mathbb{W}^{\top}(i)P_1(i)\mathbb{W}(i)W(i) & W^{\top}(i)\mathbb{W}^{\top}(i)P_1(i)\mathbb{W}(i) \\ \mathbb{W}^{\top}(i)P_1(i)\mathbb{W}(i)W(i) & \mathbb{W}^{\top}(i)P_1(i)\mathbb{W}(i) \end{bmatrix} \\
&= \begin{bmatrix} W^{\top}(i) & 0 \\ 0 & \mathbb{I} \end{bmatrix} \begin{bmatrix} \mathbb{W}^{\top}(i)P_1(i)\mathbb{W}(i) & \mathbb{W}^{\top}(i)P_1(i)\mathbb{W}(i) \\ \mathbb{W}^{\top}(i)P_1(i)\mathbb{W}(i) & \mathbb{W}^{\top}(i)P_1(i)\mathbb{W}(i) \end{bmatrix} \\
&\qquad\qquad\qquad\qquad\qquad\qquad \times \begin{bmatrix} W(i) & 0 \\ 0 & \mathbb{I} \end{bmatrix}.
\end{aligned}
$$

Using now all the previous algebraic manipulations, the stability condition for the closed-loop system becomes

$$
\begin{bmatrix} \widehat{\mathcal{M}_1(i)} & \mathcal{M}_2(i) \\ \mathcal{M}_2^{\top}(i) & \mathcal{M}_3(i) \end{bmatrix} < 0,
$$

with

$$
\begin{aligned}
\widehat{\mathcal{M}_1(i)} =\ & \mathcal{M}_1(i) + \sum_{j=1}^{N} \lambda_{ij} W^{\top}(i)\left[P_2(i)P_3^{-1}(i)P_3(j) - P_2^{\top}(j)\right] \\
& \times P_3^{-1}(j)\left[P_2(i)P_3^{-1}(i)P_3(j) - P_2^{\top}(j)\right]^{\top}W(i) \\
& + W^{\top}(i)\mathbb{W}^{\top}(i)P_1(i)\mathbb{W}(i)W(i), \\
\mathcal{M}_1(i) =\ & A(i)W(i) + W^{\top}(i)A^{\top}(i) - B(i)K_C(i)P_3^{-1}(i)P_2^{\top}(i)W(i) \\
& - W^{\top}(i)P_2(i)P_3^{-1}(i)K_C^{\top}(i)B^{\top}(i) \\
& + \sum_{j=1}^{N} \lambda_{ij} W^{\top}(i)W^{-1}(j)W(i), \\
\mathcal{M}_2(i) =\ & A(i) + W^{\top}(i)A^{\top}(i)P_1(i) \\
& + W^{\top}(i)C_y^{\top}(i)K_B^{\top}(i)P_2^{\top}(i) \\
& - W^{\top}(i)P_2(i)P_3^{-1}(i)K_C^{\top}(i)B^{\top}(i)P_1(i)
\end{aligned}
$$

$$-W^\top(i)P_2(i)P_3^{-1}(i)K_A^\top(i)P_2^\top(i)$$

$$+\sum_{j=1}^{N}\lambda_{ij}W^\top(i)\left[P_1(j)-P_2(i)P_3^{-1}(i)P_2^\top(j)\right]^\top$$

$$+W^\top(i)\mathbb{W}^\top(i)P_1(i)\mathbb{W}(i),$$

$$\mathcal{M}_3(i) = P_1(i)A(i)+P_2(i)K_B(i)C_y(i)+A^\top(i)P_1(i)$$

$$+C_y^\top(i)K_B^\top(i)P_2^\top(i)+\sum_{j=1}^{N}\lambda_{ij}P_1(j)$$

$$+\mathbb{W}^\top(i)P_1(i)\mathbb{W}(i).$$

Since

$$\sum_{j=1}^{N}\lambda_{ij}W^\top(i)\left[P_2(i)P_3^{-1}(i)P_3(j)-P_2^\top(j)\right]P_3^{-1}(j)$$

$$\times\left[P_2(i)P_3^{-1}(i)P_3(j)-P_2^\top(j)\right]^\top W(i)\geq 0,$$

and $W^\top(i)\mathbb{W}^\top(i)P_1(i)\mathbb{W}(i)W(i)\geq 0$, we get the following equivalent condition:

$$\begin{bmatrix}\mathcal{M}_1(i) & \mathcal{M}_2(i)\\ \mathcal{M}_2^\top(i) & \mathcal{M}_3(i)\end{bmatrix}<0.$$

Letting

$$P(i)=\begin{bmatrix} X(i) & Y^{-1}(i)-X(i)\\ Y^{-1}(i)-X(i) & X(i)-Y^{-1}(i)\end{bmatrix},$$

where $X(i)>0$ and $Y(i)>0$ are symmetric and positive-definite matrices for each $i\in\mathcal{S}$, that is,

$$\begin{aligned}P_1(i) &= X(i),\\ P_2(i) &= Y^{-1}(i)-X(i),\\ P_3(i) &= X(i)-Y^{-1}(i),\end{aligned}$$

implies $W(i)=\left[P_1(i)-P_2(i)P_3^{-1}(i)P_2^\top(i)\right]^{-1}=Y(i)$ and $P_3^{-1}(i)P_2^\top(i)=-\mathbb{I}$.

If we define $\mathcal{K}_B(i)$, and $\mathcal{K}_C(i)$ by

$$\begin{aligned}\mathcal{K}_B(i) &= P_2(i)K_B(i)=\left(Y^{-1}(i)-X(i)\right)K_B(i),\\ \mathcal{K}_C(i) &= -K_C(i)P_3^{-1}(i)P_2^\top(i)W(i)=K_C(i)Y(i),\end{aligned}$$

and we use all the previous algebraic manipulations, we get

$$\begin{bmatrix}\mathcal{M}_1(i) & \mathcal{M}_2(i)\\ \mathcal{M}_2^\top(i) & \mathcal{M}_3(i)\end{bmatrix}<0,$$

with

$$
\begin{aligned}
\mathcal{M}_1(i) = \; & A(i)Y(i) + Y^\top(i)A^\top(i) + B(i)\mathcal{K}_C(i) \\
& + \mathcal{K}_C^\top(i)B^\top(i) + \sum_{j=1}^{N} \lambda_{ij} Y^\top(i)Y^{-1}(j)Y(i), \\
\mathcal{M}_2(i) = \; & A(i) + Y^\top(i)A^\top(i)X(i) + Y^\top(i)C_y^\top(i)\mathcal{K}_B^\top(i) \\
& + \mathcal{K}_C^\top(i)B^\top(i)X(i) + Y^\top(i)K_A^\top(i)\left[Y^{-1}(i) - X(i)\right]^\top \\
& + \sum_{j=1}^{N} \lambda_{ij} Y^\top(i)Y^{-1}(j) \\
& + Y^\top(i)\mathbb{W}^\top(i)X(i)\mathbb{W}(i), \\
\mathcal{M}_3(i) = \; & X(i)A(i) + \mathcal{K}_B(i)C_y(i) + A^\top(i)X(i) \\
& + C_y^\top(i)\mathcal{K}_B^\top(i) + \sum_{j=1}^{N} \lambda_{ij} X(j) + \mathbb{W}^\top(i)X(i)\mathbb{W}(i).
\end{aligned}
$$

Using the expression of the controller given by

$$
\begin{cases}
K_A(i) = \left[X(i) - Y^{-1}(i)\right]^{-1}\left[A^\top(i) + X(i)A(i)Y(i)\right. \\
\qquad + X(i)B(i)\mathcal{K}_C(i) + \mathcal{K}_B(i)C_y(i)Y(i) + \mathbb{W}^\top(i)X(i)\mathbb{W}(i)Y(i) \\
\qquad \left. + \sum_{j=1}^{N} \lambda_{ij} Y^{-1}(j)Y(i)\right]Y^{-1}(i), \\
K_B(i) = \left[Y^{-1}(i) - X(i)\right]^{-1}\mathcal{K}_B(i), \\
K_C(i) = \mathcal{K}_C(i)Y^{-1}(i),
\end{cases}
$$

we have $\mathcal{M}_2(i) = 0$. This implies that the stability condition is equivalent to the following conditions:

$$
\begin{aligned}
\mathcal{M}_1(i) &< 0, \\
\mathcal{M}_3(i) &< 0,
\end{aligned}
$$

which gives

$$
A(i)Y(i) + Y^\top(i)A^\top(i) + B(i)\mathcal{K}_C(i) + \mathcal{K}_C^\top(i)B^\top(i) \\
+ \sum_{j=1}^{N} \lambda_{ij} Y^\top(i)Y^{-1}(j)Y(i) < 0,
$$

$$
X(i)A(i) + \mathcal{K}_B(i)C_y(i) + A^\top(i)X(i) + C_y^\top(i)\mathcal{K}_B^\top(i) \\
+ \sum_{j=1}^{N} \lambda_{ij} X(j) + \mathbb{W}^\top(i)X(i)\mathbb{W}(i) < 0.
$$

Notice that

$$\sum_{j=1}^{N} \lambda_{ij} Y^{\top}(i) Y^{-1}(j) Y(i) = \lambda_{ii} Y(i) + \mathcal{S}_i(Y) \mathcal{Y}_i^{-1}(Y) \mathcal{S}_i^{\top}(Y),$$

with

$$\mathcal{S}_i(Y) = \left[\sqrt{\lambda_{i1}} Y(i), \cdots, \sqrt{\lambda_{ii-1}} Y(i), \sqrt{\lambda_{ii+1}} Y(i), \right.$$
$$\left. \cdots, \sqrt{\lambda_{iN}} Y(i) \right],$$
$$\mathcal{Y}_i(Y) = \operatorname{diag} \left[Y(1), \cdots, Y(i-1), Y(i+1), \cdots, Y(N) \right].$$

Using this, the previous stability conditions become

$$\left[\begin{bmatrix} A(i)Y(i) + Y^{\top}(i)A^{\top}(i) + B(i)\mathcal{K}_C(i) \\ +\mathcal{K}_C^{\top}(i)B^{\top}(i) + \lambda_{ii}Y(i) \end{bmatrix} \quad \mathcal{S}_i(Y) \\ \mathcal{S}_i^{\top}(Y) \qquad\qquad -\mathcal{Y}_i(Y) \right] < 0,$$

$$X(i)A(i) + \mathcal{K}_B(i)C_y(i) + A^{\top}(i)X(i)$$

$$+C_y^{\top}(i)\mathcal{K}_B^{\top}(i) + \sum_{j=1}^{N} \lambda_{ij} X(j) + \mathbb{W}^{\top}(i)X(i)\mathbb{W}(i) < 0.$$

Finally, notice that

$$U^{\top}(i)V^{\top}(i)P(i)V(i)U(i) = \begin{bmatrix} Y(i) & \mathbb{I} \\ \mathbb{I} & X(i) \end{bmatrix} > 0.$$

The results of the previous algebraic manipulations are summarized by the following theorem.

Theorem 26. *System (3.51) is stochastically stable if and only if for every $i \in \mathscr{S}$, the following LMIs are feasible for some symmetric and positive-definite matrices $X = (X(1), \cdots, X(N)) > 0$, and $Y = (Y(1), \cdots, Y(N)) > 0$, and matrices $\mathcal{K}_B = (\mathcal{K}_B(1), \cdots, \mathcal{K}_B(N))$, and $\mathcal{K}_C = (\mathcal{K}_C(1), \cdots, \mathcal{K}_C(N))$:*

$$\left[\begin{bmatrix} A(i)Y(i) + Y^{\top}(i)A^{\top}(i) \\ +B(i)\mathcal{K}_C(i) \\ +\mathcal{K}_C^{\top}(i)B^{\top}(i) + \lambda_{ii}Y(i) \end{bmatrix} \quad \mathcal{S}_i(Y) \\ \mathcal{S}_i^{\top}(Y) \qquad\qquad -\mathcal{Y}_i(Y) \right] < 0, \tag{3.54}$$

$$X(i)A(i) + \mathcal{K}_B(i)C_y(i) + A^{\top}(i)X(i) + C_y^{\top}(i)\mathcal{K}_B^{\top}(i)$$

$$+ \sum_{j=1}^{N} \lambda_{ij} X(j) + \mathbb{W}^{\top}(i)X(i)\mathbb{W}(i) < 0, \tag{3.55}$$

$$\begin{bmatrix} Y(i) & \mathbb{I} \\ \mathbb{I} & X(i) \end{bmatrix} > 0, \tag{3.56}$$

with

$$\mathcal{S}_i(Y) = \left[\sqrt{\lambda_{i1}}Y(i), \cdots, \sqrt{\lambda_{ii-1}}Y(i), \sqrt{\lambda_{ii+1}}Y(i),\right.$$
$$\left.\cdots, \sqrt{\lambda_{iN}}Y(i)\right]$$
$$\mathcal{Y}_i(Y) = \text{diag}\left[Y(1), \cdots, Y(i-1), Y(i+1), \cdots, Y(N)\right].$$

Furthermore, the dynamic output-feedback controller is given by

$$\begin{cases} K_A(i) = \left[X(i) - Y^{-1}(i)\right]^{-1} \left[A^\top(i) + X(i)A(i)Y(i)\right. \\ +X(i)B(i)\mathcal{K}_C(i) + \mathcal{K}_B(i)C_y(i)Y(i) + \mathbb{W}^\top(i)X(i)\mathbb{W}(i)Y(i) \\ \left.+ \sum_{j=1}^N \lambda_{ij}Y^{-1}(j)Y(i)\right]Y^{-1}(i), \\ K_B(i) = \left[Y^{-1}(i) - X(i)\right]^{-1}\mathcal{K}_B(i), \\ K_C(i) = \mathcal{K}_C(i)Y^{-1}(i). \end{cases} \qquad (3.57)$$

Example 35. To illustrate the results developed, let us consider the two-mode system of Example 34 with all the uncertainties equal to zero and

$$\mathbb{W}(1) = \begin{bmatrix} 0.10 & 0.0 \\ 0.0 & 0.10 \end{bmatrix}, \quad \mathbb{W}(2) = \begin{bmatrix} 0.20 & 0.0 \\ 0.0 & 0.20 \end{bmatrix}.$$

Solving the LMIs (3.54)–(3.56), we get

$$X(1) = \begin{bmatrix} 155.3339 & 0.0000 \\ 0.0000 & 155.3339 \end{bmatrix}, \qquad X(2) = \begin{bmatrix} 156.7142 & -0.0000 \\ -0.0000 & 156.7142 \end{bmatrix},$$

$$Y(1) = \begin{bmatrix} 85.5704 & -0.0000 \\ -0.0000 & 85.5704 \end{bmatrix}, \qquad Y(2) = \begin{bmatrix} 76.2129 & 0.0000 \\ 0.0000 & 76.2129 \end{bmatrix},$$

$$\mathcal{K}_B(1) = \begin{bmatrix} -241.0167 & -479.2866 \\ 541.4202 & -241.0167 \end{bmatrix}, \quad \mathcal{K}_B(2) = \begin{bmatrix} -53.2470 & -7.5294 \\ 7.5294 & -45.4113 \end{bmatrix},$$

$$\mathcal{K}_C(1) = 10^3 \cdot \begin{bmatrix} -0.1707 & 3.3137 \\ -3.2795 & -0.1707 \end{bmatrix}, \quad \mathcal{K}_C(2) = \begin{bmatrix} -45.2177 & -10.8637 \\ 10.8637 & -41.4071 \end{bmatrix}.$$

The corresponding controller gains are given by

$$K_A(1) = \begin{bmatrix} -2.5462 & 35.1425 \\ -34.7425 & -2.5462 \end{bmatrix}, \quad K_B(1) = \begin{bmatrix} 1.5517 & 3.0858 \\ -3.4858 & 1.5517 \end{bmatrix},$$

$$K_C(1) = \begin{bmatrix} -1.9946 & 38.7254 \\ -38.3254 & -1.9946 \end{bmatrix}, \quad K_A(2) = \begin{bmatrix} -1.1327 & -0.6906 \\ 0.6906 & -1.0827 \end{bmatrix},$$

$$K_B(2) = \begin{bmatrix} 0.3398 & 0.0480 \\ -0.0480 & 0.2898 \end{bmatrix}, \quad K_C(2) = \begin{bmatrix} -0.5933 & -0.1425 \\ 0.1425 & -0.5433 \end{bmatrix}.$$

Using the results of this theorem, the system of this example is stochastically stable under the output feedback controller with the computed gains.

Let us now consider that the state equation contain Wiener process external disturbances and norm-bounded uncertainties. In this case the state equation becomes

$$\begin{cases} dx(t) = A(r(t),t)x(t)dt + B(r(t),t)u(t) + \mathbb{W}(r(t))x(t)d\omega(t), \\ y(t) = C_y(r(t))x(t), \\ x(0) = x_0. \end{cases} \quad (3.58)$$

Combining these dynamics with those of the controller, we get the following:

$$d\eta(t) = \begin{bmatrix} A(r(t)) & B(r(t))K_C(r(t)) \\ K_B(r(t))C_y(r(t)) & K_A(r(t)) \end{bmatrix} \eta(t)dt$$
$$+ \begin{bmatrix} D_A(r(t))F_A(r(t),t)E_A(r(t)) & \begin{bmatrix} D_B(r(t))F_B(r(t),t) \\ \times E_B(r(t))K_C(r(t)) \end{bmatrix} \\ \begin{bmatrix} K_B(r(t))D_{C_y}(r(t)) \\ \times F_{C_y}(r(t),t)E_{C_y}(r(t)) \end{bmatrix} & 0 \end{bmatrix} \eta(t)dt$$
$$+ \begin{bmatrix} \mathbb{W}(r(t)) & 0 \\ 0 & 0 \end{bmatrix} \eta(t)d\omega(t), \quad (3.59)$$

with

$$\eta(t) = \begin{bmatrix} x(t) \\ x_c(t) \end{bmatrix}.$$

These dynamics can be rewritten as follows:

$$d\eta(t) = \left[\tilde{A}(r(t)) + \Delta\tilde{A}(r(t),t) \right] \eta(t)dt + \tilde{\mathbb{W}}(r(t))\eta(t)d\omega(t),$$

with

$$\Delta\tilde{A}(r(t),t) = \Delta\tilde{A}_A(r(t),t) + \Delta\tilde{B}_B(r(t),t) + \Delta\tilde{C}_{C_y}(r(t),t),$$
$$\tilde{\mathbb{W}}(r(t)) = \begin{bmatrix} \mathbb{W}(r(t)) & 0 \\ 0 & 0 \end{bmatrix},$$

where

$$\Delta\tilde{A}_A(r(t),t) = \begin{bmatrix} D_A(r(t))F_A(r(t),t)E_A(r(t)) & 0 \\ 0 & 0 \end{bmatrix},$$
$$\Delta\tilde{B}_B(r(t),t) = \begin{bmatrix} 0 & D_B(r(t))F_B(r(t),t)E_B(r(t))K_C(r(t)) \\ 0 & 0 \end{bmatrix},$$
$$\Delta\tilde{C}_{C_y}(r(t),t) = \begin{bmatrix} 0 & 0 \\ K_B(r(t))D_{C_y}(r(t))F_{C_y}(r(t),t)E_{C_y}(r(t)) & 0 \end{bmatrix}.$$

Notice that

$$\Delta \tilde{A}_A(r(t), t) = \begin{bmatrix} D_A(r(t)) & 0 \\ 0 & 0 \end{bmatrix} \begin{bmatrix} F_A(r(t), t) & 0 \\ 0 & 0 \end{bmatrix} \begin{bmatrix} E_A(r(t)) & 0 \\ 0 & 0 \end{bmatrix}$$

$$= \tilde{D}_A(r(t)) \tilde{F}_A(r(t), t) \tilde{E}_A(r(t)),$$

$$\Delta \tilde{B}_B(r(t), t) = \begin{bmatrix} 0 & D_B(r(t)) \\ 0 & 0 \end{bmatrix} \begin{bmatrix} 0 & 0 \\ 0 & F_B(r(t), t) \end{bmatrix}$$

$$\times \begin{bmatrix} 0 & 0 \\ 0 & E_B(r(t)) K_C(r(t)) \end{bmatrix}$$

$$= \tilde{D}_B(r(t)) \tilde{F}_B(r(t), t) \tilde{E}_B(r(t)),$$

$$\Delta \tilde{C}_{C_y}(r(t), t) = \begin{bmatrix} 0 & 0 \\ 0 & K_B(r(t)) D_{C_y}(r(t)) \end{bmatrix} \begin{bmatrix} 0 & 0 \\ 0 & F_{C_y}(r(t), t) \end{bmatrix}$$

$$\times \begin{bmatrix} 0 & 0 \\ E_{C_y}(r(t)) & 0 \end{bmatrix}$$

$$= \tilde{D}_{C_y}(r(t)) \tilde{F}_{C_y}(r(t), t) \tilde{E}_{C_y}(r(t)).$$

Let us return to the extended dynamics to study their stability. Based on the results of Chapter 2, these dynamics are stable if there exists a set of symmetric and positive-definite matrices $P = (P(1), \cdots, P(N)) > 0$ such that the following holds for each $i \in \mathscr{S}$:

$$\tilde{A}^\top(i, t) P(i) + P(i) \tilde{A}(i, t) + \tilde{\mathbb{W}}^\top(i) P(i) \tilde{\mathbb{W}}(i) + \sum_{j=1}^{N} \lambda_{ij} P(j) < 0.$$

Using the expression of $\tilde{A}(i, t)$, we get

$$\tilde{A}^\top(i) P(i) + P(i) \tilde{A}(i) + P(i) \Delta \tilde{A}_A(i, t) + \Delta \tilde{A}_A^\top(i, t) P(i)$$

$$+ P(i) \Delta \tilde{B}_B(i, t) + \Delta \tilde{B}_B^\top(i, t) P(i) + P(i) \Delta \tilde{C}_{C_y}(i, t)$$

$$+ \Delta \tilde{C}_{C_y}^\top(i, t) P(i) + \tilde{\mathbb{W}}^\top(i) P(i) \tilde{\mathbb{W}}(i) + \sum_{j=1}^{N} \lambda_{ij} P(j) < 0.$$

Based on Lemma 7 in Appendix A, we have

$$P(i) \Delta \tilde{A}_A(i, t) + \Delta \tilde{A}_A^\top(i, t) P(i)$$

$$\leq \tilde{\varepsilon}_A^{-1}(i) P(i) \tilde{D}_A(i) \tilde{D}_A^\top(i) P(i) + \tilde{\varepsilon}_A(i) \tilde{E}_A^\top(i) \tilde{E}_A(i),$$

$$P(i) \Delta \tilde{B}_B(i, t) + \Delta \tilde{B}_B^\top(i, t) P(i)$$

$$\leq \tilde{\varepsilon}_B^{-1}(i) P(i) \tilde{D}_B(i) \tilde{D}_B^\top(i) P(i) + \tilde{\varepsilon}_B(i) \tilde{E}_B^\top(i) \tilde{E}_B(i),$$

$$P(i) \Delta \tilde{C}_{C_y}(i, t) + \Delta \tilde{C}_{C_y}^\top(i, t) P(i)$$

$$\leq \tilde{\varepsilon}_{C_y}^{-1}(i) P(i) \tilde{D}_{C_y}(i) \tilde{D}_{C_y}^\top(i) P(i) + \tilde{\varepsilon}_{C_y}(i) \tilde{E}_{C_y}^\top(i) \tilde{E}_{C_y}(i).$$

Using this and the Schur complement, the stability condition becomes

$$\begin{bmatrix} \tilde{J}(i) & P(i)\tilde{D}_A(i) & P(i)\tilde{D}_B(i) & P(i)\tilde{D}_{C_y}(i) \\ \tilde{D}_A^\top(i)P(i) & -\tilde{\varepsilon}_A(i)\mathbb{I} & 0 & 0 \\ \tilde{D}_B^\top(i)P(i) & 0 & -\tilde{\varepsilon}_B(i)\mathbb{I} & 0 \\ \tilde{D}_{C_y}^\top(i)P(i) & 0 & 0 & -\tilde{\varepsilon}_{C_y}(i)\mathbb{I} \end{bmatrix} < 0,$$

with

$$\tilde{J}(i) = \tilde{A}^\top(i)P(i) + P(i)\tilde{A}(i) + \tilde{\mathbb{W}}^\top(i)P(i)\tilde{\mathbb{W}}(i)$$
$$+ \sum_{j=1}^{N} \lambda_{ij}P(j) + \tilde{\varepsilon}_A(i)\tilde{E}_A^\top(i)\tilde{E}_A(i)$$
$$+\tilde{\varepsilon}_B(i)\tilde{E}_B^\top(i)\tilde{E}_B(i) + \tilde{\varepsilon}_B(i)\tilde{E}_{C_y}^\top(i)\tilde{E}_{C_y}(i).$$

Again using the Schur complement, we get

$$J(i) + P(i) \begin{bmatrix} \tilde{D}_A(i) & \tilde{D}_B(i) & \tilde{D}_{C_y}(i) \end{bmatrix} \Upsilon^{-1}(i) \begin{bmatrix} \tilde{D}_A^\top(i) \\ \tilde{D}_B^\top(i) \\ \tilde{D}_{C_y}^\top(i) \end{bmatrix} P(i)$$

$$+ \begin{bmatrix} \tilde{E}_A^\top(i) & \tilde{E}_B^\top(i) & \tilde{E}_{C_y}^\top(i) \end{bmatrix} \Upsilon(i) \begin{bmatrix} \tilde{E}_A(i) \\ \tilde{E}_B(i) \\ \tilde{E}_{C_y}(i) \end{bmatrix} < 0,$$

with

$$J(i) = \tilde{A}^\top(i)P(i) + P(i)\tilde{A}(i) + \tilde{\mathbb{W}}^\top(i)P(i)\tilde{\mathbb{W}}(i) + \sum_{j=1}^{N} \lambda_{ij}P(j).$$

This set of coupled matrix inequalities that guarantee the robust stochastic stability is nonlinear in $P(i)$ and the controller gains $K_A(i)$, $K_B(i)$, and $K_C(i)$. To cast it into an LMI form, let us pre- and post-multiply by $U^\top(i)V^\top(i)$ and $V(i)U(i)$, respectively, as we did previously. Before multiplying, notice that $U^\top(i)V^\top(i)J(i)V(i)U(i)$ has already been computed and we do not need to compute it again. For the two other terms we have

$$U^\top(i)V^\top(i) \begin{bmatrix} \tilde{E}_A^\top(i) & \tilde{E}_B^\top(i) & \tilde{E}_{C_y}^\top(i) \end{bmatrix} \Upsilon(i)$$
$$\times \begin{bmatrix} \tilde{E}_A(i) \\ \tilde{E}_B(i) \\ \tilde{E}_{C_y}(i) \end{bmatrix} V(i)U(i)$$
$$= U^\top(i)V^\top(i) \begin{bmatrix} \tilde{\varepsilon}_A(i)\tilde{E}_A^\top(i)\tilde{E}_A(i) \\ +\tilde{\varepsilon}_B(i)\tilde{E}_B^\top(i)\tilde{E}_B(i) \\ +\tilde{\varepsilon}_{C_y}(i)\tilde{E}_{C_y}^\top(i)\tilde{E}_{C_y}(i) \end{bmatrix} V(i)U(i)$$
$$= \tilde{\varepsilon}_A(i) \begin{bmatrix} W^\top(i)E_A^\top(i)E_A(i)W(i) & W^\top(i)E_A^\top(i)E_A(i) \\ E_A^\top(i)E_A(i)W(i) & E_A^\top(i)E_A(i) \end{bmatrix}$$

$$
+\tilde{\varepsilon}_B(i)
\begin{bmatrix}
\begin{bmatrix}
W^\top(i)P_2(i)P_3^{-1}(i)K_C^\top(i)E_B^\top(i) \\
\times E_B(i)K_C(i)P_3^{-1}(i)P_2^\top(i)W(i)
\end{bmatrix} & 0 \\
0 & 0
\end{bmatrix}
$$

$$
+\tilde{\varepsilon}_{C_y}
\begin{bmatrix}
W^\top(i)E_{C_y}^\top(i)E_{C_y}(i)W(i) & W^\top(i)E_{C_y}^\top(i)E_{C_y}(i) \\
E_{C_y}^\top(i)E_{C_y}(i)W(i) & E_{C_y}^\top(i)E_{C_y}(i)
\end{bmatrix}
$$

and

$$
U^\top(i)V^\top(i)P(i)\left[\tilde{D}_A(i)\ \tilde{D}_B(i)\ \tilde{D}_{C_y}(i)\right]\Upsilon^{-1}(i)
$$

$$
\times
\begin{bmatrix}
\tilde{D}_A^\top(i) \\
\tilde{D}_B^\top(i) \\
\tilde{D}_{C_y}^\top(i)
\end{bmatrix}
P(i)V(i)U(i)
$$

$$
= U^\top(i)V^\top(i)
\begin{bmatrix}
\tilde{\varepsilon}_A^{-1}(i)P(i)\tilde{D}_A(i)\tilde{D}_A^\top(i)P(i) \\
\tilde{\varepsilon}_B^{-1}(i)P(i)\tilde{D}_B(i)\tilde{D}_B^\top(i)P(i) \\
\tilde{\varepsilon}_{C_y}^{-1}(i)P(i)\tilde{D}_{C_y}(i)\tilde{D}_{C_y}^\top(i)P(i)
\end{bmatrix}
V(i)U(i).
$$

To compute the expression of this term, notice that for the first term we have

$$
U^\top(i)V^\top(i)\left[P(i)\tilde{D}_A(i)\tilde{D}_A^\top(i)P(i)\right]V(i)U(i)
$$

$$
= U^\top(i)V^\top(i)
\begin{bmatrix}
P_1(i)D_A(i)D_A^\top(i)P_1(i) \\
P_2^\top(i)D_A(i)D_A^\top(i)P_1(i)
\end{bmatrix}
$$

$$
\begin{matrix}
P_1(i)D_A(i)D_A^\top(i)P_2(i) \\
P_1(i)D_A(i)D_A^\top(i)P_2(i)
\end{matrix}
\Bigg]
V(i)U(i)
$$

$$
=
\begin{bmatrix}
W^\top(i) & -W^\top(i)P_2(i)P_3^{-1}(i) \\
\mathbb{I} & 0
\end{bmatrix}
$$

$$
\begin{bmatrix}
P_1(i)D_A(i)D_A^\top(i)P_1(i) & P_1(i)D_A(i)D_A^\top(i)P_2(i) \\
P_2^\top(i)D_A(i)D_A^\top(i)P_1(i) & P_2^\top(i)D_A(i)D_A^\top(i)P_2(i)
\end{bmatrix}
$$

$$
\begin{bmatrix}
W(i) & \mathbb{I} \\
-P_3^{-1}(i)P_2^\top(i)W(i) & 0
\end{bmatrix}
$$

$$
=
\begin{bmatrix}
\begin{bmatrix}
W^\top(i)P_1(i)D_A(i)D_A^\top(i) \\
\times\left[P_1(i) - P_2(i)P_3^{-1}(i)P_2^\top(i)\right]W(i) \\
-W^\top(i)P_2(i)P_3^{-1}(i)P_2^\top(i)D_A(i) \\
\times D_A^\top(i)\left[P_1(i) - P_2(i)P_3^{-1}(i)P_2^\top(i)\right]W(i)
\end{bmatrix}
\\
P_1(i)D_A(i)D_A^\top(i)\left[P_1(i) - P_2(i)P_3^{-1}(i)P_2^\top(i)\right]W(i)
\end{bmatrix}
$$

$$
\begin{bmatrix}
W^\top(i)\left[P_1(i) - P_2(i)P_3^{-1}(i)P_2^\top(i)\right]D_A(i)D_A^\top(i)P_1(i) \\
\\
\\
P_1(i)D_A(i)D_A^\top(i)P_1(i)
\end{bmatrix}
$$

$$= \begin{bmatrix} D_A(i)D_A^\top(i) & D_A(i)D_A^\top(i)P_1(i) \\ P_1(i)D_A(i)D_A^\top(i) & P_1(i)D_A(i)D_A^\top(i)P_1(i) \end{bmatrix}.$$

For the second term, we have

$$U^\top(i)V^\top(i) \begin{bmatrix} P_1(i)D_B(i)D_B^\top(i)P_1(i) \\ P_2^\top(i)D_B(i)D_B^\top(i)P_1(i) \end{bmatrix}$$

$$\begin{bmatrix} P_1(i)D_B(i)D_B^\top(i)P_2(i) \\ P_2(i)D_B(i)D_B^\top(i)P_2(i) \end{bmatrix} V(i)U(i)$$

$$= \begin{bmatrix} W^\top(i) & -W^\top(i)P_2(i)P_3^{-1}(i) \\ \mathbb{I} & 0 \end{bmatrix}$$

$$\begin{bmatrix} P_1(i)D_B(i)D_B^\top(i)P_1(i) & P_1(i)D_B(i)D_B^\top(i)P_2(i) \\ P_2^\top(i)D_B(i)D_B^\top(i)P_1(i) & P_2^\top(i)D_B(i)D_B^\top(i)P_2(i) \end{bmatrix}$$

$$\begin{bmatrix} W(i) & \mathbb{I} \\ -P_3^{-1}(i)P_2^\top(i)W(i) & 0 \end{bmatrix}$$

$$= \begin{bmatrix} \begin{bmatrix} W^\top(i)P_1(i)D_B(i)D_B^\top(i) \\ \times \left[P_1(i) - P_2(i)P_3^{-1}(i)P_2^\top(i) \right] W(i) \\ W^\top(i)P_2(i)P_3^{-1}(i)P_2^\top(i)D_B(i) \\ \times D_B^\top(i) \left[P_1(i) - P_2(i)P_3^{-1}(i)P_2^\top(i) \right] W(i) \end{bmatrix} \end{bmatrix}$$

$$P_1(i)D_B(i)D_B^\top(i) \left[P_1(i) - P_2(i)P_3^{-1}(i)P_2^\top(i) \right] W(i)$$

$$W^\top(i) \left[P_1(i) - P_2(i)P_3^{-1}(i)P_2^\top(i) \right] D_B(i)D_B^\top(i)P_1(i)$$

$$P_1(i)D_B(i)D_B^\top(i)P_1(i)$$

$$= \begin{bmatrix} D_B(i)D_B^\top(i) & D_B(i)D_B^\top(i)P_1(i) \\ P_1(i)D_B(i)D_B^\top(i) & P_1(i)D_B(i)D_B^\top(i)P_1(i) \end{bmatrix},$$

and for the third term, we have

$$U^\top(i)V^\top(i)$$

$$\times \begin{bmatrix} P_2(i)K_B(i)D_{C_y}(i)D_{C_y}^\top(i)K_B^\top(i)P_2^\top(i) \\ P_2(i)K_B(i)D_{C_y}(i)D_{C_y}^\top(i)K_B^\top(i)P_2^\top(i) \end{bmatrix}$$

$$\begin{bmatrix} P_2(i)K_B(i)D_{C_y}(i)D_{C_y}^\top(i)K_B^\top(i)P_3(i) \\ P_3(i)K_B(i)D_{C_y}(i)D_{C_y}^\top(i)K_B^\top(i)P_3(i) \end{bmatrix}$$

$$\times V(i)U(i)$$

$$= \begin{bmatrix} w^\top(i) & -w^\top(i)P_2(i)P_3^{-1}(i) \\ \mathbb{I} & 0 \end{bmatrix}$$

$$\begin{bmatrix} P_2(i)K_B(i)D_{C_y}(i)D_{C_y}^\top(i)K_B^\top(i)P_2^\top(i) \\ P_3(i)K_B(i)D_{C_y}(i)D_{C_y}^\top(i)K_B^\top(i)P_2^\top(i) \end{bmatrix}$$

$$\left. \begin{array}{cc} P_2(i)K_B(i)D_{C_y}(i)D_{C_y}^\top(i)K_B^\top(i)P_3(i) \\ P_3(i)K_B(i)D_{C_y}(i)D_{C_y}^\top(i)K_B^\top(i)P_3(i) \end{array} \right]$$

$$\left[\begin{array}{cc} w(i) & \mathbb{I} \\ -P_3^{-1}(i)P_2^\top(i)w(i) & 0 \end{array} \right]$$

$$= \left[\begin{array}{cc} 0 & 0 \\ 0 & P_2(i)K_B(i)D_{C_y}(i)D_{C_y}^\top(i)K_B^\top(i)P_2^\top(i) \end{array} \right].$$

Taking into account all these computations, we have

$$U^\top(i)V^\top(i) \left[\begin{array}{c} \tilde{\varepsilon}_A^{-1}(i)P(i)\tilde{D}_A(i)\tilde{D}_A^\top(i)P(i) \\ \tilde{\varepsilon}_B^{-1}(i)P(i)\tilde{D}_B(i)\tilde{D}_B^\top(i)P(i) \\ \tilde{\varepsilon}_{C_y}^{-1}(i)P(i)\tilde{D}_{C_y}(i)\tilde{D}_{C_y}^\top(i)P(i) \end{array} \right] V(i)U(i)$$

$$= \tilde{\varepsilon}_A^{-1}(i) \left[\begin{array}{cc} D_A(i)D_A^\top(i) & D_A(i)D_A^\top(i)P_1(i) \\ P_1(i)D_A(i)D_A^\top(i) & P_1(i)D_A(i)D_A^\top(i)P_1(i) \end{array} \right]$$

$$+ \tilde{\varepsilon}_B^{-1}(i) \left[\begin{array}{cc} D_B(i)D_B^\top(i) & D_B(i)D_B^\top(i)P_1(i) \\ P_1(i)D_B(i)D_B^\top(i) & P_1(i)D_B(i)D_B^\top(i)P_1(i) \end{array} \right]$$

$$+ \tilde{\varepsilon}_{C_y}^{-1}(i) \left[\begin{array}{cc} 0 & 0 \\ 0 & P_2(i)K_B(i)D_{C_y}(i)D_{C_y}^\top(i)K_B^\top(i)P_2^\top(i) \end{array} \right].$$

Using all these transformations, the previous stochastic stability condition for the extended dynamics become

$$\left[\begin{array}{cc} \widehat{\mathcal{M}_1(i)} & \mathcal{M}_2(i) \\ \mathcal{M}_2^\top(i) & \mathcal{M}_3(i) \end{array} \right]$$

$$+ \tilde{\varepsilon}_A^{-1}(i) \left[\begin{array}{cc} D_A(i)D_A^\top(i) & D_A(i)D_A^\top(i)P_1(i) \\ P_1(i)D_A(i)D_A^\top(i) & P_1(i)D_A(i)D_A^\top(i)P_1(i) \end{array} \right]$$

$$+ \tilde{\varepsilon}_B^{-1}(i) \left[\begin{array}{cc} D_B(i)D_B^\top(i) & D_B(i)D_B^\top(i)P_1(i) \\ P_1(i)D_B(i)D_B^\top(i) & P_1(i)D_B(i)D_B^\top(i)P_1(i) \end{array} \right]$$

$$+ \tilde{\varepsilon}_{C_y}^{-1}(i) \left[\begin{array}{cc} 0 & 0 \\ 0 & P_2(i)K_B(i)D_{C_y}(i)D_{C_y}^\top(i)K_B^\top(i)P_2^\top(i) \end{array} \right]$$

$$+ \tilde{\varepsilon}_A(i) \left[\begin{array}{cc} W^\top(i)E_A^\top(i)E_A(i)W(i) & W^\top(i)E_A^\top(i)E_A(i) \\ E_A^\top(i)E_A(i)W(i) & E_A^\top(i)E_A(i) \end{array} \right]$$

$$+ \tilde{\varepsilon}_B(i) \left[\begin{array}{cc} \left[\begin{array}{c} W^\top(i)P_2(i)P_3^{-1}(i)K_C^\top(i)E_B^\top(i) \\ \times E_B(i)K_C(i)P_3^{-1}(i)P_2^\top(i)W(i) \end{array} \right] & 0 \\ 0 & 0 \end{array} \right]$$

$$+ \tilde{\varepsilon}_{C_y}(i) \left[\begin{array}{cc} W^\top(i)E_{C_y}^\top(i)E_{C_y}(i)W(i) & W^\top(i)E_{C_y}^\top(i)E_{C_y}(i) \\ E_{C_y}^\top(i)E_{C_y}(i)W(i) & E_{C_y}^\top(i)E_{C_y}(i) \end{array} \right] < 0,$$

with

$$\mathcal{M}_1(i) = A(i)W(i) + W^\top(i)A^\top(i) - B(i)K_C(i)P_3^{-1}(i)P_2^\top(i)W(i)$$

$$-W^\top(i)P_2(i)P_3^{-1}(i)K_C^\top(i)B^\top(i),$$

$$\widehat{\mathcal{M}}_1(i) = \mathcal{M}_1(i) + \sum_{j=1}^{N} \lambda_{ij} W^\top(i) \left[P_2(i)P_3^{-1}(i)P_3(j) - P_2^\top(j) \right]$$
$$\times P_3^{-1}(j) \left[P_2(i)P_3^{-1}(i)P_3(j) - P_2^\top(j) \right] W(i)$$
$$+ W^\top(i)\mathbb{W}^\top(i)P_1(i)\mathbb{W}(i)W(i).$$

The last inequality can be rewritten as follows:

$$\begin{bmatrix} \widehat{H_1(i)} & H_2(i) \\ H_2^\top(i) & H_3(i) \end{bmatrix} < 0,$$

with

$$\widehat{H_1(i)} = \widehat{\mathcal{M}_1(i)} + \tilde{\varepsilon}_A^{-1}(i)D_A(i)D_A^\top(i) + \tilde{\varepsilon}_B^{-1}(i)D_B(i)D_B^\top(i)$$
$$+ \tilde{\varepsilon}_A W^\top(i)E_A^\top(i)E_A(i)W(i)$$
$$+ \tilde{\varepsilon}_B W^\top(i)P_2(i)P_3^{-1}(i)K_C^\top(i)E_B^\top(i)E_B(i)K_C(i)P_3^{-1}(i)P_2^\top(i)W(i)$$
$$+ \tilde{\varepsilon}_{C_y}(i)W^\top(i)E_{C_y}^\top(i)E_{C_y}(i)W(i),$$

$$H_2(i) = \mathcal{M}_2(i) + \tilde{\varepsilon}_A^{-1}(i)D_A(i)D_A^\top(i)P_1(i) + \tilde{\varepsilon}_B^{-1}(i)D_B(i)D_B^\top(i)P_1(i)$$
$$+ \tilde{\varepsilon}_A(i)W^\top(i)E_A^\top(i)E_A(i) + \tilde{\varepsilon}_{C_y}(i)W^\top(i)E_{C_y}^\top(i)E_{C_y}(i)$$
$$+ W^\top(i)\mathbb{W}^\top(i)P_1(i)\mathbb{W}(i),$$

$$H_3(i) = \mathcal{M}_3(i) + \tilde{\varepsilon}_A^{-1}(i)P_1(i)D_A(i)D_A^\top(i)P_1(i)$$
$$+ \tilde{\varepsilon}_B^{-1}(i)P_1(i)D_B(i)D_B^\top(i)P_1(i)$$
$$+ \tilde{\varepsilon}_{C_y}^{-1}(i)P_2(i)K_B(i)D_{C_y}(i)D_{C_y}^\top(i)K_B(i)P_2(i)$$
$$+ \tilde{\varepsilon}_A(i)E_A^\top(i)E_A(i) + \tilde{\varepsilon}_{C_y}(i)E_{C_y}^\top(i)E_{C_y}(i) + \mathbb{W}^\top(i)P_1(i)\mathbb{W}(i),$$

where $\widehat{\mathcal{M}_1(i)}$, $\mathcal{M}_1(i)$, and $\mathcal{M}_3(i)$ keep the same definition as before.

If we define $\mathcal{K}_B(i)$ and $\mathcal{K}_C(i)$ as we did previously, that is,

$$P(i) = \begin{bmatrix} X(i) & Y^{-1}(i) - X(i) \\ Y^{-1}(i) - X(i) & X(i) - Y^{-1}(i) \end{bmatrix},$$
$$\mathcal{K}_B(i) = P_2(i)K_B(i) = \left[Y^{-1}(i) - X(i) \right] K_B(i),$$
$$\mathcal{K}_C(i) = -K_C(i)P_3^{-1}(i)P_2^\top(i)W(i) = K_C(i)Y(i),$$

and we follow the same steps as before after choosing the controller gains as follows:

$$
\left\{
\begin{aligned}
K_A(i) &= \left[X(i) - Y^{-1}(i)\right]^{-1}\left[A^\top(i) + X(i)A(i)Y(i)\right. \\
&\quad + X(i)B(i)\mathcal{K}_C(i) + \mathcal{K}_B(i)C_y(i)Y(i) + \mathbb{W}^\top(i)X(i)\mathbb{W}(i)Y(i) \\
&\quad + \sum_{j=1}^{N}\lambda_{ij}Y^{-1}(j)Y(i) \\
&\quad + \tilde{\varepsilon}_A^{-1}(i)X(i)D_A(i)D_A^\top(i) + \tilde{\varepsilon}_B^{-1}(i)X(i)D_B(i)D_B^\top(i) \\
&\quad + \tilde{\varepsilon}_A(i)E_A^\top(i)E_A(i)Y(i) \\
&\quad \left. + \tilde{\varepsilon}_{C_y}(i)E_{C_y}^\top(i)E_{C_y}(i)Y(i)\right]Y^{-1}(i), \\
K_B(i) &= \left[Y^{-1}(i) - X(i)\right]^{-1}\mathcal{K}_B(i), \\
K_C(i) &= \mathcal{K}_C(i)Y^{-1}(i),
\end{aligned}
\right.
$$

we get

$$
\mathcal{H}_1(i) < 0,
$$
$$
\mathcal{H}_2(i) < 0,
$$

with

$$
\begin{aligned}
\mathcal{H}_1(i) =\ & A(i)Y(i) + Y^\top(i)A^\top(i) + B(i)\mathcal{K}_C(i) + \mathcal{K}_C^\top(i)B^\top(i) \\
& + \sum_{j=1}^{N}\lambda_{ij}Y(i)Y^{-1}(j)Y(i) + \tilde{\varepsilon}_A^{-1}(i)D_A(i)D_A^\top(i) \\
& + \tilde{\varepsilon}_B^{-1}(i)D_B(i)D_B^\top(i) + \tilde{\varepsilon}_A Y^\top(i)E_A^\top(i)E_A(i)Y(i) \\
& + \tilde{\varepsilon}_B \mathcal{K}_C^\top(i)E_B^\top(i)E_B(i)\mathcal{K}_C(i) \\
& + \tilde{\varepsilon}_{C_y}(i)Y^\top(i)E_{C_y}^\top(i)E_{C_y}(i)Y(i), \\
\mathcal{H}_2(i) =\ & X(i)A(i) + A^\top(i)X(i) + \mathcal{K}_B(i)C_y(i) + \mathcal{K}_B^\top(i)C_y^\top(i) \\
& + \sum_{j=1}^{N}\lambda_{ij}X(j) + \mathbb{W}^\top(i)X(i)\mathbb{W}(i) \\
& + \tilde{\varepsilon}_A^{-1}(i)X(i)D_A(i)D_A^\top(i)X(i) + \tilde{\varepsilon}_B^{-1}(i)X(i)D_B(i)D_B^\top(i)X(i) \\
& + \tilde{\varepsilon}_{C_y}^{-1}(i)\mathcal{K}_B(i)D_{C_y}(i)D_{C_y}^\top(i)\mathcal{K}_B^\top(i) + \tilde{\varepsilon}_A(i)E_A^\top(i)E_A(i) \\
& + \tilde{\varepsilon}_{C_y}(i)E_{C_y}^\top(i)E_{C_y}(i).
\end{aligned}
$$

Using the fact that

$$
\sum_{j=1}^{N}\lambda_{ij}Y(i)Y^{-1}(j)Y(i) = \lambda_{ii}Y(i) + \mathcal{S}_i(Y)\mathcal{Y}_i^{-1}(Y)\mathcal{S}_i^\top(Y),
$$

where $\mathcal{S}_i(Y)$ and $\mathcal{Y}_i(Y)$ keep the same definitions as before, these two matrix inequalities are equivalent to the following LMIs:

$$
\begin{bmatrix}
J_{\mathcal{H}_1}(i) & Y^\top(i)E_A^\top(i) & \mathcal{K}_C^\top(i)E_B^\top(i) \\
E_A(i)Y(i) & -\tilde{\varepsilon}_A^{-1}(i)\mathbb{I} & 0 \\
E_B(i)\mathcal{K}_C(i) & 0 & -\tilde{\varepsilon}_B^{-1}(i)\mathbb{I} \\
E_{C_y}(i)Y(i) & 0 & 0 \\
\mathcal{S}_i^\top(Y) & 0 & 0
\end{bmatrix}
$$

$$\left.\begin{array}{cc} Y^{\top}(i)E_{C_y}^{\top}(i) & S_i(Y) \\ 0 & 0 \\ 0 & 0 \\ \tilde{\varepsilon}_{C_y}^{-1}(i)\mathbb{I} & 0 \\ 0 & -\mathcal{Y}_i(Y) \end{array}\right] < 0,$$

$$\begin{bmatrix} J_{\mathcal{H}_2}(i) & X(i)D_A(i) & X(i)D_B(i) & \mathcal{K}_B(i)D_{C_y}(i) \\ D_A^{\top}(i)X(i) & -\tilde{\varepsilon}_A(i)\mathbb{I} & 0 & 0 \\ D_B^{\top}(i)X(i) & 0 & -\tilde{\varepsilon}_B(i)\mathbb{I} & 0 \\ D_{C_y}^{\top}(i)\mathcal{K}_B^{\top}(i) & 0 & 0 & -\tilde{\varepsilon}_{C_y}(i)\mathbb{I} \end{bmatrix} < 0,$$

with

$$\begin{aligned} J_{\mathcal{H}_1}(i) =\ & A(i)Y(i) + Y^{\top}(i)A^{\top}(i) + B(i)\mathcal{K}_C(i) + \mathcal{K}_C^{\top}(i)B^{\top}(i) \\ & + \lambda_{ii}Y(i) + \tilde{\varepsilon}_A^{-1}(i)D_A(i)D_A^{\top}(i) + \tilde{\varepsilon}_B^{-1}(i)D_B(i)D_B^{\top}(i), \\ J_{\mathcal{H}_2}(i) =\ & X(i)A(i) + A^{\top}(i)X(i) + \mathcal{K}_B(i)C_y(i) + \mathcal{K}_B^{\top}(i)C_y^{\top}(i) \\ & + \sum_{j=1}^{N} \lambda_{ij}X(j) + \mathbb{W}^{\top}(i)X(i)\mathbb{W}(i) + \tilde{\varepsilon}_A(i)E_A^{\top}(i)E_A(i) \\ & + \tilde{\varepsilon}_{C_y}(i)E_{C_y}^{\top}(i)E_{C_y}(i). \end{aligned}$$

The results of all these developments are summarized by the following theorem.

Theorem 27. *Let $\tilde{\varepsilon}_A = (\tilde{\varepsilon}_A(1), \cdots, \tilde{\varepsilon}_A(N))$, $\tilde{\varepsilon}_B = (\tilde{\varepsilon}_B(1), \cdots, \tilde{\varepsilon}_B(N))$, and $\tilde{\varepsilon}_{C_y} = (\tilde{\varepsilon}_{C_y}(1), \cdots, \tilde{\varepsilon}_{C_y}(N))$ be a given set of positive scalars. System (3.51) is stochastically stable if and only if for every $i \in \mathscr{S}$, the following set of coupled LMIs is feasible for some symmetric and positive-definite matrices $X = (X(1), \cdots, X(N)) > 0$, and $Y = (Y(1), \cdots, Y(N)) > 0$, and matrices $\mathcal{K}_B = (\mathcal{K}_B(1), \cdots, \mathcal{K}_B(N))$, and $\mathcal{K}_C = (\mathcal{K}_C(1), \cdots, \mathcal{K}_C(N))$:*

$$\begin{bmatrix} J_{\mathcal{H}_1}(i) & Y^{\top}(i)E_A^{\top}(i) & \mathcal{K}_C^{\top}(i)E_B^{\top}(i) \\ E_A(i)Y(i) & -\tilde{\varepsilon}_A^{-1}(i)\mathbb{I} & 0 \\ E_B(i)\mathcal{K}_C(i) & 0 & -\tilde{\varepsilon}_B^{-1}(i)\mathbb{I} \\ E_{C_y}(i)Y(i) & 0 & 0 \\ S_i^{\top}(Y) & 0 & 0 \end{bmatrix}$$

$$\left.\begin{array}{cc} Y^{\top}(i)E_{C_y}^{\top}(i) & S_i(Y) \\ 0 & 0 \\ 0 & 0 \\ \tilde{\varepsilon}_{C_y}^{-1}(i)\mathbb{I} & 0 \\ 0 & -\mathcal{Y}_i(Y) \end{array}\right] < 0, \qquad (3.60)$$

$$\begin{bmatrix} J_{\mathcal{H}_2}(i) & X(i)D_A(i) & X(i)D_B(i) & \mathcal{K}_B(i)D_{C_y}(i) \\ D_A^{\top}(i)X(i) & -\tilde{\varepsilon}_A(i)\mathbb{I} & 0 & 0 \\ D_B^{\top}(i)X(i) & 0 & -\tilde{\varepsilon}_B(i)\mathbb{I} & 0 \\ D_{C_y}^{\top}(i)\mathcal{K}_B^{\top}(i) & 0 & 0 & -\tilde{\varepsilon}_{C_y}(i)\mathbb{I} \end{bmatrix} < 0, \qquad (3.61)$$

$$\begin{bmatrix} Y(i) & \mathbb{I} \\ \mathbb{I} & X(i) \end{bmatrix} > 0, \qquad (3.62)$$

with

$$\mathcal{S}_i(Y) = \left[\sqrt{\lambda_{i1}}Y(i), \cdots, \sqrt{\lambda_{ii-1}}Y(i), \sqrt{\lambda_{ii+1}}Y(i), \cdots, \sqrt{\lambda_{iN}}Y(i) \right],$$
$$\mathcal{Y}_i(Y) = \text{diag}\left[Y(1), \cdots, Y(i-1), Y(i+1), \cdots, Y(N) \right].$$

Furthermore, the dynamic output feedback controller is given by

$$\begin{cases} K_A(i) = \left[X(i) - Y^{-1}(i) \right]^{-1} \left[A^\top(i) + X(i)A(i)Y(i) \right. \\ \qquad\quad + X(i)B(i)\mathcal{K}_C(i) + \mathcal{K}_B(i)C_y(i)Y(i) + \mathbb{W}^\top(i)X(i)\mathbb{W}(i)Y(i) \\ \qquad\quad + \sum_{j=1}^{N} \lambda_{ij}Y^{-1}(j)Y(i) \\ \qquad\quad + \tilde{\varepsilon}_A^{-1}(i)X(i)D_A(i)D_A^\top(i) + \tilde{\varepsilon}_B^{-1}(i)X(i)D_B(i)D_B^\top(i) \\ \qquad\quad + \tilde{\varepsilon}_A(i)E_A^\top(i)E_A(i)Y(i) + \tilde{\varepsilon}_{C_y}(i)E_{C_y}^\top(i)E_{C_y}(i)Y(i) \left. \right] Y^{-1}(i), \\ K_B(i) = \left[Y^{-1}(i) - X(i) \right]^{-1} \mathcal{K}_B(i), \\ K_C(i) = \mathcal{K}_C(i)Y^{-1}(i). \end{cases} \qquad (3.63)$$

Example 36. To illustrate the theoretical results of this theorem, let us consider the two-mode system of Example 34 with

$$\mathbb{W}(1) = \begin{bmatrix} 0.10 & 0.0 \\ 0.0 & 0.10 \end{bmatrix}, \quad \mathbb{W}(2) = \begin{bmatrix} 0.20 & 0.0 \\ 0.0 & 0.20 \end{bmatrix}.$$

Let us fix the positive scalars $\varepsilon_A(i)$, $\varepsilon_B(i)$, and $\varepsilon_{C_y}(i)$, $i = 1,2$ to the following values:

$$\varepsilon_A(1) = \varepsilon_A(2) = 0.50,$$
$$\varepsilon_B(1) = \varepsilon_B(2) = 0.10,$$
$$\varepsilon_{C_y}(1) = \varepsilon_{C_y}(2) = 0.10.$$

Solving the LMIs (3.60)–(3.62), we get

$$X(1) = \begin{bmatrix} 2.8024 & -1.0797 \\ -1.0797 & 0.9926 \end{bmatrix}, \qquad X(2) = \begin{bmatrix} 11.6193 & -5.7986 \\ -5.7986 & 3.3546 \end{bmatrix},$$

$$Y(1) = \begin{bmatrix} 65.7939 & -37.3834 \\ -37.3834 & 116.1613 \end{bmatrix}, \qquad Y(2) = \begin{bmatrix} 85.9922 & -29.9200 \\ -29.9200 & 81.5468 \end{bmatrix},$$

$$\mathcal{K}_B(1) = \begin{bmatrix} -119.9443 & 59.3966 \\ 61.0279 & -31.6167 \end{bmatrix}, \qquad \mathcal{K}_B(2) = \begin{bmatrix} -44.4274 & 55.3544 \\ 58.9674 & -79.5007 \end{bmatrix},$$

$$\mathcal{K}_C(1) = \begin{bmatrix} -197.9419 & 69.3903 \\ 90.4709 & -292.3150 \end{bmatrix}, \qquad \mathcal{K}_C(2) = \begin{bmatrix} -140.8357 & -59.2915 \\ 7.4807 & -96.5781 \end{bmatrix}.$$

The corresponding controller gains are given by

$$K_A(1) = \begin{bmatrix} -35.4105 & 14.4776 \\ 25.5308 & -16.6710 \end{bmatrix}, \qquad K_B(1) = \begin{bmatrix} 33.1375 & -15.4353 \\ -25.5087 & 15.1304 \end{bmatrix},$$

$$K_C(1) = \begin{bmatrix} -3.2664 & -0.4538 \\ -0.0670 & -2.5380 \end{bmatrix}, \qquad K_A(2) = \begin{bmatrix} 35.7013 & -55.3475 \\ 83.4115 & -119.7534 \end{bmatrix},$$

$$K_B(2) = \begin{bmatrix} -38.0696 & 54.3072 \\ -83.7884 & 118.1440 \end{bmatrix},$$

$$K_C(2) = \begin{bmatrix} -2.1675 & -1.5223 \\ -0.3727 & -1.3211 \end{bmatrix}.$$

Based on the results of this theorem, the system of this example is stochastically stable under the output feedback controller with the computed gains.

3.5 Observer-Based Output Stabilization

As mentioned earlier, observer-based output feedback control is one of the alternatives to control the class of systems under consideration. Our goal in this section is to focus on the design of such a controller. We will restrict ourselves to the LMI design approach. Note that other techniques exist, mainly based on the Riccati-like equation approach.

The controller we use in this section is given by

$$\begin{cases} \dot{\nu}(t) = A(r(t))\nu(t) + B(r(t))u(t) + L(r(t))\left[C_y(r(t))\nu(t) - y(t)\right], \\ u(t) = K(r(t))\nu(t), \end{cases} \tag{3.64}$$

where $\nu(t)$ is the observer state vector, and $K(i)$ and $L(i)$, $i \in \mathscr{S}$ are constant gain matrices that have to be determined and constitute one of our main goals in this section.

Let us now assume that the controller (3.64) exists and show that it stochastically stabilizes the class of systems (3.39). For this purpose, let us define the observer error by

$$e(t) = x(t) - \nu(t). \tag{3.65}$$

Combining the nominal system dynamics and the controller dynamics we get

$$\begin{aligned} \dot{e}(t) = \ & A(r(t))x(t) + B(r(t))K(r(t))\nu(t) - A(r(t))\nu(t) - B(r(t))K(r(t))\nu(t) \\ & - L(r(t))\left[C_y(r(t))\nu(t) - y(t)\right] \\ = \ & A(r(t))\left[x(t) - \nu(t)\right] + L(r(t))C_y(r(t))\left[x(t) - \nu(t)\right] \\ = \ & \left[A(r(t)) + L(r(t))C_y(r(t))\right]e(t). \end{aligned}$$

Using the system dynamics and the error dynamics, we get the following:

$$\begin{bmatrix} \dot{x}(t) \\ \dot{e}(t) \end{bmatrix} = \begin{bmatrix} A(r(t)) + B(r(t))K(r(t)) & -B(r(t))K(r(t)) \\ 0 & A(r(t)) + L(r(t))C_y(r(t)) \end{bmatrix}$$
$$\times \begin{bmatrix} x(t) \\ e(t) \end{bmatrix}.$$

The block diagram of the closed-loop system under the observer-based output feedback control is represented by Figure 3.3.

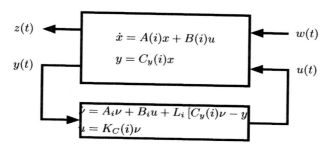

Fig. 3.3. Observer-based output feedback stabilization block diagram (nominal system).

The following theorem states the stability result of the closed loop of the class of systems we are considering under output feedback control.

Theorem 28. Let $K = (K(1), \cdots, K(N))$ and $L = (L(1), \cdots, L(N))$ be given sets of constant matrices. If there exist sets of symmetric and positive-definite matrices $P = (P(1), \cdots, P(N)) > 0$ and $Q = (Q(1), \cdots, Q(N)) > 0$ satisfying the following set of coupled LMIs for every $i \in \mathscr{S}$:

$$\begin{bmatrix} \mathscr{J}_P & P(i)B(i) \\ B^\top(i)P(i) & -\mathbb{I} \end{bmatrix} < 0,$$

$$\begin{bmatrix} \mathscr{J}_Q & K^\top(i) \\ K(i) & -\mathbb{I} \end{bmatrix} < 0,$$

where

$$\mathscr{J}_P = A^\top(i)P(i) + P(i)A(i) + K^\top(i)B^\top(i)P(i) + P(i)B(i)K(i)$$
$$+ \sum_{j=1}^{N} \lambda_{ij} P(j),$$

$$\mathscr{J}_Q = A^\top(i)Q(i) + Q(i)A(i) + Q(i)L(i)C_y(i) + C_y^\top(i)L^\top(i)Q(i)$$
$$+ \sum_{j=1}^{N} \lambda_{ij} Q(j),$$

then (3.39) is stochastically stable.

Proof: Let us consider the Lyapunov functional candidate with the following form:

$$V(x(t), e(t), r(t) = i) = x^\top(t)P(i)x(t) + e^\top(t)Q(i)e(t)$$

$$= \begin{bmatrix} x^\top(t) & e^\top(t) \end{bmatrix} \begin{bmatrix} P(i) & 0 \\ 0 & Q(i) \end{bmatrix} \begin{bmatrix} x(t) \\ e(t) \end{bmatrix},$$

where $P(i) > 0$ and $Q(i) > 0$ are symmetric and positive-definite matrices.

Let $\mathscr{L}$ be the infinitesimal generator of the process $\{(x(t), e(t), r(t)), t \geq 0\}$. Then, the expression of the infinitesimal operator acting on $V(.)$ and emanating from the point (x, i) at time t, where $x(t) = x$ and $r(t) = i$ for $i \in \mathscr{S}$, is given by

$$\mathscr{L}V(x(t), e(t), i) = \dot{x}^\top(t)P(i)x(t) + x^\top(t)P(i)\dot{x}(t) + \dot{e}^\top(t)Q(i)e(t)$$

$$+ e^\top(t)Q(i)\dot{e}(t) + \sum_{j=1}^{N} \lambda_{ij} \left[x^\top(t)P(j)x(t) + e^\top(t)Q(j)e(t) \right]$$

$$= \left[[A(i) + B(i)K(i)]x(t) - B(i)K(i)e(t) \right]^\top P(i)x(t)$$

$$+ x^\top(t)P(i) \left[[A(i) + B(i)K(i)]x(t) - B(i)K(i)e(t) \right]$$

$$+ \left[[A(i) + L(i)C_y(i)]e(t) \right]^\top Q(i)e(t) + e^\top(t)Q(i) \left[A(i) + L(i)C_y(i) \right]e(t)$$

$$+ \sum_{j=1}^{N} \lambda_{ij} \left[x^\top(t)P(j)x(t) + e^\top(t)Q(j)e(t) \right]$$

$$= x^\top(t) \left[A^\top(i)P(i) + P(i)A(i) + K^\top(i)B^\top(i)P(i) + P(i)B(i)K(i) \right.$$

$$\left. + \sum_{j=1}^{N} \lambda_{ij}P(j) \right]x(t) - 2x^\top(t)P(i)B(i)K(i)e(t)$$

$$+ e^\top(t) \left[A^\top(i)Q(i) + Q(i)A(i) + Q(i)L(i)C_y(i) + C_y^\top(i)L^\top(i)Q(i) \right.$$

$$\left. + \sum_{j=1}^{N} \lambda_{ij}Q(j) \right]e(t).$$

Note from Lemma 2 (Appendix A) that

$$-2x^\top(t)P(i)B(i)K(i)e(t) \leq x^\top(t)P(i)B(i)B^\top(i)P(i)x(t)$$

$$+ e^\top(t)K^\top(i)K(i)e(t),$$

we get

$$\mathscr{L}V(x(t), e(t), i) \leq x^\top(t) \left[A^\top(i)P(i) + P(i)A(i) + K^\top(i)B^\top(i)P(i) \right.$$

$$+ P(i)B(i)K(i) + P(i)B(i)B^\top(i)P(i) + \sum_{j=1}^{N} \lambda_{ij}P(j)\Big]x(t)$$

$$+ e^\top(t)\Big[A^\top(i)Q(i) + Q(i)A(i) + Q(i)L(i)C_y(i)$$

$$+ C_y^\top(i)L^\top(i)Q(i) + K^\top(i)K(i) + \sum_{j=1}^{N} \lambda_{ij}Q(j)\Big]e(t) = \big[\,x^\top(t)\ e^\top(t)\,\big]$$

$$\times \begin{bmatrix} \mathscr{J}_P(i) + P(i)B(i)B^\top(i)P(i) & 0 \\ 0 & \mathscr{J}_Q(i) + K^\top(i)K(i) \end{bmatrix} \begin{bmatrix} x(t) \\ e(t) \end{bmatrix}$$

$$= \big[\,x^\top(t)\ e^\top(t)\,\big]\, \Gamma(i) \begin{bmatrix} x(t) \\ e(t) \end{bmatrix}.$$

Therefore, since $\Gamma(i) < 0$ for all $i \in \mathscr{S}$, we obtain

$$\mathscr{L}V(x(t), e(t), i) \leq -\min_{j \in \mathscr{S}}\{\lambda_{min}[-\Gamma(j)]\}\,\big[\,x^\top(t)\ e^\top(t)\,\big] \begin{bmatrix} x(t) \\ e(t) \end{bmatrix}.$$

Combining this with the Dynkin's formula, we get

$$\mathbb{E}\,[V(x(t), e(t), i)] - \mathbb{E}\,[V(x(0), e(0), r_0)]$$

$$= \mathbb{E}\left[\int_0^t \mathscr{L}V(x(s), e(s), r(s))ds|(r_0, x(0), e(0))\right]$$

$$\leq -\min_{j \in \mathscr{S}}\{\lambda_{min}[-\Gamma(j)]\}\mathbb{E}\left[\int_0^t \big[\,x^\top(s)\ e^\top(s)\,\big] \begin{bmatrix} x(s) \\ e(s) \end{bmatrix} ds|(r_0, x(0), e(0))\right],$$

which gives in turn

$$\min_{j \in \mathscr{S}}\{\lambda_{min}[-\Gamma(j)]\}\mathbb{E}\left[\int_0^t \big[\,x^\top(s)\ e^\top(s)\,\big] \begin{bmatrix} x(s) \\ e(s) \end{bmatrix} ds|(r_0, x(0), e(0))\right]$$

$$\leq \mathbb{E}\,[V(x(0), e(0), r_0)].$$

This implies that the following relation holds for all $t \geq 0$:

$$\mathbb{E}\left[\int_0^t \big[\,x^\top(s)\ e^\top(s)\,\big] \begin{bmatrix} x(s) \\ e(s) \end{bmatrix} ds|(r_0, x(0), e(0))\right]$$

$$\leq \frac{\mathbb{E}\,[V(x(0), e(0), r_0)]}{\min_{j \in \mathscr{S}}\{\lambda_{min}[-\Gamma(j)]\}}.$$

This completes the proof of Theorem 28.   □

Remark 13. This theorem shows that the proposed control will stochastically stabilize the class of systems we are dealing with if we are able to find some matrices that solve our stated sufficient conditions. The rest of the section will focus on the design of the controller gain matrices $K(i)$ and $L(i)$, $i = 1, 2, \cdots, N$. Conditions in the LMI formalism are needed for our control design.

We are now in a position to synthesize the observer-based output feedback control of the form (3.64) that stochastically stabilizes (3.1).

Before giving the design algorithm, let us transform our stability conditions in the LMI formalism. For this purpose, notice that $\mathscr{J}_P(i)$ is nonlinear in the design parameters $P(i)$ and $K(i)$. To put it into the LMI form, let $X(i) = P^{-1}(i)$ and pre- and post-multiply $\mathscr{J}_P(i)$ by $P^{-1}(i)$ to get

$$X(i)A^\top(i) + A(i)X(i) + B(i)K(i)X(i) + X(i)K^\top(i)B^\top(i)$$

$$+ B(i)B^\top(i) + \sum_{j=1}^{N} \lambda_{ij} X(i)X^{-1}(j)X(i) < 0.$$

By letting $Y_c(i) = K(i)X(i)$ and noting the fact that

$$\sum_{j=1}^{N} \lambda_{ij} X(i)X^{-1}(j)X(i) = \lambda_{ii}X(i) + \mathcal{S}_i(X)\mathcal{X}_i^{-1}(X)\mathcal{S}_i^\top(X),$$

we have

$$\begin{bmatrix} \begin{bmatrix} X(i)A^\top(i) + Y_c^\top(i)B^\top(i) \\ + A(i)X(i) + B(i)Y_c(i) \\ + B(i)B^\top(i) + \lambda_{ii}X(i) \end{bmatrix} & \mathcal{S}_i(X) \\ \mathcal{S}_i^\top(X) & -\mathcal{X}_i(X) \end{bmatrix} < 0.$$

Let us now transform the condition $\mathscr{J}_Q(i) < 0$ in the LMI form. By letting $Y_o(i) = Q(i)L(i)$, we have

$$\mathscr{J}_Q(i) = A^\top(i)Q(i) + Q(i)A(i) + Y_o(i)C_y(i) + C_y^\top(i)Y_o^\top(i)$$

$$+ K^\top(i)K(i) + \sum_{j=1}^{N} \lambda_{ij}Q(j),$$

which gives, after using the Schur complement,

$$\begin{bmatrix} J_2(i) & K^\top(i) \\ K(i) & -\mathbb{I} \end{bmatrix} < 0,$$

where $J_2(i) = A^\top(i)Q(i) + Q(i)A(i) + Y_o(i)C_y(i) + C_y^\top(i)Y_o^\top(i) + \sum_{j=1}^{N} \lambda_{ij}Q(j)$.

The following theorem summarizes the results for the design of the observer-based output feedback controller.

Theorem 29. If there exist sets of symmetric and positive-definite matrices $X = (X(1), \cdots, X(N)) > 0$, $Q = (Q(1), \cdots, Q(N)) > 0$, and sets of matrices $Y_c = (Y_c(1), \cdots, Y_c(N))$ and $Y_o = (Y_o(1), \cdots, Y_o(N))$ satisfying the following set of coupled LMIs for each $i \in \mathscr{S}$:

$$\begin{bmatrix} J_1(i) & \mathcal{S}_i(X) \\ \mathcal{S}_i^\top(X) & -\mathcal{X}_i(X) \end{bmatrix} < 0, \tag{3.66}$$

$$\begin{bmatrix} J_2(i) & K^\top(i) \\ K(i) & -\mathbb{I} \end{bmatrix} < 0, \tag{3.67}$$

where

$$
\begin{aligned}
J_1(i) =\ & X(i)A^\top(i) + Y_c^\top(i)B^\top(i) + A(i)X(i) + B(i)Y_c(i) \\
& + B(i)B^\top(i) + \lambda_{ii}X(i), \\
J_2(i) =\ & A^\top(i)Q(i) + Q(i)A(i) + Y_o(i)C_y(i) + C_y^\top(i)Y_o^\top(i) \\
& + \sum_{j=1}^{N} \lambda_{ij}Q(j),
\end{aligned}
$$

then the controller gains that stabilize system (3.39) in the stochastic sense are given by

$$
\begin{aligned}
K(i) =\ & Y_c(i)X^{-1}(i), \tag{3.68} \\
L(i) =\ & Q^{-1}(i)Y_o(i). \tag{3.69}
\end{aligned}
$$

This theorem provides an algorithm to design a feedback controller of the form (3.64) that stabilizes system (3.39) in the stochastic sense.

Remark 14. Notice that the LMIs of this theorem are independent. Therefore, we can solve the first one to get $K(i), i = 1, 2, \cdots, N$ and then solve the second one to get $L(i), i = 1, 2, \cdots, N$.

Example 37. To show the usefulness of the results of this theorem, let us consider the two-mode system of Example 34.

Solving the LMIs (3.66)–(3.67), we get

$$
X(1) = \begin{bmatrix} 0.3408 & -0.0000 \\ -0.0000 & 0.3408 \end{bmatrix}, \qquad X(2) = \begin{bmatrix} 0.2556 & -0.0000 \\ -0.0000 & 0.2556 \end{bmatrix},
$$

$$
Y_c(1) = \begin{bmatrix} -1.0112 & -21.0264 \\ 21.1627 & -1.0112 \end{bmatrix}, \qquad Y_c(2) = \begin{bmatrix} -0.5767 & -10.8316 \\ 10.7038 & -0.5639 \end{bmatrix},
$$

$$
K_i(1) = \begin{bmatrix} -2.9672 & -61.7010 \\ 62.1010 & -2.9672 \end{bmatrix}, \qquad K_i(2) = \begin{bmatrix} -2.2563 & -42.3797 \\ 41.8797 & -2.2063 \end{bmatrix},
$$

$$
Q(1) = 10^3 \cdot \begin{bmatrix} 6.6508 & -0.0000 \\ -0.0000 & 6.6508 \end{bmatrix}, \qquad Q(2) = 10^3 \cdot \begin{bmatrix} 6.6508 & 0.0000 \\ 0.0000 & 6.6508 \end{bmatrix},
$$

$$
Y_o(1) = 10^4 \cdot \begin{bmatrix} -1.0006 & 0.1330 \\ 0.1330 & -1.0005 \end{bmatrix}, \qquad Y_o(2) = 10^3 \cdot \begin{bmatrix} -2.0087 & -1.6628 \\ -1.6627 & -1.6765 \end{bmatrix}.
$$

The corresponding controller gains are given by

$$
K(1) = \begin{bmatrix} -2.9672 & -61.7010 \\ 62.1010 & -2.9672 \end{bmatrix}, \qquad K(2) = \begin{bmatrix} -2.2563 & -42.3797 \\ 41.8797 & -2.2063 \end{bmatrix},
$$

$$
L(1) = \begin{bmatrix} -1.5044 & 0.2000 \\ 0.2000 & -1.5044 \end{bmatrix}, \qquad L(2) = \begin{bmatrix} -0.3020 & -0.2500 \\ -0.2500 & -0.2521 \end{bmatrix}.
$$

Based on the results of this theorem, the system of this example is stochastically stable under the observer-based output feedback control with the computed gains.

Let us now consider the uncertainties in the dynamics and see how we can modify (3.66) and (3.67) to design a robustly observer-based output feedback control that robustly stochastically stabilizes the class of system under study.

Let us assume that the controller (3.64) exists and show that it robustly stabilizes the class of systems (3.39). For this purpose, let us define the observer error $e(t)$ as before.

Combining the system dynamics and controller dynamics we get the following equation for the error using the same techniques as before and after adding and subtracting the two terms $B(r(t))K(r(t))x(t)$ and $\Delta B(r(t),t)K(r(t))x(t)$:

$$\dot{e}(t) = [A(r(t)) + L(r(t))C_y(r(t))]\,e(t) + [\Delta A(r(t),t) + \Delta B(r(t),t)K(r(t))]\,x(t)$$
$$-\Delta B(r(t),t)K(r(t))e(t) + L(i)\Delta C_y(i)x(t).$$

For the system dynamics, following the same steps as before, we get

$$\dot{x}(t) = \ [A(r(t)) + B(r(t))K(r(t))]\,x(t)$$
$$+ [\Delta A(r(t),t) + \Delta B(r(t),t)K(r(t))]\,x(t)$$
$$-B(r(t))K(r(t))e(t) - \Delta B(r(t),t)K(r(t))e(t).$$

Using all these developments, we get the following extended dynamics:

$$
\begin{bmatrix} \dot{x}(t) \\ \dot{e}(t) \end{bmatrix} =
\begin{bmatrix} \begin{bmatrix} A(r(t)) \\ +B(r(t))K(r(t)) \end{bmatrix} & -B(r(t))K(r(t)) \\ 0 & \begin{bmatrix} A(r(t)) \\ +L(r(t))C_y(r(t)) \end{bmatrix} \end{bmatrix}
\begin{bmatrix} x(t) \\ e(t) \end{bmatrix}
$$
$$
+ \begin{bmatrix} \Delta A(r(t),t) + \Delta B(r(t),t)K(r(t)) & -\Delta B(r(t),t)K(r(t)) \\ \begin{bmatrix} \Delta A(r(t),t) \\ +\Delta B(r(t),t)K(r(t)) \\ +L(i)\Delta C_y(i)x(t) \end{bmatrix} & -\Delta B(r(t),t)K(r(t)) \end{bmatrix}
$$
$$
\times \begin{bmatrix} x(t) \\ e(t) \end{bmatrix}.
$$

The following theorem states the first result on robust stability of the class of systems we are considering under the controller (3.64).

Theorem 30. Let $\varepsilon_A = (\varepsilon_A(1), \cdots, \varepsilon_A(N))$, $\varepsilon_B = (\varepsilon_B(1), \cdots, \varepsilon_B(N))$, $\varepsilon_K = (\varepsilon_K(1), \cdots, \varepsilon_K(N))$, $\varepsilon_e = (\varepsilon_e(1), \cdots, \varepsilon_e(N))$, $\varepsilon_Q = (\varepsilon_Q(1), \cdots, \varepsilon_Q(N))$, $\varepsilon_{BQ} = (\varepsilon_{BQ}(1), \cdots, \varepsilon_{BQ}(N))$, $\varepsilon_{C_y} = (\varepsilon_{C_y}(1), \cdots, \varepsilon_{C_y}(N))$, $\varepsilon_{BK} = (\varepsilon_{BK}(1), \cdots, \varepsilon_{BK}(N))$ be given sets of positive scalars. Let $K = (K(1), \cdots, K(N))$ and $L = [L(1), \cdots, L(N)]$ be a given sets of constant matrices. If there exist sets of symmetric and positive-definite matrices $P = $

$(P(1), \cdots, P(N)) > 0$ and $Q = (Q(1), \cdots, Q(N)) > 0$ satisfying the following set of coupled LMIs for every $i \in \mathscr{S}$ and for all admissible uncertainties:

$$
\begin{aligned}
\Gamma_1(i) = {} & [A(i) + B(i)K(i)]^\top P(i) + P(i)[A(i) + B(i)K(i)] \\
& + \sum_{j=1}^{N} \lambda_{ij} P(j) + \varepsilon_A(i)P(i)D_A(i)D_A^\top(i)P(i) + \varepsilon_A^{-1}(i)E_A^\top(i)E_A(i) \\
& + \varepsilon_B(i)P(i)D_B(i)D_B^\top(i)P(i) + \varepsilon_B^{-1}(i)K^\top(i)E_B^\top(i)E_B(i)K(i) \\
& + \varepsilon_K(i)P(i)B(i)B^\top(i)P(i) + \varepsilon_e(i)P(i)B(i)B^\top(i)P(i) \\
& + \varepsilon_{C_y}(i)E_{C_y}^\top(i)E_{C_y}(i) + \varepsilon_Q^{-1}(i)E_A^\top(i)E_A(i) \\
& + \varepsilon_{BQ}^{-1}(i)K^\top(i)E_B^\top(i)E_B(i)K(i) < 0, \quad\quad (3.70)
\end{aligned}
$$

$$
\begin{aligned}
\Gamma_2(i) = {} & [A(i) + L(i)C_y(i)]^\top Q(i) + Q(i)[A(i) + L(i)C_y(i)] + \sum_{j=1}^{N} \lambda_{ij} Q(j) \\
& + \varepsilon_K^{-1}(i)K^\top(i)K(i) + \varepsilon_e^{-1}(i)K^\top(i)E_B^\top(i)E_B(i)K(i) \\
& + \varepsilon_Q Q(i)D_A(i)D_A^\top(i)Q(i) + \varepsilon_{BQ}(i)Q(i)D_B(i)D_B^\top(i)Q(i) \\
& + \varepsilon_{BK}(i)Q(i)D_B(i)D_B^\top(i)Q(i) + \varepsilon_{BK}^{-1}(i)K^\top(i)E_B^\top(i)E_B(i)K(i) \\
& + \varepsilon_{C_y}^{-1}(i)Q(i)L(i)D_{C_y}(i)D_{C_y}^\top(i)L^\top(i)Q(i) < 0, \quad\quad (3.71)
\end{aligned}
$$

then system (3.39) is stochastically stable.

Proof: Let us consider the Lyapunov functional candidate with the following form:

$$
V(x(t), e(t), r(t) = i) = x^\top(t)P(i)x(t) + e^\top(t)Q(i)e(t),
$$

where $P(i) > 0$ and $Q(i) > 0$ are symmetric and positive-definite matrices.

Let $\mathscr{L}$ be the infinitesimal generator of the process $\{(x(t), e(t), r(t)), t \geq 0\}$. Then, the expression of the infinitesimal operator acting on $V(.)$ and emanating from the point (x, i) at time t, where $x(t) = x$ and $r(t) = i$ for $i \in \mathscr{S}$, is given by

$$
\begin{aligned}
\mathscr{L}V(x(t), e(t), i) = {} & \dot{x}^\top(t)P(i)x(t) + x^\top(t)P(i)\dot{x}(t) + \dot{e}^\top(t)Q(i)e(t) \\
& + e^\top(t)Q(i)\dot{e}(t) + \sum_{j=1}^{N} \lambda_{ij}\left[x^\top(t)P(j)x(t) + e^\top(t)Q(j)e(t)\right] \\
= {} & \Big[[A(i) + B(i)K(i) + \Delta A(i,t) + \Delta B(i,t)K(i)]\, x(t) \\
& \quad - [B(i)K(i) + \Delta B(i,t)K(i)]\, e(t)\Big]^\top P(i)x(t) \\
& + x^\top(t)P(i)\Big[[A(i) + B(i)K(i) + \Delta A(i,t) + \Delta B(i,t)K(i)]\, x(t) \\
& \quad\quad - [B(i)K(i) + \Delta B(i,t)K(i)]\, e(t)\Big]
\end{aligned}
$$

$$+\Big[\left[\varDelta A(i,t) + \varDelta B(i,t)K(i) + L(i)\varDelta C_y(i)\right]x(t) + \left[A(i) + L(i)C_y(i)\right.$$

$$-\varDelta B(i,t)K(i)\big]\,e(t)\Big]^{\top}Q(i)e(t) + e^{\top}(t)Q(i)\Big[\left[\varDelta A(i,t)\right.$$

$$+\varDelta B(i,t)K(i) + L(i)\varDelta C_y(i)\right]x(t) + \left[A(i) + L(i)C_y(i) - \varDelta B(i,t)K(i)\right]e(t)\Big]$$

$$+\sum_{j=1}^{N}\lambda_{ij}\left[x^{\top}(t)P(j)x(t) + e^{\top}(t)Q(j)e(t)\right],$$

which can be rewritten as follows:

$$\mathscr{L}V(x(t),e(t),i) = x^{\top}(t)\Big[\left[A(i) + B(i)K(i)\right]^{\top}P(i)$$

$$+P(i)\left[A(i) + B(i)K(i)\right] + \sum_{j=1}^{N}\lambda_{ij}P(j)\Big]x(t) + 2x^{\top}(t)P(i)\varDelta A(i,t)x(t)$$

$$+2x^{\top}(t)P(i)\varDelta B(i,t)K(i)x(t) - 2x^{\top}(t)P(i)B(i)K(i)e(t)$$

$$-2x^{\top}(t)P(i)\varDelta B(i,t)K(i)e(t) + e^{\top}(t)\Big[\left[A(i) + L(i)C_y(i)\right]^{\top}Q(i)$$

$$+Q(i)\left[A(i) + L(i)C_y(i)\right] + \sum_{j=1}^{N}\lambda_{ij}Q(j)\Big]e(t)$$

$$+2e^{\top}(t)Q(i)\varDelta A(i,t)x(t) + 2e^{\top}(t)Q(i)\varDelta B(i,t)K(i)x(t)$$

$$+2e^{\top}(t)Q(i)L(i)\varDelta C_y(i)x(t) - 2e^{\top}(t)Q(i)\varDelta B(i,t)K(i)e(t).$$

Using Lemma 7 from Appendix A, we have

$$2x^{\top}(t)P(i)\varDelta A(i,t)x(t) = 2x^{\top}(t)P(i)D_A(i)F_A(i,t)E_A(i)x(t)$$
$$\leq \varepsilon_A(i)x^{\top}(t)P(i)D_A(i)D_A^{\top}(i)P(i)x(t)$$
$$+\varepsilon_A^{-1}(i)x^{\top}(t)E_A^{\top}(i)E_A(i)x(t),$$

$$2x^{\top}(t)P(i)\varDelta B(i,t)K(i)x(t)$$
$$= 2x^{\top}(t)P(i)D_B(i)F_B(i,t)E_B(i)K(i)x(t)$$
$$\leq \varepsilon_B(i)x^{\top}(t)P(i)D_B(i)D_B^{\top}(i)P(i)x(t)$$
$$+\varepsilon_B^{-1}(i)x^{\top}(t)K^{\top}(i)E_B^{\top}(i)E_B(i)K(i)x(t),$$

$$-2x^{\top}(t)P(i)B(i)K(i)e(t) \leq \varepsilon_K(i)x^{\top}(t)P(i)B(i)B^{\top}(i)P(i)x(t)$$
$$+\varepsilon_K^{-1}(i)e^{\top}(t)K^{\top}(i)K(i)e(t),$$

$$-2x^{\top}(t)P(i)\varDelta B(i,t)K(i)e(t)$$

$$= -2x^\top(t)P(i)D_B(i)F_B(i,t)E_B(i)K(i)e(t)$$
$$\leq \varepsilon_e(i)x^\top(t)P(i)D_B(i)D_B^\top(i)P(i)x(t)$$
$$+\varepsilon_e^{-1}(i)e^\top(t)K^\top(i)E_B^\top(i)E_B(i)K(i)e(t),$$

$$2e^\top(t)Q(i)\Delta A(i,t)x(t) = 2e^\top(t)Q(i)D_A(i)F_A(i,t)E_A(i)x(t)$$
$$\leq \varepsilon_Q(i)e^\top(t)Q(i)D_A(i)D_A^\top(i)Q(i)e(t)$$
$$+\varepsilon_Q^{-1}(i)x^\top(t)E_A^\top(i)E_A(i)x(t),$$

$$2e^\top(t)Q(i)\Delta B(i,t)K(i)x(t)$$
$$= 2e^\top(t)Q(i)D_B(i)F_B(i,t)E_B(i)K(i)x(t)$$
$$\leq \varepsilon_{BQ}(i)e^\top(t)Q(i)D_B(i)D_B^\top(i)Q(i)e(t)$$
$$+\varepsilon_{BQ}^{-1}(i)x^\top(t)K^\top(i)E_B^\top(i)E_B(i)K(i)x(t),$$

$$2e^\top(t)Q(i)L(i)\Delta C_y(i)x(t)$$
$$= 2e^\top(t)Q(i)L(i)D_{C_y}(i)F_{C_y}(i,t)E_{C_y}(i)x(t)$$
$$\leq \varepsilon_{C_y}^{-1}(i)e^\top(t)Q(i)L(i)D_{C_y}(i)D_{C_y}^\top(i)L^\top(i)Q(i)e(t)$$
$$+\varepsilon_{C_y}(i)x^\top(t)E_{C_y}^\top(i)E_{C_y}(i)x(t),$$

$$-2e^\top(t)Q(i)\Delta B(i,t)K(i)e(t)$$
$$= -2e^\top(t)Q(i)D_B(i)F_B(i,t)E_B(i)K(i)e(t)$$
$$\leq \varepsilon_{BK}(i)e^\top(t)Q(i)D_B(i)D_B^\top(i)Q(i)e(t)$$
$$+\varepsilon_{BK}^{-1}(i)e^\top(t)K^\top(i)E_B^\top(i)E_B(i)K(i)e(t).$$

Using these relations, $\mathscr{L}V(x(t),e(t),i)$ can be rewritten as follows:

$$\mathscr{L}V(x(t),e(t),i) \leq x^\top(t)\Big[[A(i)+B(i)K(i)]^\top P(i)$$

$$+P(i)[A(i)+B(i)K(i)] + \sum_{j=1}^N \lambda_{ij}P(j) + \varepsilon_Q^{-1}(i)E_A^\top(i)E_A(i)$$

$$+\varepsilon_A(i)P(i)D_A(i)D_A^\top(i)P(i) + \varepsilon_A^{-1}(i)E_A^\top(i)E_A(i)$$
$$+\varepsilon_B(i)P(i)D_B(i)D_B^\top(i)P(i) + \varepsilon_B^{-1}(i)K^\top(i)E_B^\top(i)E_B(i)K(i)$$
$$+\varepsilon_K(i)P(i)B(i)B^\top(i)P(i) + \varepsilon_e(i)P(i)D_B(i)D_B^\top(i)P(i)$$
$$+\varepsilon_{BQ}^{-1}(i)K^\top(i)E_B^\top(i)E_B(i)K(i) + \varepsilon_{C_y}(i)E_{C_y}^\top(i)E_{C_y}(i)\Big]x(t)$$

$$+ e^\top(t) \Big[[A(i) + L(i)C_y(i)]^\top Q(i) + [A(i) + L(i)C_y(i)]$$

$$+ \sum_{j=1}^{N} \lambda_{ij} Q(j) + \varepsilon_K^{-1}(i) K^\top(i) K(i) + \varepsilon_e^{-1}(i) K^\top(i) E_B^\top(i) E_B(i) K(i)$$

$$+ \varepsilon_Q(i) Q(i) D_A(i) D_A^\top(i) Q(i) + \varepsilon_{BQ}(i) Q(i) D_B(i) D_B^\top(i) Q(i)$$
$$+ \varepsilon_{BK}(i) Q(i) D_B(i) D_B^\top(i) Q(i) + \varepsilon_{BK}^{-1}(i) K^\top(i) E_B^\top(i) E_B(i) K(i)$$
$$+ \varepsilon_{C_y}^{-1}(i) Q(i) L(i) D_{C_y}(i) D_{C_y}^\top(i) L^\top(i) Q(i) \Big] e(t),$$

which gives

$$\mathscr{L}V(x(t), e(t), i) \leq \begin{bmatrix} x^\top(t) \ e^\top(t) \end{bmatrix} \begin{bmatrix} \Gamma_1(i) & 0 \\ 0 & \Gamma_2(i) \end{bmatrix} \begin{bmatrix} x(t) \\ e(t) \end{bmatrix}$$

$$= \begin{bmatrix} x^\top(t) \ e^\top(t) \end{bmatrix} \Gamma(i) \begin{bmatrix} x(t) \\ e(t) \end{bmatrix}.$$

Therefore, we obtain

$$\mathscr{L}V(x(t), e(t), i) \leq - \min_{j \in \mathscr{S}} \{ \lambda_{min} [-\Gamma(j)] \} \begin{bmatrix} x^\top(t) \ e^\top(t) \end{bmatrix} \begin{bmatrix} x(t) \\ e(t) \end{bmatrix}.$$

The rest of the proof of Theorem 30 can be obtained by following the same steps as we did for the previous theorems. $\square$

Remark 15. In the previous theorem we showed that the proposed control will stochastically stabilize the class of systems we are dealing with if we are able to find some matrices that solve the given sufficient conditions. Our interest in the rest of the book will focus on the design of the controller gain matrices $K(i)$ and $L(i)$, $i = 1, 2, \cdots, N$. Conditions in the LMI formalism are needed for our control design.

We are now in a position to synthesize the observer-based output feedback control (3.64) that stochastically stabilizes (3.39). For this purpose, let us transform our stability conditions in the LMI formalism. Pre- and post-multiply (3.70) by $P^{-1}(i)$ to get

$$P^{-1}(i) [A(i) + B(i)K(i)]^\top + [A(i) + B(i)K(i)] P^{-1}(i)$$
$$+ \varepsilon_A^{-1}(i) P^{-1}(i) E_A^\top(i) E_A(i) P^{-1}(i) + \varepsilon_B(i) D_B(i) D_B^\top(i)$$
$$+ \varepsilon_B^{-1}(i) P^{-1}(i) K^\top(i) E_B^\top(i) E_B(i) K(i) P^{-1}(i)$$
$$+ \varepsilon_A(i) D_A(i) D_A^\top(i) + \varepsilon_K(i) B(i) B^\top(i) + \varepsilon_e(i) D_B(i) D_B^\top(i)$$
$$+ \varepsilon_Q^{-1}(i) P^{-1}(i) E_A^\top(i) E_A(i) P^{-1}(i) + \varepsilon_{C_y}(i) P^{-1}(i) E_{C_y}^\top(i) E_{C_y}(i) P^{-1}(i)$$
$$+ \varepsilon_{BQ}^{-1}(i) P^{-1}(i) K^\top(i) E_B^\top(i) E_B(i) K(i) P^{-1}(i)$$
$$+ \sum_{j=1}^{N} \lambda_{ij} P^{-1}(i) P(j) P^{-1}(i) < 0.$$

By letting $X(i) = P^{-1}(i)$ and $Y(i) = K(i)X(i)$, we have

$$
\begin{aligned}
&X(i)A^\top(i) + Y^\top(i)B^\top(i) + A(i)X(i) + B(i)Y(i) \\
&+\varepsilon_A^{-1}(i)X(i)E_A^\top(i)E_A(i)X(i) + \varepsilon_B(i)D_B(i)D_B^\top(i) \\
&+\varepsilon_A(i)D_A(i)D_A^\top(i) + \varepsilon_B^{-1}(i)Y^\top(i)E_B^\top(i)E_B(i)Y(i) \\
&\qquad +\varepsilon_K(i)B(i)B^\top(i) + \varepsilon_e(i)D_B(i)D_B^\top(i) \\
&+\varepsilon_Q^{-1}(i)X(i)E_A^\top(i)E_A(i)X(i) + \varepsilon_{C_y}(i)X(i)E_{C_y}^\top(i)E_{C_y}(i)X(i) \\
&+\varepsilon_{BQ}^{-1}(i)Y^\top(i)E_B^\top(i)E_B(i)Y(i) + \sum_{j=1}^{N}\lambda_{ij}X(i)X^{-1}(j)X(i) < 0. \quad (3.72)
\end{aligned}
$$

Notice that

$$
\begin{aligned}
&\varepsilon_A^{-1}(i)X(i)E_A^\top(i)E_A(i)X(i) + \varepsilon_{C_y}(i)X(i)E_{C_y}^\top(i)E_{C_y}(i)X(i) \\
&+\varepsilon_Q^{-1}(i)X(i)E_A^\top(i)E_A(i)X(i) + \varepsilon_{BQ}^{-1}(i)X(i)E_B^\top(i)E_B(i)X(i) \\
&\qquad = X(i)\left[E_A^\top(i)\ E_A^\top(i)\ E_B^\top(i)\ E_{C_y}^\top(i) \right] \\
&\times \left[\mathrm{diag}(\varepsilon_A(i)\mathbb{I}, \varepsilon_Q(i)\mathbb{I}, \varepsilon_{BQ}(i)\mathbb{I}), \varepsilon_{C_y}(i)\mathbb{I} \right]^{-1}
\begin{bmatrix} E_A(i) \\ E_A(i) \\ E_B(i) \\ E_{C_y}(i) \end{bmatrix} X(i) \\
&\qquad = X(i)\mathcal{F}^\top(i)\mathcal{D}^{-1}(i)\mathcal{F}(i)X(i),
\end{aligned}
$$

and

$$
\sum_{j=1}^{N}\lambda_{ij}X(i)X^{-1}(j)X(i) = \lambda_{ii}X(i) + S_i(X)\mathcal{X}_i^{-1}S_i^\top(X),
$$

with

$$
\begin{aligned}
\mathcal{F}^\top(i) =&\ \left[E_A^\top(i)\ E_A^\top(i)\ E_B^\top(i)\ E_{C_y}^\top(i) \right], \\
\mathcal{D}(i) =&\ \mathrm{diag}(\varepsilon_A(i)\mathbb{I}, \varepsilon_Q(i)\mathbb{I}, \varepsilon_{BQ}(i)\mathbb{I}, \varepsilon_{C_y}(i)\mathbb{I}), \\
S_i(X) =&\ \left[\sqrt{\lambda_{i1}}X(i), \cdots, \sqrt{\lambda_{ii-1}}X(i), \sqrt{\lambda_{ii+1}}X(i), \right. \\
&\ \left. \cdots, \sqrt{\lambda_{iN}}X(i) \right], \\
\mathcal{X}_i(X) =&\ \mathrm{diag}(X(1), \cdots, X(i-1), X(i+1), \cdots, X(N)).
\end{aligned}
$$

The condition (3.72) becomes

$$
\begin{aligned}
&\mathcal{F}_c(i) + X(i)\mathcal{F}^\top(i)\mathcal{D}^{-1}(i)\mathcal{F}(i)X(i) \\
&+\varepsilon_B^{-1}(i)Y^\top(i)E_B^\top(i)E_B(i)Y(i) \\
&+S_i(X)\mathcal{X}_i^{-1}S_i^\top(X) < 0,
\end{aligned}
$$

with

$$
\begin{aligned}
\mathcal{F}_c(i) = \ & X(i)A^\top(i) + Y^\top(i)B^\top(i) + A(i)X(i) + B(i)Y(i) \\
& + \varepsilon_A(i)D_A(i)D_A^\top(i) + \varepsilon_B(i)D_B(i)D_B^\top(i) \\
& + \varepsilon_K(i)B(i)B^\top(i) + \varepsilon_e(i)D_B(i)D_B^\top(i) + \lambda_{ii}X(i).
\end{aligned}
$$

Using the Schur complement, we get

$$
\begin{bmatrix}
\mathcal{F}_c(i) & X(i)\mathcal{F}^\top(i) & Y^\top(i)E_B^\top(i) & S_i(X) \\
\mathcal{F}(i)X(i) & -\mathcal{D}(i) & 0 & 0 \\
E_B(i)Y(i) & 0 & -\varepsilon_B(i)\mathbb{I} & 0 \\
S_i^\top(X) & 0 & 0 & -\mathcal{X}_i(X)
\end{bmatrix} < 0.
$$

Let us now transform (3.71) in the LMI form. For this purpose we have

$$
\Gamma_2(i) = \Big[[A(i) + L(i)C_y(i)]^\top Q(i) + Q(i)[A(i) + L(i)C_y(i)]
$$

$$
+ \sum_{j=1}^N \lambda_{ij}Q(j) + \varepsilon_K^{-1}(i)K^\top(i)K(i) + \varepsilon_e^{-1}(i)K^\top(i)E_B^\top(i)E_B(i)K(i)
$$

$$
+ \varepsilon_Q(i)Q(i)D_A(i)D_A^\top(i)Q(i) + \varepsilon_{BQ}(i)Q(i)D_B(i)D_B^\top(i)Q(i)
$$

$$
+ \varepsilon_{BK}(i)Q(i)D_B(i)D_B^\top(i)Q(i) + \varepsilon_{BK}^{-1}(i)K^\top(i)E_B^\top(i)E_B(i)K(i)
$$

$$
+ \varepsilon_{C_y}^{-1}(i)Q(i)L(i)D_{C_y}(i)D_{C_y}^\top(i)L^\top(i)Q(i) \Big].
$$

By letting $Y_o(i) = Q(i)L(i)$, we have

$$
\Gamma_2(i) = A^\top(i)Q(i) + C_y^\top(i)Y_o^\top(i) + Q(i)A(i) + Y_o(i)C_y(i)
$$

$$
+ \sum_{j=1}^N \lambda_{ij}Q(j) + \varepsilon_K^{-1}(i)K^\top(i)K(i) + \varepsilon_e^{-1}(i)K^\top(i)E_B^\top(i)E_B(i)K(i)
$$

$$
+ \varepsilon_Q(i)Q(i)D_A(i)D_A^\top(i)Q(i) + \varepsilon_{BQ}(i)Q(i)D_B(i)D_B^\top(i)Q(i)
$$

$$
+ \varepsilon_{BK}(i)Q(i)D_B(i)D_B^\top(i)Q(i) + \varepsilon_{BK}^{-1}(i)K^\top(i)E_B^\top(i)E_B(i)K(i)
$$

$$
+ \varepsilon_{C_y}^{-1}(i)Y(i)D_{C_y}(i)D_{C_y}^\top(i)Y^\top(i). \tag{3.73}
$$

Notice that

$$
\varepsilon_K^{-1}(i)K^\top(i)K(i) + \varepsilon_e^{-1}(i)K^\top(i)E_B^\top(i)E_B(i)K(i)
$$

$$
+ \varepsilon_{BK}^{-1}(i)K^\top(i)E_B^\top(i)E_B(i)K(i)
$$

$$
= K^\top(i)\left[\, \mathbb{I}\ E_B^\top(i)\ E_B^\top(i) \,\right] [\mathrm{diag}(\varepsilon_K(i)\mathbb{I}, \varepsilon_e(i)\mathbb{I}, \varepsilon_{BQ}(i)\mathbb{I})]^{-1}
$$

$$
\times \begin{bmatrix} \mathbb{I} \\ E_B(i) \\ E_B(i) \end{bmatrix} K(i)
$$

$$
= K^\top(i)\mathcal{F}_o^\top(i)\mathcal{D}_o^{-1}(i)\mathcal{F}_o(i)K(i)
$$

$$\varepsilon_Q(i)Q(i)D_A(i)D_A^\top(i)Q(i) + \varepsilon_{BQ}(i)Q(i)D_B(i)D_B^\top(i)Q(i)$$
$$+\varepsilon_{BK}(i)Q(i)D_B(i)D_B^\top(i)Q(i)$$
$$= Q(i)\left[\sqrt{\varepsilon_Q(i)}D_A(i) \ \sqrt{\varepsilon_{BQ}(i)}D_B(i) \ \sqrt{\varepsilon_{BK}(i)}D_B(i) \right]$$
$$\times \begin{bmatrix} \sqrt{\varepsilon_Q(i)}D_A^\top(i) \\ \sqrt{\varepsilon_{BQ}(i)}D_B^\top(i) \\ \sqrt{\varepsilon_{BK}(i)}D_B^\top(i) \end{bmatrix} Q(i)$$
$$= Q(i)\mathcal{G}_o^\top(i)\mathcal{G}_o(i)Q(i),$$

which gives

$$\begin{aligned}
\Gamma_2(i) = \ & A^\top(i)Q(i) + Q(i)A(i) + Y_o(i)C_y(i) + C_o^\top(i)Y_o^\top(i) \\
& + Q(i)\mathcal{G}_o^\top(i)\mathcal{G}_o(i)Q(i) + K^\top(i)\mathcal{F}_o^\top(i)\mathcal{D}_o^{-1}(i)\mathcal{F}_o(i)K(i) \\
& + \sum_{j=1}^N \lambda_{ij}Q(j) + \varepsilon_{C_y}^{-1}(i)Y(i)D_{C_y}(i)D_{C_y}^\top(i)Y^\top(i),
\end{aligned}$$

with

$$\begin{aligned}
Y_o(i) = \ & Q(i)L(i), \\
\mathcal{F}_o^\top(i) = \ & \left[\mathbb{I} \ E_B^\top(i) \ E_B^\top(i) \right], \\
\mathcal{D}_o(i) = \ & \mathrm{diag}(\varepsilon_K(i)\mathbb{I}, \varepsilon_e(i)\mathbb{I}, \varepsilon_{BQ}(i)\mathbb{I}), \\
\mathcal{G}_o^\top(i) = \ & \left[\sqrt{\varepsilon_Q(i)}D_A(i) \ \sqrt{\varepsilon_{BQ}(i)}D_B(i) \ \sqrt{\varepsilon_{BK}(i)}D_B(i) \right].
\end{aligned}$$

After using the Schur complement (3.73) becomes

$$\begin{bmatrix} \# & K^\top(i)\mathcal{F}_o^\top(i) & Q(i)\mathcal{G}_o^\top(i) & Y_o(i)D_{C_y}(i) \\ \mathcal{F}_o(i)K(i) & -\mathcal{D}_o(i) & 0 & 0 \\ \mathcal{G}_o(i)Q(i) & 0 & -\mathbb{I} & 0 \\ D_{C_y}^\top(i)Y^\top(i) & 0 & 0 & -\varepsilon_{C_y}(i)\mathbb{I} \end{bmatrix} < 0,$$

where $\# = A^\top(i)Q(i) + Q(i)A(i) + Y_o(i)C(i) + C^\top(i)Y_o^\top(i) + \sum_{j=1}^N \lambda_{ij}Q(j)$.

The following theorem gives the way to design the observer-based output feedback controller.

Theorem 31. Let $\varepsilon_A = (\varepsilon_A(i), \cdots, \varepsilon_A(N))$, $\varepsilon_B = (\varepsilon_B(i), \cdots, \varepsilon_B(N))$, $\varepsilon_K = (\varepsilon_K(i), \cdots, \varepsilon_K(N))$, $\varepsilon_Q = (\varepsilon_Q(i), \cdots, \varepsilon_Q(N))$, $\varepsilon_e = (\varepsilon_e(i), \cdots, \varepsilon_e(N))$, $\varepsilon_{BQ} = (\varepsilon_{BQ}(i), \cdots, \varepsilon_{BQ}(N))$, and $\varepsilon_{BK} = (\varepsilon_{BK}(i), \cdots, \varepsilon_{BK}(N))$ be a given sets of positive scalars. If there exist symmetric and positive-definite matrices $X = (X(1), \cdots, X(N)) > 0$, $Q_o = (Q_o(1), \cdots, Q_o(N)) > 0$, and matrices $Y_c = (Y_c(1), \cdots, Y_c(N))$ and $Y_o = (Y_o(1), \cdots, Y_o(N))$ satisfying the following set of LMIs for each mode $i \in \mathscr{S}$:

$$\begin{bmatrix} \mathcal{F}_c(i) & X(i)\mathcal{F}^\top(i) & Y^\top(i)E_B^\top(i) & S_i(X) \\ \mathcal{F}(i)X(i) & -\mathcal{D}(i) & 0 & 0 \\ E_B(i)Y(i) & 0 & -\varepsilon_B(i)\mathbb{I} & 0 \\ S_i^\top(X) & 0 & 0 & -\mathcal{X}_i(X) \end{bmatrix} < 0, \qquad (3.74)$$

$$\begin{bmatrix} \# & K^\top(i)\mathcal{F}_o^\top(i) & Q(i)\mathcal{G}_o^\top(i) & Y_o(i)D_{C_y}(i) \\ \mathcal{F}_o(i)K(i) & -\mathcal{D}_o(i) & 0 & 0 \\ \mathcal{G}_o(i)Q(i) & 0 & -\mathbb{I} & 0 \\ D_{C_y}^\top(i)Y^\top(i) & 0 & 0 & -\varepsilon_{C_y}(i)\mathbb{I} \end{bmatrix} < 0, \qquad (3.75)$$

where

$$\begin{aligned} \mathcal{F}_c(i) = {}& X(i)A^\top(i) + Y^\top(i)B^\top(i) + A(i)X(i) + B(i)Y(i) \\ & + \varepsilon_A(i)D_A(i)D_A^\top(i) + \varepsilon_B(i)D_B(i)D_B^\top(i) \\ & + \varepsilon_K(i)B(i)B^\top(i) + \varepsilon_e(i)D_B(i)D_B^\top(i) + \lambda_{ii}X(i), \\ \# = {}& A^\top(i)Q(i) + Q(i)A(i) + Y_o(i)C(i) + C^\top(i)Y_o^\top(i) + \sum_{j=1}^N \lambda_{ij}Q(j), \end{aligned}$$

then the controller gains that stabilize system (3.39) in the stochastic sense are given by

$$K(i) = Y_c(i)X^{-1}(i), \qquad (3.76)$$
$$L(i) = Q_o^{-1}(i)Y_o(i). \qquad (3.77)$$

This theorem provides a procedure to design the observer-output feedback control of the form (3.64) that stabilizes system (3.39) in the robust stochastic sense. To solve the LMIs of this theorem, we can solve the first LMI to get the gain $K(i), i = 1, 2, \cdots, N$, that will be used in the second LMI to get the second set of gains $L(i), i = 1, 2, \cdots, N$.

Example 38. To show the usefulness of the theoretical results developed in this theorem, let us consider the two-mode system of Example 34.

The positive constant $\varepsilon_A(i)$ and $\varepsilon_B(i)$, $i = 1, 2$ are fixed to the following values:

$$\begin{aligned} \varepsilon_A(1) = {}& 0.10, \\ \varepsilon_B(1) = {}& \varepsilon_K(1) = \varepsilon_Q(1) = \varepsilon_e(1) = \varepsilon_{BQ}(1) = \varepsilon_{BK}(1) = 0.50, \\ \varepsilon_A(2) = {}& 0.10, \\ \varepsilon_B(2) = {}& \varepsilon_K(2) = \varepsilon_Q(2) = \varepsilon_e(2) = \varepsilon_{BQ}(2) = \varepsilon_{BK}(2) = 0.50. \end{aligned}$$

Solving the LMIs (3.74)–(3.75), we get

$$X(1) = \begin{bmatrix} 0.3344 & -0.0117 \\ -0.0117 & 0.3459 \end{bmatrix}, \qquad X(2) = \begin{bmatrix} 0.2968 & -0.0094 \\ -0.0094 & 0.2966 \end{bmatrix},$$

$$Y_c(1) = \begin{bmatrix} -0.9783 & 0.0314 \\ 0.0969 & -0.9987 \end{bmatrix}, \qquad Y_c(2) = \begin{bmatrix} -0.4623 & -0.1517 \\ -0.0303 & -0.4487 \end{bmatrix},$$

$$K_i(1) = \begin{bmatrix} -2.9254 & -0.0084 \\ 0.1889 & -2.8807 \end{bmatrix}, \qquad K_i(2) = \begin{bmatrix} -1.5752 & -0.5612 \\ -0.1500 & -1.5178 \end{bmatrix},$$

$$Q(1) = \begin{bmatrix} 35.8239 & -11.8485 \\ -11.8485 & 18.0388 \end{bmatrix}, \qquad Q(2) = \begin{bmatrix} 25.1454 & -12.7588 \\ -12.7588 & 31.9395 \end{bmatrix},$$

$$Y_o(1) = \begin{bmatrix} -48.5105 & 18.9195 \\ 17.5438 & -66.1793 \end{bmatrix}, \quad Y_o(2) = \begin{bmatrix} -36.3759 & -13.9821 \\ -13.6054 & 11.3450 \end{bmatrix}.$$

This gives the following gains for the desired controller:

$$K(1) = \begin{bmatrix} -2.9254 & -0.0084 \\ 0.1889 & -2.8807 \end{bmatrix}, \quad K(2) = \begin{bmatrix} -1.5752 & -0.5612 \\ -0.1500 & -1.5178 \end{bmatrix},$$

$$L(1) = \begin{bmatrix} -1.3190 & -0.8755 \\ 0.1062 & -4.2438 \end{bmatrix}, \quad L(2) = \begin{bmatrix} -2.0855 & -0.4714 \\ -1.2590 & 0.1669 \end{bmatrix}.$$

Based on the results of this theorem, the system of this example is stochastically stable under the observer-based output feedback control with the computed gains.

Let us now consider that the dynamics are perturbed by a Wiener process. We want to stochastically stabilize the new dynamics and design the observer-based output feedback controller that will accomplish this goal. Assume that the new dynamics are given by the following:

$$\begin{cases} dx(t) = A(r(t),t)x(t)dt + B(r(t),t)u(t)dt + \mathbb{W}(r(t))x(t)d\omega(t), \\ y(t) = C_y(r(t),t)x(t), \\ x(0) = x_0, \end{cases} \tag{3.78}$$

where all the variables and the matrices keep the same meaning as before.

Let the desired controller be given by the following form:

$$\begin{cases} d\nu(t) = A(r(t))\nu(t)dt + B(r(t))u(t)dt + L(r(t))\left[C_y(r(t))\nu(t) - y(t)\right]dt \\ \qquad + \mathbb{W}(r(t))\nu(t)d\omega(t), \nu(0) = 0, \\ u(t) = K(r(t))\nu(t), \end{cases} \tag{3.79}$$

where the matrices $K(r(t))$ and $L(r(t))$ are design parameters that have to be determined.

Let us define as before the estimation error $e(t)$ and rewrite the dynamics (3.78) and those of the error using the state variable $\eta(t) = \begin{bmatrix} x(t) \\ e(t) \end{bmatrix}$.

For the dynamics (3.78), we have

$$dx(t) = [A(r(t)) + \Delta A(r(t),t)]x(t)dt + [B(r(t)) + \Delta B(r(t),t)]u(t)dt$$
$$+ \mathbb{W}(r(t))x(t)d\omega(t).$$

Replacing the control $u(t)$ by its expression and by adding and subtracting $B(r(t))K(r(t))x(t)dt$ and $\Delta B(r(t))K(r(t))x(t)dt$, we get

$$dx(t) = [A(r(t)) + B(r(t))K(r(t))]x(t)dt - B(r(t))K(r(t))[x(t) - \nu(t)]dt$$
$$- \Delta B(r(t),t)K(r(t))[x(t) - \nu(t)]u(t)dt + \Delta A(r(t),t)x(t)dt$$
$$+ \Delta B(r(t),t)K(r(t))x(t)dt + \mathbb{W}(r(t))x(t)d\omega(t)$$

$$
\begin{aligned}
= &\ [A(r(t)) + B(r(t))K(r(t))]\, x(t)dt + \big[\Delta A(r(t), t) \\
&\ + \Delta B(r(t), t)K(r(t))\big]x(t)dt - B(r(t))K(r(t))e(t)dt \\
&\ - \Delta B(r(t), t)K(r(t))e(t)dt + \mathbb{W}(r(t))x(t)d\omega(t).
\end{aligned}
$$

For the error dynamics, we have

$$
\begin{aligned}
de(t) = &\ dx(t) - d\nu(t) = [A(r(t)) + \Delta A(r(t), t)]\, x(t)dt \\
&\ + [B(r(t)) + \Delta B(r(t), t)]\, u(t)dt + \mathbb{W}(r(t))x(t)d\omega(t) \\
&\ - A(r(t))\nu(t)dt - B(r(t))K(r(t))\nu(t)dt \\
&\ - L(r(t))\, [C_y(r(t))\nu(t) - C_y(r(t))x(t) - \Delta C_y(r(t), t)]\, dt \\
&\ - \mathbb{W}(r(t))\nu(t)d\omega(t) \\
= &\ [A(r(t)) + L(r(t))C_y(r(t))]\, e(t)dt + \big[\Delta A(r(t), t) \\
&\ + \Delta B(r(t), t)K(r(t)) + L(r(t))\Delta C_y(r(t), t)\big]\, x(t)dt \\
&\ - \Delta B(r(t), t)K(r(t))e(t)dt + \mathbb{W}(r(t))e(t)d\omega(t).
\end{aligned}
$$

By letting $\eta(t)$, $\tilde{A}(r(t))$, $\Delta\tilde{A}(r(t), t)$, and $\tilde{\mathbb{W}}(r(t))$ be defined as follows:

$$
\eta(t) = \begin{bmatrix} x(t) \\ e(t) \end{bmatrix},
$$

$$
\tilde{A}(r(t)) = \begin{bmatrix} A(r(t)) + B(r(t))K(r(t)) & -B(r(t))K(r(t)) \\ 0 & A(r(t)) + L(r(t))C_y(r(t)) \end{bmatrix},
$$

$$
\Delta\tilde{A}(r(t), t) = \begin{bmatrix} \Delta A(r(t), t) + \Delta B(r(t))K(r(t)) & -\Delta B(r(t), t)K(r(t)) \\ \begin{bmatrix} \Delta A(r(t), t) \\ +\Delta B(r(t))K(r(t)) \\ +L(r(t))\Delta C_y(r(t)) \end{bmatrix} & -\Delta B(r(t), t)K(r(t)) \end{bmatrix},
$$

$$
\tilde{\mathbb{W}}(r(t)) = \begin{bmatrix} \mathbb{W}(r(t)) & 0 \\ 0 & \mathbb{W}(r(t)) \end{bmatrix},
$$

the extended dynamics becomes:

$$
d\eta(t) = \left[\tilde{A}(r(t)) + \Delta\tilde{A}(r(t), t)\right]\eta(t)dt + \tilde{\mathbb{W}}(r(t))\eta(t)d\omega(t). \tag{3.80}
$$

As we did previously for systems without external disturbance, let us start with the study of the nominal class of systems and see how we can design a stochastically stabilizing observer-based output feedback control and then take care of the effects of the uncertainties on the system's matrices.

The following result shows that if there exist sets of gains $K = (K(1), \cdots , K(N))$ and $L = (L(1), \cdots , L(N))$, then the observer-based output feedback control will stochastically stabilize the class of systems disturbed by a Wiener process (nominal systems) if some appropriate conditions are satisfied.

Theorem 32. *Let $K(i)$, $i = 1, 2, \cdots , N$ be a given set of gains. If there exist sets of symmetric and positive-definite matrices $P = (P(1), \cdots , P(N)) > 0$*

and $Q = (Q(1), \cdots, Q(N)) > 0$ such that the following LMIs hold for each $i \in \mathscr{S}$:

$$\begin{bmatrix} \mathscr{R}_P(i) & P(i)B(i) \\ B^\top(i)P(i) & -\mathbb{I} \end{bmatrix} < 0, \tag{3.81}$$

$$\begin{bmatrix} \mathscr{R}_Q(i) & K^\top(i) \\ K(i) & -\mathbb{I} \end{bmatrix} < 0, \tag{3.82}$$

where

$$\begin{aligned} \mathscr{R}_P(i) = & \ [A(i) + B(i)K(i)]^\top P(i) + P(i)[A(i) + B(i)K(i)] \\ & + \sum_{j=1}^{N} \lambda_{ij} P(j) + \mathbb{W}^\top(i)P(i)\mathbb{W}(i), \\ \mathscr{R}_Q(i) = & \ [A(i) + L(i)C_y(i)]^\top Q(i) + Q(i)[A(i) + L(i)C_y(i)] \\ & + \sum_{j=1}^{N} \lambda_{ij} Q(j) + \mathbb{W}^\top(i)Q(i)\mathbb{W}(i), \end{aligned}$$

then the closed-loop system is stochastically stable under the controller (3.79).

Proof: Let us consider the Lyapunov candidate function to be given by the following:

$$\begin{aligned} V(x(t), e(t), r(t) = i) = & \ x^\top(t)P(i)x(t) + e^\top(t)Q(i)e(t) \\ = & \ \begin{bmatrix} x^\top(t) & e^\top(t) \end{bmatrix} \begin{bmatrix} P(i) & 0 \\ 0 & Q(i) \end{bmatrix} \begin{bmatrix} x(t) \\ e(t) \end{bmatrix} \\ = & \ \eta^\top(t)\mathbb{P}(i)\eta(t), \end{aligned}$$

where the matrices $P(i) > 0$ and $Q(i) > 0$ are symmetric and positive-definite matrices.

Using Theorem 82 from Appendix A, we get

$$\mathscr{L}V(x(t), e(t), i) = \left[\tilde{A}(i)\eta(t) \right]^\top V_\eta(x(t), e(t), i)$$

$$+ \sum_{j=1}^{N} \lambda_{ij} V(x(t), e(t), j) + \frac{1}{2} tr\left(\eta^\top(t)\tilde{\mathbb{W}}^\top(i)V_{\eta\eta}(x(t), e(t), i)\tilde{\mathbb{W}}(i)\eta(t) \right).$$

Using the fact that

$$V_\eta(x(t), e(t), i) = 2 \begin{bmatrix} P(i) & 0 \\ 0 & Q(i) \end{bmatrix} \begin{bmatrix} x(t) \\ e(t) \end{bmatrix} = 2\mathbb{P}(i)\eta(t)$$

$$V_{\eta\eta}(x(t), e(t), i) = 2 \begin{bmatrix} P(i) & 0 \\ 0 & Q(i) \end{bmatrix} = 2\mathbb{P}(i),$$

we obtain the following:

$$\mathscr{L}V(x(t), e(t), i) = 2\eta^\top(t)\tilde{A}^\top(i)\mathbb{P}(i)\eta(t) + \sum_{j=1}^{N} \lambda_{ij}\eta^\top(t)\mathbb{P}(j)\eta(t)$$
$$+\eta^\top(t)\tilde{\mathbb{W}}^\top(i)\mathbb{P}(i)\tilde{\mathbb{W}}(i)\eta(t),$$

which can be rewritten as

$$\mathscr{L}V(x(t), e(t), i) = \eta^\top(t)\Big[\tilde{A}^\top(i)\mathbb{P}(i) + \mathbb{P}(i)\tilde{A}(i) + \sum_{j=1}^{N} \lambda_{ij}\mathbb{P}(j)$$
$$+\tilde{\mathbb{W}}^\top(i)\mathbb{P}(i)\tilde{\mathbb{W}}(i)\Big]\eta(t)$$
$$= \eta^\top(t)\Psi(i)\eta(t),$$

with

$$\Psi(i) = \tilde{A}^\top(i)\mathbb{P}(i) + \mathbb{P}(i)\tilde{A}(i) + \sum_{j=1}^{N} \lambda_{ij}\mathbb{P}(j) + \tilde{\mathbb{W}}^\top(i)\mathbb{P}(i)\tilde{\mathbb{W}}(i).$$

Notice that

$$\mathbb{P}(i)\tilde{A}(i) = \begin{bmatrix} P(i) & 0 \\ 0 & Q(i) \end{bmatrix} \begin{bmatrix} \begin{bmatrix} A(i) \\ +B(i)K(i) \end{bmatrix} & -B(i)K(i) \\ 0 & \begin{bmatrix} A(i) \\ +L(i)C_y(i) \end{bmatrix} \end{bmatrix}$$
$$= \begin{bmatrix} \begin{bmatrix} P(i)[A(i) \\ +B(i)K(i)] \end{bmatrix} & -P(i)B(i)K(i) \\ 0 & \begin{bmatrix} Q(i)[A(i) \\ +L(i)C_y(i)] \end{bmatrix} \end{bmatrix},$$

$$\tilde{A}^\top(i)\mathbb{P}(i) = \begin{bmatrix} \begin{bmatrix} [A(i) \\ +B(i)K(i)]^\top P(i) \end{bmatrix} & 0 \\ -K^\top(i)B^\top(i)P(i) & \begin{bmatrix} [A(i) \\ +L(i)C_y(i)]^\top Q(i) \end{bmatrix} \end{bmatrix},$$

$$\tilde{\mathbb{W}}^\top(i)\mathbb{P}(i)\tilde{\mathbb{W}}(i) = \begin{bmatrix} W^\top(i) & 0 \\ 0 & W^\top(i) \end{bmatrix} \begin{bmatrix} P(i) & 0 \\ 0 & Q(i) \end{bmatrix}$$
$$\times \begin{bmatrix} \mathbb{W}(i) & 0 \\ 0 & \mathbb{W}(i) \end{bmatrix}$$
$$= \begin{bmatrix} \mathbb{W}^\top(i)P(i)\mathbb{W}(i) & 0 \\ 0 & \mathbb{W}^\top(i)Q(i)\mathbb{W}(i) \end{bmatrix}.$$

Using these expressions, we have

$$\Psi(i) = \begin{bmatrix} \tilde{J}_P(i) & -P(i)B(i)K(i) \\ -K^\top(i)B^\top(i)P(i) & \tilde{J}_Q(i) \end{bmatrix}$$

$$+ \begin{bmatrix} \sum_{j=1}^{N} \lambda_{ij} P(j) & 0 \\ 0 & \sum_{j=1}^{N} \lambda_{ij} Q(j) \end{bmatrix}$$

$$= \begin{bmatrix} \tilde{J}_P(i) + \sum_{j=1}^{N} \lambda_{ij} P(j) & 0 \\ 0 & \tilde{J}_Q(i) + \sum_{j=1}^{N} \lambda_{ij} Q(j) \end{bmatrix}$$

$$+ \begin{bmatrix} 0 & -P(i)B(i)K(i) \\ -K^\top(i)B^\top(i)P(i) & 0 \end{bmatrix},$$

where

$$\tilde{J}_P(i) = [A(i) + B(i)K(i)]^\top P(i) + P(i)[A(i) + B(i)K(i)]$$
$$+ \mathbb{W}^\top(i)P(i)\mathbb{W}(i),$$

$$\tilde{J}_Q(i) = Q(i)[A(i) + L(i)C_y(i)] + [A(i) + L(i)C_y(i)]^\top Q(i)$$
$$+ \mathbb{W}^\top(i)Q(i)\mathbb{W}(i).$$

Notice that

$$\begin{bmatrix} 0 & -P(i)B(i)K(i) \\ -K^\top(i)B^\top(i)P(i) & 0 \end{bmatrix}$$

$$= \begin{bmatrix} 0 & -P(i)B(i)K(i) \\ 0 & 0 \end{bmatrix} + \begin{bmatrix} 0 & 0 \\ -K^\top(i)B^\top(i)P(i) & 0 \end{bmatrix}$$

and

$$\begin{bmatrix} 0 & -P(i)B(i)K(i) \\ 0 & 0 \end{bmatrix} = \begin{bmatrix} 0 & -P(i)B(i) \\ 0 & 0 \end{bmatrix} \begin{bmatrix} 0 & K(i) \\ 0 & 0 \end{bmatrix}.$$

Using Lemma 3 from Appendix A, we get

$$\begin{bmatrix} 0 & -P(i)B(i)K(i) \\ 0 & 0 \end{bmatrix} + \begin{bmatrix} 0 & 0 \\ -K^\top(i)B^\top(i)P(i) & 0 \end{bmatrix}$$

$$\leq \begin{bmatrix} P(i)B(i)B^\top(i)P(i) & 0 \\ 0 & K^\top(i)K(i) \end{bmatrix}.$$

Using conditions (3.81)–(3.82), we conclude that $\Psi(i) < 0$ for all $i \in \mathscr{S}$ and therefore

$$\mathscr{L}V(x(t), e(t), i) \leq - \min_{j \in \mathscr{S}} \{ \lambda_{min} [-\Psi(j)] \} \begin{bmatrix} x^\top(t) & e^\top(t) \end{bmatrix} \begin{bmatrix} x(t) \\ e(t) \end{bmatrix}.$$

The rest of the proof of this theorem can be obtained in a similar way as before. □

Let us now return to the determination of the controller parameters $K = (K(1), \cdots, K(N))$ and $L = (L(1), \cdots, L(N))$. Let us transform the conditions (3.81) and (3.82), starting with (3.81). Let $X(i) = P^{-1}(i)$. Pre- and post-multiplying the left-hand side of this condition by $X(i)$ after using the Schur complement, we get the following condition, which implies the previous one:

$$X(i)\left[A(i)+B(i)K(i)\right]^{\top}+\left[A(i)+B(i)K(i)\right]X(i)+B(i)B^{\top}(i)$$

$$+\sum_{j=1}^{N}\lambda_{ij}X(i)X^{-1}(j)X(i)+X(i)\mathbb{W}^{\top}(i)X^{-1}(i)\mathbb{W}(i)X(i)<0.$$

This inequality can be rewritten as follows:

$$X(i)A^{\top}(i)+X(i)K^{\top}(i)B^{\top}(i)+A(i)X(i)$$

$$+B(i)K(i)X(i)+B(i)B^{\top}(i)+\sum_{j=1}^{N}\lambda_{ij}X(i)X^{-1}(j)X(i)$$

$$+X(i)\mathbb{W}^{\top}(i)X^{-1}(i)\mathbb{W}(i)X(i)<0,$$

which, after letting $Y_c(i)=K(i)X(i)$ and using the Schur complement and the fact that $\sum_{j=1}^{N}\lambda_{ij}X(i)X^{-1}(j)X(i)=\lambda_{ii}X(i)+\mathcal{S}_i(X)\mathcal{X}^{-1}(X)\mathcal{S}_i^{\top}(X)$ gives

$$\begin{bmatrix} \mathscr{R}_X(i) & X(i)\mathbb{W}^{\top}(i) & \mathcal{S}_i(X) \\ \mathbb{W}(i)X(i) & -X(i) & 0 \\ \mathcal{S}_i^{\top}(X) & 0 & -\mathcal{X}(X) \end{bmatrix}<0,$$

with

$$\mathscr{R}_X(i)=X(i)A^{\top}(i)+Y_c^{\top}(i)B^{\top}(i)+A(i)X(i)+B(i)Y_c(i)$$
$$+B(i)B^{\top}(i)+\lambda_{ii}X(i).$$

For (3.82), we have

$$A^{\top}(i)Q(i)+Q(i)A(i)+Q(i)L(i)C_y(i)+C_y^{\top}(i)L^{\top}(i)Q(i)$$

$$+\sum_{j=1}^{N}\lambda_{ij}Q(j)+\mathbb{W}^{\top}(i)Q(i)\mathbb{W}(i)+K^{\top}(i)K(i)<0.$$

Letting $Y_o(i)=Q(i)L(i)$ and using the Schur complement, we get

$$\begin{bmatrix} \mathscr{R}_Q(i) & K^{\top}(i) \\ K(i) & -\mathbb{I} \end{bmatrix}<0,$$

with

$$\mathscr{R}_Q(i)=A^{\top}(i)Q(i)+Q(i)A(i)+Y_o(i)C_y(i)$$
$$+C_y^{\top}(i)Y_o^{\top}(i)+\sum_{j=1}^{N}\lambda_{ij}Q(j)+\mathbb{W}^{\top}(i)Q(i)\mathbb{W}(i).$$

The following theorem summarizes the results of this development.

Theorem 33. *If there exist sets of symmetric and positive-definite matrices* $X = (X(1), \cdots, X(N)) > 0$ *and* $Q = (Q(1), \cdots, Q(N)) > 0$ *and matrices* $Y_c = (Y_c(1), \cdots, Y_c(N))$ *and* $Y_o = (Y_o(1), \cdots, Y_o(N))$ *such that the following set of coupled LMIs holds for each* $i \in \mathscr{S}$:

$$\begin{bmatrix} \mathscr{R}_X(i) & X(i)\mathbb{W}^\top(i) & \mathcal{S}_i(X) \\ \mathbb{W}(i)X(i) & -X(i) & 0 \\ \mathcal{S}_i^\top(X) & 0 & -\mathcal{X}(X) \end{bmatrix} < 0, \tag{3.83}$$

$$\begin{bmatrix} \mathscr{R}_Q(i) & K^\top(i) \\ K(i) & -\mathbb{I} \end{bmatrix} < 0, \tag{3.84}$$

with

$$\begin{aligned} \mathscr{R}_X(i) = \; & X(i)A^\top(i) + Y_c^\top(i)B^\top(i) + A(i)X(i) \\ & + B(i)Y_c(i) + B(i)B^\top(i) + \lambda_{ii}X(i), \\ \mathscr{R}_Q(i) = \; & A^\top(i)Q(i) + Q(i)A(i) + Y_o(i)C_y(i) \\ & + C_y^\top(i)Y_o^\top(i) + \sum_{j=1}^N \lambda_{ij}Q(j) + \mathbb{W}^\top(i)Q(i)\mathbb{W}(i), \end{aligned}$$

then the controller gains that stochastically stabilize system (3.78) are given by

$$K(i) = Y_c(i)X^{-1}(i), \tag{3.85}$$

$$L(i) = Q^{-1}(i)Y_o(i). \tag{3.86}$$

Example 39. To show the usefulness of the theoretical results developed in this theorem, let us consider the two-mode system of the previous example.

Solving the LMIs (3.83)–(3.84), we get

$$X(1) = \begin{bmatrix} 18.1321 & -0.0132 \\ -0.0132 & 18.2444 \end{bmatrix}, \qquad X(2) = \begin{bmatrix} 16.0151 & -0.0141 \\ -0.0141 & 16.2159 \end{bmatrix},$$

$$Y_c(1) = \begin{bmatrix} -11.9152 & -272.2874 \\ 279.5776 & -11.6643 \end{bmatrix}, \qquad Y_c(2) = 10^3 \cdot \begin{bmatrix} -0.0326 & -2.9349 \\ 2.9266 & -0.0317 \end{bmatrix},$$

$$K_i(1) = \begin{bmatrix} -0.6680 & -14.9249 \\ 15.4185 & -0.6282 \end{bmatrix}, \qquad K_i(2) = \begin{bmatrix} -2.1920 & -180.9908 \\ 182.7414 & -1.7957 \end{bmatrix},$$

$$Q(1) = 10^4 \cdot \begin{bmatrix} 3.7464 & -0.0000 \\ -0.0000 & 3.7464 \end{bmatrix}, \qquad Q(2) = 10^4 \cdot \begin{bmatrix} 3.7464 & 0.0000 \\ 0.0000 & 3.7464 \end{bmatrix},$$

$$Y_o(1) = 10^4 \cdot \begin{bmatrix} -5.6386 & 0.7493 \\ 0.7493 & -5.6386 \end{bmatrix}, \qquad Y_o(2) = 10^4 \cdot \begin{bmatrix} -1.3123 & -0.9741 \\ -0.9741 & -0.9931 \end{bmatrix}.$$

This gives the following gains for the desired controller:

$$K(1) = \begin{bmatrix} -0.6680 & -14.9249 \\ 15.4185 & -0.6282 \end{bmatrix}, \qquad K(2) = \begin{bmatrix} -2.1920 & -180.9908 \\ 182.7414 & -1.7957 \end{bmatrix},$$

$$L(1) = \begin{bmatrix} -1.5051 & 0.2000 \\ 0.2000 & -1.5051 \end{bmatrix}, \qquad L(2) = \begin{bmatrix} -0.3503 & -0.2600 \\ -0.2600 & -0.2651 \end{bmatrix}.$$

Using the results of this theorem, the system of this example is stochastically stable under the observer-based output feedback control with the computed gains.

Let us now consider the effects of the uncertainties on the dynamics and establish the corresponding results that permit us to design the robust stabilizing controller of the form (3.79). Before proceeding with the design of such a controller, let us assume that some sets of gains characterize the controller and show that they stabilize our class of systems. The following theorem gives such results.

Theorem 34. Let $\varepsilon_A = (\varepsilon_A(1), \cdots, \varepsilon_A(N))$, $\varepsilon_B = (\varepsilon_B(1), \cdots, \varepsilon_B(N))$, $\varepsilon_Q = (\varepsilon_Q(1), \cdots, \varepsilon_Q(N))$, $\varepsilon_{Q_1} = (\varepsilon_{Q_1}(1), \cdots, \varepsilon_{Q_1}(N))$, $\varepsilon_{Q_2} = (\varepsilon_{Q_2}(1), \cdots, \varepsilon_{Q_2}(N))$, and $\varepsilon_{P_1} = (\varepsilon_{P_1}(1), \cdots, \varepsilon_{P_1}(N))$ be sets of given positive scalars. If there exist sets of symmetric and positive-definite matrices $P = (P(1), \cdots, P(N)) > 0$ and $Q = (Q(1), \cdots, Q(N)) > 0$ such that the following set of coupled LMIs holds for each $i \in \mathscr{S}$:

$$\begin{bmatrix} \mathscr{J}_P(i) & P(i)D_A(i) & P(i)D_B(i) & P(i)D_B(i) \\ D_A^\top(i)P(i) & -\varepsilon_A^{-1}(i)\mathbb{I} & 0 & 0 \\ D_B^\top(i)P(i) & 0 & -\varepsilon_B^{-1}(i)\mathbb{I} & 0 \\ D_B^\top(i)P(i) & 0 & 0 & -\varepsilon_{P_1}^{-1}(i)\mathbb{I} \\ E_B(i)K(i) & 0 & 0 & 0 \\ E_B(i)K(i) & 0 & 0 & 0 \\ B^\top(i)P(i) & 0 & 0 & 0 \end{bmatrix}$$

$$\begin{matrix} K^\top(i)E_B^\top(i) & K^\top(i)E_B^\top(i) & P(i)B(i) \\ 0 & 0 & 0 \\ 0 & 0 & 0 \\ 0 & 0 & 0 \\ -\varepsilon_B(i)\mathbb{I} & 0 & 0 \\ 0 & -\varepsilon_{Q_2}(i)\mathbb{I} & 0 \\ 0 & 0 & -\mathbb{I} \end{matrix} \Bigg] < 0, \qquad (3.87)$$

$$\begin{bmatrix} \mathscr{J}_Q(i) & Q(i)D_A(i) & Q(i)D_B(i) & Q(i)D_B(i) \\ D_A^\top(i)Q(i) & -\varepsilon_{Q_1}^{-1}(i)\mathbb{I} & 0 & 0 \\ D_B^\top(i)Q(i) & 0 & -\varepsilon_{Q_2}^{-1}(i)\mathbb{I} & 0 \\ D_B^\top(i)Q(i) & 0 & 0 & -\varepsilon_Q^{-1}(i)\mathbb{I} \\ E_B(i)K(i) & 0 & 0 & 0 \\ E_B(i)K(i) & 0 & 0 & 0 \\ K(i) & 0 & 0 & 0 \\ D_{C_y}^\top(i)L^\top(i)Q(i) & 0 & 0 & 0 \end{bmatrix}$$

$$\begin{bmatrix} K^\top(i)E_B^\top(i) & K^\top(i)E_B^\top(i) & K^\top(i) & Q(i)L(i)D_{C_y}(i) \\ 0 & 0 & 0 & 0 \\ 0 & 0 & 0 & 0 \\ 0 & 0 & 0 & 0 \\ -\varepsilon_{P_1}(i)\mathbb{I} & 0 & 0 & 0 \\ 0 & -\varepsilon_Q(i)\mathbb{I} & 0 & 0 \\ 0 & 0 & -\mathbb{I} & 0 \\ 0 & 0 & 0 & -\varepsilon_{C_y}(i)\mathbb{I} \end{bmatrix} < 0, \qquad (3.88)$$

where

$$\begin{aligned}
\mathscr{I}_P(i) = {}& [A(i) + B(i)K(i)]^\top P(i) + P(i)[A(i) + B(i)K(i)] \\
& + \sum_{j=1}^{N} \lambda_{ij} P(j) + \mathbb{W}^\top(i)P(i)\mathbb{W}(i) + \varepsilon_A^{-1}(i)E_A^\top(i)E_A(i) \\
& + \varepsilon_{C_y}(i)E_{C_y}^\top(i)E_{C_y}(i) + \varepsilon_{Q_1}^{-1}(i)E_A^\top(i)E_A(i), \\
\mathscr{I}_Q(i) = {}& [A(i) + L(i)C_y(i)]^\top Q(i) + Q(i)[A(i) + L(i)C_y(i)] \\
& + \sum_{j=1}^{N} \lambda_{ij} Q(j) + \mathbb{W}^\top(i)Q(i)\mathbb{W}(i),
\end{aligned}$$

then the closed-loop system is stochastically stable under the controller (3.79).

Proof: Considering the same Lyapunov candidate function as the previous theorem and following the same steps, we get

$$\begin{aligned}
\mathscr{L}V(x(t),e(t),i) = {}& 2\eta^\top(t)\tilde{A}^\top(i)\mathbb{P}(i)\eta(t) + 2\eta^\top(t)\varDelta\tilde{A}^\top(i,t)\mathbb{P}(i)\eta(t) \\
& + \sum_{j=1}^{N} \lambda_{ij}\eta^\top(t)\mathbb{P}(j)\eta(t) + \eta^\top(t)\tilde{\mathbb{W}}^\top(i)\mathbb{P}(i)\tilde{\mathbb{W}}(i)\eta(t),
\end{aligned}$$

which can be rewritten as

$$\begin{aligned}
\mathscr{L}V(x(t),e(t),i) = {}& \eta^\top(t)\Big[\tilde{A}^\top(i)\mathbb{P}(i) + \mathbb{P}(i)\tilde{A}(i) + \sum_{j=1}^{N} \lambda_{ij}\mathbb{P}(j) \\
& + \varDelta\tilde{A}^\top(i,t)\mathbb{P}(i) + \mathbb{P}(i)\varDelta\tilde{A}(i,t) \\
& + \tilde{\mathbb{W}}^\top(i)\mathbb{P}(i)\tilde{\mathbb{W}}(i)\Big]\eta(t).
\end{aligned}$$

Notice that

$$\mathbb{P}(i)\varDelta\tilde{A}(i,t) = \begin{bmatrix} P(i) & 0 \\ 0 & Q(i) \end{bmatrix}$$

$$\times \begin{bmatrix} \varDelta A(i,t) + \varDelta B(i,t)K(i) & -\varDelta B(i,t)K(i) \\ \varDelta A(i,t) + \varDelta B(i,t)K(i) + L(i)\varDelta C_y(i,t) & -\varDelta B(i,t)K(i) \end{bmatrix}$$

$$
= \begin{bmatrix} P(i)\Delta A(i,t) + P(i)\Delta B(i,t)K(i) & -P(i)\Delta B(i,t)K(i) \\ \begin{bmatrix} Q(i)\Delta A(i,t) \\ +Q(i)\Delta B(i,t)K(i) \\ +Q(i)L(i)\Delta C_y(i,t) \end{bmatrix} & -Q(i)\Delta B(i,t)K(i) \end{bmatrix},
$$

$$
\Delta \tilde{A}^\top(i,t)\mathbb{P}(i)
$$

$$
= \begin{bmatrix} \begin{bmatrix} \Delta A^\top(i,t)P(i) \\ +K^\top(i)\Delta B^\top(i,t)P(i) \end{bmatrix} & \begin{bmatrix} \Delta A^\top(i,t)Q(i) \\ +K^\top(i)\Delta B^\top(i,t)Q(i) \\ \Delta C_{C_y}^\top(i)L^\top(i)Q(i) \end{bmatrix} \\ -K^\top(i)\Delta B^\top(i,t)P(i) & -K^\top(i)\Delta B^\top(i,t)Q(i) \end{bmatrix},
$$

and

$$
\begin{bmatrix} P(i)\Delta A(i,t) & 0 \\ 0 & 0 \end{bmatrix} = \begin{bmatrix} P(i)D_A(i)F_A(i,t)E_A(i) & 0 \\ 0 & 0 \end{bmatrix}
$$
$$
= \begin{bmatrix} P(i)D_A(i) & 0 \\ 0 & 0 \end{bmatrix} \begin{bmatrix} F_A(i,t) & 0 \\ 0 & 0 \end{bmatrix} \begin{bmatrix} E_A(i) & 0 \\ 0 & 0 \end{bmatrix},
$$

$$
\begin{bmatrix} 0 & 0 \\ Q(i)\Delta A(i,t) & 0 \end{bmatrix} = \begin{bmatrix} 0 & 0 \\ Q(i)D_A(i)F_A(i,t)E_A(i) & 0 \end{bmatrix}
$$
$$
= \begin{bmatrix} 0 & 0 \\ 0 & Q(i)D_A(i) \end{bmatrix} \begin{bmatrix} 0 & 0 \\ 0 & F_A(i,t) \end{bmatrix} \begin{bmatrix} 0 & 0 \\ E_A(i) & 0 \end{bmatrix},
$$

$$
\begin{bmatrix} P(i)\Delta B(i,t)K(i) & 0 \\ 0 & 0 \end{bmatrix} = \begin{bmatrix} P(i)D_B(i)F_B(i,t)E_B(i)K(i) & 0 \\ 0 & 0 \end{bmatrix}
$$
$$
= \begin{bmatrix} P(i)D_B(i) & 0 \\ 0 & 0 \end{bmatrix} \begin{bmatrix} F_B(i,t) & 0 \\ 0 & 0 \end{bmatrix} \begin{bmatrix} E_B(i)K(i) & 0 \\ 0 & 0 \end{bmatrix},
$$

$$
\begin{bmatrix} 0 & 0 \\ Q(i)\Delta B(i,t)K(i) & 0 \end{bmatrix} = \begin{bmatrix} 0 & 0 \\ Q(i)D_B(i)F_B(i,t)E_B(i)K(i) & 0 \end{bmatrix}
$$
$$
= \begin{bmatrix} 0 & 0 \\ 0 & Q(i)D_B(i) \end{bmatrix} \begin{bmatrix} 0 & 0 \\ 0 & F_B(i,t) \end{bmatrix} \begin{bmatrix} 0 & 0 \\ E_B(i)K(i) & 0 \end{bmatrix},
$$

$$
\begin{bmatrix} 0 & 0 \\ Q(i)L(i)\Delta C_y(i,t) & 0 \end{bmatrix} = \begin{bmatrix} 0 & 0 \\ Q(i)L(i)D_{C_y}(i)F_{C_y}(i,t)E_{C_y}(i) & 0 \end{bmatrix}
$$
$$
= \begin{bmatrix} 0 & 0 \\ 0 & Q(i)L(i)D_{C_y}(i) \end{bmatrix} \begin{bmatrix} 0 & 0 \\ 0 & F_{C_y}(i,t) \end{bmatrix} \begin{bmatrix} 0 & 0 \\ E_{C_y}(i) & 0 \end{bmatrix},
$$

$$\begin{bmatrix} 0 & -P(i)\Delta B(i,t)K(i) \\ 0 & 0 \end{bmatrix} = \begin{bmatrix} 0 & -P(i)D_B(i)F_B(i,t)E_B(i)K(i) \\ 0 & 0 \end{bmatrix}$$

$$= \begin{bmatrix} 0 & -P(i)D_B(i) \\ 0 & 0 \end{bmatrix} \begin{bmatrix} 0 & 0 \\ 0 & F_B(i,t) \end{bmatrix} \begin{bmatrix} 0 & 0 \\ 0 & E_B(i)K(i) \end{bmatrix},$$

$$\begin{bmatrix} 0 & 0 \\ 0 & -Q(i)\Delta B(i,t)K(i) \end{bmatrix} = \begin{bmatrix} 0 & 0 \\ 0 & -Q(i)D_B(i)F_B(i,t)E_B(i)K(i) \end{bmatrix}$$

$$= \begin{bmatrix} 0 & 0 \\ 0 & -Q(i)D_B(i) \end{bmatrix} \begin{bmatrix} 0 & 0 \\ 0 & F_B(i,t) \end{bmatrix} \begin{bmatrix} 0 & 0 \\ 0 & E_B(i)K(i) \end{bmatrix}.$$

Using Lemma 7 from Appendix A, we get

$$\begin{bmatrix} P(i)\Delta A(i,t) & 0 \\ 0 & 0 \end{bmatrix} + \begin{bmatrix} \Delta A^\top(i,t)P(i) & 0 \\ 0 & 0 \end{bmatrix}$$

$$\le \varepsilon_A(i) \begin{bmatrix} P(i)D_A(i) & 0 \\ 0 & 0 \end{bmatrix} \begin{bmatrix} D_A^\top(i)P(i) & 0 \\ 0 & 0 \end{bmatrix}$$

$$+ \varepsilon_A^{-1}(i) \begin{bmatrix} E_A^\top(i) & 0 \\ 0 & 0 \end{bmatrix} \begin{bmatrix} E_A(i) & 0 \\ 0 & 0 \end{bmatrix}$$

$$= \begin{bmatrix} \varepsilon_A(i)P(i)D_A(i)D_A^\top(i)P(i) + \varepsilon_A^{-1}(i)E_A^\top(i)E_A(i) & 0 \\ 0 & 0 \end{bmatrix},$$

$$\begin{bmatrix} P(i)\Delta B(i,t)K(i) & 0 \\ 0 & 0 \end{bmatrix} + \begin{bmatrix} K^\top(i)\Delta B^\top(i,t)P(i) & 0 \\ 0 & 0 \end{bmatrix}$$

$$\le \varepsilon_B(i) \begin{bmatrix} P(i)D_B(i) & 0 \\ 0 & 0 \end{bmatrix} \begin{bmatrix} D_B^\top(i)P(i) & 0 \\ 0 & 0 \end{bmatrix}$$

$$+ \varepsilon_B^{-1}(i) \begin{bmatrix} K^\top(i)E_B^\top(i) & 0 \\ 0 & 0 \end{bmatrix} \begin{bmatrix} E_B(i)K(i) & 0 \\ 0 & 0 \end{bmatrix}$$

$$= \begin{bmatrix} \varepsilon_B(i)P(i)D_B(i)D_B^\top(i)P(i) + \varepsilon_B^{-1}(i)K^\top(i)E_B^\top(i)E_B(i)K(i) & 0 \\ 0 & 0 \end{bmatrix},$$

$$\begin{bmatrix} 0 & 0 \\ Q(i)\Delta A(i,t) & 0 \end{bmatrix} + \begin{bmatrix} 0 & \Delta A^\top(i,t)Q(i) \\ 0 & 0 \end{bmatrix}$$

$$\le \varepsilon_{Q_1}(i) \begin{bmatrix} 0 & 0 \\ 0 & Q(i)D_A(i)D_A^\top(i)Q(i) \end{bmatrix}$$

$$+ \varepsilon_{Q_1}^{-1}(i) \begin{bmatrix} E_A^\top(i)E_A(i) & 0 \\ 0 & 0 \end{bmatrix}$$

$$= \begin{bmatrix} \varepsilon_{Q_1}^{-1}(i)E_A^\top(i)E_A(i) & 0 \\ 0 & \varepsilon_{Q_1}(i)Q(i)D_A(i)D_A^\top(i)Q(i) \end{bmatrix},$$

$$\begin{bmatrix} 0 & 0 \\ Q(i)\Delta B(i,t)K(i) & 0 \end{bmatrix} + \begin{bmatrix} 0 & K^\top(i)\Delta B^\top(i,t)Q(i) \\ 0 & 0 \end{bmatrix}$$

$$\leq \varepsilon_{Q_2}(i) \begin{bmatrix} 0 & 0 \\ 0 & Q(i)D_B(i)D_B^\top(i)Q(i) \end{bmatrix}$$

$$+ \varepsilon_{Q_2}^{-1}(i) \begin{bmatrix} K^\top(i)E_B^\top(i)E_B(i)K(i) & 0 \\ 0 & 0 \end{bmatrix}$$

$$= \begin{bmatrix} \varepsilon_{Q_2}^{-1}(i)K^\top(i)E_B^\top(i)E_B(i)K(i) & 0 \\ 0 & \varepsilon_{Q_2}(i)Q(i)D_B(i)D_B^\top(i)Q(i) \end{bmatrix},$$

$$\begin{bmatrix} 0 & 0 \\ Q(i)L(i)\Delta C_y(i,t) & 0 \end{bmatrix} + \begin{bmatrix} 0 & \Delta C_y^\top(i,t)L^\top(i)Q(i) \\ 0 & 0 \end{bmatrix}$$

$$\leq \varepsilon_{C_y}^{-1}(i) \begin{bmatrix} 0 & 0 \\ 0 & Q(i)L(i)D_{C_y}(i)D_{C_y}^\top(i)L^\top(i)Q(i) \end{bmatrix}$$

$$+ \varepsilon_{C_y}(i) \begin{bmatrix} E_{C_y}^\top(i)E_{C_y}(i) & 0 \\ 0 & 0 \end{bmatrix}$$

$$= \begin{bmatrix} \varepsilon_{C_y}(i)E_{C_y}^\top(i)E_{C_y}(i) & 0 \\ 0 & \varepsilon_{C_y}^{-1}(i)Q(i)L(i)D_{C_y}(i)D_{C_y}^\top(i)L^\top(i)Q(i) \end{bmatrix},$$

$$\begin{bmatrix} 0 & -P(i)\Delta B(i,t)K(i) \\ 0 & 0 \end{bmatrix} + \begin{bmatrix} 0 & 0 \\ -K^\top(i)\Delta B^\top(i,t)P(i) & \end{bmatrix}$$

$$\leq \varepsilon_{P_1}(i) \begin{bmatrix} P(i)D_B(i)D_B^\top(i)P(i) & 0 \\ 0 & 0 \end{bmatrix}$$

$$+ \varepsilon_{P_1}^{-1}(i) \begin{bmatrix} 0 & 0 \\ 0 & K^\top(i)E_B^\top(i)E_B(i)K(i) \end{bmatrix}$$

$$= \begin{bmatrix} \varepsilon_{P_1}(i)P(i)D_B(i)D_B^\top(i)P(i) & 0 \\ 0 & \varepsilon_{P_1}^{-1}(i)K^\top(i)E_B^\top(i)E_B(i)K(i) \end{bmatrix},$$

$$\begin{bmatrix} 0 & 0 \\ 0 & -Q(i)\Delta B(i,t)K(i) \end{bmatrix} + \begin{bmatrix} 0 & 0 \\ 0 & -K^\top(i)\Delta B^\top(i,t)Q(i) \end{bmatrix}$$

$$\leq \varepsilon_Q(i) \begin{bmatrix} 0 & 0 \\ 0 & Q(i)D_B(i)D_B^\top(i)Q(i) \end{bmatrix}$$

$$+ \varepsilon_Q^{-1}(i) \begin{bmatrix} 0 & 0 \\ 0 & K^\top(i)E_B^\top(i)E_B(i)K(i) \end{bmatrix}$$

$$= \begin{bmatrix} 0 & 0 \\ 0 & \begin{array}{c} \varepsilon_Q(i)Q(i)D_B(i)D_B^\top(i)Q(i) \\ + \varepsilon_Q^{-1}(i)K^\top(i)E_B^\top(i)E_B(i)K(i) \end{array} \end{bmatrix}.$$

Using all these transformations, we obtain

$$\mathscr{L}V(x(t), e(t), i) \leq \eta^\top(t)\Psi(i)\eta(t),$$

with

$$\Phi(i) = \tilde{A}^\top(i)\mathbb{P}(i) + \mathbb{P}(i)\tilde{A}(i) + \sum_{j=1}^{N} \lambda_{ij}\mathbb{P}(j) + \tilde{\mathbb{W}}^\top(i)\mathbb{P}(i)\tilde{\mathbb{W}}(i)$$

$$+ \begin{bmatrix} \varepsilon_A(i)P(i)D_A(i)D_A^\top(i)P(i) + \varepsilon_A^{-1}(i)E_A^\top(i)E_A(i) & 0 \\ 0 & 0 \end{bmatrix}$$

$$+ \begin{bmatrix} \begin{bmatrix} \varepsilon_B(i)P(i)D_B(i)D_B^\top(i)P(i) \\ +\varepsilon_B^{-1}(i)K^\top(i)E_B^\top(i)E_B(i)K(i) \end{bmatrix} & 0 \\ 0 & 0 \end{bmatrix}$$

$$+ \begin{bmatrix} \varepsilon_{Q_1}^{-1}(i)E_A^\top(i)E_A(i) & 0 \\ 0 & \varepsilon_{Q_1}(i)Q(i)D_A(i)D_A^\top(i)Q(i) \end{bmatrix}$$

$$+ \begin{bmatrix} \varepsilon_{Q_2}^{-1}(i)K^\top(i)E_B^\top(i)E_B(i)K(i) & 0 \\ 0 & \varepsilon_{Q_2}(i)Q(i)D_B(i)D_B^\top(i)Q(i) \end{bmatrix}$$

$$+ \begin{bmatrix} \varepsilon_{C_y}(i)E_{C_y}^\top(i)E_{C_y}(i) & 0 \\ 0 & \varepsilon_{C_y}^{-1}(i)Q(i)L(i)D_{C_y}(i)D_{C_y}^\top(i)L^\top(i)Q(i) \end{bmatrix}$$

$$+ \begin{bmatrix} \varepsilon_{P_1}(i)P(i)D_B(i)D_B^\top(i)P(i) & 0 \\ 0 & \varepsilon_{P_1}^{-1}(i)K^\top(i)E_B^\top(i)E_B(i)K(i) \end{bmatrix}$$

$$+ \begin{bmatrix} 0 & 0 \\ 0 & \begin{bmatrix} \varepsilon_Q(i)Q(i)D_B(i)D_B^\top(i)Q(i) \\ +\varepsilon_Q^{-1}(i)K^\top(i)E_B^\top(i)E_B(i)K(i) \end{bmatrix} \end{bmatrix}.$$

Notice that

$$\mathbb{P}(i)\tilde{A}(i) = \begin{bmatrix} P(i) & 0 \\ 0 & Q(i) \end{bmatrix} \begin{bmatrix} \begin{bmatrix} A(i) \\ +B(i)K(i) \end{bmatrix} & -B(i)K(i) \\ 0 & \begin{bmatrix} A(i) \\ +L(i)C_y(i) \end{bmatrix} \end{bmatrix}$$

$$= \begin{bmatrix} \begin{bmatrix} P(i)\,[A(i) \\ +B(i)K(i)] \end{bmatrix} & -P(i)B(i)K(i) \\ 0 & \begin{bmatrix} Q(i)\,[A(i) \\ +L(i)C_y(i)] \end{bmatrix} \end{bmatrix},$$

$$\tilde{A}^\top(i)\mathbb{P}(i) = \begin{bmatrix} \begin{bmatrix} [A(i) \\ +B(i)K(i)]^\top P(i) \end{bmatrix} & 0 \\ -K^\top(i)B^\top(i)P(i) & \begin{bmatrix} [A(i) \\ +L(i)C_y(i)]^\top Q(i) \end{bmatrix} \end{bmatrix},$$

$$\tilde{\mathbb{W}}^\top(i)\mathbb{P}(i)\tilde{\mathbb{W}}(i) = \begin{bmatrix} \mathbb{W}^\top(i) & 0 \\ 0 & \mathbb{W}^\top(i) \end{bmatrix} \begin{bmatrix} P(i) & 0 \\ 0 & Q(i) \end{bmatrix}$$

$$\times \begin{bmatrix} \mathbb{W}(i) & 0 \\ 0 & \mathbb{W}(i) \end{bmatrix}$$

$$= \begin{bmatrix} \mathbb{W}^{\top}(i)P(i)\mathbb{W}(i) & 0 \\ 0 & \mathbb{W}^{\top}(i)Q(i)\mathbb{W}(i) \end{bmatrix}.$$

Using these expressions, we have

$$\Phi(i) = \begin{bmatrix} \bar{J}_P(i) & -P(i)B(i)K(i) \\ -K^{\top}(i)B^{\top}(i)P(i) & \bar{J}_Q(i) \end{bmatrix}$$

$$+ \begin{bmatrix} \sum_{j=1}^{N} \lambda_{ij} P(j) & 0 \\ 0 & \sum_{j=1}^{N} \lambda_{ij} Q(j) \end{bmatrix}$$

$$= \begin{bmatrix} \bar{J}_P(i) + \sum_{j=1}^{N} \lambda_{ij} P(j) & 0 \\ 0 & \bar{J}_Q(i) + \sum_{j=1}^{N} \lambda_{ij} Q(j) \end{bmatrix}$$

$$+ \begin{bmatrix} 0 & -P(i)B(i)K(i) \\ -K^{\top}(i)B^{\top}(i)P(i) & 0 \end{bmatrix},$$

where

$$
\begin{aligned}
\bar{J}_P(i) =\ & [A(i) + B(i)K(i)]^{\top} P(i) + P(i)[A(i) + B(i)K(i)] \\
& + \mathbb{W}^{\top}(i)P(i)\mathbb{W}(i) \\
& + \varepsilon_A(i)P(i)D_A(i)D_A^{\top}(i)P(i) + \varepsilon_A^{-1}(i)E_A^{\top}(i)E_A(i) \\
& + \varepsilon_B(i)P(i)D_B(i)D_B^{\top}(i)P(i) \\
& + \varepsilon_B^{-1}(i)K^{\top}(i)E_B^{\top}(i)E_B(i)K(i) \\
& + \varepsilon_{Q_1}^{-1}(i)E_A^{\top}(i)E_A(i) \\
& + \varepsilon_{Q_2}^{-1}(i)K^{\top}(i)E_B^{\top}(i)E_B(i)K(i) \\
& + \varepsilon_{P_1}(i)P(i)D_B(i)D_B^{\top}(i)P(i) + \varepsilon_{C_y}(i)E_{C_y}^{\top}(i)E_{C_y}(i),
\end{aligned}
$$

$$
\begin{aligned}
\bar{J}_Q(i) =\ & Q(i)[A(i) + L(i)C_y(i)] + [A(i) + L(i)C_y(i)]^{\top} Q(i) \\
& + \mathbb{W}^{\top}(i)Q(i)\mathbb{W}(i) \\
& + \varepsilon_{Q_1}(i)Q(i)D_A(i)D_A^{\top}(i)Q(i) \\
& + \varepsilon_{Q_2}(i)Q(i)D_B(i)D_B^{\top}(i)Q(i) \\
& + \varepsilon_{P_1}^{-1}(i)K^{\top}(i)E_B^{\top}(i)E_B(i)K(i) \\
& + \varepsilon_Q(i)Q(i)D_B(i)D_B^{\top}(i)Q(i) \\
& + \varepsilon_Q^{-1}(i)K^{\top}(i)E_B^{\top}(i)E_B(i)K(i) \\
& + \varepsilon_{C_y}^{-1}(i)Q(i)L(i)D_{C_y}(i)D_{C_y}^{\top}(i)L^{\top}(i)Q(i).
\end{aligned}
$$

Notice that

$$\begin{bmatrix} 0 & -P(i)B(i)K(i) \\ 0 & 0 \end{bmatrix} = \begin{bmatrix} 0 & -P(i)B(i) \\ 0 & 0 \end{bmatrix} \begin{bmatrix} 0 & K(i) \\ 0 & 0 \end{bmatrix}.$$

Using Lemma 2 from Appendix A, we get

$$\begin{bmatrix} 0 & -P(i)B(i)K(i) \\ 0 & 0 \end{bmatrix} + \begin{bmatrix} 0 & 0 \\ -K^{\top}(i)B^{\top}(i)P(i) & 0 \end{bmatrix}$$

$$\leq \begin{bmatrix} P(i)B(i)B^\top(i)P(i) & 0 \\ 0 & K^\top(i)K(i) \end{bmatrix}.$$

letting $\mathcal{J}_P(i)$ and $\mathcal{J}_Q(i)$ be defined as follows:

$$\begin{aligned}
\mathcal{J}_P(i) = &\ [A(i) + B(i)K(i)]\, P(i) + P(i)\,[A(i) + B(i)K(i)]^\top \\
&+ \mathbb{W}^\top(i)P(i)\mathbb{W}(i) + \sum_{ij} \lambda_{ij} P(j) + \varepsilon_A^{-1}(i)E_A^\top(i)E_A(i) \\
&+ \varepsilon_{C_y}(i)E_{C_y}^\top(i)E_{C_y}(i) + \varepsilon_{Q_1}^{-1}(i)E_A^\top(i)E_A(i) \\
\mathcal{J}_Q(i) = &\ Q(i)\,[A(i) + L(i)C_y(i)] + [A(i) + L(i)C_y(i)]^\top Q(i) \\
&+ \mathbb{W}^\top(i)Q(i)\mathbb{W}(i) + \sum_{j=1}^N \lambda_{ij} Q(j),
\end{aligned}$$

and taking care of all these transformations, we get

$$\begin{bmatrix}
\mathcal{J}_P(i) & P(i)D_A(i) & P(i)D_B(i) & P(i)D_B(i) \\
D_A^\top(i)P(i) & -\varepsilon_A^{-1}\mathbb{I} & 0 & 0 \\
D_B^\top(i)P(i) & 0 & -\varepsilon_B^{-1}\mathbb{I} & 0 \\
D_B^\top(i)P(i) & 0 & 0 & -\varepsilon_{P_1}^{-1}\mathbb{I} \\
E_B(i)K(i) & 0 & 0 & 0 \\
E_B(i)K(i) & 0 & 0 & 0 \\
B^\top(i)P(i) & 0 & 0 & 0
\end{bmatrix}$$

$$\left.\begin{bmatrix}
K^\top(i)E_B^\top(i) & K^\top(i)E_B^\top(i) & P(i)B(i) \\
0 & 0 & 0 \\
0 & 0 & 0 \\
0 & 0 & 0 \\
-\varepsilon_B\mathbb{I} & 0 & 0 \\
0 & -\varepsilon_{Q_2}\mathbb{I} & 0 \\
0 & 0 & -\mathbb{I}
\end{bmatrix}\right] < 0,$$

$$\begin{bmatrix}
\mathcal{J}_Q(i) & Q(i)D_A(i) & Q(i)D_B(i) & Q(i)D_B(i) \\
D_A^\top(i)Q(i) & -\varepsilon_{Q_1}^{-1}\mathbb{I} & 0 & 0 \\
D_B^\top(i)Q(i) & 0 & -\varepsilon_{Q_2}^{-1}\mathbb{I} & 0 \\
D_B^\top(i)Q(i) & 0 & 0 & -\varepsilon_Q^{-1}\mathbb{I} \\
E_B(i)K(i) & 0 & 0 & 0 \\
E_B(i)K(i) & 0 & 0 & 0 \\
K(i) & 0 & 0 & 0 \\
D_{C_y}^\top(i)L^\top(i)Q(i) & 0 & 0 & 0
\end{bmatrix}$$

$$
\left.
\begin{matrix}
K^\top(i)E_B^\top(i) & K^\top(i)E_B^\top(i) & K^\top(i) & Q(i)L(i)D_{C_y}(i) \\
0 & 0 & 0 & 0 \\
0 & 0 & 0 & 0 \\
0 & 0 & 0 & 0 \\
-\varepsilon_{P_1}\mathbb{I} & 0 & 0 & 0 \\
0 & -\varepsilon_Q\mathbb{I} & 0 & 0 \\
0 & 0 & -\mathbb{I} & 0 \\
0 & 0 & 0 & -\varepsilon_{C_y}(i)\mathbb{I}
\end{matrix}
\right] < 0,
$$

which are both negative-definite by hypothesis.

Using (3.87)–(3.88), we conclude that $\Phi(i) < 0$ for all $i \in \mathscr{S}$ and therefore

$$
\mathscr{L}V(x(t), e(t), i) \le - \min_{j \in \mathscr{S}} \{\lambda_{min}[-\Phi(j)]\} \begin{bmatrix} x^\top(t) & e^\top(t) \end{bmatrix} \begin{bmatrix} x(t) \\ e(t) \end{bmatrix}.
$$

The rest of the proof of this theorem can be obtained using the same steps used in the proof of Theorem 32. □

Let us now return to the determination of the controller parameters $K = (K(1), \cdots, K(N))$ and $L = (L(1), \cdots, L(N))$. Transform the conditions (3.87) and (3.88), starting with (3.87). Let $X(i) = P^{-1}(i)$. Pre- and post-multiplying the left-hand side of the equivalence of this condition (using the Schur complement) by $X(i)$ gives the following condition, which implies the previous one:

$$
X(i)[A(i) + B(i)K(i)]^\top + [A(i) + B(i)K(i)]X(i)
$$
$$
+ X(i)\mathbb{W}^\top(i)X^{-1}(i)\mathbb{W}(i)X(i)
$$
$$
+ \varepsilon_A(i)D_A(i)D_A^\top(i) + \varepsilon_A^{-1}(i)X(i)E_A^\top(i)E_A(i)X(i)
$$
$$
+ \varepsilon_B(i)D_B(i)D_B^\top(i) + \varepsilon_B^{-1}(i)X(i)K^\top(i)E_B^\top(i)E_B(i)K(i)X(i)
$$
$$
+ \varepsilon_{Q_1}^{-1}(i)X(i)E_A^\top(i)E_A(i)X(i)
$$
$$
+ \varepsilon_{Q_2}^{-1}(i)X(i)K^\top(i)E_B^\top(i)E_B(i)K(i)X(i) + \varepsilon_{C_y}(i)X(i)E_{C_y}^\top(i)E_{C_y}(i)X(i)
$$
$$
+ \varepsilon_{P_1}(i)D_B(i)D_B^\top(i) + \sum_{j=1}^{N}\lambda_{ij}X(i)X^{-1}(j)X(i) + B(i)B^\top(i) < 0.
$$

This inequality can be rewritten as follows:

$$
X(i)A^\top(i) + X(i)K^\top(i)B^\top(i) + A(i)X(i) + B(i)K(i)X(i)
$$
$$
+ B(i)B^\top(i) + \varepsilon_A(i)D_A(i)D_A^\top(i) + \varepsilon_A^{-1}(i)X(i)E_A^\top(i)E_A(i)X(i)
$$
$$
+ \varepsilon_B(i)D_B(i)D_B^\top(i) + \varepsilon_B^{-1}(i)X(i)K^\top(i)E_B^\top(i)E_B(i)K(i)X(i)
$$
$$
+ \varepsilon_{Q_1}^{-1}(i)X(i)E_A^\top(i)E_A(i)X(i) + \varepsilon_{C_y}(i)X(i)E_{C_y}^\top(i)E_{C_y}(i)X(i)
$$
$$
+ \varepsilon_{Q_2}^{-1}(i)X(i)K^\top(i)E_B^\top(i)E_B(i)K(i)X(i) + \varepsilon_{P_1}(i)D_B(i)D_B^\top(i)
$$
$$
+ \sum_{j=1}^{N}\lambda_{ij}X(i)X^{-1}(j)X(i) + X(i)\mathbb{W}^\top(i)X^{-1}(i)\mathbb{W}(i)X(i) < 0,
$$

which, after letting $Y_c(i) = K(i)X(i)$ and using the Schur complement and the fact that $\sum_{j=1}^{N} \lambda_{ij} X(i) X^{-1}(j) X(i) = \lambda_{ii} X(i) + S_i(X) \mathcal{X}^{-1}(X) S_i^\top(X)$, gives

$$
\begin{bmatrix}
\mathscr{R}_X(i) & X(i)E_A^\top(i) & X(i)E_A^\top(i) & Y_c^\top(i)E_B^\top(i) \\
E_A(i)X(i) & -\varepsilon_A(i)\mathbb{I} & 0 & 0 \\
E_A(i)X(i) & 0 & -\varepsilon_{Q_1}(i)\mathbb{I} & 0 \\
E_B(i)Y_c(i) & 0 & 0 & -\varepsilon_B(i)\mathbb{I} \\
E_B(i)Y_c(i) & 0 & 0 & 0 \\
\mathbb{W}(i)X(i) & 0 & 0 & 0 \\
E_{C_y}(i)X(i) & 0 & 0 & 0 \\
S_i^\top(X) & 0 & 0 & 0
\end{bmatrix}
$$

$$
\begin{bmatrix}
Y_c^\top(i)E_B^\top(i) & X(i)\mathbb{W}^\top(i) & X(i)E_{C_y}^\top(i) & S_i(X) \\
0 & 0 & 0 & 0 \\
0 & 0 & 0 & 0 \\
0 & 0 & 0 & 0 \\
-\varepsilon_{Q_2}(i)\mathbb{I} & 0 & 0 & 0 \\
0 & -X(i) & 0 & 0 \\
0 & 0 & -\mathbb{I} & 0 \\
0 & 0 & 0 & -\mathcal{X}(X)
\end{bmatrix} < 0,
$$

with

$$
\begin{aligned}
\mathscr{R}_X(i) =\ & X(i)A^\top(i) + Y_c^\top(i)B^\top(i) + A(i)X(i) + B(i)Y_c(i) \\
& + B(i)B^\top(i) + \lambda_{ii}X(i) \\
& + \varepsilon_A(i)D_A(i)D_A^\top(i) + \varepsilon_B(i)D_B(i)D_B^\top(i) \\
& + \varepsilon_{P_1}(i)D_B(i)D_B^\top(i).
\end{aligned}
$$

For (3.88), following the same steps as for the inequality (3.87), we have

$$
\begin{aligned}
& Q(i)\left[A(i) + L(i)C_y(i)\right] + \left[A(i) + L(i)C_y(i)\right]^\top Q(i) + \mathbb{W}^\top(i)Q(i)\mathbb{W}(i) \\
& + \varepsilon_{Q_1}(i)Q(i)D_A(i)D_A^\top(i)Q(i) + \varepsilon_{Q_2}(i)Q(i)D_B(i)D_B^\top(i)Q(i) \\
& + \varepsilon_{P_1}^{-1}(i)K^\top(i)E_B^\top(i)E_B(i)K(i) + \varepsilon_Q(i)Q(i)D_B(i)D_B^\top(i)Q(i) \\
& + \varepsilon_Q^{-1}(i)K^\top(i)E_B^\top(i)E_B(i)K(i) + K^\top(i)K(i) + \sum_{j=1}^{N}\lambda_{ij}Q(j) \\
& + \varepsilon_{C_y}^{-1}(i)Q(i)L(i)D_{C_y}(i)D_{C_y}^\top(i)L^\top(i)Q(i) < 0.
\end{aligned}
$$

Letting $Y_o(i) = Q(i)L(i)$ and using the Schur complement, we get

$$
\begin{bmatrix}
\mathscr{R}_Q(i) & Q(i)D_A(i) & Q(i)D_B(i) & Q(i)D_B(i) \\
D_A^\top(i)Q(i) & -\varepsilon_{Q_1}^{-1}(i)\mathbb{I} & 0 & 0 \\
D_B^\top(i)Q(i) & 0 & -\varepsilon_{Q_2}^{-1}(i)\mathbb{I} & 0 \\
D_A^\top(i)Q(i) & 0 & 0 & -\varepsilon_Q^{-1}(i)\mathbb{I} \\
E_B(i)K(i) & 0 & 0 & 0 \\
E_B(i)K(i) & 0 & 0 & 0 \\
K(i) & 0 & 0 & 0 \\
D_{C_y}^\top(i)Y_o^\top(i) & 0 & 0 & 0
\end{bmatrix}
$$

$$
\begin{bmatrix}
K^\top(i)E_B^\top(i) & K^\top(i)E_B^\top(i) & K^\top(i) & Y_o(i)D_{c_y}(i) \\
0 & 0 & 0 & 0 \\
0 & 0 & 0 & 0 \\
0 & 0 & 0 & 0 \\
-\varepsilon_{P_1}(i)\mathbb{I} & 0 & 0 & 0 \\
0 & -\varepsilon_Q(i)\mathbb{I} & 0 & 0 \\
0 & 0 & -\mathbb{I} & 0 \\
0 & 0 & 0 & -\varepsilon_{C_y}(i)\mathbb{I}
\end{bmatrix} < 0,
$$

with

$$
\mathscr{R}_Q(i) = A^\top(i)Q(i) + Q(i)A(i) + Y_o(i)C_y(i) + C_y^\top(i)Y_o^\top(i)
$$

$$
+ \sum_{j=1}^N \lambda_{ij} Q(j) + \mathbb{W}^\top(i)Q(i)\mathbb{W}(i).
$$

The following theorem summarizes the results of this development.

Theorem 35. *Let* $\varepsilon_A = (\varepsilon_A(1), \cdots, \varepsilon_A(N))$, $\varepsilon_B = (\varepsilon_B(1), \cdots, \varepsilon_B(N))$, $\varepsilon_Q = (\varepsilon_Q(1), \cdots, \varepsilon_Q(N))$, $\varepsilon_{Q_1} = (\varepsilon_{Q_1}(1), \cdots, \varepsilon_{Q_1}(N))$, $\varepsilon_{Q_2} = (\varepsilon_{Q_2}(1), \cdots, \varepsilon_{Q_2}(N))$, *and* $\varepsilon_{P_1} = (\varepsilon_{P_1}(1), \cdots, \varepsilon_{P_1}(N))$ *be sets of given positive scalars. If there exist sets of symmetric and positive-definite matrices* $X = (X(1), \cdots, X(N)) > 0$ *and* $Q = (Q(1), \cdots, Q(N)) > 0$ *and matrices* $Y_c = (Y_c(1), \cdots, Y_c(N))$ *and* $Y_o = (Y_o(1), \cdots, Y_o(N))$ *such that the following set of coupled LMIs holds for each* $i \in \mathscr{S}$:

$$
\begin{bmatrix}
\mathscr{R}_X(i) & X(i)E_A^\top(i) & X(i)E_A^\top(i) & Y_c^\top(i)E_B^\top(i) \\
E_A(i)X(i) & -\varepsilon_A(i)\mathbb{I} & 0 & 0 \\
E_A(i)X(i) & 0 & -\varepsilon_{Q_1}(i)\mathbb{I} & 0 \\
E_B(i)Y_c(i) & 0 & 0 & -\varepsilon_B(i)\mathbb{I} \\
E_B(i)Y_c(i) & 0 & 0 & 0 \\
\mathbb{W}(i)X(i) & 0 & 0 & 0 \\
E_{C_y}(i)X(i) & 0 & 0 & 0 \\
\mathcal{S}_i^\top(X) & 0 & 0 & 0
\end{bmatrix}
$$

$$\left[\begin{array}{cccc}
Y_c^\top(i)E_B^\top(i) & X(i)\mathbb{W}^\top(i) & X(i)E_{C_y}^\top(i) & \mathcal{S}_i(X) \\
0 & 0 & 0 & 0 \\
0 & 0 & 0 & 0 \\
0 & 0 & 0 & 0 \\
-\varepsilon_{Q_2}(i)\mathbb{I} & 0 & 0 & 0 \\
0 & -X(i) & 0 & \\
0 & 0 & -\mathbb{I} & 0 \\
0 & 0 & 0 & -\mathcal{X}(X)
\end{array}\right] < 0, \qquad (3.89)$$

$$\left[\begin{array}{cccc}
\mathscr{R}_Q(i) & Q(i)D_A(i) & Q(i)D_B(i) & Q(i)D_B(i) \\
D_A^\top(i)Q(i) & -\varepsilon_{Q_1}^{-1}(i)\mathbb{I} & 0 & 0 \\
D_B^\top(i)Q(i) & 0 & -\varepsilon_{Q_2}^{-1}(i)\mathbb{I} & 0 \\
D_A^\top(i)Q(i) & 0 & 0 & -\varepsilon_Q^{-1}(i)\mathbb{I} \\
E_B(i)K(i) & 0 & 0 & 0 \\
E_B(i)K(i) & 0 & 0 & 0 \\
K(i) & 0 & 0 & 0 \\
D_{C_y}^\top(i)Y_o^\top(i) & 0 & 0 & 0
\end{array}\right.$$

$$\left.\begin{array}{cccc}
K^\top(i)E_B^\top(i) & K^\top(i)E_B^\top(i) & K^\top(i) & Y_o(i)D_{c_y}(i) \\
0 & 0 & 0 & 0 \\
0 & 0 & 0 & 0 \\
0 & 0 & 0 & 0 \\
-\varepsilon_{P_1}(i)\mathbb{I} & 0 & 0 & 0 \\
0 & -\varepsilon_Q(i)\mathbb{I} & 0 & 0 \\
0 & 0 & -\mathbb{I} & 0 \\
0 & 0 & 0 & -\varepsilon_{C_y}(i)\mathbb{I}
\end{array}\right] < 0, \qquad (3.90)$$

with

$$\begin{aligned}
\mathscr{R}_X(i) =\ & X(i)A^\top(i) + Y_c^\top(i)B^\top(i) + A(i)X(i) + B(i)Y_c(i) \\
& + B(i)B^\top(i) + \lambda_{ii}X(i) \\
& + \varepsilon_A(i)D_A(i)D_A^\top(i) + \varepsilon_B(i)D_B(i)D_B^\top(i) \\
& + \varepsilon_{P_1}(i)D_B(i)D_B^\top(i), \\
\mathscr{R}_Q(i) =\ & A^\top(i)Q(i) + Q(i)A(i) + Y_o(i)C_y(i) + C_y^\top(i)Y_o^\top(i) \\
& + \sum_{j=1}^{N} \lambda_{ij}Q(j) + \mathbb{W}^\top(i)Q(i)\mathbb{W}(i),
\end{aligned}$$

then the controller gains that stochastically stabilize system (3.78) are given by

$$K(i) = Y_c(i)X^{-1}(i), \qquad (3.91)$$
$$L(i) = Q^{-1}(i)Y_o(i). \qquad (3.92)$$

Example 40. To show the usefulness of the developed results in this theorem, let us consider the two-mode system of the example considered before with the same data.

Fix the required positive scalars to the following values:

$$\begin{aligned}
\varepsilon_A(1) &= 0.10, \\
\varepsilon_B(1) &= \varepsilon_Q(1) = \varepsilon_{Q1}(1) = \varepsilon_{Q2}(1) = \varepsilon_P(1) = 0.50, \\
\varepsilon_A(2) &= 0.10, \\
\varepsilon_B(2) &= \varepsilon_Q(2) = \varepsilon_{Q1}(2) = \varepsilon_{Q2}(2) = \varepsilon_P(2) = 0.50.
\end{aligned}$$

Solving the LMIs (3.89)–(3.90), we get

$$X(1) = \begin{bmatrix} 0.2375 & -0.0032 \\ -0.0032 & 0.2538 \end{bmatrix}, \qquad X(2) = \begin{bmatrix} 0.2119 & -0.0010 \\ -0.0010 & 0.2220 \end{bmatrix},$$

$$Y_c(1) = \begin{bmatrix} -1.5026 & -0.0270 \\ 0.0156 & -1.4673 \end{bmatrix}, \qquad Y_c(2) = \begin{bmatrix} -1.0779 & -0.1456 \\ -0.0421 & -1.0979 \end{bmatrix},$$

$$Q(1) = \begin{bmatrix} 65.6278 & -24.0341 \\ -24.0341 & 29.3642 \end{bmatrix}, \qquad Q(2) = \begin{bmatrix} 42.7741 & -26.7655 \\ -26.7655 & 57.0332 \end{bmatrix},$$

$$Y_o(1) = \begin{bmatrix} -87.1069 & 37.2563 \\ 37.2085 & -121.1409 \end{bmatrix}, \qquad Y_o(2) = \begin{bmatrix} -146.6626 & 29.8761 \\ 30.2071 & -46.8421 \end{bmatrix}.$$

The corresponding gains for the considered controller are given by

$$K(1) = \begin{bmatrix} -6.3302 & -0.1874 \\ -0.0131 & -5.7818 \end{bmatrix}, \qquad K(2) = \begin{bmatrix} -5.0911 & -0.6797 \\ -0.2228 & -4.9460 \end{bmatrix},$$

$$L(1) = \begin{bmatrix} -1.2327 & -1.3468 \\ 0.2582 & -5.2278 \end{bmatrix}, \qquad L(2) = \begin{bmatrix} -4.3851 & 0.2613 \\ -1.5283 & -0.6987 \end{bmatrix}.$$

Based on the results of this theorem, the system of this example is stochastically stable under the observer-based output feedback control with the computed gains.

3.6 Stabilization with Constant Gains

Earlier in this chapter we supposed complete access to the system mode. In reality, this is hard to achieve and we must estimate the mode to continue to apply the previous results. In this section, we relax this assumption and try to synthesize controllers that do not require knowledge of the mode.

Let us start by designing a constant gain state feedback controller for the nominal system. The structure for this controller is given by

$$u = \mathcal{K} x(t), \tag{3.93}$$

where $\mathcal{K}$ is a constant gain that we have to determine.

Using the nominal system dynamics (3.1) and the expression of the controller, we get the following expression for the closed-loop system:

$$\dot{x}(t) = [A(r(t)) + B(r(t))\mathcal{K}] x(t).$$

Theorem 36. *Let $\mathcal{K}$ be a given gain matrix. If there exists a symmetric and positive-definite matrix $P > 0$ such that the following set of LMIs holds for each $i \in \mathscr{S}$:*

$$A^\top(i)P + \mathcal{K}^\top B^\top(i)P + PA(i) + PB(i)\mathcal{K} < 0, \qquad (3.94)$$

then the nominal system (3.1) is stochastically stable under the state feedback controller with constant gain.

Proof: Let $P > 0$ be a symmetric and positive-definite matrix and definite the Lyapunov candidate function as follows:

$$V(x_t, r(t) = i) = x^\top(t)Px(t).$$

Let $\mathscr{L}$ denote the infinitesimal generator of the Markov process $(x(t), r(t))$. The expression of the infinitesimal operator acting on $V(.)$ and emanating from the point (x, i) at time t, where $x(t) = x$ and $r(t) = i$ for $i \in \mathscr{S}$, is given by

$$\mathscr{L}V(x(t), i) = \dot{x}^\top(t)Px(t) + x^\top(t)P\dot{x}(t) + \sum_{j=1}^{N} \lambda_{ij} x^\top(t)Px(t).$$

Using the fact that $\sum_{j=1}^{N} \lambda_{ij} = 0$ for every $i \in \mathscr{S}$, we get

$$\begin{aligned}
\mathscr{L}V(x(t), i) &= \dot{x}^\top(t)Px(t) + x^\top(t)P\dot{x}(t) \\
&= \left[[A(i) + B(i)\mathcal{K}]\, x(t)\right]^\top Px(t) + x^\top(t)P\left[A(i) + B(i)\mathcal{K}\right]x(t) \\
&= x^\top(t)\left[A^\top(i)P + \mathcal{K}^\top B^\top(i)P + PA(i) + P\mathcal{K}B(i)\right]x(t) \\
&= x^\top(t)\Lambda(i)x(t),
\end{aligned}$$

with $\Lambda(i) = A^\top(i)P + \mathcal{K}^\top B^\top(i)P + PA(i) + PB(i)\mathcal{K}$.

Using (3.94) we get

$$\mathscr{L}V(x(t), i) \leq -\min_{i \in \mathscr{S}} \left\{\lambda_{min}\left(-\Lambda(i)\right)\right\} x^\top(t)x(t).$$

The rest of the proof of Theorem 36 can be obtained following the same steps as before. □

From this theorem, it is possible to stochastically stabilize the class of systems we are considering if condition (3.94) is satisfied. The P we need is constant and does not depend on the mode i.

To determine the gain $\mathcal{K}$, transform the condition (3.94). For this purpose, let $X = P^{-1}$ and pre- and post-multiply this inequality by X to give:

$$XA^\top(i) + X\mathcal{K}^\top B^\top(i) + A(i)X + B(i)\mathcal{K}X < 0.$$

Letting $K = \mathcal{K}X$, we get

$$XA^\top(i) + K^\top B^\top(i) + A(i)X + B(i)K < 0.$$

If this LMI is feasible, the controller gain is given by

$$\mathscr{K} = KX^{-1} = KP. \tag{3.95}$$

The controller gain can be determined by solving the LMI of the following theorem.

Theorem 37. *If there exist a symmetric and positive-definite matrix $X > 0$ and a constant gain K such that the following set of LMIs holds for each $i \in \mathscr{S}$:*

$$XA^\top(i) + A(i)X + K^\top B^\top(i) + B(i)K < 0, \tag{3.96}$$

then the state feedback controller with the gain $\mathscr{K} = KX^{-1}$ stochastically stabilizes the nominal system.

Example 41. Let us consider the system with two modes considered in the previous example. Notice that the system is instable in mode 1 and it is stochastically instable. Solving the LMI (3.96), we get

$$X = \begin{bmatrix} 0.4791 & 0.1129 \\ 0.1129 & 0.4698 \end{bmatrix},$$

which is a symmetric and positive-definite matrix. Using (3.96) gives the following constant gain:

$$\mathscr{K} = \begin{bmatrix} 9.8277 & -51.0015 \\ 51.1597 & -14.6804 \end{bmatrix}.$$

With this controller, the closed-loop state equation becomes

$$\dot{x}(t) = A_{cl}(i)x(t),$$

with

$$A_{cl}(i) = \begin{cases} \begin{bmatrix} 10.8277 & -51.5015 \\ 51.2597 & -13.6804 \end{bmatrix}, \text{when } i = 1, \\[4mm] \begin{bmatrix} 9.6277 & -51.5015 \\ 51.6597 & -14.9304 \end{bmatrix}, \text{otherwise.} \end{cases} \tag{3.97}$$

The standard conditions for stochastic stability can be summarized as follows: If there exists a set of symmetric and positive-definite matrices $P = (P(1), P(2)) > 0$ such that the following holds for each $r(t) = i \in \mathscr{S}$:

$$A_{cl}^{\top}(i)P(i) + P(i)A_{cl}(i) + \sum_{j=1}^{N} \lambda_{ij}P(j) < 0,$$

then the closed-loop state equation is stochastically stable.

Using these conditions, we get the following matrices:

$$P(1) = \begin{bmatrix} 8.5974 & -2.0276 \\ -2.0276 & 8.4410 \end{bmatrix}, \quad P(2) = \begin{bmatrix} 5.5621 & -0.4028 \\ -0.4028 & 5.6338 \end{bmatrix},$$

which are both symmetric and positive-definite and therefore the closed-loop system is stochastically stable under the constant gain state feedback controller.

If the uncertainties are acting on the dynamics, the previous results can be extended to handle this case. In fact, if we replace $A(i)$ and $B(i)$ by $A(i) + \Delta A(i,t)$ and $B(i) + \Delta B(i,t)$, respectively, in the previous condition, we get for every $i \in \mathscr{S}$:

$$XA^{\top}(i) + A(i)X + K^{\top}B^{\top}(i) + B(i)K + X\Delta A^{\top}(i,t)$$
$$+\Delta A(i,t)X + K^{\top}\Delta B^{\top}(i,t) + \Delta B(i,t)K < 0.$$

Using Lemma 7 from Appendix A, we get

$$X\Delta A^{\top}(i,t) + \Delta A(i,t)X \leq \varepsilon_A(i)D_A(i)D_A^{\top}(i)$$
$$+\varepsilon_A^{-1}(i)XE_A^{\top}(i)E_A(i)X,$$
$$K^{\top}\Delta B^{\top}(i,t) + \Delta B(i,t)K \leq \varepsilon_B(i)D_B(i)D_B^{\top}(i)$$
$$+\varepsilon_B^{-1}(i)K^{\top}E_B^{\top}(i)E_B(i)K.$$

Based on these inequalities we need to have the following to guarantee the stochastic stability of the closed-loop state equation:

$$XA^{\top}(i) + A(i)X + K^{\top}B^{\top}(i) + B(i)K$$
$$+\varepsilon_A(i)D_A(i)D_A^{\top}(i) + \varepsilon_B(i)D_B(i)D_B^{\top}(i)$$
$$+\varepsilon_A^{-1}(i)XE_A^{\top}(i)E_A(i)X$$
$$+\varepsilon_B^{-1}(i)K^{\top}E_B^{\top}(i)E_B(i)K < 0,$$

which implies the previous one.

Using the Schur complement, we get

$$\begin{bmatrix} \mathscr{J}(i) & XE_A^{\top}(i) & K^{\top}E_B^{\top}(i) \\ E_A(i)X & -\varepsilon_A(i)\mathbb{I} & 0 \\ E_B(i)K & 0 & -\varepsilon_B(i)\mathbb{I} \end{bmatrix} < 0, \tag{3.98}$$

with

$$\mathscr{J}(i) = XA^{\top}(i) + A(i)X + K^{\top}B^{\top}(i) + B(i)K$$

$$+\varepsilon_A(i)D_A(i)D_A^\top(i) + \varepsilon_B(i)D_B(i)D_B^\top(i).$$

The following theorem determines the controller with constant gain that robustly stochastically stabilizes the class of systems we are studying.

Theorem 38. *If there exist symmetric and positive-definite matrices $X > 0$ and sets of positive constants $\varepsilon_A = (\varepsilon_A(1), \cdots, \varepsilon_A(N))$ and $\varepsilon_B = (\varepsilon_B(1), \cdots, \varepsilon_B(N))$ and a constant gain K such the following sets of LMIs (3.98) hold for every $i \in \mathscr{S}$, then the stabilizing controller gain is $\mathscr{K} = KX^{-1}$.*

Example 42. To design a robust stabilizing controller with constant gain, let us consider again the two-mode system of the previous example.

Letting $\varepsilon_A(1) = \varepsilon_A(2) = 0.5$ and $\varepsilon_B(1) = \varepsilon_B(2) = 0.1$ and solving the LMIs (3.98), we get

$$X = \begin{bmatrix} 0.2641 & 0.0897 \\ 0.0897 & 0.2001 \end{bmatrix}, \quad K = \begin{bmatrix} -0.5365 & -0.0496 \\ -0.0315 & -0.5139 \end{bmatrix},$$

which gives the following gain:

$$\mathscr{K} = \begin{bmatrix} -2.2975 & 0.7827 \\ 0.8891 & -2.9678 \end{bmatrix}.$$

Using the results of this theorem, the system of this example is stochastically stable under the state feedback controller with the computed constant gain.

Let us now consider the effects of external Wiener process disturbance to see how we can design a controller with constant gain that stochastically and/or robustly stochastically stabilizes the system.

Based on the results of Chapter 2, the free nominal system will be stochastically stable if there exists a symmetric and positive-definite matrix $P > 0$, such that the following holds for each $i \in \mathscr{S}$:

$$A^\top(i)P + PA(i) + \mathbb{W}^\top(i)P\mathbb{W}(i) < 0.$$

Plugging the controller expression into the system dynamics, we get the following dynamics for the closed loop:

$$dx(t) = [A(r(t)) + B(r(t))\mathscr{K}]\,x(t)dt + \mathbb{W}(i)x(t)d\omega(t).$$

Replacing $A(i)$ by $A(i) + B(i)\mathscr{K}$, we get the following condition for stochastic stability of the closed-loop state equation for every $i \in \mathscr{S}$:

$$[A(i) + B(i)\mathscr{K}]^\top P + P[A(i) + B(i)\mathscr{K}] + \mathbb{W}^\top(i)P\mathbb{W}(i) < 0.$$

Notice that this inequality is nonlinear in the design parameters $P > 0$ and $\mathscr{K}$. Let $X = P^{-1}$. To put it into LMI form, pre- and post-multiply this inequality by X. After some simple algebraic manipulations, we get

$$XA^\top(i) + X\mathscr{K}^\top B^\top(i) + A(i)X + B(i)\mathscr{K}X$$
$$+ XW^\top(i)P^{-1}W(i)P < 0.$$

Letting $K = \mathscr{K}X$, we obtain the following LMI:

$$\begin{bmatrix} \mathscr{J}(i) & XW^\top(i) \\ W(i)X & -X \end{bmatrix} < 0, \tag{3.99}$$

where

$$\mathscr{J}(i) = XA^\top(i) + XA(i) + K^\top B^\top(i) + B(i)K.$$

The following theorem summarizes the results of this development.

Theorem 39. *If there exist a symmetric and positive-definite matrix $X > 0$ and a constant gain K such that the following set of LMIs (3.99) holds for each $i \in \mathscr{S}$, then the state feedback controller with the gain $\mathscr{K} = KX^{-1}$ stochastically stabilizes the system.*

Example 43. To show the usefulness of the results of this theorem, let us consider the two-mode system of the previous example. Solving the LMI (3.99), we get

$$X = \begin{bmatrix} 11.7639 & 1.8477 \\ 1.8477 & 11.6092 \end{bmatrix}, \quad K = \begin{bmatrix} -15.5182 & 19.8940 \\ -19.0505 & -16.6443 \end{bmatrix},$$

which is a symmetric and positive-definite matrix. (3.95) gives the following constant gain:

$$\mathscr{K} = \begin{bmatrix} -1.6290 & 1.9729 \\ -1.4300 & -1.2061 \end{bmatrix}.$$

With this controller, the closed-loop state equation becomes

$$\dot{x}(t) = A_{cl}(r(t))x(t),$$

with

$$A_{cl}(i) = \begin{cases} \begin{bmatrix} -0.6290 & 1.4729 \\ -1.3300 & -0.2061 \end{bmatrix}, \text{when } r(t) = 1, \\ \\ \begin{bmatrix} -1.8290 & 1.4729 \\ -0.9300 & -1.4561 \end{bmatrix}, \text{otherwise.} \end{cases}$$

The standard conditions for stochastic stability can be summarized as follows: If there exists a set of symmetric and positive-definite matrices $P = (P(1), P(2)) > 0$ such that the following holds for each $i \in \mathscr{S}$:

$$A_{cl}^{\top}(i)P(i) + P(i)A_{cl}(i) + \sum_{j=1}^{N} \lambda_{ij}P(j) < 0,$$

then the closed-loop system is stochastically stable.

Using these conditions, we get the following matrices:

$$P(1) = \begin{bmatrix} 23.2198 & -1.0277 \\ -1.0277 & 31.4927 \end{bmatrix}, \quad P(2) = \begin{bmatrix} 13.4892 & 1.7971 \\ 1.7971 & 17.1656 \end{bmatrix},$$

which are both symmetric and positive-definite. Therefore the closed-loop system is stochastically stable under the constant gain state feedback controller.

If both the external Wiener process and the uncertainties are acting on the systems, similar results can be obtained. In fact, if we replace $A(i)+B(i)\mathscr{K}$ by $A(i)+\Delta A(i,t)+B(i)\mathscr{K}+\Delta B(i,t)\mathscr{K}$, we get the following required condition that guarantees robust stochastic stability:

$$[A(i) + \Delta A(i,t) + B(i)\mathscr{K} + \Delta B(i,t)\mathscr{K}]^{\top} P$$
$$+P[A(i) + \Delta A(i,t) + B(i)\mathscr{K} + \Delta B(i,t)\mathscr{K}]$$
$$+\mathbb{W}^{\top}(i)P\mathbb{W}(i) < 0.$$

Notice that this inequality is nonlinear in the design parameters $P > 0$ and $\mathscr{K}$. Let $X = P^{-1}$. To put it into LMI form, let us pre- and post-multiply this inequality by X. After some simple algebraic manipulations, we get for every $i \in \mathscr{S}$:

$$XA^{\top}(i) + X\mathscr{K}^{\top}B^{\top}(i) + A(i)X + B(i)\mathscr{K}X$$
$$+X\mathbb{W}^{\top}(i)P^{-1}\mathbb{W}(i)X + X\Delta A^{\top}(i,t) + \Delta A(i,t)X$$
$$+X\mathscr{K}^{\top}\Delta B^{\top}(i,t) + \Delta B(i,t)\mathscr{K}X < 0.$$

Using Lemma 7, we can transform the term $X\Delta A^{\top}(i,t) + \Delta A(i,t)X$ and $X\mathscr{K}^{\top}\Delta B^{\top}(i,t) + \Delta B(i,t)\mathscr{K}X$ as before, and after letting $K = \mathscr{K}X$, we obtain the required LMI that guarantees the stochastic stability of the closed-loop:

$$\begin{bmatrix} \mathscr{J}(i) & XE^{\top}(i) & K^{\top}E_B^{\top}(i) & X\mathbb{W}^{\top}(i) \\ E_A(i)X & -\varepsilon_A(i)\mathbb{I} & 0 & 0 \\ E_B(i)K & 0 & -\varepsilon_B(i)\mathbb{I} & 0 \\ \mathbb{W}(i)P & 0 & 0 & -X \end{bmatrix} < 0, \qquad (3.100)$$

where

$$\mathscr{J}(i) = XA^{\top}(i) + A(i)X + K^{\top}B^{\top}(i) + B(i)K$$
$$+\varepsilon_A(i)D_A(i)D_A^{\top}(i) + \varepsilon_B(i)D_B(i)D_B^{\top}(i).$$

The following theorem summarizes the results of this development.

Theorem 40. *If there exist a symmetric and positive-definite matrix $X > 0$ and sets of positive constants $\varepsilon_A = (\varepsilon_A(1), \cdots, \varepsilon_A(N))$ and $\varepsilon_B = (\varepsilon_B(1), \cdots, \varepsilon_B(N))$ and a constant gain K such the following sets of LMIs (3.100) hold for every $i \in \mathscr{S}$ then the stabilizing controller gain is $\mathscr{K} = KX^{-1}$.*

Example 44. To design a robust stabilizing controller with constant gain, let us consider again the system with two modes of the previous example. Letting $\varepsilon_A(1) = \varepsilon_A(2) = 0.5$ and $\varepsilon_B(1) = \varepsilon_B(2) = 0.1$ and solving the LMIs (3.100), we get

$$P = \begin{bmatrix} 0.4208 & 0.0578 \\ 0.0578 & 0.4165 \end{bmatrix}, \quad K = \begin{bmatrix} -0.5702 & 0.0343 \\ 0.0044 & -0.6076 \end{bmatrix},$$

which gives the following gain:

$$\mathscr{K} = \begin{bmatrix} -1.3931 & 0.2755 \\ 0.2147 & -1.4885 \end{bmatrix}.$$

Using the results of this theorem, the system of this example is stochastically stable under the state feedback controller with the computed constant gain.

3.7 Case Study

To end this chapter let us consider the system described in Chapter 1 involving the VTOL helicopter. With the same data given in Chapter 1 and assuming that we have access to the state vector, let us design a state feedback controller using the developed design method. Solving the set of coupled LMIs, we get

$$X(1) = \begin{bmatrix} 0.6065 & 0.0053 & 0.0872 & 0.0395 \\ 0.0053 & 0.5756 & 0.0103 & 0.0031 \\ 0.0872 & 0.0103 & 0.6253 & -0.0854 \\ 0.0395 & 0.0031 & -0.0854 & 0.5060 \end{bmatrix},$$

$$X(2) = \begin{bmatrix} 1.1272 & 0.0239 & 0.2684 & 0.2290 \\ 0.0239 & 1.0095 & 0.0275 & 0.0129 \\ 0.2684 & 0.0275 & 1.1807 & -0.2519 \\ 0.2290 & 0.0129 & -0.2519 & 0.7662 \end{bmatrix},$$

$$X(3) = \begin{bmatrix} 1.1347 & 0.0128 & 0.1733 & 0.2183 \\ 0.0128 & 1.0560 & 0.0264 & 0.0342 \\ 0.1733 & 0.0264 & 1.2140 & -0.2585 \\ 0.2183 & 0.0342 & -0.2585 & 1.0168 \end{bmatrix},$$

$$Y(1) = \begin{bmatrix} -0.0436 & -0.0125 & -0.0838 & 0.1196 \\ -0.0396 & -0.0673 & -0.0 & -0.2067 \end{bmatrix},$$

$$Y(2) = \begin{bmatrix} -0.0966 & 0.1326 & -0.0591 & -0.0611 \\ -0.1242 & -0.0567 & 0.0 & -0.4055 \end{bmatrix},$$

$$Y(3) = \begin{bmatrix} -0.0154 & 0.1039 & -0.1804 & 0.6468 \\ -0.1239 & -0.0234 & 0.0 & -0.0982 \end{bmatrix},$$

which gives the following gains:

$$K(1) = \begin{bmatrix} -0.0731 & -0.0207 & -0.0925 & 0.2266 \\ -0.0301 & -0.1135 & -0.0505 & -0.4139 \end{bmatrix},$$

$$K(2) = \begin{bmatrix} -0.0578 & 0.1354 & -0.0580 & -0.0838 \\ 0.0412 & -0.0460 & -0.1330 & -0.5846 \end{bmatrix},$$

$$K(3) = \begin{bmatrix} -0.1448 & 0.0782 & 0.0125 & 0.6678 \\ -0.0939 & -0.0185 & -0.0025 & -0.0764 \end{bmatrix}.$$

We can also assume that we do not have access to the mode and try to design a constant gain state feedback controller. Using the previous results and solving the set of LMIs, we have

$$X = \begin{bmatrix} 3.3304 & -1.0031 & 0.1220 & 0.6575 \\ -1.0031 & 1.4556 & 0.2100 & -0.4049 \\ 0.1220 & 0.2100 & 1.4461 & -0.5585 \\ 0.6575 & -0.4049 & -0.5585 & 0.9409 \end{bmatrix},$$

$$Y = \begin{bmatrix} 0.1386 & 0.1416 & 0.2689 & 0.2778 \\ -0.0621 & 0.1817 & 0.4079 & -0.2551 \end{bmatrix},$$

which gives the following constant gain:

$$K = \begin{bmatrix} -0.0465 & 0.1876 & 0.4156 & 0.6552 \\ 0.0186 & 0.0719 & 0.2235 & -0.1205 \end{bmatrix}.$$

If we assume that the state vector is not accessible, we can design an output feedback controller using the previous results. Using the same data as before and solving the appropriate set of coupled LMIs gives

$$X(1) = \begin{bmatrix} 9.7282 & -1.8814 & -0.2079 & -1.9571 \\ -1.8814 & 4.7466 & 1.9035 & 3.6740 \\ -0.2079 & 1.9035 & 9.2696 & 1.4797 \\ -1.9571 & 3.6740 & 1.4797 & 14.7762 \end{bmatrix},$$

$$X(2) = \begin{bmatrix} 8.8208 & -1.4447 & 0.3358 & 1.1365 \\ -1.4447 & 4.2255 & 1.4165 & 4.9098 \\ 0.3358 & 1.4165 & 9.4633 & 0.4707 \\ 1.1365 & 4.9098 & 0.4707 & 10.5495 \end{bmatrix},$$

$$X(3) = \begin{bmatrix} 9.7866 & -3.3633 & 1.0574 & -0.4175 \\ -3.3633 & 6.1659 & 2.7704 & 4.5104 \\ 1.0574 & 2.7704 & 7.5558 & -1.2452 \\ -0.4175 & 4.5104 & -1.2452 & 8.9236 \end{bmatrix},$$

$$Y(1) = \begin{bmatrix} 6.1898 & 0.0671 & 0.8070 & 0.3487 \\ 0.0671 & 5.8269 & 0.0888 & 0.0155 \\ 0.8070 & 0.0888 & 6.2532 & -0.7316 \\ 0.3487 & 0.0155 & -0.7316 & 5.2439 \end{bmatrix},$$

$$Y(2) = \begin{bmatrix} 10.4897 & 0.2056 & 2.4237 & 1.9123 \\ 0.2056 & 9.5823 & 0.2373 & 0.0815 \\ 2.4237 & 0.2373 & 11.0087 & -2.1617 \\ 1.9123 & 0.0815 & -2.1617 & 7.2361 \end{bmatrix},$$

$$Y(3) = \begin{bmatrix} 10.3650 & 0.1051 & 1.4839 & 1.7359 \\ 0.1051 & 9.9009 & 0.1995 & 0.2453 \\ 1.4839 & 0.1995 & 11.1810 & -2.1769 \\ 1.7359 & 0.2453 & -2.1769 & 9.3320 \end{bmatrix},$$

$$\mathcal{K}_B(1) = \begin{bmatrix} -3.3800 & 5.1499 \\ -5.4839 & 0.3679 \\ 1.5367 & 0.0 \\ -14.1434 & 13.5104 \end{bmatrix},$$

$$\mathcal{K}_B(2) = \begin{bmatrix} -4.1110 & 2.5550 \\ -3.9626 & 1.0908 \\ 1.7776 & 0.0 \\ -4.8024 & 16.8858 \end{bmatrix},$$

$$\mathcal{K}_B(3) = \begin{bmatrix} -3.5742 & 2.2258 \\ -5.4298 & 1.4920 \\ 2.5429 & -0.0 \\ -15.4107 & 17.4410 \end{bmatrix},$$

$$\mathcal{K}_C(1) = \begin{bmatrix} -0.3879 & -0.2185 & -0.8007 & 1.0865 \\ -0.3573 & -0.7192 & -0.0 & -2.2098 \end{bmatrix},$$

$$\mathcal{K}_C(2) = \begin{bmatrix} -0.8189 & 1.0843 & -0.5558 & -0.6340 \\ -1.0365 & -0.5550 & -0.0 & -3.8356 \end{bmatrix},$$

$$\mathcal{K}_C(3) = \begin{bmatrix} -0.1996 & 0.8602 & -1.5629 & 5.8799 \\ -1.0439 & -0.2591 & 0.0 & -0.9351 \end{bmatrix},$$

which gives the following gains:

$$K_A(1) = \begin{bmatrix} -0.8071 & 0.6479 & -0.7407 & 0.1657 \\ -0.9743 & -0.8361 & -1.0124 & -0.6135 \\ 0.8275 & -0.0130 & 0.0909 & -1.6445 \\ -0.8736 & 1.1252 & 0.1745 & 1.1515 \end{bmatrix},$$

$$K_B(1) = \begin{bmatrix} 0.7175 & -0.6640 \\ 1.0119 & 0.5559 \\ -0.5036 & 0.0556 \\ 0.8608 & -1.1607 \end{bmatrix},$$

$$K_C(1) = \begin{bmatrix} -0.0608 & -0.0358 & -0.0965 & 0.1979 \\ -0.0266 & -0.1213 & -0.0446 & -0.4255 \end{bmatrix},$$

$$K_A(2) = \begin{bmatrix} -1.0678 & -1.1790 & -1.0286 & -1.9246 \\ -2.7181 & -6.1764 & -1.5655 & -5.4328 \\ 1.0969 & -0.1029 & -0.3614 & -1.2342 \\ 0.7875 & 4.3755 & 1.7833 & 4.4348 \end{bmatrix},$$

$$K_B(2) = \begin{bmatrix} 1.0047 & 1.2911 \\ 2.4371 & 5.7491 \\ -0.5630 & -0.7171 \\ -0.7801 & -4.4594 \end{bmatrix},$$

$$K_C(2) = \begin{bmatrix} -0.0492 & 0.1165 & -0.0606 & -0.0940 \\ 0.0344 & -0.0508 & -0.1192 & -0.5742 \end{bmatrix},$$

$$K_A(3) = \begin{bmatrix} -0.2155 & -3.5038 & -0.1381 & -3.8405 \\ 0.9598 & -10.5379 & 0.8146 & -10.2005 \\ 0.0594 & 5.5874 & -1.0472 & 4.1746 \\ -2.2909 & 7.7417 & -1.1495 & 7.8591 \end{bmatrix},$$

$$K_B(3) = \begin{bmatrix} 0.1085 & 3.6175 \\ -0.8997 & 10.1916 \\ 0.3646 & -5.6498 \\ 2.2687 & -7.8583 \end{bmatrix},$$

$$K_C(3) = \begin{bmatrix} -0.1298 & 0.0720 & 0.0033 & 0.6531 \\ -0.0856 & -0.0231 & -0.0047 & -0.0848 \end{bmatrix}.$$

For the observer-based output feedback controller, using the same data as before and solving the appropriate set of coupled LMIs gives

$$X(1) = \begin{bmatrix} 76.2103 & 0.6638 & 10.9612 & 4.9596 \\ 0.6638 & 72.3326 & 1.2929 & 0.3932 \\ 10.9612 & 1.2929 & 78.5798 & -10.7318 \\ 4.9596 & 0.3932 & -10.7318 & 63.5807 \end{bmatrix},$$

$$X(2) = \begin{bmatrix} 141.6507 & 3.0038 & 33.7306 & 28.7827 \\ 3.0038 & 126.8566 & 3.4561 & 1.6248 \\ 33.7306 & 3.4561 & 148.3684 & -31.6521 \\ 28.7827 & 1.6248 & -31.6521 & 96.2770 \end{bmatrix},$$

$$X(3) = \begin{bmatrix} 142.5915 & 1.6080 & 21.7749 & 27.4260 \\ 1.6080 & 132.6978 & 3.3168 & 4.2958 \\ 21.7749 & 3.3168 & 152.5544 & -32.4800 \\ 27.4260 & 4.2958 & -32.4800 & 127.7716 \end{bmatrix},$$

$$Y_c(1) = \begin{bmatrix} -5.6210 & -6.4418 & -5.9411 & 15.0267 \\ -5.2413 & -6.1006 & 0.0 & -25.9682 \end{bmatrix},$$

$$Y_c(2) = \begin{bmatrix} -12.2878 & 13.0699 & -2.8444 & -7.6732 \\ -15.8768 & -3.7312 & 0.0 & -50.9584 \end{bmatrix},$$

$$Y_c(3) = \begin{bmatrix} -2.0774 & 7.4119 & -18.0888 & 81.2837 \\ -15.8380 & -1.2188 & -0.0 & -12.3391 \end{bmatrix},$$

which gives the gains

$$K(1) = \begin{bmatrix} -0.0842 & -0.0890 & -0.0298 & 0.2384 \\ -0.0339 & -0.0809 & -0.0505 & -0.4138 \end{bmatrix},$$

$$K(2) = \begin{bmatrix} -0.0707 & 0.1061 & -0.0198 & -0.0669 \\ 0.0387 & -0.0192 & -0.1330 & -0.5843 \end{bmatrix},$$

$$K(3) = \begin{bmatrix} -0.1529 & 0.0345 & 0.0472 & 0.6798 \\ -0.0959 & -0.0055 & -0.0025 & -0.0764 \end{bmatrix}.$$

Using this set of gains and solving the other set of coupled LMIs gives

$$Q(1) = \begin{bmatrix} 1.3964 & 0.2046 & 0.0651 & 0.0539 \\ 0.2046 & 2.4239 & -0.0332 & 0.2200 \\ 0.0651 & -0.0332 & 1.3019 & -0.1052 \\ 0.0539 & 0.2200 & -0.1052 & 0.7997 \end{bmatrix},$$

$$Q(2) = \begin{bmatrix} 1.4429 & 0.1126 & 0.0149 & 0.1668 \\ 0.1126 & 2.0044 & 0.0530 & 0.5171 \\ 0.0149 & 0.0530 & 1.3584 & -0.2004 \\ 0.1668 & 0.5171 & -0.2004 & 0.6454 \end{bmatrix},$$

$$Q(3) = \begin{bmatrix} 1.4050 & 0.0540 & 0.0987 & 0.0798 \\ 0.0540 & 1.9969 & 0.0673 & 0.5242 \\ 0.0987 & 0.0673 & 1.2275 & -0.1072 \\ 0.0798 & 0.5242 & -0.1072 & 0.6142 \end{bmatrix},$$

$$Y_o(1) = \begin{bmatrix} -0.6682 & 2.6114 \\ -1.9517 & 3.1193 \\ 0.3768 & -0.0 \\ -2.4647 & -0.4393 \end{bmatrix},$$

$$Y_o(2) = \begin{bmatrix} -0.5672 & 1.9127 \\ -0.9708 & 3.4482 \\ 0.2759 & -0.0 \\ -1.6452 & -0.5542 \end{bmatrix},$$

$$Y_o(3) = \begin{bmatrix} -0.6580 & 3.2837 \\ -2.7291 & 3.4712 \\ 0.3580 & 0.0 \\ -3.0845 & -0.7045 \end{bmatrix},$$

which gives the following set of gains:

$$L(1) = \begin{bmatrix} -0.2932 & 1.7356 \\ -0.5153 & 1.2313 \\ 0.0555 & -0.1380 \\ -2.9133 & -1.0232 \end{bmatrix},$$

$$L(2) = \begin{bmatrix} -0.0873 & 1.5375 \\ 0.2466 & 2.5532 \\ -0.2174 & -0.6325 \\ -2.7919 & -3.4984 \end{bmatrix},$$

$$L(3) = \begin{bmatrix} -0.0873 & 1.5375 \\ 0.2466 & 2.5532 \\ -0.2174 & -0.6325 \\ -2.7919 & -3.4984 \end{bmatrix}.$$

3.8 Notes

This chapter dealt with the stochastic stabilization problem and the robustness of the class of piecewise deterministic systems. Under some appropriate assumptions, different types of controllers like state feedback, output feedback, and observer-based output feedback have been studied and LMI design approaches have been developed for the nominal systems and also for systems with norm-bounded uncertainties. The results we developed can easily be solved using any LMI Toolbox such as Matlab or Scilab. Most of the results in this chapter are based on the work of the author and his coauthors. The three stabilization techniques require complete access to the mode and the state vector of the system at time t. We have also relaxed the dependence of the gain on the mode and we developed design approach for a state feedback controller with constant gain.

4

$\mathscr{H}_\infty$ Control Problem

In the stabilization chapter we discussed the design of controllers that guarantee the stochastic stability of the closed loop for nominal and uncertain dynamical systems belonging to the class of piecewise deterministic systems we are considering in this book. In practice we are interested in more than stability and its robustness; for instance, designing a controller that rejects the effect of external disturbance that may act on the system dynamics. Among the controllers we can use to reach this goal, the linear quadratic regulator is a good candidate for stabilizing and rejecting the effect of disturbances for the class of piecewise deterministic systems. Unfortunately this approach requires special assumptions on these external disturbances that should be Gaussian with some given statistical properties that are difficult to satisfy.

An alternative that copes with the limitation of the linear quadratic regulator was proposed to synthesize controllers to stabilize systems subject to arbitrary external disturbances with finite energy or finite average power, and simultaneously guarantee the disturbance rejection with some desired level.

In this chapter, we study the design of controllers for the class of piecewise deterministic systems that guarantee disturbance rejection and ensure stochastic stability. The robustness problem of this class of systems is also discussed.

The rest of the chapter is organized as follows. In Section 4.1, the $\mathscr{H}_\infty$ control problem is stated and the effectiveness definitions given. Section 4.2 deals with the state feedback $\mathscr{H}_\infty$ stabilization problem and its robustness. In the Section 4.3, the output feedback $\mathscr{H}_\infty$ stabilization problem is covered. The robust stabilization of this controller is also described. Section 4.4 treats the observer-based output feedback $\mathscr{H}_\infty$ stabilization and its robustness. Section 4.5 treats the $\mathscr{H}_\infty$ stabilization for stochastic switching systems with multiplicative noise. All the results developed in this chapter are in the LMI framework and are illustrated by simple examples.

4.1 Problem Statement

Let us consider a dynamical system defined in a probability space $(\Omega, \mathcal{F}, \mathbb{P})$ and assume that its dynamics are described by the following differential equations:

$$\begin{cases} \dot{x}(t) = A(r(t),t)x(t) + B(r(t),t)u(t) + B_\omega(r(t))\omega(t), x(0) = x_0, \\ y(t) = C_y(r(t),t)x(t) + D_y(r(t),t)u(t) + B_y(r(t))\omega(t), \\ z(t) = C_z(r(t),t)x(t) + D_z(r(t),t)u(t) + B_z(r(t))\omega(t), \end{cases} \quad (4.1)$$

where $x(t) \in \mathbb{R}^n$ is the state vector; $u(t) \in \mathbb{R}^m$ is the control vector; $y(t) \in \mathbb{R}^p$ is the measured output; $z(t) \in \mathbb{R}^q$ is the controlled output; and $\omega(t) \in \mathbb{R}^l$ is the system external disturbance. The matrices $A(r(t),t)$, $B(r(t),t)$, $C_y(r(t),t)$, $D_y(r(t),t)$, $C_z(r(t),t)$, and $D_z(r(t),t)$ are given by

$$\begin{cases} A(r(t),t) = A(r(t)) + D_A(r(t))F_A(r(t),t)E_A(r(t)), \\ B(r(t),t) = B(r(t)) + D_B(r(t))F_B(r(t),t)E_B(r(t)), \\ C_y(r(t),t) = C_y(r(t)) + D_{C_y}(r(t))F_{C_y}(r(t),t)E_{C_y}(r(t)), \\ C_z(r(t),t) = C_z(r(t)) + D_{C_z}(r(t))F_{C_z}(r(t),t)E_{C_z}(r(t)), \\ D_y(r(t),t) = D_y(r(t)) + D_{D_y}(r(t))F_{D_y}(r(t),t)E_{D_y}(r(t)), \\ D_z(r(t),t) = D_z(r(t)) + D_{D_z}(r(t))F_{D_z}(r(t),t)E_{D_z}(r(t)), \end{cases}$$

where $A(r(t))$, $B(r(t))$, $B_\omega(r(t))$, $C_y(r(t))$, $D_y(r(t))$, $B_y(r(t))$, $C_z(r(t))$, $D_z(r(t))$, $B_z(r(t))$, $D_A(r(t))$, $E_A(r(t))$, $D_B(r(t))$, $E_B(r(t))$, $D_{C_y}(r(t))$, $E_{C_y}(r(t))$, $D_{C_z}(r(t))$, $E_{C_z}(r(t))$, $D_{D_y}(r(t)), E_{D_y}(r(t))$, $D_{D_z}(r(t))$, and $E_{D_z}(r(t))$ are known real matrices with appropriates dimensions. The matrices $F_A(r(t),t)$, $F_B(r(t),t)$, $F_{C_y}(r(t),t)$, $F_{D_y}(r(t),t)$, $F_{C_z}(r(t),t)$, and $F_{D_z}(r(t),t)$ are time-varying unknown matrices satisfying the following for every $i \in \mathscr{S}$:

$$\begin{cases} F_A^\top(i,t)F_A(i,t) \leq \mathbb{I}, \\ F_B^\top(i,t)F_B(i,t) \leq \mathbb{I}, \\ F_{C_y}^\top(i,t)F_{C_y}(i,t) \leq \mathbb{I}, \\ F_{D_y}^\top(i,t)F_{D_y}(i,t) \leq \mathbb{I}, \\ F_{C_z}^\top(i,t)F_{C_z}(i,t) \leq \mathbb{I}, \\ F_{D_z}^\top(i,t)F_{D_z}(i,t) \leq \mathbb{I}. \end{cases}$$

The Markov process $\{r(t), t \geq 0\}$ besides taking values in the finite set $\mathscr{S}$ represents the switching between the different modes where we assume the behavior is described by the following probability transitions:

$$\mathbb{P}\left[r(t+h) = j | r(t) = i\right]$$
$$= \begin{cases} \lambda_{ij}h + o(h) & \text{when } r(t) \text{ jumps from } i \text{ to } j, \\ 1 + \lambda_{ii}h + o(h) & \text{otherwise,} \end{cases} \quad (4.2)$$

where λ_{ij} is the transition rate from mode i to mode j with $\lambda_{ij} \geq 0$ when $i \neq j$ and $\lambda_{ii} = -\sum_{j=1, j\neq i}^{N} \lambda_{ij}$ and $o(h)$ is such that $\lim_{h\to 0} \frac{o(h)}{h} = 0$.

The system disturbance $\omega(t)$ is assumed to belong to $\mathcal{L}_2[0, \infty)$, which means that the following holds:

$$\int_0^\infty \omega^\top(t)\omega(t)dt < \infty. \tag{4.3}$$

This implies that the disturbance has finite energy.

In the rest of this chapter we deal with the design of controllers that stochastically stabilize closed-loop systems and guarantee the disturbance rejection with a certain level $\gamma > 0$. We also discuss the design of robust controllers that guarantee the same goal. Mathematically we are concerned with the design of a controller that guarantees the following for all $\omega \in \mathcal{L}_2[0, \infty)$:

$$\|z(t)\|_2 < \gamma \left[\|\omega(t)\|_2^2 + M(x_0, r_0)\right]^{\frac{1}{2}},$$

where $\gamma > 0$ is a prescribed level of disturbance rejection to be achieved; x_0 and r_0 are the initial conditions of the state vector and the mode, respectively, at time $t = 0$; and $M(x_0, r_0)$ is a constant that depends on the initial conditions (x_0, r_0).

Let us begin this chapter by developing results for the nominal system, when all uncertainties are equal to zero. Before proceeding let us define the different concepts we will use for this purpose.

Definition 5. *Let $\gamma > 0$ be a given positive constant. System (4.1) with $u(t) \equiv 0$ is said to be stochastically stable with γ-disturbance attenuation if there exists a constant $M(x_0, r_0)$ with $M(0, r_0) = 0$, for all $r_0 \in \mathscr{S}$, such that the following holds:*

$$\|z\|_2 \triangleq \left[\mathbb{E}\int_0^\infty z^\top(t)z(t)dt|(x_0, r_0)\right]^{1/2} \leq \gamma \left[\|\omega\|_2^2 + M(x_0, r_0)\right]^{\frac{1}{2}}. \tag{4.4}$$

Definition 6. *System (4.1) with $u(t) \equiv 0$ is said to be internally mean square quadratically stable (MSQS) if there exists a set of symmetric and positive-definite matrices $P = (P(1), \cdots, P(N)) > 0$ satisfying the following for every $i \in \mathscr{S}$:*

$$A^\top(i)P(i) + P(i)A(i) + \sum_{j=1}^N \lambda_{ij}P(j) < 0. \tag{4.5}$$

By Definition 1 it is obvious that system (4.1) is internally MSQS when $\omega(t) \equiv 0$, (4.1) being free of input disturbance. Likewise, we can give the following definitions:

Definition 7. *System (4.1) with $u(t) \equiv 0$ is said to be internally SS (MES) if it is SS (MES) when $\omega(t) \equiv 0$.*

Definition 8. *System (4.1) is said to be stabilizable with γ-disturbance rejection in the SS (MES, MSQS) sense if there exists a control law such that the closed-loop system under this control law is SS (MES, MSQS) and satisfies (4.4).*

For uncertain dynamics we have similar definitions that are summarized as follows:

Definition 9. *Let $\gamma > 0$ be a given positive constant. System (4.1) with $u(t) \equiv 0$ is said to be robustly stochastically stable with γ-disturbance attenuation if there exists a constant $M(x_0, r_0)$ with $M(0, r_0) = 0$ for all $r_0 \in \mathscr{S}$, such that the following holds for all admissible uncertainties:*

$$\|z\|_2 \triangleq \left[\mathbb{E} \int_0^\infty z^\top(t) z(t) dt | (x_0, r_0) \right]^{\frac{1}{2}} \leq \gamma \left[\|\omega\|_2^2 + M(x_0, r_0) \right]^{\frac{1}{2}}. \quad (4.6)$$

Definition 10. *System (4.1) with $u(t) \equiv 0$ is said to be internally robust mean square quadratically stable (RMSQS) if there exists a set of symmetric and positive-definite matrices $P = (P(1), \cdots, P(N)) > 0$ satisfying the following for every $i \in \mathscr{S}$ and for all admissible uncertainties:*

$$A^\top(i, t) P(i) + P(i) A(i, t) + \sum_{j=1}^N \lambda_{ij} P(j) < 0. \quad (4.7)$$

By Definition 1 it is obvious that system (4.1) is internally RMSQS when $\omega(t) \equiv 0$, (4.1) being free of input disturbance. Likewise, we can give the following definitions:

Definition 11. *System (4.1) with $u(t) \equiv 0$ is said to be internally RSS (RMES) if it is RSS (RMES) when $\omega(t) \equiv 0$.*

Definition 12. *System (4.1) is said to be robust stabilizable with γ-disturbance rejection in the RSS (RMES, RMSQS) sense if there exists a control law such that the closed-loop system under this control law is RSS (RMES, RMSQS) and satisfies (4.6).*

The following theorem shows that when $\omega(t) \not\equiv 0$, internal MSQS implies stochastic stability.

Theorem 41. *If system (4.1) with $u(t) \equiv 0$ is internally MSQS, then it is stochastically stable.*

Proof: To prove this theorem, let us consider a candidate Lyapunov function defined as follows:

$$V(x(t), r(t)) = x^\top(t) P(r(t)) x(t),$$

where $P(i) > 0$ is symmetric and positive-definite matrix for every $i \in \mathscr{S}$.

As before, the infinitesimal operator $\mathscr{L}$ of the Markov process $\{(x(t), r(t)), t \geq 0\}$ acting on $V(x(t), r(t))$ and emanating from the point (x, i) at time t, where $x(t) = x$ and $r(t) = i$ for $i \in \mathscr{S}$, is given by:

$$\mathscr{L}V(x(t), i) = \dot{x}^{\top}(t)P(i)x(t) + x^{\top}(t)P(i)\dot{x}(t) + \sum_{j=1}^{N}\lambda_{ij}x^{\top}(t)P(j)x(t)$$

$$= x^{\top}(t)\left[A^{\top}(i)P(i) + P(i)A(i) + \sum_{j=1}^{N}\lambda_{ij}P(j)\right]x(t)$$

$$+2x^{\top}(t)P(i)B_{\omega}(i)\omega(t).$$

Using Lemma 7 in Appendix A, we get the following for any $\varepsilon_w(i) > 0$:

$$2x^{\top}(t)P(i)B_{\omega}(i)\omega(t) \leq \varepsilon_w^{-1}(i)x^{\top}(t)P(i)B_{\omega}(i)B_{\omega}^{\top}(i)P(i)x(t)$$

$$+\varepsilon_w(i)\omega^{\top}(t)\omega(t).$$

Combining this with the expression of $\mathscr{L}V(x(t), i)$ yields

$$\mathscr{L}V(x(t), i) \leq x^{\top}(t)\left[A^{\top}(i)P(i) + P(i)A(i) + \sum_{j=1}^{N}\lambda_{ij}P(j)\right]x(t)$$

$$+\varepsilon_w^{-1}(i)x^{\top}(t)P(i)B_{\omega}(i)B_{\omega}^{\top}(i)P(i)x(t) + \varepsilon_w(i)\omega^{\top}(t)\omega(t)$$

$$= x^{\top}(t)\left[A^{\top}(i)P(i) + P(i)A(i) + \sum_{j=1}^{N}\lambda_{ij}P(j)\right]x(t)$$

$$+x^{\top}(t)\left[\varepsilon_w^{-1}(i)P(i)B_{\omega}(i)B_{\omega}^{\top}(i)P(i)\right]x(t) + \varepsilon_w(i)\omega^{\top}(t)\omega(t),$$

$$= x^{\top}(t)\Xi(i)x(t) + \varepsilon_w(i)\omega^{\top}(t)\omega(t), \quad (4.8)$$

with

$$\Xi(i) = A^{\top}(i)P(i) + P(i)A(i) + \sum_{j=1}^{N}\lambda_{ij}P(j)$$

$$+\varepsilon_w^{-1}(i)P(i)B_{\omega}(i)B_{\omega}^{\top}(i)P(i).$$

Based on Dynkin's formula, we get the following:

$$\mathbb{E}\left[V(x(t), i) - V(x_0, r_0)\right] = \mathbb{E}\left[\int_0^t \mathscr{L}V(x(s), r(s))ds|x_0, r_0\right],$$

which combined with (4.8) yields

$$\mathbb{E}\left[V(x(t), i) - V(x_0, r_0)\right] \leq \mathbb{E}\left[\int_0^t x^{\top}(s)\Xi(r(s))x(s)ds|x_0, r_0\right]$$

$$+\varepsilon_w(i)\int_0^t \omega^\top(s)\omega(s)ds. \qquad (4.9)$$

Since $V(x(t),i)$ is nonnegative, (4.9) implies

$$\mathbb{E}[V(x(t),i)] + \mathbb{E}\left[\int_0^t x^\top(s)[-\Xi(r(s))]x(s)ds|x_0,r_0\right]$$
$$\leq V(x_0,r_0) + \varepsilon_w(i)\int_0^t \omega^\top(s)\omega(s)ds,$$

which yields

$$\min_{i\in\mathscr{S}}\{\lambda_{min}(-\Xi(i))\}\mathbb{E}\left[\int_0^t x^\top(s)x(s)ds\right] \leq \mathbb{E}\left[\int_0^t x^\top(s)[-\Xi(r(s))]x(s)ds\right]$$
$$\leq V(x_0,r_0) + \varepsilon_w(i)\int_0^\infty \omega^\top(s)\omega(s)ds.$$

This proves that system (4.1) is stochastically stable.   □

Let us now establish what conditions should be satisfied if we want (4.1) be stochastically stable with γ-disturbance rejection. The following theorem gives such conditions.

Theorem 42. *Let γ be a given positive constant. If there exists a set of symmetric and positive-definite matrices $P = (P(1),\cdots,P(N)) > 0$ such that the following set of coupled LMIs holds for every $i \in \mathscr{S}$:*

$$\begin{bmatrix} J_0(i) & \begin{bmatrix} C_z^\top(i)B_z(i) \\ +P(i)B_\omega(i) \end{bmatrix} \\ \begin{bmatrix} B_z^\top(i)C_z(i) \\ +B_\omega^\top(i)P(i) \end{bmatrix} & B_z^\top(i)B_z(i) - \gamma^2\mathbb{I} \end{bmatrix} < 0, \qquad (4.10)$$

where $J_0(i) = A^\top(i)P(i)+P(i)A(i)+\sum_{j=1}^N \lambda_{ij}P(j)+C_z^\top(i)C_z(i)$, then system (4.1) with $u(t) \equiv 0$ is stochastically stable and satisfies the following:

$$\|z\|_2 \leq \left[\gamma^2\|\omega\|_2^2 + x_0^\top P(r_0)x_0\right]^{\frac{1}{2}}, \qquad (4.11)$$

which means that the system with $u_t = 0$ for all $t \geq 0$ is stochastically stable with γ-disturbance attenuation.

Proof: From (4.10) and using the Schur complement, we get the following inequality:

$$A^\top(i)P(i) + P(i)A(i) + \sum_{j=1}^N \lambda_{ij}P(j) + C_z^\top(i)C_z(i) < 0,$$

which implies the following since $C_z^\top(i)C_z(i) \geq 0$:

$$A^{\top}(i)P(i) + P(i)A(i) + \sum_{j=1}^{N} \lambda_{ij} P(j) < 0.$$

Based on Definition 6, this proves that the system under study is internally MSQS. Using Theorem 41, we conclude that system (4.1) with $u(t) \equiv 0$ is stochastically stable.

Let us now prove that (4.11) is satisfied. To this end, let us define the following performance function:

$$J_T = \mathbb{E}\left[\int_0^T [z^{\top}(t)z(t) - \gamma^2 \omega^{\top}(t)\omega(t)]dt\right].$$

To prove (4.11) it suffices to establish that J_∞ is bounded, that is,

$$J_\infty \leq V(x_0, r_0) = x_0^{\top} P(r_0) x_0.$$

Notice that for $V(x(t), i) = x^{\top}(t)P(i)x(t)$, we have

$$\mathscr{L}V(x(t), i) = x^{\top}(t)\left[A^{\top}(i)P(i) + P(i)A(i) + \sum_{j=1}^{N} \lambda_{ij} P(j)\right]x(t)$$
$$+ x^{\top}(t)P(i)B_\omega(i)\omega(t) + \omega^{\top}(t)B_\omega^{\top}(i)P(i)x(t),$$

and

$$z^{\top}(t)z(t) - \gamma^2 \omega(t)\omega(t)$$
$$= [C_z(i)x(t) + B_z(i)\omega(t)]^{\top}[C_z(i)x(t) + B_z(i)\omega(t)] - \gamma^2 \omega(t)\omega(t)$$
$$= x^{\top}(t)C_z^{\top}(i)C_z(i)x(t) + x^{\top}(t)C_z^{\top}(i)B_z(i)\omega(t)$$
$$+ \omega^{\top}(t)B_z^{\top}(i)C_z(i)x(t) + \omega^{\top}(t)B_z^{\top}(i)B_z(i)\omega(t) - \gamma^2 \omega^{\top}(t)\omega(t),$$

which implies the following equality:

$$z^{\top}(t)z(t) - \gamma^2 \omega^{\top}(t)\omega(t) + \mathscr{L}V(x(t), r(t)) = \eta^{\top}(t)\Theta(r(t))\eta(t),$$

with

$$\Theta(i) = \begin{bmatrix} J_0(i) & \begin{bmatrix} C_z^{\top}(i)B_z(i) \\ +P(i)B_\omega(i) \end{bmatrix} \\ \begin{bmatrix} B_z^{\top}(i)C_z(i) \\ +B_\omega^{\top}(i)P(i) \end{bmatrix} & B_z^{\top}(i)B_z(i) - \gamma^2 \mathbb{I} \end{bmatrix}$$
$$\eta^{\top}(t) = \begin{bmatrix} x^{\top}(t) & \omega^{\top}(t) \end{bmatrix}.$$

Therefore,

$$J_T = \mathbb{E}\left[\int_0^T [z^{\top}(t)z(t) - \gamma^2 \omega^{\top}(t)\omega(t) + \mathscr{L}V(x(t), r(t))]dt\right]$$

$$-\mathbb{E}\left[\int_0^T \mathscr{L}V(x(t), r(t))]dt\right].$$

Using now Dynkin's formula,

$$\mathbb{E}\left[\int_0^T \mathscr{L}V(x(t), r(t))dt|x_0, r_0\right] = \mathbb{E}[V(x(T), r(T))] - V(x_0, r_0),$$

we get

$$J_T = \mathbb{E}\left[\int_0^T \eta^\top(t)\Theta(r(t))\eta(t)dt\right] - \mathbb{E}[V(x(T), r(T))] + V(x_0, r_0).$$

Since $\Theta(i) < 0$ and $\mathbb{E}[V(x(T), r(T))] \geq 0$, (4.12) implies the following:

$$J_T \leq V(x_0, r_0),$$

which yields $J_\infty \leq V(x_0, r_0)$, i.e., $\|z\|_2^2 - \gamma^2\|\omega\|_2^2 \leq x_0^\top P(r_0)x_0$.

This gives the desired results:

$$\|z\|_2 \leq \left[\gamma^2\|\omega\|_2^2 + x_0^\top P(r_0)x_0\right]^{\frac{1}{2}}.$$

This ends the proof of the theorem.   □

Example 45. To illustrate the effectiveness of the results of this theorem, let us consider a two-mode system with the following data:

- mode #1:

$$A(1) = \begin{bmatrix} -0.5 & 1.0 \\ 0.3 & -2.5 \end{bmatrix}, \quad B(1) = \begin{bmatrix} 1.0 & 0.0 \\ 0.0 & 1.0 \end{bmatrix},$$

$$B_w(1) = \begin{bmatrix} 1.0 & 0.0 \\ 0.0 & 1.0 \end{bmatrix}, \quad C_z(1) = \begin{bmatrix} 1.0 & 0.0 \\ 0.0 & 1.0 \end{bmatrix},$$

$$B_z(1) = \begin{bmatrix} 1.0 & 0.0 \\ 0.0 & 1.0 \end{bmatrix}, \quad D_z(1) = \begin{bmatrix} 1.0 & 0.0 \\ 0.0 & 1.0 \end{bmatrix},$$

- mode #2:

$$A(2) = \begin{bmatrix} -1.0 & 0.1 \\ 0.2 & -2.0 \end{bmatrix}, \quad B(2) = \begin{bmatrix} 1.0 & 0.0 \\ 0.0 & 1.0 \end{bmatrix},$$

$$B_w(2) = \begin{bmatrix} 1.0 & 0.0 \\ 0.0 & 1.0 \end{bmatrix}, \quad C_z(2) = \begin{bmatrix} 1.0 & 0.0 \\ 0.0 & 1.0 \end{bmatrix},$$

$$B_z(2) = \begin{bmatrix} 1.0 & 0.0 \\ 0.0 & 1.0 \end{bmatrix}, \quad D_z(2) = \begin{bmatrix} 1.0 & 0.0 \\ 0.0 & 1.0 \end{bmatrix}.$$

The switching between the two modes is assumed to be described by the following transition matrix:

$$\Lambda = \begin{bmatrix} -2.0 & 2.0 \\ 3.0 & -3.0 \end{bmatrix}.$$

The desired disturbance level for this example is fixed to $\gamma = 5.0$. Solving the LMI (4.10), we get

$$P(1) = \begin{bmatrix} 11.8533 & 2.5811 \\ 2.5811 & 6.4475 \end{bmatrix}, P(2) = \begin{bmatrix} 10.9411 & 1.3425 \\ 1.3425 & 6.6235 \end{bmatrix}.$$

Based on the results of the theorem we conclude that the system is stochastically stable and guarantees the disturbance rejection of the desired level γ.

Let us return to the dynamics (4.1) and consider now that the uncertainties are not equal to zero. In this case the system with $u(t) = 0$ for all $t \geq 0$ is internally mean square quadratically stable if there exists a set of symmetric and positive-definite matrices $P = (P(1), \cdots, P(N)) > 0$, such that the following holds for all admissible uncertainties and for every $i \in \mathscr{S}$:

$$A^\top(i,t)P(i) + P(i)A(i,t) + \sum_{j=1}^{N} \lambda_{ij} P(j) < 0.$$

This condition is useless since it contains the uncertainties $F_A(i,t)$. Let us transform it into a useful one that can be used to check the robust stability. If we use the expression of $A(i,t)$, we get

$$A^\top(i)P(i) + P(i)A(i) + \sum_{j=1}^{N} \lambda_{ij} P(j) + P(i)D_A(i)F_A(i,t)E_A(i)$$
$$+ E_A^\top(i)F_A^\top(i,t)D_A^\top(i)P(i) < 0.$$

Using now Lemma 7 in Appendix A, the previous inequality will be satisfied if the following holds:

$$A^\top(i)P(i) + P(i)A(i) + \sum_{j=1}^{N} \lambda_{ij} P(j) + \varepsilon_A(i)E_A^\top(i)E_A(i)$$
$$+ \varepsilon_A^{-1}(i)P(i)D_A(i)D_A^\top(i)P(i) < 0,$$

with $\varepsilon_A(i) > 0$ for all $i \in \mathscr{S}$.

Using the Schur complement we get the desired condition:

$$\begin{bmatrix} J_0(i) & P(i)D_A(i) \\ D_A^\top(i)P(i) & -\varepsilon_A(i)\mathbb{I} \end{bmatrix} < 0, \tag{4.12}$$

with $J_0(i) = A^\top(i)P(i) + P(i)A(i) + \sum_{j=1}^N \lambda_{ij}P(j) + \varepsilon_A(i)E_A^\top(i)E_A(i)$.

The results of this development are summarized by the following theorem.

Theorem 43. *If there exist a set of symmetric and positive-definite matrices $P = (P(1), \cdots, P(N)) > 0$ and a set of positive scalars $\varepsilon_A = (\varepsilon_A(1), \cdots, \varepsilon_A(N))$ such that the following set of coupled LMIs (4.12) holds for every $i \in \mathscr{S}$ and for all admissible uncertainties, then system (4.1) with $u(t) = 0$ for all $t \geq 0$ is internally mean square quadratically stable.*

The following result shows that if (4.1) is internally mean square stochastically stable for all admissible uncertainties, it is also robustly stochastically stable.

Theorem 44. *Let $\omega(.) \in \mathscr{L}_2[0, \infty)$. If the system (4.1) with $u(t) = 0$ for all $t \geq 0$ is internally mean square stochastically stable for all admissible uncertainties, it is also stochastically stable.*

Proof: The proof of this theorem is similar to the one of the previous theorem and the details are omitted $\qquad\square$

Theorem 45. *If there exists a set of symmetric and positive-definite matrices $P = (P(1), \cdots, P(N)) > 0$ such that the following set of coupled LMIs holds for every $i \in \mathscr{S}$ and for all admissible uncertainties:*

$$\begin{bmatrix} J_u(i) & \begin{bmatrix} C_z^\top(i,t)B_z(i) \\ +P(i)B_w(i) \end{bmatrix} \\ \begin{bmatrix} B^\top(i)C_z(i,t) \\ +B_w^\top(i)P(i) \end{bmatrix} & B_z^\top(i)B_z(i) - \gamma^2 \mathbb{I} \end{bmatrix} < 0, \tag{4.13}$$

with $J_u(i) = A^\top(i,t)P(i) + P(i)A(i,t) + \sum_{j=1}^N \lambda_{ij}P(j) + C_z^\top(i,t)C_z(i,t)$, then system (4.1) with $u(t) = 0$ for all $t \geq 0$ is robustly stochastically stable and satisfies the disturbance rejection of level γ,

$$\|z\|_2 \leq \left[\gamma^2 \|\omega\|_2^2 + x_0^\top P(r_0)x_0\right]^{\frac{1}{2}}. \tag{4.14}$$

Proof: Start by proving that if the LMI (4.13) is satisfied then it is implied that the system is robustly stochastically stable. Notice that if the LMI (4.13) is satisfied for every $i \in \mathscr{S}$, then it is implied that

$$A^\top(i,t)P(i) + P(i)A(i,t) + \sum_{j=1}^N \lambda_{ij}P(j) + C_z^\top(i,t)C_z(i,t) < 0,$$

and since $C_z^\top(i,t)C_z(i,t) \geq 0$ for all $i \in \mathscr{S}$ and for all t, then we get

$$A^\top(i,t)P(i) + P(i)A(i,t) + \sum_{j=1}^N \lambda_{ij}P(j) < 0,$$

which implies that the system is mean square quadratically stable.

Let us prove also that the LMI (4.13) implies that the system satisfies the disturbance rejection of level γ. For this purpose, choose a Lyapunov function $V(x(t), r(t))$ defined as

$$V(x(t), r(t)) = x^\top(t)P(r(t))x(t),$$

where $P(i) > 0$, $i \in \mathscr{S}$, is symmetric and positive-definite matrix.

The infinitesimal operator $\mathscr{L}$ of the Markov process $\{(x(t), r(t)), t \geq 0\}$ acting on $V(.)$ and emanating from the point (x, i) at time t, where $x(t) = x$ and $r(t) = i$ for $i \in \mathscr{S}$ is given by

$$\mathscr{L}V(x(t), i) = x^\top(t)\left[A^\top(i, t)P(i) + P(i)A(i, t) + \sum_{j=1}^{N}\lambda_{ij}P(j)\right]x(t)$$
$$+ 2x^\top(t)P(i)B_w(i)\omega(t).$$

Let us define the following performance function:

$$J_T = \mathbb{E}\left[\int_0^T [z^\top(t)z(t) - \gamma^2\omega^\top(t)\omega(t)]dt\right].$$

To prove (4.14), it suffices to establish that J_∞ is bounded for all admissible uncertainties, that is,

$$J_\infty \leq V(x_0, r_0) = x_0^\top P(r_0)x_0.$$

Notice that

$$\mathscr{L}V(x(t), i) = x^\top(t)\left[A^\top(i, t)P(i) + P(i)A(i, t) + \sum_{j=1}^{N}\lambda_{ij}P(j)\right]x(t)$$
$$+ x^\top(t)P(i)B_\omega(i)\omega(t) + \omega^\top(t)B_\omega^\top(i)P(i)x(t),$$

and

$$z^\top(t)z(t) - \gamma^2\omega(t)\omega(t)$$
$$= [C_z(i, t)x(t) + B_z(i)\omega(t)]^\top [C_z(i, t)x(t) + B_z(i)\omega(t)]$$
$$- \gamma^2\omega(t)\omega(t)$$
$$= x^\top(t)C_z^\top(i, t)C_z(i, t)x(t) + x^\top(t)C_z^\top(i, t)B_z(i)\omega(t)$$
$$+ \omega^\top(t)B_z^\top(i)C_z(i, t)x(t)$$
$$+ \omega^\top(t)B_z^\top(i)B_z(i)\omega(t) - \gamma^2\omega^\top(t)\omega(t),$$

which implies

$$z^\top(t)z(t) - \gamma^2\omega^\top(t)\omega(t) + \mathscr{L}V(x(t), i) = \eta^\top(t)\Theta_u(i)\eta(t),$$

with

$$
\Theta_u(i) = \begin{bmatrix} J_u(i) & \begin{bmatrix} C_z^\top(i,t)B_z(i) \\ +P(i)B_\omega(i) \end{bmatrix} \\ \begin{bmatrix} B_z^\top(i)C_z(i,t) \\ +B_\omega^\top(i)P(i) \end{bmatrix} & B_z^\top(i)B_z(i) - \gamma^2\mathbb{I} \end{bmatrix},
$$

$$
\eta^\top(t) = \begin{bmatrix} x^\top(t) \ \omega^\top(t) \end{bmatrix}.
$$

Therefore,

$$
J_T = \mathbb{E}\left[\int_0^T \left[z^\top(t)z(t) - \gamma^2\omega^\top(t)\omega(t) + \mathscr{L}V(x(t),r(t))\right]dt\right]
$$
$$
-\mathbb{E}\left[\int_0^T \mathscr{L}V(x(t),r(t))dt\right].
$$

From Dynkin's formula, we get

$$
\mathbb{E}\left[\int_0^T \mathscr{L}V(x(t),r(t))dt|x_0,r_0\right] = \mathbb{E}\left[V(x(T),r(T))\right] - V(x_0,r_0),
$$

which implies

$$
J_T = \mathbb{E}\left[\int_0^T \eta^\top(t)\Theta_u(r(t))\eta(t)dt\right] - \mathbb{E}[V(x(T),r(T))] + V(x_0,r_0). \quad (4.15)
$$

Since $\Theta_u(i) < 0$ and $\mathbb{E}[V(x(T),r(T))] \geq 0$, (4.15) implies the following:

$$
J_T \leq V(x_0,r_0),
$$

which yields $J_\infty \leq V(x_0,r_0)$, i.e., $\|z\|_2^2 - \gamma^2\|\omega\|_2^2 \leq x_0^\top P(r_0)x_0$.

This gives the desired results:

$$
\|z\|_2 \leq \left[\gamma^2\|\omega\|_2^2 + x_0^\top P(r_0)x_0\right]^{\frac{1}{2}},
$$

which gives (4.14). This ends the proof of the theorem. $\quad\square$

The LMI of this theorem is useless since it depends on the uncertainties. Let us transform it to get an equivalent LMI condition that does not depend on the system uncertainties, which we can use easily to check if a given system is robustly stochastically stable. For this purpose, notice that

$$
\begin{bmatrix} \begin{bmatrix} A^\top(i,t)P(i) \\ +P(i)A(i,t) \\ +\sum_{j=1}^N \lambda_{ij}P(j) \\ +C_z^\top(i,t)C_z(i,t) \end{bmatrix} & \begin{bmatrix} C_z^\top(i,t)B_z(i) \\ +P(i)B_\omega(i) \end{bmatrix} \\ \begin{bmatrix} B_z^\top(i)C_z(i,t) \\ +B_\omega^\top(i)P(i) \end{bmatrix} & B_z^\top(i)B_z(i) - \gamma^2\mathbb{I} \end{bmatrix} =
$$

$$\begin{bmatrix} J_1(i,t) & P(i)B_\omega(i) \\ B_\omega^\top(i)P(i) & -\gamma^2\mathbb{I} \end{bmatrix} + \begin{bmatrix} C_z^\top(i,t) \\ B_z^\top(i) \end{bmatrix}\begin{bmatrix} C_z(i,t) & B_z(i) \end{bmatrix}.$$

Using the Schur complement we show that this is equivalent to the following inequality:

$$\begin{bmatrix} J_1(i,t) & P(i)B_\omega(i) & C_z^\top(i,t) \\ B_\omega^\top(i)P(i) & -\gamma^2\mathbb{I} & B_z^\top(i) \\ C_z(i,t) & B_z(i) & -\mathbb{I} \end{bmatrix} < 0.$$

Using the expressions of $A(i,t)$ and $C_z(i,t)$ we get

$$\begin{bmatrix} J_1(i) & P(i)B_\omega(i) & C_z^\top(i) \\ B_\omega^\top(i)P(i) & -\gamma^2\mathbb{I} & B_z^\top(i) \\ C_z(i) & B_z(i) & -\mathbb{I} \end{bmatrix}$$

$$+ \begin{bmatrix} E_A^\top(i)F_A^\top(i,t)D_A^\top(i)P(i) & 0 & 0 \\ 0 & 0 & 0 \\ 0 & 0 & 0 \end{bmatrix}$$

$$+ \begin{bmatrix} P(i)D_A(i)F_A(i,t)E_A(i) & 0 & 0 \\ 0 & 0 & 0 \\ 0 & 0 & 0 \end{bmatrix}$$

$$+ \begin{bmatrix} 0 & 0 & E_{C_z}^\top(i)F_{C_z}^\top(i,t)D_{C_z}^\top(i) \\ 0 & 0 & 0 \\ 0 & 0 & 0 \end{bmatrix}$$

$$+ \begin{bmatrix} 0 & 0 & 0 \\ 0 & 0 & 0 \\ D_{C_z}(i)F_{C_z}(i,t)E_{C_z}(i) & 0 & 0 \end{bmatrix} < 0,$$

with $J_1(i) = A^\top(i)P(i) + P(i)A(i) + \sum_{j=1}^N \lambda_{ij}P(j)$.
    Notice that

$$\begin{bmatrix} E_A^\top(i)F_A^\top(i,t)D_A^\top(i)P(i) & 0 & 0 \\ 0 & 0 & 0 \\ 0 & 0 & 0 \end{bmatrix}$$

$$= \begin{bmatrix} E_A^\top(i) & 0 & 0 \\ 0 & 0 & 0 \\ 0 & 0 & 0 \end{bmatrix}\begin{bmatrix} F_A^\top(i,t) & 0 & 0 \\ 0 & 0 & 0 \\ 0 & 0 & 0 \end{bmatrix}\begin{bmatrix} D_A^\top(i)P(i) & 0 & 0 \\ 0 & 0 & 0 \\ 0 & 0 & 0 \end{bmatrix},$$

$$\begin{bmatrix} P(i)D_A(i)F_A(i,t)E_A(i) & 0 & 0 \\ 0 & 0 & 0 \\ 0 & 0 & 0 \end{bmatrix}$$

$$
= \begin{bmatrix} P(i)D_A(i) & 0 & 0 \\ 0 & 0 & 0 \\ 0 & 0 & 0 \end{bmatrix} \begin{bmatrix} F_A(i,t) & 0 & 0 \\ 0 & 0 & 0 \\ 0 & 0 & 0 \end{bmatrix} \begin{bmatrix} E_A(i) & 0 & 0 \\ 0 & 0 & 0 \\ 0 & 0 & 0 \end{bmatrix},
$$

$$
\begin{bmatrix} 0 & 0 & E_{C_z}^\top(i)F_{C_z}^\top(i,t)D_{C_z}^\top(i) \\ 0 & 0 & 0 \\ 0 & 0 & 0 \end{bmatrix}
$$

$$
= \begin{bmatrix} 0 & 0 & E_{C_z}^\top(i) \\ 0 & 0 & 0 \\ 0 & 0 & 0 \end{bmatrix} \begin{bmatrix} 0 & 0 & 0 \\ 0 & 0 & 0 \\ 0 & 0 & F_{C_z}^\top(i,t) \end{bmatrix} \begin{bmatrix} 0 & 0 & 0 \\ 0 & 0 & 0 \\ 0 & 0 & D_{C_z}^\top(i) \end{bmatrix},
$$

and

$$
\begin{bmatrix} 0 & 0 & 0 \\ 0 & 0 & 0 \\ D_{C_z}(i)F_{C_z}(i,t)E_{C_z}(i) & 0 & 0 \end{bmatrix}
$$

$$
= \begin{bmatrix} 0 & 0 & 0 \\ 0 & 0 & 0 \\ 0 & 0 & D_{C_z}(i) \end{bmatrix} \begin{bmatrix} 0 & 0 & 0 \\ 0 & 0 & 0 \\ 0 & 0 & F_{C_z}(i,t) \end{bmatrix} \begin{bmatrix} 0 & 0 & 0 \\ 0 & 0 & 0 \\ E_{C_z}(i) & 0 & 0 \end{bmatrix}.
$$

Using Lemma 7 in Appendix A, we get

$$
\begin{bmatrix} P(i)D_A(i)F_A(i,t)E_A(i) & 0 & 0 \\ 0 & 0 & 0 \\ 0 & 0 & 0 \end{bmatrix}
$$

$$
+ \begin{bmatrix} E_A^\top(i)F_A^\top(i,t)D_A^\top(i)P(i) & 0 & 0 \\ 0 & 0 & 0 \\ 0 & 0 & 0 \end{bmatrix}
$$

$$
\leq \varepsilon_A^{-1}(i) \begin{bmatrix} P(i)D_A(i) & 0 & 0 \\ 0 & 0 & 0 \\ 0 & 0 & 0 \end{bmatrix} \begin{bmatrix} D_A^\top(i)P(i) & 0 & 0 \\ 0 & 0 & 0 \\ 0 & 0 & 0 \end{bmatrix}
$$

$$
+ \varepsilon_A(i) \begin{bmatrix} E_A^\top(i) & 0 & 0 \\ 0 & 0 & 0 \\ 0 & 0 & 0 \end{bmatrix} \begin{bmatrix} E_A(i) & 0 & 0 \\ 0 & 0 & 0 \\ 0 & 0 & 0 \end{bmatrix}
$$

$$
= \begin{bmatrix} \varepsilon_A^{-1}(i)P(i)D_A(i)D_A^\top(i)P(i) + \varepsilon_A(i)E_A^\top(i)E_A(i) & 0 & 0 \\ 0 & 0 & 0 \\ 0 & 0 & 0 \end{bmatrix},
$$

$$
\begin{bmatrix} 0 & 0 & E_{C_z}^\top(i)F_{C_z}^\top(i,t)D_{C_z}^\top(i) \\ 0 & 0 & 0 \\ 0 & 0 & 0 \end{bmatrix}
$$

$$
+ \begin{bmatrix} 0 & 0 & 0 \\ 0 & 0 & 0 \\ D_{C_z}(i)F_{C_z}(i,t)E_{C_z}(i) & 0 & 0 \end{bmatrix}
$$

$$\leq \varepsilon_{C_z}^{-1}(i) \begin{bmatrix} 0 & 0 & E_{C_z}^\top(i) \\ 0 & 0 & 0 \\ 0 & 0 & 0 \end{bmatrix} \begin{bmatrix} 0 & 0 & 0 \\ 0 & 0 & 0 \\ E_{C_z}(i,t) & 0 & 0 \end{bmatrix}$$

$$+ \varepsilon_{C_z}(i) \begin{bmatrix} 0 & 0 & 0 \\ 0 & 0 & 0 \\ 0 & 0 & D_{C_z}(i) \end{bmatrix} \begin{bmatrix} 0 & 0 & 0 \\ 0 & 0 & 0 \\ 0 & 0 & D_{C_z}^\top(i) \end{bmatrix}$$

$$= \begin{bmatrix} \varepsilon_{C_z}^{-1}(i) E_{C_z}^\top(i) E_{C_z}(i,t) & 0 & 0 \\ 0 & 0 & 0 \\ 0 & 0 & \varepsilon_{C_z}(i) D_{C_z}(i) D_{C_z}^\top(i) \end{bmatrix}.$$

Using these transformations, we get

$$\begin{bmatrix} J_n(i) & P(i)B_\omega(i) & C_z^\top(i) \\ B_\omega^\top(i)P(i) & -\gamma^2 \mathbb{I} & B_z^\top(i) \\ C_z(i) & B_z(i) & -\mathbb{I} \end{bmatrix}$$

$$+ \begin{bmatrix} \varepsilon_A^{-1}(i)P(i)D_A(i)D_A^\top(i)P(i) + \varepsilon_A(i)E_A^\top(i)E_A(i) & 0 & 0 \\ 0 & 0 & 0 \\ 0 & 0 & 0 \end{bmatrix}$$

$$+ \begin{bmatrix} \varepsilon_{C_z}^{-1}(i)E_{C_z}^\top(i)E_{C_z}(i,t) & 0 & 0 \\ 0 & 0 & 0 \\ 0 & 0 & \varepsilon_{C_z}(i)D_{C_z}(i)D_{C_z}^\top(i) \end{bmatrix} < 0,$$

with $J_n(i) = A^\top(i)P(i) + P(i)A(i) + \sum_{j=1}^N \lambda_{ij}P(j)$.
Let $J_m(i)$, $\mathcal{W}_n(i)$, and $\mathcal{T}_m(i)$ be defined as

$$J_m(i) = J_n(i) + \varepsilon_A(i)E_A^\top(i)E_A(i),$$
$$\mathcal{W}_m(i) = \operatorname{diag}\left[\varepsilon_A(i)\mathbb{I}, \varepsilon_{C_z}(i)\mathbb{I}\right],$$
$$\mathcal{T}_m(i) = \left(P(i)D_A(i), E_{C_z}^\top(i)\right),$$

and using the Schur complement we get the equivalent inequality:

$$\begin{bmatrix} J_m(i) & P(i)B_\omega(i) & C_z^\top(i) & \mathcal{T}_m(i) \\ B_\omega^\top(i)P(i) & -\gamma^2 \mathbb{I} & B_z^\top(i) & 0 \\ C_z(i) & B_z(i) & -\mathbb{I} + \varepsilon_{C_z}(i)D_{C_z}(i)D_{C_z}^\top(i) & 0 \\ \mathcal{T}_m^\top(i) & 0 & 0 & -\mathcal{W}_m(i) \end{bmatrix} < 0. \qquad (4.16)$$

The following theorem summarizes the results of this development.

Theorem 46. *Let γ be a positive constant. If there exist a set of symmetric and positive-definite matrices $P = (P(1), \cdots, P(N)) > 0$ and sets of positive scalars $\varepsilon_A = (\varepsilon_A(1), \cdots, \varepsilon_A(N))$ and $\varepsilon_{C_z} = (\varepsilon_{C_z}(1), \cdots, \varepsilon_{C_z}(N))$ such that* *the following set of coupled LMIs (4.16) holds for every $i \in \mathscr{S}$ and for all admissible uncertainties, then the system (4.1) is robustly stochastically stable and moreover the system satisfies the disturbance rejection of level γ.*

Example 46. To illustrate the effectiveness of the results of this theorem, let us consider the two-mode system of Example 45 with the following extra data:

- mode #1:

$$D_A(1) = \begin{bmatrix} 0.10 \\ 0.20 \end{bmatrix}, \quad E_A(1) = \begin{bmatrix} 0.20 & 0.10 \end{bmatrix},$$

$$D_B(1) = \begin{bmatrix} 0.10 \\ 0.20 \end{bmatrix}, \quad E_B(1) = \begin{bmatrix} 0.20 & 0.10 \end{bmatrix},$$

$$D_{C_z}(1) = \begin{bmatrix} 0.10 \\ 0.20 \end{bmatrix}, \quad E_{C_z}(1) = \begin{bmatrix} 0.20 & 0.10 \end{bmatrix},$$

$$D_{D_z}(1) = \begin{bmatrix} 0.10 \\ 0.20 \end{bmatrix}, \quad E_{D_z}(1) = \begin{bmatrix} 0.20 & 0.10 \end{bmatrix},$$

- mode #2:

$$D_A(2) = \begin{bmatrix} 0.13 \\ 0.10 \end{bmatrix}, \quad E_A(2) = \begin{bmatrix} 0.10 & 0.20 \end{bmatrix},$$

$$D_B(2) = \begin{bmatrix} 0.13 \\ 0.10 \end{bmatrix}, \quad E_B(2) = \begin{bmatrix} 0.10 & 0.20 \end{bmatrix},$$

$$D_{C_z}(2) = \begin{bmatrix} 0.13 \\ 0.10 \end{bmatrix}, \quad E_{C_z}(2) = \begin{bmatrix} 0.10 & 0.20 \end{bmatrix},$$

$$D_{D_z}(2) = \begin{bmatrix} 0.13 & 0.10 \end{bmatrix}, \quad E_{D_z}(2) = \begin{bmatrix} 0.10 & 0.20 \end{bmatrix}.$$

Solving the LMI (4.16), we get

$$P(1) = \begin{bmatrix} 2.7536 & 0.4831 \\ 0.4831 & 3.2025 \end{bmatrix}, \quad P(2) = \begin{bmatrix} 2.2407 & 0.1031 \\ 0.1031 & 3.2327 \end{bmatrix}.$$

Based on the results of this theorem, we conclude that the system is stochastically stable and satisfies the disturbance rejection of level $\gamma = 2.7901$.

4.2 State Feedback Stabilization

In this section the structure of the controller we consider is given by the following form:

$$u(t) = K(i)x(t), \tag{4.17}$$

where $x(t)$ is the state vector and $K(i)$, $i \in \mathscr{S}$ is a design parameter with an appropriate dimension that has to be chosen. In this section we assume the complete access to the state vector and to the mode at each time t.

Let us drop the uncertainties from the dynamics and see how we can design a controller of the form (4.17). Plugging the expression of the controller in the dynamics (4.1), we get

$$\begin{cases} \dot{x}(t) = \bar{A}(i)x(t) + B_w(i)\omega(t), \\ z(t) = \bar{C}_z(i)x(t) + B_z(i)\omega(t), \end{cases} \tag{4.18}$$

where $\bar{A}(i) = A(i) + B(i)K(i)$ and $\bar{C}_z(i) = C_z(i) + D_z(i)K(i)$.

Using the results of Theorem 42 we get the following results for the stochastic stability and disturbance rejection of level $\gamma > 0$ for the closed-loop dynamics.

Theorem 47. *Let γ be a given positive constant and $K = (K(1), \cdots, K(N))$ be a set of given gains. If there exists a set of symmetric and positive-definite matrices $P = (P(1), \cdots, P(N)) > 0$ such that the following set of coupled LMIs holds for every $i \in \mathscr{S}$:*

$$\begin{bmatrix} \bar{J}_0(i) & \begin{bmatrix} \bar{C}_z^\top(i)B_z(i) \\ +P(i)B_\omega(i) \end{bmatrix} \\ \begin{bmatrix} B_z^\top(i)\bar{C}_z(i) \\ +B_\omega^\top(i)P(i) \end{bmatrix} & B_z^\top(i)B_z(i) - \gamma^2\mathbb{I} \end{bmatrix} < 0, \tag{4.19}$$

with $\bar{J}_0(i) = \bar{A}^\top(i)P(i) + P(i)\bar{A}(i) + \sum_{j=1}^N \lambda_{ij}P(j) + \bar{C}_z^\top(i)\,\bar{C}_z(i)$, then system (4.1) is stochastically stable under the controller (4.17) and satisfies

$$\|z\|_2 \leq \left[\gamma^2\|\omega\|_2^2 + x_0^\top P(r_0)x_0\right]^{\frac{1}{2}}, \tag{4.20}$$

which means that the system is stochastically stable with γ-disturbance attenuation.

To synthesize the controller gain, let us transform the LMI (4.19) into a form that can be used easily to compute this gain for every mode $i \in \mathscr{S}$. Notice that

$$\begin{bmatrix} \bar{J}_0(i) & \begin{bmatrix} \bar{C}_z^\top(i)B_z(i) \\ +P(i)B_\omega(i) \end{bmatrix} \\ \begin{bmatrix} B_z^\top(i)\bar{C}_z(i) \\ +B_\omega^\top(i)P(i) \end{bmatrix} & B_z^\top(i)B_z(i) - \gamma^2\mathbb{I} \end{bmatrix} =$$

$$\begin{bmatrix} \bar{J}_1(i) & P(i)B_\omega(i) \\ B_\omega^\top(i)P(i) & -\gamma^2\mathbb{I} \end{bmatrix}$$

$$+ \begin{bmatrix} \bar{C}_z^\top(i) \\ B_z^\top(i) \end{bmatrix} \begin{bmatrix} \bar{C}_z(i) & B_z(i) \end{bmatrix},$$

with $\bar{J}_1(i) = \bar{A}^\top(i)P(i) + P(i)\bar{A}(i) + \sum_{j=1}^N \lambda_{ij}P(j)$.

Using the Schur complement we show that (4.19) is equivalent to the following inequality:

$$\begin{bmatrix} \bar{J}_1(i) & P(i)B_\omega(i) & \bar{C}_z^\top(i) \\ B_\omega^\top(i)P(i) & -\gamma^2\mathbb{I} & B_z^\top(i) \\ \bar{C}_z(i) & B_z(i) & -\mathbb{I} \end{bmatrix} < 0.$$

Since $\bar{A}(i)$ is nonlinear in $K(i)$ and $P(i)$, the previous inequality is also nonlinear. Therefore it cannot be solved using existing linear algorithms. To transform it into an LMI, let $X(i) = P^{-1}(i)$. As we did many times previously, let us pre- and post-multiply this inequality by $\mathrm{diag}[X(i), \mathbb{I}, \mathbb{I}]$, which gives

$$\begin{bmatrix} \bar{J}_X(i) & B_\omega(i) & X(i)\bar{C}_z^\top(i) \\ B_\omega^\top(i) & -\gamma^2\mathbb{I} & B_z^\top(i) \\ \bar{C}_z(i)X(i) & B_z(i) & -\mathbb{I} \end{bmatrix} < 0,$$

with $\bar{J}_X(i) = X(i)\bar{A}^\top(i) + \bar{A}(i)X(i) + \sum_{j=1}^N \lambda_{ij}X(i)X^{-1}(j)X(i)$.

Notice that

$$X(i)\bar{A}^\top(i) + \bar{A}(i)X(i) = \begin{aligned}& X(i)A^\top(i) + A(i)X(i) + Y^\top(i)B^\top(i) \\ & + B(i)Y(i),\end{aligned}$$

$$\sum_{j=1}^N \lambda_{ij}X(i)X^{-1}(j)X(i) = \lambda_{ii}X(i) + \mathcal{S}_i(X)\mathcal{X}_i^{-1}(X)\mathcal{S}_i^\top(X),$$

$$X(i)\left[C_z(i) + D_z(i)K(i)\right]^\top = X(i)C_z^\top(i) + Y^\top(i)D_z^\top(i),$$

where $Y(i) = K(i)X(i)$, and $\mathcal{S}_i(X)$ and $\mathcal{X}_i(X)$ are defined as

$$\mathcal{S}_i(X) = \left[\sqrt{\lambda_{i1}}X(i), \cdots, \sqrt{\lambda_{ii-1}}X(i), \sqrt{\lambda_{ii+1}}X(i),\right.$$
$$\left. \cdots, \sqrt{\lambda_{iN}}X(i)\right],$$
$$\mathcal{X}_i(X) = \mathrm{diag}\left[X(1), \cdots, X(i-1), X(i+1), \cdots, Y(N)\right].$$

Using the Schur complement implies that the previous inequality is equivalent to the following:

$$\begin{bmatrix} J(i) & B_\omega(i) & \begin{bmatrix} X(i)C_z^\top(i) \\ +Y^\top(i)D_z^\top(i) \end{bmatrix} & \mathcal{S}_i(X) \\ B_\omega^\top(i) & -\gamma^2\mathbb{I} & B_z^\top(i) & 0 \\ \begin{bmatrix} C_z(i)X(i) \\ +D_z(i)Y(i) \end{bmatrix} & B_z(i) & -\mathbb{I} & 0 \\ \mathcal{S}_i^\top(X) & 0 & 0 & -\mathcal{X}_i(X) \end{bmatrix} < 0, \qquad (4.21)$$

with $J(i) = X(i)A^\top(i) + A(i)X(i) + Y^\top(i)B^\top(i) + B(i)Y(i) + \lambda_{ii}X(i)$.

From this discussion we get the following theorem.

Theorem 48. *Let γ be a positive constant. If there exist a set of symmetric and positive-definite matrices $X = (X(1), \cdots, X(N)) > 0$ and a set of matrices $Y = (Y(1), \cdots, Y(N))$ such that the following set of coupled LMIs*

(4.21) holds for every $i \in \mathscr{S}$, then system (4.1) under the controller (4.17) with $K(i) = Y(i)X^{-1}(i)$ is stochastically stable and, moreover, the closed-loop system satisfies the disturbance rejection of level γ.

Example 47. To illustrate the effectiveness of the results developed in this theorem, let us consider the two-mode system of Example 45 with the same data.

Solving the LMI (4.21), we get

$$X(1) = \begin{bmatrix} 2.5267 & -0.0000 \\ -0.0000 & 2.5267 \end{bmatrix}, \quad X(2) = \begin{bmatrix} 2.2378 & -0.0000 \\ -0.0000 & 2.2378 \end{bmatrix},$$

$$Y(1) = \begin{bmatrix} -4.8707 & -1.1826 \\ -2.1020 & 0.1826 \end{bmatrix}, \quad Y(2) = \begin{bmatrix} -2.7171 & 0.6162 \\ -1.2876 & -0.4792 \end{bmatrix}.$$

Based on the results of this theorem, we conclude that the system is stochastically stable under the state feedback controller given by

$$u(t) = K(i)x(t), i = 1, 2,$$

with

$$K(1) = \begin{bmatrix} -1.9277 & -0.4681 \\ -0.8319 & 0.0723 \end{bmatrix}, \quad K(2) = \begin{bmatrix} -1.2141 & 0.2754 \\ -0.5754 & -0.2141 \end{bmatrix},$$

and that satisfies the desired disturbance rejection of level $\gamma = 5.0$.

The controller that stochastically stabilizes the system and at the same time guarantees the minimum disturbance rejection is of great practical interest. This controller can be obtained by solving the following optimization problem:

$$P: \begin{cases} \min\limits_{\substack{\nu > 0, \\ X = (X(1), \cdots, X(N)) > 0, \\ Y = (Y(1), \cdots, Y(N)),}} \nu, \\ s.t. : (4.21) \text{ with } \nu = \gamma^2. \end{cases}$$

The following corollary gives the results on the design of the controller that stochastically stabilizes the system (4.1) and simultaneously guarantees the smallest disturbance rejection level.

Corollary 6. *Let $\nu > 0$, $X = (X(1), \cdots, X(N)) > 0$, and $Y = (Y(1), \cdots, Y(N))$ be the solution of the optimization problem P. Then the controller (4.17) with $K(i) = Y(i)X^{-1}(i)$ stochastically stabilizes the class of systems we are considering and, moreover, the closed-loop system satisfies the disturbance rejection of level $\sqrt{\nu}$.*

Example 48. To illustrate the theoretical results of this theorem, let us consider the two-mode system of Example 45 with the same data. Solving the problem P, we get

$$X(1) = \begin{bmatrix} 36.2448 & -0.0000 \\ -0.0000 & 36.2448 \end{bmatrix}, \qquad X(2) = \begin{bmatrix} 33.4662 & -0.0000 \\ -0.0000 & 33.4662 \end{bmatrix},$$

$$Y(1) = 10^3 \cdot \begin{bmatrix} -9.8677 & -0.0236 \\ -0.0235 & -9.7952 \end{bmatrix}, \qquad Y(2) = 10^3 \cdot \begin{bmatrix} -9.7516 & -0.0050 \\ -0.0050 & -9.7181 \end{bmatrix}.$$

The corresponding gains are given by

$$K(1) = \begin{bmatrix} -272.2522 & -0.6507 \\ -0.6493 & -270.2522 \end{bmatrix}, \qquad K(2) = \begin{bmatrix} -291.3851 & -0.1500 \\ -0.1500 & -290.3851 \end{bmatrix}.$$

Based on the results of this corollary we conclude that the system is stochastically stable and guarantees the disturbance of level $\gamma = 1.0394$.

Let us examine the dynamics (4.1) when the uncertainties are not equal to zero and see how to synthesize the state feedback controller of the form (4.17) that robustly stabilizes the class of systems we are studying while guaranteeing the desired disturbance rejection of level γ. Following the same steps for the nominal case starting from the results of Theorem 45, we get

$$\begin{bmatrix} \bar{J}_0(i,t) & P(i)B_\omega(i) & \bar{C}_z^\top(i,t) \\ B_\omega^\top(i)P(i) & -\gamma^2\mathbb{I} & B_z^\top(i) \\ \bar{C}_z(i,t) & B_z(i) & -\mathbb{I} \end{bmatrix} < 0, \qquad (4.22)$$

with

$$\bar{J}_0(i,t) = \bar{A}^\top(i,t)P(i) + P(i)\bar{A}(i,t) + \sum_{j=1}^{N}\lambda_{ij}P(j),$$

$$\bar{A}(i,t) = A(i,t) + B(i,t)K(i),$$
$$\bar{C}_z(i,t) = C_z(i,t) + D_z(i,t)K(i).$$

Using the expressions of $\bar{A}(i,t)$, $\bar{C}_z(i,t)$ and their components, we obtain the following inequality:

$$\begin{bmatrix} J_1(i) & P(i)B_\omega(i) & \begin{bmatrix} C_z^\top(i) \\ +K^\top(i)D_z^\top(i) \end{bmatrix} \\ B_\omega^\top(i)P(i) & -\gamma^2\mathbb{I} & B_z^\top(i) \\ \begin{bmatrix} D_z(i)K(i) \\ +C_z(i) \end{bmatrix} & B_z(i) & -\mathbb{I} \end{bmatrix}$$
$$+ \begin{bmatrix} E_A^\top(i)F_A^\top(i,t)D_A^\top(i)P(i) & 0 & 0 \\ 0 & 0 & 0 \\ 0 & 0 & 0 \end{bmatrix}$$

$$+ \begin{bmatrix} P(i)D_A(i)F_A(i,t)E_A(i) & 0 & 0 \\ 0 & 0 & 0 \\ 0 & 0 & 0 \end{bmatrix}$$

$$+ \begin{bmatrix} P(i)D_B(i)F_B(i,t)E_B(i)K(i) & 0 & 0 \\ 0 & 0 & 0 \\ 0 & 0 & 0 \end{bmatrix}$$

$$+ \begin{bmatrix} K^\top(i)E_B^\top(i)F_B^\top(i,t)D_B^\top(i)P(i) & 0 & 0 \\ 0 & 0 & 0 \\ 0 & 0 & 0 \end{bmatrix}$$

$$+ \begin{bmatrix} 0 & 0 & K^\top(i)E_{D_z}^\top(i)F_{D_z}^\top(i,t)D_{D_z}^\top(i) \\ 0 & 0 & 0 \\ 0 & 0 & 0 \end{bmatrix}$$

$$+ \begin{bmatrix} 0 & 0 & 0 \\ 0 & 0 & 0 \\ D_{D_z}(i)F_{D_z}(i,t)E_{D_z}(i)K(i) & 0 & 0 \end{bmatrix}$$

$$+ \begin{bmatrix} 0 & 0 & E_{C_z}^\top(i)F_{C_z}^\top(i,t)D_{C_z}^\top(i) \\ 0 & 0 & 0 \\ 0 & 0 & 0 \end{bmatrix}$$

$$+ \begin{bmatrix} 0 & 0 & 0 \\ 0 & 0 & 0 \\ D_{C_z}(i)F_{C_z}(i,t)E_{C_z}(i) & 0 & 0 \end{bmatrix} < 0,$$

with

$$J_1(i) = A^\top(i)P(i) + P(i)A(i) + K^\top(i)B^\top(i)P(i)$$
$$+ P(i)B(i)K(i) + \sum_{j=1}^{N}\lambda_{ij}P(j).$$

Notice that

$$\begin{bmatrix} E_A^\top(i)F_A^\top(i,t)D_A^\top(i)P(i) & 0 & 0 \\ 0 & 0 & 0 \\ 0 & 0 & 0 \end{bmatrix}$$

$$= \begin{bmatrix} E_A^\top(i) & 0 & 0 \\ 0 & 0 & 0 \\ 0 & 0 & 0 \end{bmatrix} \begin{bmatrix} F_A^\top(i,t) & 0 & 0 \\ 0 & 0 & 0 \\ 0 & 0 & 0 \end{bmatrix} \begin{bmatrix} D_A^\top(i)P(i) & 0 & 0 \\ 0 & 0 & 0 \\ 0 & 0 & 0 \end{bmatrix},$$

$$\begin{bmatrix} P(i)D_A(i)F_A(i,t)E_A(i) & 0 & 0 \\ 0 & 0 & 0 \\ 0 & 0 & 0 \end{bmatrix}$$

$$= \begin{bmatrix} P(i)D_A(i) & 0 & 0 \\ 0 & 0 & 0 \\ 0 & 0 & 0 \end{bmatrix} \begin{bmatrix} F_A(i,t) & 0 & 0 \\ 0 & 0 & 0 \\ 0 & 0 & 0 \end{bmatrix} \begin{bmatrix} E_A(i) & 0 & 0 \\ 0 & 0 & 0 \\ 0 & 0 & 0 \end{bmatrix},$$

$$= \begin{bmatrix} P(i)D_B(i) & 0 & 0 \\ 0 & 0 & 0 \\ 0 & 0 & 0 \end{bmatrix} \begin{bmatrix} F_B(i,t) & 0 & 0 \\ 0 & 0 & 0 \\ 0 & 0 & 0 \end{bmatrix} \begin{bmatrix} E_B(i)K(i) & 0 & 0 \\ 0 & 0 & 0 \\ 0 & 0 & 0 \end{bmatrix} = \begin{bmatrix} P(i)D_B(i)F_B(i,t)E_B(i)K(i) & 0 & 0 \\ 0 & 0 & 0 \\ 0 & 0 & 0 \end{bmatrix},$$

$$= \begin{bmatrix} K^\top(i)E_B^\top(i) & 0 & 0 \\ 0 & 0 & 0 \\ 0 & 0 & 0 \end{bmatrix} \begin{bmatrix} F_B^\top(i,t) & 0 & 0 \\ 0 & 0 & 0 \\ 0 & 0 & 0 \end{bmatrix} \begin{bmatrix} D_B^\top(i)P(i) & 0 & 0 \\ 0 & 0 & 0 \\ 0 & 0 & 0 \end{bmatrix} = \begin{bmatrix} K^\top(i)E_B^\top(i)F_B^\top(i,t)D_B^\top(i)P(i) & 0 & 0 \\ 0 & 0 & 0 \\ 0 & 0 & 0 \end{bmatrix},$$

$$= \begin{bmatrix} 0 & 0 & K^\top(i)E_{D_z}^\top(i) \\ 0 & 0 & 0 \\ 0 & 0 & 0 \end{bmatrix} \begin{bmatrix} 0 & 0 & 0 \\ 0 & 0 & 0 \\ 0 & 0 & F_{D_z}^\top(i,t) \end{bmatrix} \begin{bmatrix} 0 & 0 & 0 \\ 0 & 0 & 0 \\ 0 & 0 & D_{D_z}^\top(i) \end{bmatrix} = \begin{bmatrix} 0 & 0 & K^\top(i)E_{D_z}^\top(i)F_{D_z}^\top(i,t)D_{D_z}^\top(i) \\ 0 & 0 & 0 \\ 0 & 0 & 0 \end{bmatrix},$$

$$= \begin{bmatrix} 0 & 0 & 0 \\ 0 & 0 & 0 \\ 0 & 0 & D_{D_z}(i) \end{bmatrix} \begin{bmatrix} 0 & 0 & 0 \\ 0 & 0 & 0 \\ 0 & 0 & F_{D_z}(i,t) \end{bmatrix} \begin{bmatrix} 0 & 0 & 0 \\ 0 & 0 & 0 \\ E_{D_z}(i)K(i) & 0 & 0 \end{bmatrix} = \begin{bmatrix} 0 & 0 & 0 \\ 0 & 0 & 0 \\ D_{D_z}(i)F_{D_z}(i,t)E_{D_z}(i)K(i) & 0 & 0 \end{bmatrix},$$

$$= \begin{bmatrix} 0 & 0 & E_{C_z}^\top(i) \\ 0 & 0 & 0 \\ 0 & 0 & 0 \end{bmatrix} \begin{bmatrix} 0 & 0 & 0 \\ 0 & 0 & 0 \\ 0 & 0 & F_{C_z}^\top(i,t) \end{bmatrix} \begin{bmatrix} 0 & 0 & 0 \\ 0 & 0 & 0 \\ 0 & 0 & D_{C_z}^\top(i) \end{bmatrix} = \begin{bmatrix} 0 & 0 & E_{C_z}^\top(i)F_{C_z}^\top(i,t)D_{C_z}^\top(i) \\ 0 & 0 & 0 \\ 0 & 0 & 0 \end{bmatrix},$$

and

$$\begin{bmatrix} 0 & 0 & 0 \\ 0 & 0 & 0 \\ D_{C_z}(i)F_{C_z}(i,t)E_{C_z}(i) & 0 & 0 \end{bmatrix}$$

$$= \begin{bmatrix} 0 & 0 & 0 \\ 0 & 0 & 0 \\ 0 & 0 & D_{C_z}(i) \end{bmatrix} \begin{bmatrix} 0 & 0 & 0 \\ 0 & 0 & 0 \\ 0 & 0 & F_{C_z}(i,t) \end{bmatrix} \begin{bmatrix} 0 & 0 & 0 \\ 0 & 0 & 0 \\ E_{C_z}(i) & 0 & 0 \end{bmatrix}.$$

Using Lemma 7 in Appendix A, we get

$$\begin{bmatrix} P(i)D_A(i)F_A(i,t)E_A(i) & 0 & 0 \\ 0 & 0 & 0 \\ 0 & 0 & 0 \end{bmatrix}$$

$$+ \begin{bmatrix} E_A^\top(i)F_A^\top(i,t)D_A^\top(i)P(i) & 0 & 0 \\ 0 & 0 & 0 \\ 0 & 0 & 0 \end{bmatrix}$$

$$\leq \varepsilon_A(i) \begin{bmatrix} P(i)D_A(i) & 0 & 0 \\ 0 & 0 & 0 \\ 0 & 0 & 0 \end{bmatrix} \begin{bmatrix} D_A^\top(i)P(i) & 0 & 0 \\ 0 & 0 & 0 \\ 0 & 0 & 0 \end{bmatrix}$$

$$+ \varepsilon_A^{-1}(i) \begin{bmatrix} E_A^\top(i) & 0 & 0 \\ 0 & 0 & 0 \\ 0 & 0 & 0 \end{bmatrix} \begin{bmatrix} E_A(i) & 0 & 0 \\ 0 & 0 & 0 \\ 0 & 0 & 0 \end{bmatrix}$$

$$= \begin{bmatrix} \varepsilon_A(i)P(i)D_A(i)D_A^\top(i)P(i) + \varepsilon_A^{-1}(i)E_A^\top(i)E_A(i) & 0 & 0 \\ 0 & 0 & 0 \\ 0 & 0 & 0 \end{bmatrix},$$

$$\begin{bmatrix} P(i)D_B(i)F_B(i,t)E_B(i)K(i) & 0 & 0 \\ 0 & 0 & 0 \\ 0 & 0 & 0 \end{bmatrix}$$

$$+ \begin{bmatrix} K^\top(i)E_B^\top(i)F_B^\top(i,t)D_B^\top(i)P(i) & 0 & 0 \\ 0 & 0 & 0 \\ 0 & 0 & 0 \end{bmatrix}$$

$$\leq \varepsilon_B(i) \begin{bmatrix} P(i)D_B(i) & 0 & 0 \\ 0 & 0 & 0 \\ 0 & 0 & 0 \end{bmatrix} \begin{bmatrix} D_B^\top(i)P(i) & 0 & 0 \\ 0 & 0 & 0 \\ 0 & 0 & 0 \end{bmatrix}$$

$$+ \varepsilon_B^{-1}(i) \begin{bmatrix} K^\top(i)E_B^\top(i) & 0 & 0 \\ 0 & 0 & 0 \\ 0 & 0 & 0 \end{bmatrix} \begin{bmatrix} E_B(i)K(i) & 0 & 0 \\ 0 & 0 & 0 \\ 0 & 0 & 0 \end{bmatrix}$$

$$= \begin{bmatrix} \begin{bmatrix} \varepsilon_B(i)P(i)D_B(i)D_B^\top(i)P(i) \\ + \varepsilon_B^{-1}(i)K^\top(i)E_B^\top(i)E_B(i)K(i) \end{bmatrix} & 0 & 0 \\ 0 & 0 & 0 \\ 0 & 0 & 0 \end{bmatrix},$$

$$\begin{bmatrix} 0 & 0 & K^\top(i)E_{D_z}^\top(i)F_{D_z}^\top(i,t)D_{D_z}^\top(i) \\ 0 & 0 & 0 \\ 0 & 0 & 0 \end{bmatrix}$$

$$+ \begin{bmatrix} & 0 & 0 & 0 \\ & 0 & 0 & 0 \\ D_{D_z}(i)F_{D_z}(i,t)E_{D_z}(i)K(i) & 0 & 0 \end{bmatrix}$$

$$\leq \varepsilon_{D_z}^{-1}(i) \begin{bmatrix} 0 & 0 & K^\top(i)E_{D_z}^\top(i) \\ 0 & 0 & 0 \\ 0 & 0 & 0 \end{bmatrix} \begin{bmatrix} 0 & 0 & 0 \\ 0 & 0 & 0 \\ E_{D_z}(i)K(i) & 0 & 0 \end{bmatrix}$$

$$+ \varepsilon_{D_z}(i) \begin{bmatrix} 0 & 0 & 0 \\ 0 & 0 & 0 \\ 0 & 0 & D_{D_z}(i) \end{bmatrix} \begin{bmatrix} 0 & 0 & 0 \\ 0 & 0 & 0 \\ 0 & 0 & D_{D_z}^\top(i) \end{bmatrix}$$

$$= \begin{bmatrix} \varepsilon_{D_z}^{-1}(i)K^\top(i)E_{D_z}^\top(i)E_{D_z}(i,t)K(i) & 0 & 0 \\ 0 & 0 & 0 \\ 0 & 0 & \varepsilon_{D_z}(i)D_{D_z}(i)D_{D_z}^\top(i) \end{bmatrix},$$

$$\begin{bmatrix} 0 & 0 & E_{C_z}^\top(i)F_{C_z}^\top(i,t)D_{C_z}^\top(i) \\ 0 & 0 & 0 \\ 0 & 0 & 0 \end{bmatrix}$$

$$+ \begin{bmatrix} & 0 & 0 & 0 \\ & 0 & 0 & 0 \\ D_{C_z}(i)F_{C_z}(i,t)E_{C_z}(i) & 0 & 0 \end{bmatrix}$$

$$\leq \varepsilon_{C_z}^{-1}(i) \begin{bmatrix} 0 & 0 & E_{C_z}^\top(i) \\ 0 & 0 & 0 \\ 0 & 0 & 0 \end{bmatrix} \begin{bmatrix} 0 & 0 & 0 \\ 0 & 0 & 0 \\ E_{C_z}(i,t) & 0 & 0 \end{bmatrix}$$

$$+ \varepsilon_{C_z}(i) \begin{bmatrix} 0 & 0 & 0 \\ 0 & 0 & 0 \\ 0 & 0 & D_{C_z}(i) \end{bmatrix} \begin{bmatrix} 0 & 0 & 0 \\ 0 & 0 & 0 \\ 0 & 0 & D_{C_z}^\top(i) \end{bmatrix}$$

$$= \begin{bmatrix} \varepsilon_{C_z}^{-1}(i)E_{C_z}^\top(i)E_{C_z}(i) & 0 & 0 \\ 0 & 0 & 0 \\ 0 & 0 & \varepsilon_{C_z}(i)D_{C_z}(i)D_{C_z}^\top(i) \end{bmatrix}.$$

Using all these transformations we get

$$\begin{bmatrix} J_1(i) & P(i)B_\omega(i) & C_z^\top(i) + K^\top(i)D_z^\top(i) \\ B_\omega^\top(i)P(i) & -\gamma^2\mathbb{I} & B_z^\top(i) \\ D_z(i)K(i) + C_z(i) & B_z(i) & -\mathbb{I} \end{bmatrix}$$

$$+ \begin{bmatrix} \varepsilon_A(i)P(i)D_A(i)D_A^\top(i)P(i) + \varepsilon_A^{-1}(i)E_A^\top(i)E_A(i) & 0 & 0 \\ 0 & 0 & 0 \\ 0 & 0 & 0 \end{bmatrix}$$

$$+ \begin{bmatrix} \begin{bmatrix} \varepsilon_B(i)P(i)D_B(i)D_B^\top(i)P(i) \\ +\varepsilon_B^{-1}K^\top(i)E_B^\top(i)E_B(i)K(i) \end{bmatrix} & 0 & 0 \\ 0 & 0 & 0 \\ 0 & 0 & 0 \end{bmatrix}$$

$$+ \begin{bmatrix} \varepsilon_{D_z}^{-1}(i)K^\top(i)E_{D_z}^\top(i)E_{D_z}(i)K(i) & 0 & 0 \\ 0 & 0 & 0 \\ 0 & 0 & \varepsilon_{D_z}(i)D_{D_z}(i)D_{D_z}^\top(i) \end{bmatrix}$$

$$+ \begin{bmatrix} \varepsilon_{C_z}^{-1}(i)K^\top(i)E_{C_z}^\top(i)E_{C_z}(i,t)K(i) & 0 & 0 \\ 0 & 0 & 0 \\ 0 & 0 & \varepsilon_{C_z}(i)D_{C_z}(i)D_{C_z}^\top(i) \end{bmatrix} < 0,$$

with

$$J_1(i) = A^\top(i)P(i) + P(i)A(i) + K^\top(i)B^\top(i)P(i)$$

$$+ P(i)B(i)K(i) + \sum_{j=1}^{N} \lambda_{ij}P(j).$$

Let $J_2(i)$, $\mathcal{W}(i)$, and $\mathcal{T}(i)$ be defined as

$$J_2(i) = J_1(i) + \varepsilon_A^{-1}(i)E^\top(i)E_A(i) + \varepsilon_B^{-1}(i)K^\top(i)E_B^\top(i)E_B(i)K(i),$$

$$\mathcal{W}(i) = \text{diag}[\varepsilon_A^{-1}(i)\mathbb{I}, \varepsilon_B^{-1}(i)\mathbb{I}, \varepsilon_{C_z}(i)\mathbb{I}, \varepsilon_{D_z}(i)\mathbb{I}],$$

$$\mathcal{T}(i) = \left(P(i)D_A(i), P(i)D_B(i), K^\top(i)E_{C_z}^\top(i), K^\top(i)E_{D_z}^\top(i)\right).$$

Using the Schur complement we get the equivalent inequality:

$$\begin{bmatrix} J_2(i) & P(i)B_\omega(i) & \begin{bmatrix} C_z^\top(i) \\ +K^\top(i)D_z^\top(i) \end{bmatrix} & \mathcal{T}(i) \\ B_\omega^\top(i)P(i) & -\gamma^2\mathbb{I} & B_z^\top(i) & 0 \\ \begin{bmatrix} D_z(i)K(i) \\ +C_z(i) \end{bmatrix} & B_z(i) & -\mathcal{U}(i) & 0 \\ \mathcal{T}^\top(i) & 0 & 0 & -\mathcal{W}(i) \end{bmatrix} < 0,$$

with $\mathcal{U}(i) = \mathbb{I} - \varepsilon_{D_z}(i)D_{D_z}(i)D_{D_z}^\top(i) - \varepsilon_{C_z}(i)D_{C_z}(i)D_{C_z}^\top(i)$.

This matrix inequality is nonlinear in $P(i)$ and $K(i)$. To put it into LMI form, let $X(i) = P^{-1}(i)$. Pre- and post-multiply this matrix inequality by $\text{diag}[X(i), \mathbb{I}, \mathbb{I}, \mathbb{I}]$ to get

$$\begin{bmatrix} J_3(i) & B_\omega(i) & \begin{bmatrix} X(i)C_z^\top(i) \\ +X(i)K^\top(i)D_z^\top(i) \end{bmatrix} & X(i)\mathcal{T}(i) \\ B_\omega^\top(i) & -\gamma^2\mathbb{I} & B_z^\top(i) & 0 \\ \begin{bmatrix} D_z(i)K(i)X(i) \\ +C_z(i)X(i) \end{bmatrix} & B_z(i) & -\mathcal{U}(i) & 0 \\ \mathcal{T}^\top(i)X(i) & 0 & 0 & -\mathcal{W}(i) \end{bmatrix} < 0,$$

with

$$
\begin{aligned}
J_3(i) = \; & X(i)A^\top(i) + A(i)X(i) + X(i)K^\top(i)B^\top(i) \\
& + B(i)K(i)X(i) + \varepsilon_A^{-1}(i)X(i)E_A^\top(i)E_A(i)X(i) \\
& + \varepsilon_B^{-1}(i)X(i)K^\top(i)E_B^\top(i)E_B(i)K(i)X(i) \\
& + \sum_{j=1}^{N} \lambda_{ij} X(i)X^{-1}(j)X(i).
\end{aligned}
$$

Notice that

$$
X(i)\mathcal{T}(i) = \left(D_A(i), D_B(i), X(i)K^\top(i)E_{C_z}^\top(i), X(i)K^\top(i)E_{D_z}^\top(i) \right),
$$

and

$$
\sum_{j=1}^{N} \lambda_{ij} X(i)X^{-1}(j)X(i) = \lambda_{ii}X(i) + \mathcal{S}_i(X)\mathcal{X}^{-1}(X)\mathcal{S}_i^\top(X).
$$

Letting $Y(i) = K(i)X(i)$ and using the Schur complement we obtain

$$
\left[
\begin{array}{ccc}
\tilde{J}(i) & B_\omega(i) & \begin{bmatrix} X(i)C_z^\top(i) \\ +Y^\top(i)D_z^\top(i) \end{bmatrix} \\
B_\omega^\top(i) & -\gamma^2\mathbb{I} & B_z^\top(i) \\
\begin{bmatrix} D_z(i)Y(i) \\ +C_z(i)X(i) \end{bmatrix} & B_z(i) & -\mathcal{U}(i) \\
\mathcal{Z}^\top(i) & 0 & 0 \\
\mathcal{S}_i^\top(X) & 0 & 0
\end{array}
\right.
\tag{4.23}
$$

$$
\left.
\begin{array}{cc}
\mathcal{Z}(i) & \mathcal{S}_i(X) \\
0 & 0 \\
0 & 0 \\
-\mathcal{V}(i) & 0 \\
0 & -\mathcal{X}_i(X)
\end{array}
\right] < 0,
\tag{4.24}
$$

with

$$
\begin{aligned}
\tilde{J}(i) = \; & X(i)A^\top(i) + A(i)X(i) + Y^\top(i)B^\top(i) + B(i)Y(i) \\
& + \varepsilon_A(i)D_A(i)D_A^\top(i) + \varepsilon_B(i)D_B(i)D_B^\top(i) \\
& + \lambda_{ii}X(i), \\
\mathcal{U}(i) = \; & \mathbb{I} - \varepsilon_{D_z}(i)D_{D_z}(i)D_{D_z}^\top(i) - \varepsilon_{C_z}(i)D_{C_z}(i)D_{C_z}^\top(i), \\
\mathcal{Z}(i) = \; & \left(X(i)E_A^\top(i), Y^\top(i)E_B^\top(i), Y^\top(i)E_{C_z}^\top(i), Y^\top(i)E_{D_z}^\top(i) \right), \\
\mathcal{V}(i) = \; & \mathrm{diag}[\varepsilon_A(i)\mathbb{I}, \varepsilon_B(i)\mathbb{I}, \varepsilon_{C_z}(i)\mathbb{I}, \varepsilon_{D_z}(i)\mathbb{I}].
\end{aligned}
$$

The following theorem summarizes the results of this development.

Theorem 49. *Let γ be a positive constant. If there exist a set of symmetric and positive-definite matrices $X = (X(1), \cdots, X(N)) > 0$ and a set of matrices $Y = (Y(1), \cdots, Y(N))$ and sets of positive scalars $\varepsilon_A = (\varepsilon_A(1), \cdots, \varepsilon_A(N))$, $\varepsilon_B = (\varepsilon_B(1), \cdots, \varepsilon_B(N))$, $\varepsilon_{C_z} = (\varepsilon_{C_z}(1), \cdots, \varepsilon_{C_z}(N))$, and $\varepsilon_{D_z} = (\varepsilon_{D_z}(1), \cdots, \varepsilon_{D_z}(N))$ such that the following set of coupled LMIs (4.24) holds for every $i \in \mathscr{S}$ and for all admissible uncertainties, then the system (4.1) under the controller (4.17) with $K(i) = Y(i)X^{-1}(i)$ is stochastically stable and, moreover, the closed-loop system satisfies the disturbance rejection of level γ.*

Example 49. To show the effectiveness of the theoretical results of this theorem, let us consider the two-mode system of Example 46 with the same data.
Fix the required positive constant to the following values:

$$\varepsilon_A(1) = \varepsilon_A(2) = 0.50,$$
$$\varepsilon_B(1) = \varepsilon_B(2) = \varepsilon_C(1) = \varepsilon_C(2) = \varepsilon_D(1) = \varepsilon_D(2) = 0.10.$$

Solving the LMI (4.24), we get

$$X(1) = \begin{bmatrix} 1.2271 & 0.0800 \\ 0.0800 & 0.9253 \end{bmatrix}, \qquad X(2) = \begin{bmatrix} 1.7939 & 0.0942 \\ 0.0942 & 1.1861 \end{bmatrix},$$
$$Y(1) = \begin{bmatrix} -1.3360 & -0.2774 \\ -0.2751 & -0.6293 \end{bmatrix}, \qquad Y(2) = \begin{bmatrix} -0.2270 & -0.1418 \\ -0.1485 & 0.3717 \end{bmatrix}.$$

The corresponding gains are given by

$$K(1) = \begin{bmatrix} -1.0753 & -0.2068 \\ -0.1808 & -0.6645 \end{bmatrix}, \qquad K(2) = \begin{bmatrix} -0.1208 & -0.1099 \\ -0.0996 & 0.3213 \end{bmatrix}.$$

Based on the results of this theorem, we conclude that the system is stochastically stable and guarantees the desired disturbance rejection of level $\gamma = 5.0$.

As was done for the nominal system, we can determine the controller that stochastically stabilizes the class of systems we are considering and at the same time guarantees the minimum disturbance rejection by solving the following optimization problem:

$$\text{Pu}: \begin{cases} \min\limits_{\nu > 0,} & \nu, \\ \varepsilon_A = (\varepsilon_A(1), \cdots, \varepsilon_A(N)) > 0, \\ \varepsilon_B = (\varepsilon_B(1), \cdots, \varepsilon_B(N)) > 0, \\ \varepsilon_{C_z} = (\varepsilon_{D_z}(1), \cdots, \varepsilon_{C_z}(N)) > 0, \\ \varepsilon_{D_z} = (\varepsilon_{D_z}(1), \cdots, \varepsilon_{D_z}(N)) > 0, \\ X = (X(1), \cdots, X(N)) > 0, \\ Y = (Y(1), \cdots, Y(N)), \\ s.t.: (4.24) \text{ with } \nu = \gamma^2. \end{cases}$$

The following corollary summarizes the results on the design of the controller that stochastically stabilizes the system (4.1) and simultaneously guarantees the smallest disturbance rejection level.

Corollary 7. *Let $\nu > 0$, $\varepsilon_A = (\varepsilon_A(1), \cdots, \varepsilon_A(N)) > 0$, $\varepsilon_B = (\varepsilon_B(1), \cdots, \varepsilon_B(N)) > 0$, $\varepsilon_{C_z} = (\varepsilon_{C_z}(1), \cdots, \varepsilon_{C_z}(N)) > 0$, $\varepsilon_{D_z} = (\varepsilon_{D_z}(1), \cdots, \varepsilon_{D_z}(N)) > 0$, $X = (X(1), \cdots, X(N)) > 0$, and $Y = (Y(1), \cdots, Y(N))$ be the solution of the optimization problem Pu. Then the controller (4.17) with $K(i) = Y(i)X^{-1}(i)$ stochastically stabilizes the class of systems we are considering and, moreover, the closed-loop system satisfies the disturbance rejection of level $\sqrt{\nu}$.*

Example 50. To illustrate the effectiveness of the results developed in this theorem, let us consider the two-mode system of Example 49 with the same data.

The required positive scalars are fixed to the following values:

$$\varepsilon_A(1) = \varepsilon_A(2) = 0.50,$$
$$\varepsilon_B(1) = \varepsilon_B(2) = \varepsilon_C(1) = \varepsilon_C(2) = \varepsilon_D(1) = \varepsilon_D(2) = 0.10.$$

Solving the problem Pu gives

$$X(1) = \begin{bmatrix} 0.8525 & 0.0321 \\ 0.0321 & 0.7363 \end{bmatrix}, \qquad X(2) = \begin{bmatrix} 1.0584 & 0.0414 \\ 0.0414 & 0.8281 \end{bmatrix},$$
$$Y(1) = \begin{bmatrix} -1.5699 & -0.1755 \\ -0.1659 & -1.2116 \end{bmatrix}, \qquad Y(2) = \begin{bmatrix} -0.9827 & -0.0893 \\ -0.0812 & -0.5942 \end{bmatrix},$$

which gives the following gains:

$$K(1) = \begin{bmatrix} -1.8356 & -0.1584 \\ -0.1329 & -1.6397 \end{bmatrix}, \qquad K(2) = \begin{bmatrix} -0.9261 & -0.0615 \\ -0.0488 & -0.7150 \end{bmatrix}.$$

Based on the results of this corollary, we conclude that the system is stochastically stable and assures a disturbance rejection level equal to $\gamma = 1.0926$.

Let us now discuss the design of a nonfragile controller that robustly stochastically stabilizes the class of systems we are considering and simultaneously guarantees the desired disturbance rejection of level γ. For this purpose, let the matrices $B(i,t)$, $D_y(i,t)$, and $D_z(i,t)$ be free of uncertainties, that is,

$$B(i,t) = B(i),$$
$$D_y(i,t) = D_y(i),$$
$$D_z(i,t) = D_z(i).$$

Following the steps of Chapter 3, the nonfragile controller gain we will be synthesizing has the following form:

$$K(i,t) = K(i) + \Delta K(i,t), \tag{4.25}$$

with $\Delta K(i,t)$ given by

$$\Delta K(i,t) = \rho(i)F_K(i,t)K(i),$$

where $\rho(i)$ is an uncertain real parameter indicating the measure of non-fragility against controller gain variations and $F_K(i,t)$ is the uncertainty that will be supposed to satisfy the following for every $i \in \mathscr{S}$:

$$F_K^\top(i,t)F_K(i,t) \leq \mathbb{I}.$$

Plugging the expression of the state feedback controller with the gain given by (4.25) in the dynamics (4.1), we obtain the following closed-loop dynamics:

$$\begin{cases} \dot{x}(t) = [A(i,t) + B(i)\left[K(i) + \rho(i)F_K(i,t)K(i)\right]] x(t) + B_w(i)\omega(t), \\ y(t) = [C_y(i,t) + D_y(i)\left[K(i) + \rho(i)F_K(i,t)K(i)\right]] x(t) + B_y(i)\omega(t), \\ z(t) = [C_z(i,t) + D_z(i)\left[K(i) + \rho(i)F_K(i,t)K(i)\right]] x(t) + B_z(i)\omega(t). \end{cases}$$

These dynamics will be robustly stochastically stable and have the desired disturbance rejection of level γ if the following holds for every $i \in \mathscr{S}$ and for all admissible uncertainties:

$$\begin{bmatrix} \bar{J}_0(i,t) & P(i)B_w(i) & \bar{C}_z^\top(i,t) \\ B_w^\top(i)P(i) & -\gamma^2\mathbb{I} & B_z^\top(i) \\ \bar{C}_z(i,t) & B_z(i) & -\mathbb{I} \end{bmatrix} < 0, \qquad (4.26)$$

with

$$\bar{J}_0(i,t) = \bar{A}^\top(i,t)P(i) + P(i)\bar{A}(i,t) + \sum_{j=1}^{N}\lambda_{ij}P(j),$$

$$\bar{A}(i,t) = A(i,t) + B(i)\left[K(i) + \rho(i)F_K(i,t)K(i)\right],$$

$$\bar{C}_z(i,t) = C_z(i,t) + D_z(i)\left[K(i) + \rho(i)F_K(i,t)K(i)\right].$$

The controller gain we are designing is assumed to have the following form:

$$K(i) = \varrho(i)B^\top(i)P(i),$$

where $P(i) > 0$ is a design parameter that is symmetric and positive-definite matrix for every $i \in \mathscr{S}$ that is the solution of (4.26).

Using the expression of $A(i,t)$ we get

$$\bar{A}^\top(i,t)P(i) + P(i)\bar{A}(i,t) + \sum_{j=1}^{N}\lambda_{ij}P(j) =$$
$$A^\top(i)P(i) + P(i)A(i) + P(i)D_A(i)F_A(i,t)E_A(i)$$
$$+E_A^\top(i,t)F_A^\top(i,t)D_A^\top(i)P(i)$$
$$+K^\top(i)B^\top(i)P(i) + P(i)B(i)K(i)$$
$$+\rho(i)P(i)B(i)F_K(i,t)K(i)$$

$$+\rho(i)K^\top(i)F_K^\top(i,t)B^\top(i)P(i) + \sum_{j=1}^N \lambda_{ij}P(j),$$

and

$$\bar{C}_z(i,t) = C_z(i) + D_{C_z}(i)F_{C_z}(i,t)E_{C_z}(i)$$
$$+D_z(i)\left[\varrho(i)B^\top(i)P(i) + \rho(i)\varrho(i)F_K(i,t)B^\top(i)P(i)\right]$$
$$= C_z(i) + D_{C_z}(i)F_{C_z}(i,t)E_{C_z}(i)$$
$$+\varrho(i)D_z(i)B^\top(i)P(i) + \rho(i)\varrho(i)D_z(i)F_K(i,t)B^\top(i)P(i).$$

From Lemma 7 in Appendix A, we get

$$P(i)D_A(i)F_A(i,t)E_A(i) + E_A^\top(i,t)F_A^\top(i,t)D_A^\top(i)P(i)$$
$$\leq \varepsilon_A(i)P(i)D_A(i)D_A^\top(i)P(i) + \varepsilon_A^{-1}(i)E_A^\top(i)E_A(i),$$

$$\rho(i)\varrho(i)P(i)B(i)F_K(i,t)B^\top(i)P(i)$$
$$+\rho(i)\varrho(i)P(i)B(i)F_K^\top(i,t)B^\top(i)P(i)$$
$$\leq \varepsilon_K^{-1}(i)\rho(i)\varrho^2(i)P(i)B(i)B^\top(i)P(i)$$
$$+\varepsilon_K(i)\rho(i)P(i)B(i)F_K^\top(i,t)F_K(i,t)B^\top(i)P(i)$$
$$\leq \varepsilon_K^{-1}(i)\rho(i)\varrho^2(i)P(i)B(i)B^\top(i)P(i)$$
$$+\varepsilon_K(i)\rho(i)P(i)B(i)B^\top(i)P(i),$$

and

$$\begin{bmatrix} 0 & 0 & E_{C_z}^\top(i)F_{C_z}^\top(i)D_{C_z}^\top(i) \\ 0 & 0 & 0 \\ 0 & 0 & 0 \end{bmatrix}$$
$$+\begin{bmatrix} 0 & & 0 & 0 \\ 0 & & 0 & 0 \\ D_{C_z}(i)F_{C_z}(i,t)E_{C_z}(i) & 0 & 0 \end{bmatrix}$$
$$\leq \varepsilon_{C_z}^{-1}(i)\begin{bmatrix} 0 & 0 & E_{C_z}^\top(i) \\ 0 & 0 & 0 \\ 0 & 0 & 0 \end{bmatrix}\begin{bmatrix} 0 & & 0 & 0 \\ 0 & & 0 & 0 \\ E_{C_z}(i) & 0 & 0 \end{bmatrix}$$
$$+\varepsilon_{C_z}(i)\begin{bmatrix} 0 & 0 & 0 \\ 0 & 0 & 0 \\ 0 & 0 & D_{C_z}(i) \end{bmatrix}\begin{bmatrix} 0 & 0 & 0 \\ 0 & 0 & 0 \\ 0 & 0 & D_{C_z}^\top(i) \end{bmatrix}$$
$$=\begin{bmatrix} \varepsilon_{C_z}^{-1}(i)E_{C_z}^\top(i)E_{C_z}(i) & 0 & 0 \\ 0 & 0 & 0 \\ 0 & 0 & \varepsilon_{C_z}(i)D_{C_z}(i)D_{C_z}^\top(i) \end{bmatrix},$$

$$\begin{bmatrix} 0 & 0 & \rho(i)\varrho(i)P(i)B(i)F_K^\top(i,t)D_{D_z}^\top(i) \\ 0 & 0 & 0 \\ 0 & 0 & 0 \end{bmatrix}$$

$$+ \begin{bmatrix} & 0 & 0 & 0 \\ & 0 & 0 & 0 \\ \rho(i)\varrho(i)D_z(i)F_K(i,t)B^\top(i)P(i) & 0 & 0 \end{bmatrix}$$

$$\leq \rho(i)\varepsilon_{D_z}^{-1}(i) \begin{bmatrix} 0 & 0 & \varrho(i)P(i)B(i) \\ 0 & 0 & 0 \\ 0 & 0 & 0 \end{bmatrix} \begin{bmatrix} 0 & 0 & 0 \\ 0 & 0 & 0 \\ \varrho(i)B^\top(i)P(i) & 0 & 0 \end{bmatrix}$$

$$+ \rho(i)\varepsilon_{D_z}(i) \begin{bmatrix} 0 & 0 & 0 \\ 0 & 0 & 0 \\ 0 & 0 & D_{D_z}(i) \end{bmatrix} \begin{bmatrix} 0 & 0 & 0 \\ 0 & 0 & 0 \\ 0 & 0 & D_{D_z}^\top(i) \end{bmatrix}$$

$$= \begin{bmatrix} \begin{bmatrix} \varepsilon_{D_z}^{-1}(i)\rho(i)\varrho^2(i)P(i) \\ \times B(i)B^\top(i)P(i) \end{bmatrix} & 0 & 0 \\ 0 & 0 & 0 \\ 0 & 0 & \begin{bmatrix} \rho(i)\varepsilon_{D_z}(i)D_{D_z}(i) \\ \times D_{D_z}^\top(i) \end{bmatrix} \end{bmatrix}.$$

Using all these transformations the previous matrix inequality becomes

$$\begin{bmatrix} J(i) & P(i)B_\omega(i) \\ B_\omega^\top(i)P(i) & -\gamma^2\mathbb{I} \\ \varrho(i)D_z(i)B^\top(i)P(i)+C_z(i) & B_z(i) \\ \\ C_z^\top(i)+\varrho(i)P(i)B(i)D_z^\top(i) \\ B_z^\top(i) \\ -\mathcal{U}_\rho(i) \end{bmatrix} < 0,$$

with

$$J(i) = A^\top(i)P(i) + P(i)A(i) + P(i)B(i)K(i)$$
$$+ K^\top(i)B^\top(i)P(i) + \sum_{j=1}^{N}\lambda_{ij}P(j)$$
$$+ \varepsilon_A(i)P(i)D_A(i)D_A^\top(i)P(i) + \varepsilon_A^{-1}(i)E_A^\top(i)E_A(i)$$
$$+ \varepsilon_K^{-1}(i)\rho(i)\varrho^2(i)P(i)B(i)B^\top(i)P(i)$$
$$+ \varepsilon_K(i)\rho(i)P(i)B(i)B^\top(i)P(i)$$
$$+ \varepsilon_{D_z}^{-1}(i)\rho(i)\varrho^2(i)P(i)B(i)B^\top(i)P(i) + \varepsilon_{C_z}^{-1}(i)E_{C_z}^\top(i)E_{C_z}(i)$$
$$\mathcal{U}_\rho(i) = \mathbb{I} - \varepsilon_{C_z}(i)D_{C_z}(i)D_{C_z}^\top(i) - \rho(i)\varepsilon_{D_z}(i)D_{D_z}(i)D_{D_z}^\top(i).$$

Letting

$$\mathcal{Z}(i) = \left[E_A^\top(i), \varrho(i)P(i)B(i), E_{C_z}^\top(i), \varrho(i)P(i)B(i) \right],$$
$$\mathcal{N}(i) = \text{diag}\left[\varepsilon_A(i)\mathbb{I}, \frac{\varepsilon_K(i)}{\rho(i)}\mathbb{I}, \varepsilon_{C_z}(i)\mathbb{I}, \frac{\varepsilon_{D_z}(i)}{\rho(i)}\mathbb{I} \right],$$

and using the Schur complement we get

$$
\begin{bmatrix}
\bar{J}(i) & P(i)B_\omega(i) \\
B_\omega^\top(i)P(i) & -\gamma^2\mathbb{I} \\
\varrho(i)D_z(i)B^\top(i)P(i) + C_z(i) & B_z(i) \\
\mathcal{Z}^\top(i) & 0 \\
C_z^\top(i) + \varrho(i)P(i)B(i)D_z^\top(i) & \mathcal{Z}(i) \\
B_z^\top(i) & 0 \\
-\mathcal{U}_\rho(i) & 0 \\
0 & -\mathcal{N}(i)
\end{bmatrix} < 0,
$$

where

$$
\bar{J}(i) = A^\top(i)P(i) + P(i)A(i) + P(i)B(i)K(i)
$$
$$
+ K^\top(i)B^\top(i)P(i) + \sum_{j=1}^{N} \lambda_{ij}P(j)
$$
$$
+ \varepsilon_A(i)P(i)D_A(i)D_A^\top(i)P(i) + \varepsilon_K(i)\rho(i)P(i)B(i)B^\top(i)P(i).
$$

This inequality is nonlinear in $\varrho(i)$ and $P(i)$. To cast it into an LMI, let $X(i) = P^{-1}(i)$ and pre- and post-multiply the previous inequality matrix by $\mathrm{diag}\,[X(i), \mathbb{I}, \mathbb{I}, \mathbb{I}]$ to imply

$$
\begin{bmatrix}
\tilde{J}(i) & B_\omega(i) \\
B_\omega^\top(i) & -\gamma^2\mathbb{I} \\
\varrho(i)D_z(i)B^\top(i) + C_z(i)X(i) & B_z(i) \\
\mathcal{L}^\top(i)X(i) & 0 \\
X(i)C_z^\top(i) + \varrho(i)B(i)D_z^\top(i) & X(i)\mathcal{L}(i) \\
B_z^\top(i) & 0 \\
-\mathcal{U}_\rho(i) & 0 \\
0 & -\mathcal{N}(i)
\end{bmatrix} < 0, \qquad (4.27)
$$

with

$$
\tilde{J}(i) = X(i)A^\top(i) + A(i)X(i) + B(i)K(i)X(i)
$$
$$
+ X(i)K^\top(i)B^\top(i) + \sum_{j=1}^{N} \lambda_{ij}X(i)X^{-1}(j)X(i)
$$
$$
+ \varepsilon_A(i)D_A(i)D_A^\top(i) + \varepsilon_K(i)\rho(i)B(i)B^\top(i).
$$

Letting $Y(i) = K(i)X(i)$ and noticing that

$$
X(i)\mathcal{L}(i) = \left[X(i)E_A^\top(i), \varrho(i)B(i), X(i)E_{C_z}^\top(i), \varrho(i)B(i) \right]
$$
$$
\sum_{j=1}^{N} \lambda_{ij}X(i)X^{-1}(j)X(i) = \lambda_{ii}X(i) + \mathcal{S}_i(X)\mathcal{X}_i^{-1}(X)\mathcal{S}_i^\top(X),
$$

and after using the Schur complement again, we get

$$
\begin{bmatrix}
\hat{J}(i) & B_\omega(i) \\
B_\omega^\top(i) & -\gamma^2\mathbb{I} \\
\varrho(i)D_z(i)B^\top(i) + C_z(i)X(i) & B_z(i) \\
\mathcal{U}^\top(i) & 0 \\
\mathcal{S}_i^\top(X) & 0
\end{bmatrix}
$$

$$
\begin{bmatrix}
X(i)C_z^\top(i) + \varrho(i)B(i)D_z^\top(i) & \mathcal{U}(i) & \mathcal{S}_i(X) \\
B_z^\top(i) & 0 & 0 \\
-\mathcal{U}_\rho(i) & 0 & 0 \\
0 & -\mathcal{W}(i) & 0 \\
0 & 0 & -\mathcal{X}_i(X)
\end{bmatrix} < 0, \qquad (4.28)
$$

with

$$
\begin{aligned}
\hat{J}(i) &= X(i)A^\top(i) + A(i)X(i) + B(i)Y(i) + Y^\top(i)B^\top(i) \\
&\quad + \lambda_{ii}X(i) + \varepsilon_A(i)D_A(i)D_A^\top(i) + \varepsilon_K(i)\rho(i)B(i)B^\top(i), \\
\mathcal{U}(i) &= \left[X(i)E_A^\top(i), X(i)E_{C_z}^\top(i), \varrho(i)B(i), \varrho(i)B(i)\right], \\
\mathcal{W}(i) &= \mathrm{diag}\left[\varepsilon_A(i)\mathbb{I}, \varepsilon_{C_z}(i)\mathbb{I}, \frac{\varepsilon_{D_z}(i)}{\rho(i)}\mathbb{I}, \frac{\varepsilon_K(i)}{\rho(i)}\mathbb{I}\right].
\end{aligned}
$$

The following theorem summarizes the results of this development.

Theorem 50. *Let γ be a positive constant. If there exist a set of symmetric and positive-definite matrices $X = (X(1), \cdots, X(N)) > 0$ and a set of matrices $Y = (Y(1), \cdots, Y(N))$ and positive scalars $\varepsilon_A = (\varepsilon_A(1), \cdots, \varepsilon_A(N))$, $\varepsilon_{C_z} = (\varepsilon_{C_z}(1), \cdots, \varepsilon_{C_z}(N))$, $\varepsilon_{D_z} = (\varepsilon_{D_z}(1), \cdots, \varepsilon_{D_z}(N))$, and $\varepsilon_K = (\varepsilon_K(1), \cdots, \varepsilon_K(N))$ such that the following set of coupled LMIs (4.28) holds for every $i \in \mathscr{S}$ and for all admissible uncertainties, then the system (4.1) under the controller (4.17) with $K(i) = Y(i)X^{-1}(i)$ is stochastically stable and, moreover, the closed-loop system satisfies the disturbance rejection of level γ.*

Example 51. To show the effectiveness of the results of this theorem, let us consider the two-mode system of Example 49 with the same data except that $\Delta B(i, t) = 0$ for every $i \in \mathscr{S}$.

Let us fix all the required positive scalars to the following values:

$$
\begin{aligned}
\varepsilon_A(1) &= \varepsilon_A(2) = 0.5, \\
\varepsilon_C(1) &= \varepsilon_C(2) = \varepsilon_D(1) = \varepsilon_D(2) = \varepsilon_K(1) = \varepsilon_K(2) = 0.1, \\
\rho(1) &= \rho(2) = 0.5.
\end{aligned}
$$

The desired disturbance rejection level is fixed to $\gamma = 5.0$. Solving the problem Pw, we get

$$
X(1) = \begin{bmatrix} 5.3344 & -0.4466 \\ -0.4466 & 5.6761 \end{bmatrix}, \quad X(2) = \begin{bmatrix} 5.0491 & -0.4032 \\ -0.4032 & 4.8075 \end{bmatrix},
$$

$$Y(1) = \begin{bmatrix} -11.1933 & 15.4484 \\ -28.0308 & 2.7894 \end{bmatrix}, \quad Y(2) = \begin{bmatrix} 0.1564 & 6.3403 \\ -10.8011 & 3.6176 \end{bmatrix},$$

$$\varrho(1) = -5.4060, \qquad\qquad \varrho(2) = -4.8388.$$

The corresponding controller gains are given by

$$K(1) = \begin{bmatrix} -1.8828 & 2.5735 \\ -5.2482 & 0.0785 \end{bmatrix}, \quad K(2) = \begin{bmatrix} 0.1372 & 1.3303 \\ -2.0932 & 0.5770 \end{bmatrix}.$$

Based on the results of this theorem, we conclude that the system is stochastically stable and guarantees the desired disturbance level.

As done previously, we can determine a nonfragile controller that stochastically stabilizes the class of systems we are considering and simultaneously guarantees the minimum disturbance rejection by solving the following optimization problem:

$$\text{Pw}: \begin{cases} \min\limits_{\substack{\nu>0, \\ \varepsilon_A=(\varepsilon_A(1),\cdots,\varepsilon_A(N))>0, \\ \varepsilon_B=(\varepsilon_B(1),\cdots,\varepsilon_B(N))>0, \\ \varepsilon_{C_z}=(\varepsilon_{D_z}(1),\cdots,\varepsilon_{D_z}(N))>0, \\ \varepsilon_{D_z}=(\varepsilon_{D_z}(1),\cdots,\varepsilon_{D_z}(N))>0, \\ X=(X(1),\cdots,X(N))>0, \\ Y=(Y(1),\cdots,Y(N)), }} \nu, \\ s.t. : (4.28) \text{ with } \nu = \gamma^2. \end{cases} \tag{4.29}$$

The following corollary summarizes the results of the design of the controller that stochastically stabilizes the system (4.1) and simultaneously guarantees the smallest disturbance rejection level.

Corollary 8. *Let $\nu > 0$, $\varepsilon_A = (\varepsilon_A(1), \cdots, \varepsilon_A(N)) > 0$, $\varepsilon_{C_z} = (\varepsilon_{C_z}(1), \cdots, \varepsilon_{C_z}(N)) > 0$, $\varepsilon_{D_z} = (\varepsilon_{D_z}(1), \cdots, \varepsilon_{D_z}(N)) > 0$, $\varepsilon_K = (\varepsilon_K(1), \cdots, \varepsilon_K(N)) > 0$, $X = (X(1), \cdots, X(N)) > 0$, and $Y = (Y(1), \cdots, Y(N))$ be the solution of the optimization problem Pw. Then the controller (4.17) with $K(i) = Y(i)X^{-1}(i)$ stochastically stabilizes the class of systems we are considering and, moreover, the closed-loop system satisfies the disturbance rejection of level $\sqrt{\nu}$.*

Example 52. To show the effectiveness of the results of this theorem, let us consider the two-mode system of the previous example with the same data.
The required positive scalars are fixed to the following values:

$$\varepsilon_A(1) = \varepsilon_A(2) = 0.5,$$
$$\varepsilon_C(1) = \varepsilon_C(2) = \varepsilon_D(1) = \varepsilon_D(2) = \varepsilon_K(1) = \varepsilon_K(2) = 0.1,$$
$$\rho(1) = \rho(2) = 0.5.$$

Solving the problem Pw, we get

$$X(1) = \begin{bmatrix} 16.0649 & -0.0112 \\ -0.0112 & 16.0720 \end{bmatrix}, \qquad X(2) = \begin{bmatrix} 15.2067 & -0.0220 \\ -0.0220 & 15.1748 \end{bmatrix},$$

$$Y(1) = \begin{bmatrix} -164.2228 & -51.6292 \\ -57.0415 & -66.0119 \end{bmatrix}, \quad Y(2) = \begin{bmatrix} -35.4198 & -22.8874 \\ -20.8337 & -49.6442 \end{bmatrix},$$

$$\varrho(1) = -17.0442, \qquad\qquad\qquad \varrho(2) = -16.1671.$$

The corresponding controller gains are given by

$$K(1) = \begin{bmatrix} -10.2247 & -3.2195 \\ -3.5536 & -4.1097 \end{bmatrix}, \quad K(2) = \begin{bmatrix} -2.3314 & -1.5116 \\ -1.3748 & -3.2735 \end{bmatrix}.$$

Based on the results of this corollary, we conclude that the system is stochastically stable and guarantees a disturbance level equal to $\gamma = 1.0043$.

Notice that in designing the state feedback controller, we have always assumed complete access to the system mode. This assumption may not be practically valid in some cases and an alternative is required. In the rest of this section we focus on the design of the state feedback controller that does not require access to the system mode. The structure of this controller is given by the following expression:

$$u(t) = \mathscr{K} x(t), \tag{4.30}$$

where $\mathscr{K}$ is a constant gain to be determined.

Plugging the controller's expression into the system dynamics we get

$$\dot{x}(t) = \bar{A}(r(t))x(t) + B_w(r(t))\omega(t),$$
$$z(t) = \bar{C}_z(r(t))x(t) + B_z(r(t))\omega(t),$$

where $\bar{A}(r(t)) = A(r(t)) + B(r(t))\mathscr{K}$, $\bar{C}_z(r(t)) = C_z(r(t)) + D_z(r(t))\mathscr{K}$.

Based on what we developed earlier, we have the following results.

Theorem 51. *Let γ be a given positive constant and $\mathscr{K}$ a given constant gain. If there exists a symmetric and positive-definite matrix $P > 0$ such that the following set of LMIs holds for every $i \in \mathscr{S}$:*

$$\begin{bmatrix} \bar{J}_0(i) & \bar{C}_z(i)B_z(i) + PB_w(i) \\ B_z^\top(i)\bar{C}_z(i) + B_w^\top(i)P & B_z^\top(i)B_z(i) - \gamma^2\mathbb{I} \end{bmatrix} < 0, \tag{4.31}$$

where $\bar{J}_0(i) = \bar{A}^\top(i)P + P\bar{A}(i) + \bar{C}_z^\top(i)\,\bar{C}_z(i)$, then system (4.1) is stochastically stable and satisfies the following:

$$\|z\|_2 \le \left[\gamma^2\|\omega\|_2^2 + x^\top(0)Px(0)\right]^{\frac{1}{2}}. \tag{4.32}$$

Proof: The proof of this theorem is direct and the detail is omitted.   □

To synthesize the controller gain, let us transform the LMI (4.31) into a form that can be used easily to compute this gain. Notice that

$$\begin{bmatrix} \bar{A}^\top(i)P + P\bar{A}(i) + \bar{C}_z^\top(i)\bar{C}_z(i) & \bar{C}_z^\top(i)B_z(i) + PB_\omega(i) \\ B_z^\top(i)\bar{C}_z(i) + B_\omega^\top(i)P & B_z^\top(i)B_z(i) - \gamma^2\mathbb{I} \end{bmatrix} =$$
$$\begin{bmatrix} \bar{A}^\top(i)P + P\bar{A}(i) & PB_\omega(i) \\ B_\omega^\top(i)P & -\gamma^2\mathbb{I} \end{bmatrix} + \begin{bmatrix} \bar{C}_z^\top(i) \\ B_z^\top(i) \end{bmatrix} \begin{bmatrix} \bar{C}_z(i) & B_z(i) \end{bmatrix}.$$

Using the Schur complement we show that (4.31) is equivalent to the following inequality:

$$\begin{bmatrix} \bar{A}^\top(i)P + P\bar{A}(i) & PB_\omega(i) & \bar{C}_z^\top(i) \\ B_\omega^\top(i)P & -\gamma^2\mathbb{I} & B_z^\top(i) \\ \bar{C}_z(i) & B_z(i) & -\mathbb{I} \end{bmatrix} < 0.$$

Since $\bar{A}(i)$ is nonlinear in $\mathscr{K}$ and P, the previous inequality is then nonlinear and therefore it cannot be solved using existing linear algorithms. To transform it into an LMI, let $X = P^{-1}$. As done many times previously, pre- and post-multiply this inequality by $\mathrm{diag}[X, \mathbb{I}, \mathbb{I}]$, which gives

$$\begin{bmatrix} X\bar{A}^\top(i) + \bar{A}(i)X & B_\omega(i) & X\bar{C}_z^\top(i) \\ B_\omega^\top(i) & -\gamma^2\mathbb{I} & B_z^\top(i) \\ \bar{C}_z(i)X & B_z(i) & -\mathbb{I} \end{bmatrix} < 0.$$

Notice that

$$X\bar{A}^\top(i) + \bar{A}(i)X = XA^\top(i) + A(i)X + Y^\top B^\top(i) + B(i)Y,$$
$$X(i)\left[C_z(i) + D_z(i)K(i)\right]^\top = X(i)C_z^\top(i) + Y^\top(i)D_z^\top(i),$$

where $Y = \mathscr{K}X$.

Using the Schur complement again implies that the previous inequality is equivalent to the following:

$$\begin{bmatrix} J(i) & B_\omega(i) & XC_z^\top(i) + Y^\top D_z^\top(i) \\ B_\omega^\top(i) & -\gamma^2\mathbb{I} & B_z^\top(i) \\ C_z(i)X + D_z(i)Y & B_z(i) & -\mathbb{I} \end{bmatrix} < 0, \qquad (4.33)$$

with $J(i) = XA^\top(i) + A(i)X + Y^\top B^\top(i) + B(i)Y$.

From this discussion and using the expression of X, we get the following theorem.

Theorem 52. *Let γ be a positive constant. If there exist a symmetric and positive-definite matrix $X > 0$ and a matrix Y such that the following set of LMIs (4.33) holds for every $i \in \mathscr{S}$, then the system (4.1) under the controller (4.30) with $\mathscr{K} = YX^{-1}$ is stochastically stable and, moreover, the closed-loop system satisfies the disturbance rejection of level γ.*

Example 53. Let us in this example consider the two-mode system of Example 45 with the same data. Solving the LMI of the previous theorem when $\gamma = 5.0$ we get

$$X = \begin{bmatrix} 9.5507 & 1.1766 \\ 1.1766 & 4.5472 \end{bmatrix}, \quad Y = \begin{bmatrix} -6.7818 & -0.8451 \\ -0.8451 & -2.8332 \end{bmatrix},$$

which gives the following gain:

$$\mathcal{K} = \begin{bmatrix} -0.7098 & -0.0022 \\ -0.0121 & -0.6199 \end{bmatrix}.$$

With this controller, the closed-loop state equation becomes

$$\dot{x}(t) = A_{cl}(i)x(t),$$

with

$$A_{cl}(i) = \begin{cases} \begin{bmatrix} -1.2098 & 0.9978 \\ 0.2879 & -3.1199 \end{bmatrix}, \text{when } i = 1, \\[2em] \begin{bmatrix} -1.7098 & 0.0978 \\ 0.1879 & -2.6199 \end{bmatrix}, \text{otherwise.} \end{cases}$$

The standard conditions for stochastic stability can be summarized as follows: If there exists a set of symmetric and positive-definite matrices $P = (P(1), P(2)) > 0$ such that the following holds for each $r(t) = i \in \mathcal{S}$:

$$A_{cl}^{\top}(i)P(i) + P(i)A_{cl}(i) + \sum_{j=1}^{N} \lambda_{ij} P(j) < 0,$$

then the closed-loop system is stochastically stable.

Using these conditions, we get the following matrices:

$$P(1) = \begin{bmatrix} 12.9296 & 2.3139 \\ 2.3139 & 6.2671 \end{bmatrix}, \quad P(2) = \begin{bmatrix} 11.2888 & -0.1445 \\ -0.1445 & 8.0984 \end{bmatrix},$$

which are both symmetric and positive-definite matrices and therefore the closed-loop system is stochastically stable under the constant gain state feedback controller.

From the practical point of view, the controller that stochastically stabilizes the system and simultaneously guarantees the minimum disturbance rejection is of great interest. This controller can be obtained by solving the following optimization problem:

$$\text{Pc} : \begin{cases} \min_{\nu > 0,} \nu, \\ \phantom{\min_{\nu}} X > 0, \\ \phantom{\min_{\nu}} Y, \\ s.t. : (4.33) \text{ with } \nu = \gamma^2. \end{cases} \tag{4.34}$$

The following corollary gives the results of the design of the controller that stochastically stabilizes the system (4.1) and simultaneously guarantees the smallest disturbance rejection level.

Corollary 9. *Let $\nu > 0$, $X > 0$, and Y be the solution of the optimization problem Pc. Then the controller (4.30) with $\mathscr{K} = YX^{-1}$ stochastically stabilizes the class of systems we are considering and, moreover, the closed-loop system satisfies the disturbance rejection of level $\sqrt{\nu}$.*

Example 54. In this example let us consider the same system of the previous example with the same data. Solving the LMI of the previous theorem iteratively, we get

$$X = \begin{bmatrix} 3.2560 & 0.3312 \\ 0.3312 & 1.8966 \end{bmatrix}, \quad Y = \begin{bmatrix} -4.4808 & -0.4168 \\ -0.4168 & -2.8857 \end{bmatrix},$$

which gives the following gain:

$$\mathscr{K} = \begin{bmatrix} -1.3783 & 0.0210 \\ 0.0273 & -1.5263 \end{bmatrix}.$$

With this controller, the closed-loop state equation becomes

$$\dot{x}(t) = A_{cl}(i)x(t),$$

with

$$A_{cl}(i) = \begin{cases} \begin{bmatrix} -1.8783 & 1.0210 \\ 0.3273 & -4.0263 \end{bmatrix}, \text{ when } i = 1, \\[3ex] \begin{bmatrix} -2.3783 & 0.1210 \\ 0.2273 & -3.5263 \end{bmatrix}, \text{ otherwise.} \end{cases}$$

The standard conditions for stochastic stability can be summarized as follows: If there exists a set of symmetric and positive-definite matrices $P = (P(1), P(2)) > 0$ such that the following holds for each $r(t) = i \in \mathscr{S}$:

$$A_{cl}^{\top}(i)P(i) + P(i)A_{cl}(i) + \sum_{j=1}^{N} \lambda_{ij}P(j) < 0,$$

then the closed-loop system is stochastically stable.

Using these conditions, we get the following matrices:

$$P(1) = \begin{bmatrix} 7.2981 & 1.0301 \\ 1.0301 & 3.9112 \end{bmatrix}, \quad P(2) = \begin{bmatrix} 7.0007 & -0.3456 \\ -0.3456 & 5.7152 \end{bmatrix},$$

which are both symmetric and positive-definite matrices and therefore the closed-loop system is stochastically stable under the constant gain state feedback controller. The corresponding minimal disturbance rejection level γ is equal to 1.0043.

Let us now focus on the design of a robust state feedback controller with constant gain. Assume that the dynamics (4.1) have uncertainties and see how we can design a controller of the form (4.30) that robustly stabilizes the class of systems we are studying and at the same time guarantees the desired disturbance rejection of level γ. Following the same steps for the nominal case starting from the results of Theorem 52, we get

$$\begin{bmatrix} \bar{J}_0(i,t) & PB_\omega(i) & \bar{C}_z^\top(i,t) \\ B_\omega^\top(i)P & -\gamma^2\mathbb{I} & B_z^\top(i) \\ \bar{C}_z(i,t) & B_z(i) & -\mathbb{I} \end{bmatrix} < 0, \tag{4.35}$$

with

$$\begin{aligned} \bar{J}_0(i,t) &= \bar{A}^\top(i,t)P + P\bar{A}(i,t), \\ \bar{A}(i,t) &= A(i,t) + B(i,t)\mathscr{K}, \\ \bar{C}_z(i,t) &= C_z(i,t) + D_z(i,t)\mathscr{K}. \end{aligned}$$

Using the expressions of $\bar{A}(i,t)$, $\bar{C}_z(i,t)$, and their components, we obtain the following inequality:

$$\begin{bmatrix} J_1(i) & PB_\omega(i) & \begin{bmatrix} C_z^\top(i) \\ +\mathscr{K}^\top D_z^\top(i) \end{bmatrix} \\ B_\omega^\top(i)P & -\gamma^2\mathbb{I} & B_z^\top(i) \\ \begin{bmatrix} D_z(i)\mathscr{K} \\ +C_z(i) \end{bmatrix} & B_z(i) & -\mathbb{I} \end{bmatrix}$$

$$+ \begin{bmatrix} E_A^\top(i)F_A^\top(i,t)D_A^\top(i)P & 0 & 0 \\ 0 & 0 & 0 \\ 0 & 0 & 0 \end{bmatrix}$$

$$+ \begin{bmatrix} PD_A(i)F_A(i,t)E_A(i) & 0 & 0 \\ 0 & 0 & 0 \\ 0 & 0 & 0 \end{bmatrix}$$

$$+ \begin{bmatrix} PD_B(i)F_B(i,t)E_B(i)\mathscr{K} & 0 & 0 \\ 0 & 0 & 0 \\ 0 & 0 & 0 \end{bmatrix}$$

$$+ \begin{bmatrix} \mathscr{K}^\top E_B^\top(i)F_B^\top(i,t)D_B^\top(i)P & 0 & 0 \\ 0 & 0 & 0 \\ 0 & 0 & 0 \end{bmatrix}$$

$$+ \begin{bmatrix} 0 & 0 & \mathscr{K}^\top E_{D_z}^\top(i)F_{D_z}^\top(i,t)D_{D_z}^\top(i) \\ 0 & 0 & 0 \\ 0 & 0 & 0 \end{bmatrix}$$

$$+ \begin{bmatrix} 0 & 0 & 0 \\ 0 & 0 & 0 \\ D_{D_z}(i)F_{D_z}(i,t)E_{D_z}(i)\mathscr{K} & 0 & 0 \end{bmatrix}$$

$$+ \begin{bmatrix} 0 & 0 & E_{C_z}^\top(i)F_{C_z}^\top(i,t)D_{C_z}^\top(i) \\ 0 & 0 & 0 \\ 0 & 0 & 0 \end{bmatrix}$$

$$+ \begin{bmatrix} 0 & 0 & 0 \\ 0 & 0 & 0 \\ D_{C_z}(i)F_{C_z}(i,t)E_{C_z}(i) & 0 & 0 \end{bmatrix} < 0,$$

with $J_1(i) = A^\top(i)P + PA(i) + \mathscr{K}^\top B^\top(i)P + PB(i)\mathscr{K}$.

Notice that

$$\begin{bmatrix} E_A^\top(i)F_A^\top(i,t)D_A^\top(i)P & 0 & 0 \\ 0 & 0 & 0 \\ 0 & 0 & 0 \end{bmatrix}$$

$$= \begin{bmatrix} E_A^\top(i) & 0 & 0 \\ 0 & 0 & 0 \\ 0 & 0 & 0 \end{bmatrix} \begin{bmatrix} F_A^\top(i,t) & 0 & 0 \\ 0 & 0 & 0 \\ 0 & 0 & 0 \end{bmatrix} \begin{bmatrix} D_A^\top(i)P & 0 & 0 \\ 0 & 0 & 0 \\ 0 & 0 & 0 \end{bmatrix},$$

$$\begin{bmatrix} PD_A(i)F_A(i,t)E_A(i) & 0 & 0 \\ 0 & 0 & 0 \\ 0 & 0 & 0 \end{bmatrix}$$

$$= \begin{bmatrix} PD_A(i) & 0 & 0 \\ 0 & 0 & 0 \\ 0 & 0 & 0 \end{bmatrix} \begin{bmatrix} F_A(i,t) & 0 & 0 \\ 0 & 0 & 0 \\ 0 & 0 & 0 \end{bmatrix} \begin{bmatrix} E_A(i) & 0 & 0 \\ 0 & 0 & 0 \\ 0 & 0 & 0 \end{bmatrix},$$

$$\begin{bmatrix} PD_B(i)F_B(i,t)E_B(i)\mathscr{K} & 0 & 0 \\ 0 & 0 & 0 \\ 0 & 0 & 0 \end{bmatrix}$$

$$= \begin{bmatrix} PD_B(i) & 0 & 0 \\ 0 & 0 & 0 \\ 0 & 0 & 0 \end{bmatrix} \begin{bmatrix} F_B(i,t) & 0 & 0 \\ 0 & 0 & 0 \\ 0 & 0 & 0 \end{bmatrix} \begin{bmatrix} E_B(i)\mathscr{K} & 0 & 0 \\ 0 & 0 & 0 \\ 0 & 0 & 0 \end{bmatrix},$$

$$\begin{bmatrix} \mathscr{K}^\top E_B^\top(i)F_B^\top(i,t)D_B^\top(i)P & 0 & 0 \\ 0 & 0 & 0 \\ 0 & 0 & 0 \end{bmatrix}$$

$$= \begin{bmatrix} \mathscr{K}^\top E_B^\top(i) & 0 & 0 \\ 0 & 0 & 0 \\ 0 & 0 & 0 \end{bmatrix} \begin{bmatrix} F_B^\top(i,t) & 0 & 0 \\ 0 & 0 & 0 \\ 0 & 0 & 0 \end{bmatrix} \begin{bmatrix} D_B^\top(i)P & 0 & 0 \\ 0 & 0 & 0 \\ 0 & 0 & 0 \end{bmatrix},$$

$$\begin{bmatrix} 0 & 0 & \mathscr{K}^\top E_{D_z}^\top(i)F_{D_z}^\top(i,t)D_{D_z}^\top(i) \\ 0 & 0 & 0 \\ 0 & 0 & 0 \end{bmatrix}$$

$$= \begin{bmatrix} 0 & 0 & \mathscr{K}^\top E_{D_z}^\top(i) \\ 0 & 0 & 0 \\ 0 & 0 & 0 \end{bmatrix} \begin{bmatrix} 0 & 0 & 0 \\ 0 & 0 & 0 \\ 0 & 0 & F_{D_z}^\top(i,t) \end{bmatrix} \begin{bmatrix} 0 & 0 & 0 \\ 0 & 0 & 0 \\ 0 & 0 & D_{D_z}^\top(i) \end{bmatrix},$$

$$\begin{bmatrix} & 0 & & 0 & 0 \\ & 0 & & 0 & 0 \\ D_{D_z}(i)F_{D_z}(i,t)E_{D_z}(i)\mathscr{K} & 0 & 0 \end{bmatrix}$$
$$= \begin{bmatrix} 0 & 0 & 0 \\ 0 & 0 & 0 \\ 0 & 0 & D_{D_z}(i) \end{bmatrix} \begin{bmatrix} 0 & 0 & 0 \\ 0 & 0 & 0 \\ 0 & 0 & F_{D_z}(i,t) \end{bmatrix} \begin{bmatrix} & 0 & & 0 & 0 \\ & 0 & & 0 & 0 \\ E_{D_z}(i)\mathscr{K} & 0 & 0 \end{bmatrix},$$

$$\begin{bmatrix} 0 & 0 & E_{C_z}^\top(i)F_{C_z}^\top(i,t)D_{C_z}^\top(i) \\ 0 & 0 & 0 \\ 0 & 0 & 0 \end{bmatrix}$$
$$= \begin{bmatrix} 0 & 0 & E_{C_z}^\top(i) \\ 0 & 0 & 0 \\ 0 & 0 & 0 \end{bmatrix} \begin{bmatrix} 0 & 0 & 0 \\ 0 & 0 & 0 \\ 0 & 0 & F_{C_z}^\top(i,t) \end{bmatrix} \begin{bmatrix} 0 & 0 & 0 \\ 0 & 0 & 0 \\ 0 & 0 & D_{C_z}^\top(i) \end{bmatrix},$$

and

$$\begin{bmatrix} & 0 & & 0 & 0 \\ & 0 & & 0 & 0 \\ D_{C_z}(i)F_{C_z}(i,t)E_{C_z}(i) & 0 & 0 \end{bmatrix}$$
$$= \begin{bmatrix} 0 & 0 & 0 \\ 0 & 0 & 0 \\ 0 & 0 & D_{C_z}(i) \end{bmatrix} \begin{bmatrix} 0 & 0 & 0 \\ 0 & 0 & 0 \\ 0 & 0 & F_{C_z}(i,t) \end{bmatrix} \begin{bmatrix} & 0 & & 0 & 0 \\ & 0 & & 0 & 0 \\ E_{C_z}(i) & 0 & 0 \end{bmatrix}.$$

Using now Lemma 7 in Appendix A, we get

$$\begin{bmatrix} PD_A(i)F_A(i,t)E_A(i) & 0 & 0 \\ 0 & 0 & 0 \\ 0 & 0 & 0 \end{bmatrix}$$
$$+ \begin{bmatrix} E_A^\top(i)F_A^\top(i,t)D_A^\top(i)P & 0 & 0 \\ 0 & 0 & 0 \\ 0 & 0 & 0 \end{bmatrix}$$
$$\leq \varepsilon_A(i) \begin{bmatrix} PD_A(i) & 0 & 0 \\ 0 & 0 & 0 \\ 0 & 0 & 0 \end{bmatrix} \begin{bmatrix} D_A^\top(i)P & 0 & 0 \\ 0 & 0 & 0 \\ 0 & 0 & 0 \end{bmatrix}$$
$$+ \varepsilon_A^{-1}(i) \begin{bmatrix} E_A^\top(i) & 0 & 0 \\ 0 & 0 & 0 \\ 0 & 0 & 0 \end{bmatrix} \begin{bmatrix} E_A(i) & 0 & 0 \\ 0 & 0 & 0 \\ 0 & 0 & 0 \end{bmatrix}$$
$$= \begin{bmatrix} \varepsilon_A(i)PD_A(i)D_A^\top(i)P + \varepsilon_A^{-1}(i)E_A^\top(i)E_A(i) & 0 & 0 \\ 0 & 0 & 0 \\ 0 & 0 & 0 \end{bmatrix},$$

$$\begin{bmatrix} PD_B(i)F_B(i,t)E_B(i)\mathscr{K} & 0 & 0 \\ 0 & 0 & 0 \\ 0 & 0 & 0 \end{bmatrix}$$

$$+ \begin{bmatrix} \mathscr{K}^\top E_B^\top(i)F_B^\top(i,t)D_B^\top(i)P & 0 & 0 \\ 0 & 0 & 0 \\ 0 & 0 & 0 \end{bmatrix}$$

$$\leq \varepsilon_B(i) \begin{bmatrix} PD_B(i) & 0 & 0 \\ 0 & 0 & 0 \\ 0 & 0 & 0 \end{bmatrix} \begin{bmatrix} D_B^\top(i)P & 0 & 0 \\ 0 & 0 & 0 \\ 0 & 0 & 0 \end{bmatrix}$$

$$+ \varepsilon_B^{-1}(i) \begin{bmatrix} \mathscr{K}^\top E_B^\top(i) & 0 & 0 \\ 0 & 0 & 0 \\ 0 & 0 & 0 \end{bmatrix} \begin{bmatrix} E_B(i)\mathscr{K} & 0 & 0 \\ 0 & 0 & 0 \\ 0 & 0 & 0 \end{bmatrix}$$

$$= \begin{bmatrix} \varepsilon_B(i)PD_B(i)D_B^\top(i)P + \varepsilon_B^{-1}(i)\mathscr{K}^\top E_B^\top(i)E_B(i)\mathscr{K} & 0 & 0 \\ 0 & 0 & 0 \\ 0 & 0 & 0 \end{bmatrix},$$

$$\begin{bmatrix} 0 & 0 & \mathscr{K}^\top E_{D_z}^\top(i)F_{D_z}^\top(i,t)D_{D_z}^\top(i) \\ 0 & 0 & 0 \\ 0 & 0 & 0 \end{bmatrix}$$

$$+ \begin{bmatrix} 0 & & 0 & 0 \\ 0 & & 0 & 0 \\ D_{D_z}(i)F_{D_z}(i,t)E_{D_z}(i)\mathscr{K} & & 0 & 0 \end{bmatrix}$$

$$\leq \varepsilon_{D_z}^{-1}(i) \begin{bmatrix} 0 & 0 & \mathscr{K}^\top E_{D_z}^\top(i) \\ 0 & 0 & 0 \\ 0 & 0 & 0 \end{bmatrix} \begin{bmatrix} 0 & & 0 & 0 \\ 0 & & 0 & 0 \\ E_{D_z}(i)\mathscr{K} & & 0 & 0 \end{bmatrix}$$

$$+ \varepsilon_{D_z}(i) \begin{bmatrix} 0 & 0 & 0 \\ 0 & 0 & 0 \\ 0 & 0 & D_{D_z}(i) \end{bmatrix} \begin{bmatrix} 0 & 0 & 0 \\ 0 & 0 & 0 \\ 0 & 0 & D_{D_z}^\top(i) \end{bmatrix}$$

$$= \begin{bmatrix} \varepsilon_{D_z}^{-1}(i)\mathscr{K}^\top E_{D_z}^\top(i)E_{D_z}(i)\mathscr{K} & 0 & 0 \\ 0 & 0 & 0 \\ 0 & 0 & \varepsilon_{D_z}(i)D_{D_z}(i)D_{D_z}^\top(i) \end{bmatrix},$$

$$\begin{bmatrix} 0 & 0 & E_{C_z}^\top(i)F_{C_z}^\top(i,t)D_{C_z}^\top(i) \\ 0 & 0 & 0 \\ 0 & 0 & 0 \end{bmatrix}$$

$$+ \begin{bmatrix} 0 & & 0 & 0 \\ 0 & & 0 & 0 \\ D_{C_z}(i)F_{C_z}(i,t)E_{C_z}(i) & & 0 & 0 \end{bmatrix}$$

$$\leq \varepsilon_{C_z}^{-1}(i) \begin{bmatrix} 0 & 0 & E_{C_z}^{\top}(i) \\ 0 & 0 & 0 \\ 0 & 0 & 0 \end{bmatrix} \begin{bmatrix} 0 & 0 & 0 \\ 0 & 0 & 0 \\ E_{C_z}(i,t) & 0 & 0 \end{bmatrix}$$

$$+\varepsilon_{C_z}(i) \begin{bmatrix} 0 & 0 & 0 \\ 0 & 0 & 0 \\ 0 & 0 & D_{C_z}(i) \end{bmatrix} \begin{bmatrix} 0 & 0 & 0 \\ 0 & 0 & 0 \\ 0 & 0 & D_{C_z}^{\top}(i) \end{bmatrix}$$

$$= \begin{bmatrix} \varepsilon_{C_z}^{-1}(i) E_{C_z}^{\top}(i) E_{C_z}(i) & 0 & 0 \\ 0 & 0 & 0 \\ 0 & 0 & \varepsilon_{C_z}(i) D_{C_z}(i) D_{C_z}^{\top}(i) \end{bmatrix}.$$

Using these transformations we get

$$\begin{bmatrix} J_1(i) & PB_\omega(i) & C_z^{\top}(i) + \mathcal{K}^{\top} D_z^{\top}(i) \\ B_\omega^{\top}(i)P & -\gamma^2\mathbb{I} & B_z^{\top}(i) \\ D_z(i)\mathcal{K} + C_z(i) & B_z(i) & -\mathbb{I} \end{bmatrix}$$

$$+ \begin{bmatrix} \varepsilon_A(i)PD_A(i)D_A^{\top}(i)P + \varepsilon_A^{-1}(i)E_A^{\top}(i)E_A(i) & 0 & 0 \\ 0 & 0 & 0 \\ 0 & 0 & 0 \end{bmatrix}$$

$$+ \begin{bmatrix} \varepsilon_B(i)PD_B(i)D_B^{\top}(i)P + \varepsilon_B^{-1}\mathcal{K}^{\top} E_B^{\top}(i)E_B(i)\mathcal{K} & 0 & 0 \\ 0 & 0 & 0 \\ 0 & 0 & 0 \end{bmatrix}$$

$$+ \begin{bmatrix} \varepsilon_{D_z}^{-1}(i)\mathcal{K}^{\top} E_{D_z}^{\top}(i)E_{D_z}(i)\mathcal{K} & 0 & 0 \\ 0 & 0 & 0 \\ 0 & 0 & \varepsilon_{D_z}(i)D_{D_z}(i)D_{D_z}^{\top}(i) \end{bmatrix}$$

$$+ \begin{bmatrix} \varepsilon_{C_z}^{-1}(i)E_{C_z}^{\top}(i)E_{C_z}(i) & 0 & 0 \\ 0 & 0 & 0 \\ 0 & 0 & \varepsilon_{C_z}(i)D_{C_z}(i)D_{C_z}^{\top}(i) \end{bmatrix} < 0,$$

with $J_1(i) = A^{\top}(i)P + PA(i) + \mathcal{K}^{\top} B^{\top}(i)P + PB(i)\mathcal{K}$.
Let $J_2(i)$, $\mathcal{W}(i)$, and $\mathcal{T}(i)$ be defined as

$$J_2(i) = J_1(i) + \varepsilon_A^{-1}(i)E_A^{\top}(i)E_A(i) + \varepsilon_B^{-1}(i)\mathcal{K}^{\top} E_B^{\top}(i)E_B(i)\mathcal{K},$$
$$\mathcal{W}(i) = \text{diag}[\varepsilon_A^{-1}(i)\mathbb{I}, \varepsilon_B^{-1}(i)\mathbb{I}, \varepsilon_{C_z}(i)\mathbb{I}, \varepsilon_{D_z}(i)\mathbb{I}],$$
$$\mathcal{T}(i) = \left(PD_A(i), PD_B(i), E_{C_z}^{\top}(i), \mathcal{K}^{\top} E_{D_z}^{\top}(i)\right).$$

Using the Schur complement we get the equivalent inequality:

$$\begin{bmatrix} J_2(i) & PB_\omega(i) & \begin{bmatrix} C_z^{\top}(i) \\ +\mathcal{K}^{\top} D_z^{\top}(i) \end{bmatrix} & \mathcal{T}(i) \\ B_\omega^{\top}(i)P & -\gamma^2\mathbb{I} & B_z^{\top}(i) & 0 \\ \begin{bmatrix} D_z(i)\mathcal{K} \\ +C_z(i) \end{bmatrix} & B_z(i) & -\mathcal{U}(i) & 0 \\ \mathcal{T}^{\top}(i) & 0 & 0 & -\mathcal{W}(i) \end{bmatrix} < 0,$$

with $\mathcal{U}(i) = \mathbb{I} - \varepsilon_{D_z}(i)D_{D_z}(i)D_{D_z}^\top(i) - \varepsilon_{C_z}(i)D_{C_z}(i)D_{C_z}^\top(i)$.

This matrix inequality is nonlinear in P and $\mathscr{K}$. To put it into LMI form, let $X = P^{-1}$. Pre- and post-multiply this matrix inequality by $\mathrm{diag}[X,\mathbb{I},\mathbb{I},\mathbb{I}]$ to get

$$
\begin{bmatrix}
J_3(i) & B_\omega(i) & \begin{bmatrix} XC_z^\top(i) \\ +XK^\top(i)D_z^\top(i) \end{bmatrix} & XT(i) \\
B_\omega^\top(i) & -\gamma^2\mathbb{I} & B_z^\top(i) & 0 \\
\begin{bmatrix} D_z(i)\mathscr{K}X \\ +C_z(i)X \end{bmatrix} & B_z(i) & -\mathcal{U}(i) & 0 \\
T^\top(i)X & 0 & 0 & -\mathcal{W}(i)
\end{bmatrix} < 0,
$$

with

$$
\begin{aligned}
J_3(i) =\ & XA^\top(i) + A(i)X + X\mathscr{K}^\top B^\top(i) \\
& + B(i)\mathscr{K}X + \varepsilon_A^{-1}(i)XE_A^\top(i)E_A(i)X(i) \\
& + \varepsilon_B^{-1}(i)X\mathscr{K}^\top E_B^\top(i)E_B(i)\mathscr{K}X.
\end{aligned}
$$

Notice that

$$
XT(i) = \left(D_A(i), D_B(i), XE_{C_z}^\top(i), X\mathscr{K}^\top E_{D_z}^\top(i) \right).
$$

Letting $Y = \mathscr{K}X$ and using the Schur complement we obtain

$$
\begin{bmatrix}
\tilde{J}(i) & B_\omega(i) & \begin{bmatrix} XC_z^\top(i) \\ +Y^\top D_z^\top(i) \end{bmatrix} & \mathcal{Z}(i) \\
B_\omega^\top(i) & -\gamma^2\mathbb{I} & B_z^\top(i) & 0 \\
\begin{bmatrix} D_z(i)Y \\ +C_z(i)X \end{bmatrix} & B_z(i) & -\mathcal{U}(i) & 0 \\
\mathcal{Z}^\top(i) & 0 & 0 & -\mathcal{V}(i)
\end{bmatrix} < 0, \tag{4.36}
$$

with

$$
\begin{aligned}
\tilde{J}(i) =\ & XA^\top(i) + A(i)X + Y^\top B^\top(i) + B(i)Y \\
& + \varepsilon_A(i)D_A(i)D_A^\top(i) + \varepsilon_B(i)D_B(i)D_B^\top(i), \\
\mathcal{U}(i) =\ & \mathbb{I} - \varepsilon_{D_z}(i)D_{D_z}(i)D_{D_z}^\top(i) - \varepsilon_{C_z}(i)D_{C_z}(i)D_{C_z}^\top(i), \\
\mathcal{Z}(i) =\ & \left[XE_A^\top(i), Y^\top E_B^\top(i), X^\top E_{C_z}^\top(i), Y^\top E_{D_z}^\top(i) \right], \\
\mathcal{V}(i) =\ & \mathrm{diag}[\varepsilon_A(i)\mathbb{I}, \varepsilon_B(i)\mathbb{I}, \varepsilon_{C_z}(i)\mathbb{I}, \varepsilon_{D_z}(i)\mathbb{I}].
\end{aligned}
$$

The following theorem summarizes the results of this development.

Theorem 53. *Let γ be a positive constant. If there exist a symmetric and positive-definite matrix $X > 0$ and a matrix Y and sets of positive scalars $\varepsilon_A = (\varepsilon_A(1), \cdots, \varepsilon_A(N))$, $\varepsilon_B = (\varepsilon_B(1), \cdots, \varepsilon_B(N))$, $\varepsilon_{C_z} = (\varepsilon_{C_z}(1), \cdots, \varepsilon_{C_z}(N))$, and $\varepsilon_{D_z} = (\varepsilon_{D_z}(1), \cdots, \varepsilon_{D_z}(N))$ such that the following set of LMIs (4.36) holds for every $i \in \mathscr{S}$ and for all admissible uncertainties, then the system (4.1) under the controller (4.30) with $\mathscr{K} = YX^{-1}$ is stochastically stable and, moreover, the closed-loop system satisfies the disturbance rejection of level γ.*

Example 55. To show the effectiveness of the theoretical results of this theorem, let us consider the two-mode system of Example 49 with the same data. Let us fix the required positive constant to the following values:

$$\varepsilon_A(1) = \varepsilon_A(2) = 0.50,$$
$$\varepsilon_B(1) = \varepsilon_B(2) = \varepsilon_C(1) = \varepsilon_C(2) = \varepsilon_D(1) = \varepsilon_D(2) = 0.10.$$

Solving the LMI (4.36) with $\gamma = 5.0$, we get

$$X = \begin{bmatrix} 3.7967 & -2.8740 \\ -2.8740 & 4.5911 \end{bmatrix}, \quad Y = \begin{bmatrix} -3.5141 & 2.8775 \\ 2.8017 & -3.8579 \end{bmatrix}.$$

The corresponding gains are given by

$$\mathcal{K} = \begin{bmatrix} -0.8575 & 0.0900 \\ 0.1936 & -0.7191 \end{bmatrix}.$$

Based on the results of this theorem, we conclude that the system is stochastically stable and guarantees the desired disturbance rejection level $\gamma = 5.0$.

With this controller, the closed-loop state equation becomes

$$\dot{x}(t) = A_{cl}(i)x(t),$$

with

$$A_{cl}(i) = \begin{cases} \begin{bmatrix} -1.3575 & 1.0900 \\ 0.4936 & -3.2191 \end{bmatrix}, \text{when } i = 1, \\ \\ \begin{bmatrix} -1.8575 & 0.1900 \\ 0.3936 & -2.7191 \end{bmatrix}, \text{otherwise.} \end{cases}$$

The standard conditions for stochastic stability can be summarized as follows: If there exists a set of symmetric and positive-definite matrices $P = (P(1), P(2)) > 0$ such that the following holds for each $r(t) = i \in \mathcal{S}$:

$$A_{cl}^{\top}(i)P(i) + P(i)A_{cl}(i) + \sum_{j=1}^{N} \lambda_{ij}P(j) < 0,$$

then the closed-loop system is stochastically stable.

Using these conditions, we get the following matrices:

$$P(1) = \begin{bmatrix} 11.7556 & 2.5739 \\ 2.5739 & 6.0528 \end{bmatrix}, \quad P(2) = \begin{bmatrix} 10.3809 & 0.2804 \\ 0.2804 & 7.7345 \end{bmatrix},$$

which are both symmetric and positive-definite matrices and therefore the closed-loop system is stochastically stable under the constant gain state feedback controller.

As done for nominal system, we can determine the controller that stochastically stabilizes the class of systems we are considering and simultaneously guarantees the minimum disturbance rejection by solving the following optimization problem:

$$
\text{Pu} : \begin{cases} \min\limits_{\substack{\nu>0, \\ \varepsilon_A=(\varepsilon_A(1),\cdots,\varepsilon_A(N))>0, \\ \varepsilon_B=(\varepsilon_B(1),\cdots,\varepsilon_B(N))>0, \\ \varepsilon_{C_z}=(\varepsilon_{D_z}(1),\cdots,\varepsilon_{C_z}(N))>0, \\ \varepsilon_{D_z}=(\varepsilon_{D_z}(1),\cdots,\varepsilon_{D_z}(N))>0, \\ X>0, \\ Y,}} \nu, \\[2pt] s.t. : (4.16) \quad \text{with} \quad \nu = \gamma^2. \end{cases}
$$

The following corollary summarizes the results of the design of the controller that stochastically stabilizes the system (4.1) and simultaneously guarantees the smallest disturbance rejection level.

Corollary 10. *Let $\nu > 0$, $\varepsilon_A = (\varepsilon_A(1), \cdots, \varepsilon_A(N)) > 0$, $\varepsilon_B = (\varepsilon_B(1), \cdots, \varepsilon_B(N)) > 0$, $\varepsilon_{C_z} = (\varepsilon_{C_z}(1), \cdots, \varepsilon_{C_z}(N)) > 0$, $\varepsilon_{D_z} = (\varepsilon_{D_z}(1), \cdots, \varepsilon_{D_z}(N)) > 0$, $X > 0$, and Y be the solution of the optimization problem Pu. Then the controller (4.30) with $\mathcal{K} = YX^{-1}$ stochastically stabilizes the class of systems we are considering and, moreover, the closed-loop system satisfies the disturbance rejection of level $\sqrt{\nu}$.*

Example 56. To illustrate the effectiveness of the results developed in this theorem, let us consider the two-mode system of the previous example with the same data.

The required positive scalars are fixed to the following values:

$$\varepsilon_A(1) = \varepsilon_A(2) = 0.50,$$
$$\varepsilon_B(1) = \varepsilon_B(2) = \varepsilon_C(1) = \varepsilon_C(2) = \varepsilon_D(1) = \varepsilon_D(2) = 0.10.$$

Solving the problem Pu gives

$$
X = \begin{bmatrix} 28.8798 & -38.7485 \\ -38.7485 & 55.0593 \end{bmatrix}, \quad Y = \begin{bmatrix} -29.9278 & 38.8262 \\ 38.7400 & -56.0388 \end{bmatrix},
$$

which gives the following gains:

$$
\mathcal{K} = \begin{bmatrix} -1.6168 & -0.4327 \\ -0.4333 & -1.3227 \end{bmatrix}.
$$

Based on the results of this theorem, we conclude that the system is stochastically stable and assures a disturbance rejection level equal to $\gamma = 1.0043$.

With this controller, the closed-loop state equation becomes

$$\dot{x}(t) = A_{cl}(i)x(t),$$

with

$$A_{cl}(i) = \begin{cases} \begin{bmatrix} -2.1168 & 0.5673 \\ -0.1333 & -3.8227 \end{bmatrix}, \text{when } i = 1, \\[3ex] \begin{bmatrix} -2.6168 & -0.3327 \\ -0.2333 & -3.3227 \end{bmatrix}, \text{otherwise.} \end{cases}$$

The standard conditions for stochastic stability can be summarized as follows: If there exists a set of symmetric and positive-definite matrices $P = (P(1), P(2)) > 0$ such that the following holds for each $r(t) = i \in \mathscr{S}$:

$$A_{cl}^\top(i)P(i) + P(i)A_{cl}(i) + \sum_{j=1}^{N} \lambda_{ij}P(j) < 0,$$

then the closed-loop system is stochastically stable.

Using these conditions, we get the following matrices:

$$P(1) = \begin{bmatrix} 6.2321 & 0.0531 \\ 0.0531 & 3.9069 \end{bmatrix}, \quad P(2) = \begin{bmatrix} 6.2483 & -1.0435 \\ -1.0435 & 5.9999 \end{bmatrix},$$

which are both symmetric and positive-definite matrices and therefore the closed-loop system is stochastically stable under the constant gain state feedback controller.

4.3 Output Feedback Stabilization

As mentioned earlier in Chapter 3, state feedback stabilization is sometimes difficult to use since complete access to the state vector is not always possible for reasons such as lack of technology or inadequate budget. To overcome this difficulty, we use output feedback control or observer-based output feedback control. In this section, we focus on the design of output feedback control and see how to compute the gains of this controller. The structure we use is given by the following dynamics:

$$\begin{cases} \dot{x}_c(t) = K_A(r(t))x_c(t) + K_B(r(t))y(t), x_c(0) = 0, \\ u(t) = K_C(r(t))x_c(t), \end{cases} \tag{4.37}$$

where $x_c(t)$ is the state of the controller and $K_A(i)$, $K_B(i)$, and $K_C(i)$ are design parameters to be determined.

For simplicity in development, we assume that in the dynamics (4.1), the matrices $D_y(r(t))$ and $B_y(r(t))$ are always equals to zero for each mode.

Using (4.1) and (4.37), we get the following dynamics for the extended system:

$$\dot{\eta}(t) = \begin{bmatrix} A(r(t)) & B(r(t))K_C(r(t)) \\ K_B(r(t))C_y(r(t)) & K_A(r(t)) \end{bmatrix} \eta(t) + \begin{bmatrix} B_w(r(t)) \\ 0 \end{bmatrix} \omega(t)$$

$$= \tilde{A}(r(t))\eta(t) + \tilde{B}_w(r(t))\omega(t),$$

with

$$\eta(t) = \begin{bmatrix} x(t) \\ x_c(t) \end{bmatrix},$$

$$\tilde{A}(r(t)) = \begin{bmatrix} A(r(t)) & B(r(t))K_C(r(t)) \\ K_B(r(t))C_y(r(t)) & K_A(r(t)) \end{bmatrix},$$

$$\tilde{B}_w(r(t)) = \begin{bmatrix} B_w(r(t)) \\ 0 \end{bmatrix}.$$

For the controlled output, we have

$$z(t) = \begin{bmatrix} C_z(r(t)) & D_z(r(t))K_C(r(t)) \end{bmatrix} \eta(t) + B_z(r(t))\omega(t)$$
$$= \tilde{C}_z(r(t))\eta(t) + \tilde{B}_z(r(t))\omega(t).$$

Based on the previous results of this chapter, the extended system will be stochastically stable and guarantee the disturbance rejection of level γ, if there exists a set of symmetric and positive-definite matrices $P = (P(1), \cdots, P(N)) > 0$ such that the following holds for each $i \in \mathscr{S}$:

$$\tilde{J}(i) + \tilde{C}_z^\top(i)\tilde{C}_z(i)$$
$$+ \left[\tilde{C}_z^\top(i)\tilde{B}_z(i) + P(i)\tilde{B}_w(i)\right] \left[\gamma^2\mathbb{I} - \tilde{B}_z^\top(i)\tilde{B}_z(i)\right]^{-1}$$
$$\times \left[\tilde{B}_z^\top(i)\tilde{C}_z(i) + \tilde{B}_w^\top(i)P(i)\right] < 0, \qquad (4.38)$$

with $\tilde{J}(i) = \tilde{A}^\top(i)P(i) + P(i)\tilde{A}(i) + \sum_{j=1}^N \lambda_{ij}P(j)$.

Let $P(i)$, $i \in \mathscr{S}$ be defined by

$$P(i) = \begin{bmatrix} P_1(i) & P_2(i) \\ P_2^\top(i) & P_3(i) \end{bmatrix},$$

where $P_1(i) > 0$, $P_3(i) > 0$.

Let us now define the following matrices:

$$W(i) = \left[P_1(i) - P_2(i)P_3^{-1}(i)P_2^\top(i)\right]^{-1},$$
$$U(i) = \begin{bmatrix} W(i) & \mathbb{I} \\ W(i) & 0 \end{bmatrix},$$
$$V(i) = \begin{bmatrix} \mathbb{I} & 0 \\ 0 & -P_3^{-1}(i)P_2^\top(i) \end{bmatrix}.$$

Based on these definitions, we conclude that

$$V(i)U(i) = \begin{bmatrix} W(i) & \mathbb{I} \\ -P_3^{-1}(i)P_2^\top(i)W(i) & 0 \end{bmatrix}.$$

Pre- and post-multiply the-left hand side of (4.38) by $U^\top(i)V^\top(i)$ and $V(i)U(i)$, respectively, to get

$$U^\top(i)V^\top(i)\left[\tilde{J}(i)+\tilde{C}_z^\top(i)\tilde{C}_z(i)\right]V(i)U(i)$$

$$+U^\top(i)V^\top(i)\left[\tilde{C}_z^\top(i)\tilde{B}_z(i)+P(i)\tilde{B}_w(i)\right]$$

$$\times\left[\gamma^2\mathbb{I}-\tilde{B}_z^\top(i)\tilde{B}_z(i)\right]^{-1}\left[\tilde{B}_z^\top(i)\tilde{C}_z(i)+\tilde{B}_w^\top(i)P(i)\right]V(i)U(i)<0.$$

Now compute

$$U^\top(i)V^\top(i)P(i)\tilde{A}(i)V(i)U(i)$$

and

$$U^\top(i)V^\top(i)P(j)V(i)U(i)$$

in function of the system data. The other required terms are computed later. In fact, for the first term we have

$$U^\top(i)V^\top(i)P(i)\tilde{A}(i)V(i)U(i)$$

$$=\begin{bmatrix}W^\top(i) & -W^\top(i)P_2(i)P_3^{-1}(i)\\ \mathbb{I} & 0\end{bmatrix}\begin{bmatrix}P_1(i) & P_2(i)\\ P_2^\top(i) & P_3(i)\end{bmatrix}$$

$$\times\begin{bmatrix}A(i) & B(i)K_C(i)\\ K_B(i)C_y(i) & K_A(i)\end{bmatrix}\begin{bmatrix}W(i) & \mathbb{I}\\ -P_3^{-1}(i)P_2^\top(i)W(i) & 0\end{bmatrix}$$

$$=\begin{bmatrix}Z_1(i) & Z_2(i)\\ Z_3(i) & Z_4(i)\end{bmatrix}.$$

Performing the matrices multiplication we get

$$Z_1(i)=W^\top(i)P_1(i)A(i)W(i)-W^\top(i)P_2(i)K_B(i)C_y(i)W(i)$$
$$-W^\top(i)P_2(i)P_3^{-1}(i)P_2^\top(i)A(i)W(i)$$
$$-W^\top(i)P_1(i)B(i)K_C(i)P_3^{-1}(i)P_2^\top(i)W(i)$$
$$-W^\top(i)P_2(i)K_A(i)P_3^{-1}(i)P_2^\top(i)W(i)$$
$$+W^\top(i)P_2(i)P_3^{-1}(i)P_2^\top(i)B(i)K_C(i)P_3^{-1}(i)P_2^\top(i)W(i)$$
$$+W^\top(i)P_2(i)K_A(i)P_3^{-1}(i)P_2^\top(i)W(i)+W^\top(i)P_2(i)K_B(i)C_y(i)W(i),$$

$$Z_2(i)=W^\top(i)P_1(i)A(i)-W^\top(i)P_2(i)P_3^{-1}(i)P_2^\top(i)A(i)$$
$$-W^\top(i)P_2(i)K_B(i)C_y(i)+W^\top(i)P_2(i)K_B(i)C_y(i),$$

$$Z_3(i)=P_1(i)A(i)W(i)+P_2(i)K_B(i)C_y(i)W(i)$$

$$-P_1(i)B(i)K_C(i)P_3^{-1}(i)P_2^\top(i)W(i)$$
$$-P_2(i)K_A(i)P_3^{-1}(i)P_2^\top(i)W(i),$$

$$Z_4(i) = P_1(i)A(i) + P_2(i)K_B(i)C_y(i).$$

Using some basic algebraic manipulations and

$$W(i)\left[P_1(i) - P_2(i)P_3^{-1}(i)P_2^\top(i)\right] = \mathbb{I},$$

the previous elements $Z_1(i)$, $Z_2(i)$, $Z_3(i)$, and $Z_4(i)$ become

$$
\begin{aligned}
Z_1(i) &= W^\top(i)\left[P_1(i) - P_2(i)P_3^{-1}(i)P_2^\top(i)\right]A(i)W(i) \\
&\quad -W^\top(i)\left[P_1(i) - P_2(i)P_3^{-1}(i)P_2^\top(i)\right]B(i)K_C(i) \\
&\quad\qquad\qquad\qquad\qquad\qquad \times P_3^{-1}(i)P_2^\top(i)W(i) \\
&= A(i)W(i) - B(i)K_C(i)P_3^{-1}(i)P_2^\top(i)W(i),
\end{aligned}
$$

$$Z_2(i) = W^\top(i)\left[P_1(i) - P_2(i)P_3^{-1}(i)P_2^\top(i)\right]A(i) = A(i),$$

$$
\begin{aligned}
Z_3(i) &= P_1(i)A(i)W(i) + P_2(i)K_B(i)C_y(i)W(i) \\
&\quad -P_1(i)B(i)K_C(i)P_3^{-1}(i)P_2^\top(i)W(i) \\
&\quad -P_2(i)K_A(i)P_3^{-1}(i)P_2^\top(i)W(i),
\end{aligned}
$$

$$Z_4(i) = P_1(i)A(i) + P_2(i)K_B(i)C_y(i).$$

Using all these computations, we get

$$U^\top(i)V^\top(i)P(i)\tilde{A}(i)V(i)U(i)$$

$$
= \left[
\begin{array}{cc}
\begin{bmatrix} A(i)W(i) \\ -B(i)K_C(i) \\ \times P_3^{-1}(i)P_2^\top(i)W(i) \end{bmatrix} & A(i) \\[2em]
\begin{bmatrix} P_1(i)A(i)W(i) \\ +P_2(i)K_B(i)C_y(i)W(i) \\ -P_1(i)B(i)K_C(i) \\ \times P_3^{-1}(i)P_2^\top(i)W(i) \\ -P_2(i)K_A(i) \\ \times P_3^{-1}(i)P_2^\top(i)W(i) \end{bmatrix} & \begin{bmatrix} P_1(i)A(i) \\ +P_2(i)K_B(i)C_y(i) \end{bmatrix}
\end{array}
\right].
$$

Using the fact that $U^\top(i)V^\top(i)\tilde{A}^\top(i)P(i)V(i)U(i)$ is the transpose of $U^\top(i)V^\top(i)P(i)\tilde{A}(i)V(i)U(i)$ we get

$$U^{\top}(i)V^{\top}(i)\tilde{A}^{\top}(i)P(i)V(i)U(i)$$

$$= \begin{bmatrix} \begin{bmatrix} W^{\top}(i)A^{\top}(i) \\ -W^{\top}(i)P_2(i) \\ \times P_3^{-1}(i)K_C^{\top}(i)B^{\top}(i) \end{bmatrix} & \begin{bmatrix} W^{\top}(i)A^{\top}(i)P_1(i) \\ +W^{\top}(i)C_y^{\top}(i) \\ \times K_B^{\top}(i)P_2^{\top}(i) \\ -W^{\top}(i)P_2(i)P_3^{-1}(i) \\ \times K_C^{\top}(i)B^{\top}(i)P_1(i) \\ -W^{\top}(i)P_2(i) \\ \times P_3^{-1}(i)K_A^{\top}(i)P_2^{\top}(i) \end{bmatrix} \\ A^{\top}(i) & \begin{bmatrix} A^{\top}(i)P_1(i) \\ +C_y^{\top}(i)K_B^{\top}(i)P_2^{\top}(i) \end{bmatrix} \end{bmatrix}.$$

For the term $U^{\top}(i)V^{\top}(i)P(j)V(i)U(i)$, we have

$$\begin{bmatrix} W^{\top}(i) & -W^{\top}(i)P_2(i)P_3^{-1}(i) \\ \mathbb{I} & 0 \end{bmatrix} \begin{bmatrix} P_1(j) & P_2(j) \\ P_2^{\top}(j) & P_3(j) \end{bmatrix}$$
$$\begin{bmatrix} W(i) & \mathbb{I} \\ -P_3^{-1}(i)P_2^{\top}(i)W(i) & 0 \end{bmatrix}$$

$$= \begin{bmatrix} \begin{bmatrix} W^{\top}(i)P_1(j)W(i) \\ -W^{\top}(i)P_2(i) \\ \times P_3^{-1}(i)P_2^{\top}(j)W(i) \\ -W^{\top}(i)P_2(j) \\ \times P_3^{-1}(i)P_2^{\top}(i)W(i) \\ +W^{\top}(i)P_2(i)P_3^{-1}(i) \\ \times P_3(j)P_3^{-1}(i)P_2^{\top}(i)W(i) \end{bmatrix} & \begin{bmatrix} W^{\top}(i)P_1(j) \\ -W^{\top}(i)P_2(i) \\ \times P_3^{-1}(i)P_2^{\top}(j) \end{bmatrix} \\ P_1(j)W(i) - P_2(j)P_3^{-1}(i)P_2^{\top}(i)W(i) & P_1(j) \end{bmatrix},$$

which can be rewritten as follows using the fact that $W^{-1}(j) = P_1(j) - P_2(j)P_3^{-1}(j)P_2^{\top}(j)$ and some algebraic manipulations:

$$\begin{bmatrix} \begin{bmatrix} W^{\top}(i)W^{-1}(j)W(i) \\ +W^{\top}(i)\left[P_2(i)P_3^{-1}(i)P_3(j) - P_2^{\top}(j)\right]P_3^{-1}(j) \\ \times \left[P_2(i)P_3^{-1}(i)P_3(j) - P_2^{\top}(j)\right]^{\top}W(i) \end{bmatrix} & \star \\ \left[P_1(j) - P_2(j)P_3^{-1}(i)P_2^{\top}(i)\right]W(i) & P_1(j) \end{bmatrix}.$$

For the term $U^{\top}(i)V^{\top}(i)\tilde{C}_z^{\top}(i)\tilde{C}_z(i)V(i)U(i)$, we have

$$U^{\top}(i)V^{\top}(i)\tilde{C}_z^{\top}(i)\tilde{C}_z(i)V(i)U(i)$$
$$= \begin{bmatrix} W^{\top}(i) & -W^{\top}(i)P_2(i)P_3^{-1}(i) \\ \mathbb{I} & 0 \end{bmatrix} \begin{bmatrix} C_z^{\top}(i) \\ K_C^{\top}(i)D_z^{\top}(i) \end{bmatrix}$$
$$\times \begin{bmatrix} C_z(i) & D_z(i)K_C(i) \end{bmatrix} \begin{bmatrix} W(i) & \mathbb{I} \\ -P_3^{-1}(i)P_2^{\top}(i)W(i) & 0 \end{bmatrix}$$
$$= \begin{bmatrix} W^{\top}(i) & -W^{\top}(i)P_2(i)P_3^{-1}(i) \\ \mathbb{I} & 0 \end{bmatrix}$$

$$\begin{bmatrix} C_z^\top(i)C_z(i) & C_z^\top(i)D_z(i)K_C(i) \\ K_C^\top(i)D_z^\top(i)C_z(i) & K_C^\top(i)D_z^\top(i)D_z(i)K_C(i) \end{bmatrix}$$

$$\begin{bmatrix} W(i) & \mathbb{I} \\ -P_3^{-1}(i)P_2^\top(i)W(i) & 0 \end{bmatrix}$$

$$= \begin{bmatrix} \begin{bmatrix} W^\top(i)C_z^\top(i)C_z(i)W(i) \\ -W^\top(i)C_z^\top(i)D_z(i)K_C(i)P_3^{-1}(i)P_2^\top(i)W(i) \\ -W^\top(i)P_2(i)P_3^{-1}(i)K_C^\top(i)D_z^\top(i)C_z(i)W(i) \\ +W^\top(i)P_2(i)P_3^{-1}(i)K_C^\top(i)D_z^\top(i) \\ \times D_z(i)K_C(i)P_3^{-1}(i)P_2^\top(i)W(i) \end{bmatrix} & \begin{bmatrix} C_z^\top(i)C_z(i)W(i) - C_z^\top(i)D_z(i)K_C(i) \\ \times P_3^{-1}(i)P_2^\top(i)W(i) \end{bmatrix} \\[4mm] \begin{bmatrix} W^\top(i)C_z^\top(i)C_z(i) \\ -W^\top(i)P_2(i)P_3^{-1}(i)K_C^\top(i) \\ \times D_z^\top(i)C_z(i) \end{bmatrix} & C_z^\top(i)C_z(i) \end{bmatrix}.$$

For the term $U^\top(i)V^\top(i)\tilde{C}_z^\top(i)B_z(i)$, we have

$$U^\top(i)V^\top(i)\tilde{C}_z^\top(i)B_z(i)$$

$$= \begin{bmatrix} W^\top(i) & -W^\top(i)P_2(i)P_3^{-1}(i) \\ \mathbb{I} & 0 \end{bmatrix} \begin{bmatrix} C_z^\top(i) \\ K_C^\top(i)D_z^\top(i) \end{bmatrix} B_z(i)$$

$$= \begin{bmatrix} \begin{bmatrix} W^\top(i)C_z^\top(i)B_z(i) \\ -W^\top(i)P_2(i)P_3^{-1}(i) \\ \times K_C^\top(i)D_z^\top(i)B_z(i) \end{bmatrix} \\ C_z^\top(i)B_z(i) \end{bmatrix}.$$

For the term $U^\top(i)V^\top(i)P(i)\tilde{B}_w(i)$, we have

$$U^\top(i)V^\top(i)P(i)\tilde{B}_w(i)$$

$$= \begin{bmatrix} W^\top(i) & -W^\top(i)P_2(i)P_3^{-1}(i) \\ \mathbb{I} & 0 \end{bmatrix} \begin{bmatrix} P_1(i) & P_2(i) \\ P_2^\top(i) & P_3(i) \end{bmatrix} \begin{bmatrix} B_w(i) \\ 0 \end{bmatrix}$$

$$= \begin{bmatrix} \begin{bmatrix} W^\top(i)P_1(i)B_w(i) \\ -W^\top(i)P_2(i)P_3^{-1}(i) \\ \times P_2^\top(i)B_w(i) \end{bmatrix} \\ P_1(i)B_w(i) \end{bmatrix}$$

$$= \begin{bmatrix} B_w(i) \\ P_1(i)B_w(i) \end{bmatrix}.$$

Using all the previous computations, the stability condition for the closed-loop system becomes

$$\begin{bmatrix} \widehat{\mathcal{M}_1(i)} & \mathcal{M}_2(i) \\ \mathcal{M}_2^\top(i) & \mathcal{M}_3(i) \end{bmatrix} + \begin{bmatrix} \mathcal{N}_1(i) \\ \mathcal{N}_2(i) \end{bmatrix} \mathcal{N}^{-1}(i) \begin{bmatrix} \mathcal{N}_1^\top(i) & \mathcal{N}_2^\top(i) \end{bmatrix} < 0,$$

that is,

$$\begin{bmatrix} \widehat{\mathcal{M}_1(i)} + \mathcal{N}_1(i)\mathcal{N}^{-1}(i)\mathcal{N}_1^\top(i) & \mathcal{M}_2(i) + \mathcal{N}_1(i)\mathcal{N}^{-1}(i)\mathcal{N}_2^\top(i) \\ \mathcal{M}_2^\top(i) + \mathcal{N}_2(i)\mathcal{N}^{-1}(i)\mathcal{N}_1^\top(i) & \mathcal{M}_3(i) + \mathcal{N}_2(i)\mathcal{N}^{-1}(i)\mathcal{N}_2^\top(i) \end{bmatrix} < 0,$$

with

$$\begin{aligned}
\widehat{\mathcal{M}_1(i)} =&\ \mathcal{M}_1(i) + W^\top(i)C_z^\top(i)C_z(i)W(i) \\
&+ W^\top(i)P_2(i)P_3^{-1}(i)K_C^\top(i)D_z^\top(i)D_z(i) \\
&\times K_C(i)P_3^{-1}(i)P_2^\top(i)W(i) \\
&+ \sum_{j=1}^N \lambda_{ij}W^\top(i)\left[P_2(i)P_3^{-1}(i)P_3(j) - P_2^\top(j)\right]P_3^{-1}(j) \\
&\times \left[P_2(i)P_3^{-1}(i)P_3(j) - P_2^\top(j)\right]^\top W(i), \\
\mathcal{M}_1(i) =&\ A(i)W(i) + W^\top(i)A^\top(i) \\
&- B(i)K_C(i)P_3^{-1}(i)P_2^\top(i)W(i) \\
&- W^\top(i)P_2(i)P_3^{-1}(i)K_C^\top(i)B^\top(i) \\
&+ \sum_{j=1}^N \lambda_{ij}W^\top(i)W^{-1}(j)W(i) \\
&- W^\top(i)C_z^\top(i)D_z(i)K_C(i)P_3^{-1}(i)P_2^\top(i)W(i) \\
&- W^\top(i)P_2(i)P_3^{-1}(i)K_C^\top(i)D_z^\top(i)C_z(i)W(i), \\
\mathcal{M}_2(i) =&\ A(i) + W^\top(i)A^\top(i)P_1(i) + W^\top(i)C_y^\top(i)K_B^\top(i)P_2^\top(i) \\
&- W^\top(i)P_2(i)P_3^{-1}(i)K_C^\top(i)B^\top(i)P_1(i) \\
&- W^\top(i)P_2(i)P_3^{-1}(i)K_A^\top(i)P_2^\top(i) \\
&+ \sum_{j=1}^N \lambda_{ij}W^\top(i)\left[P_1(j) - P_2(j)P_3^{-1}(i)P_2^\top(i)\right]^\top \\
&+ W^\top(i)C_z^\top(i)C_z(i) \\
&- W^\top(i)P_2(i)P_3^{-1}(i)K_C^\top(i)D_z^\top(i)C_z(i), \\
\mathcal{M}_3(i) =&\ P_1(i)A(i) + P_2(i)K_B(i)C_y(i) + A^\top(i)P_1(i) \\
&+ C_y^\top(i)K_B^\top(i)P_2^\top(i) + \sum_{j=1}^N \lambda_{ij}P_1(j) + C_z^\top(i)C_z(i), \\
\mathcal{N}(i) =&\ \gamma^2\mathbb{I} - B_z^\top(i)B_z(i), \\
\mathcal{N}_1(i) =&\ W^\top(i)C_z^\top(i)B_z(i) \\
&- W^\top(i)P_2(i)P_3^{-1}(i)K_C^\top(i)D_z^\top(i)B_z(i) + B_w(i), \\
\mathcal{N}_2(i) =&\ C_z^\top(i)B_z(i) + P_1(i)B_w(i).
\end{aligned}$$

Since

$$\sum_{j=1}^{N} \lambda_{ij} W^{\top}(i) \left[P_2(i) P_3^{-1}(i) P_3(j) - P_2^{\top}(j) \right] P_3^{-1}(j)$$

$$\times \left[P_2(i) P_3^{-1}(i) P_3(j) - P_2^{\top}(j) \right]^{\top} W(i)$$

$$+ W^{\top}(i) P_2(i) P_3^{-1}(i) K_C^{\top}(i) D_z^{\top}(i) D_z(i) K_C(i) P_3^{-1}(i) P_2^{\top}(i) W(i) \geq 0,$$

we get the following equivalent condition:

$$\begin{bmatrix} \mathcal{M}_1(i) + \mathcal{N}_1(i)\mathcal{N}^{-1}(i)\mathcal{N}_1^{\top}(i) & \mathcal{M}_2(i) + \mathcal{N}_1(i)\mathcal{N}^{-1}(i)\mathcal{N}_2^{\top}(i) \\ \mathcal{M}_2^{\top}(i) + \mathcal{N}_2(i)\mathcal{N}^{-1}(i)\mathcal{N}_1^{\top}(i) & \mathcal{M}_3(i) + \mathcal{N}_2(i)\mathcal{N}^{-1}(i)\mathcal{N}_2^{\top}(i) \end{bmatrix} < 0.$$

Letting

$$P(i) = \begin{bmatrix} X(i) & Y^{-1}(i) - X(i) \\ Y^{-1}(i) - X(i) & X(i) - Y^{-1}(i) \end{bmatrix},$$

that is,

$$P_1(i) = X(i),$$
$$P_2(i) = Y^{-1}(i) - X(i),$$
$$P_3(i) = X(i) - Y^{-1}(i),$$

implies $W(i) = \left[P_1(i) - P_2(i) P_3^{-1}(i) P_2^{\top}(i) \right]^{-1} = Y(i)$ and $P_3^{-1}(i) P_2^{\top}(i) = -\mathbb{I}$. If we define $\mathcal{K}_B(i)$ and $\mathcal{K}_C(i)$ by

$$\mathcal{K}_B(i) = P_2(i) K_B(i) = \left[Y^{-1}(i) - X(i) \right] K_B(i),$$
$$\mathcal{K}_C(i) = -K_C(i) P_3^{-1}(i) P_2^{\top}(i) W(i) = K_C(i) Y(i),$$

and we use all the previous development, we get

$$\begin{bmatrix} \mathcal{M}_1^{\star}(i) + \mathcal{N}_1(i)\mathcal{N}^{-1}(i)\mathcal{N}_1^{\top}(i) & \mathcal{M}_2(i) + \mathcal{N}_1(i)\mathcal{N}^{-1}(i)\mathcal{N}_2^{\top}(i) \\ \mathcal{M}_2^{\top}(i) + \mathcal{N}_2(i)\mathcal{N}^{-1}(i)\mathcal{N}_1^{\top}(i) & \mathcal{M}_3(i) + \mathcal{N}_2(i)\mathcal{N}^{-1}(i)\mathcal{N}_2^{\top}(i) \end{bmatrix} < 0,$$

with

$$\mathcal{M}_1^{\star}(i) = A(i)Y(i) + Y^{\top}(i)A^{\top}(i) + B(i)\mathcal{K}_C(i) + \mathcal{K}_C^{\top}(i)B^{\top}(i)$$

$$+ \sum_{j=1}^{N} \lambda_{ij} Y^{\top}(i) Y^{-1}(j) Y(i) + Y^{\top}(i) C_z^{\top}(i) D_z(i) \mathcal{K}_C(i)$$

$$+ \mathcal{K}_C^{\top}(i) D_z^{\top}(i) C_z(i) Y(i) + Y^{\top}(i) C_z^{\top}(i) C_z(i) Y(i),$$

$$\mathcal{M}_2(i) = A(i) + Y^{\top}(i) A^{\top}(i) X(i) + Y^{\top}(i) C_y^{\top}(i) \mathcal{K}_B^{\top}(i)$$

$$+ \mathcal{K}_C^{\top}(i) B^{\top}(i) X(i) - Y^{\top}(i) K_A^{\top}(i) \left[X(i) - Y^{-1}(i) \right]^{\top}$$

$$+ \sum_{j=1}^{N} \lambda_{ij} Y^{\top}(i) Y^{-1}(j) + Y^{\top}(i) C_z^{\top}(i) C_z(i) + \mathcal{K}_C^{\top}(i) D_z^{\top}(i) C_z(i),$$

$$\mathcal{M}_3(i) = X(i)A(i) + \mathcal{K}_B(i)C_y(i) + A^\top(i)X(i) + C_z^\top(i)C_z(i)$$

$$+C_y^\top(i)\mathcal{K}_B^\top(i) + \sum_{j=1}^N \lambda_{ij}X(j),$$

$$\mathcal{N}(i) = \gamma^2\mathbb{I} - B_z^\top(i)B_z(i),$$

$$\mathcal{N}_1(i) = Y^\top(i)C_z^\top(i)B_z(i) + \mathcal{K}_C^\top(i)D_z^\top(i)B_z(i) + B_w(i),$$

$$\mathcal{N}_2(i) = C_z^\top(i)B_z(i) + X(i)B_w(i).$$

Using the following expressions for the controller parameters:

$$\begin{cases} K_A(i) = \left[X(i) - Y^{-1}(i)\right]^{-1}\left[A^\top(i) + X(i)A(i)Y(i)\right.\\ \qquad\qquad +X(i)B(i)\mathcal{K}_C(i) + \mathcal{K}_B(i)C_y(i)Y(i)\\ \qquad\qquad +C_z^\top(i)C_z(i)Y(i) + C_z^\top(i)D_z(i)\mathcal{K}_C(i)\\ \qquad\qquad +\mathcal{N}_2(i)\mathcal{N}^{-1}(i)\mathcal{N}_1^\top(i)\\ \qquad\qquad \left.+\sum_{j=1}^N \lambda_{ij}Y^{-1}(j)Y(i)\right]Y^{-1}(i),\\ K_B(i) = \left[Y^{-1}(i) - X(i)\right]^{-1}\mathcal{K}_B(i),\\ K_C(i) = \mathcal{K}_C(i)Y^{-1}(i), \end{cases}$$

we have $\mathcal{M}_2(i) = 0$, which implies that the stability condition is equivalent to the following conditions:

$$\mathcal{M}_1(i) + Y^\top(i)C_z^\top(i)C_z(i)Y(i) + \mathcal{N}_1(i)\mathcal{N}^{-1}(i)\mathcal{N}_1^\top(i) < 0,$$
$$\mathcal{M}_3(i) + \mathcal{N}_2(i)\mathcal{N}^{-1}(i)\mathcal{N}_2^\top(i) < 0,$$

which gives

$$\left[A(i) + \mathcal{K}_C^\top(i)D_z^\top(i)C_z(i)\right]Y(i) + B(i)\mathcal{K}_C(i) + Y^\top(i)C_z^\top(i)C_z(i)Y(i)$$

$$+Y^\top(i)\left[A^\top(i) + C_z^\top(i)D_z(i)\mathcal{K}_C(i)\right] + \mathcal{K}_C^\top(i)B^\top(i)$$

$$+\sum_{j=1}^N \lambda_{ij}Y^\top(i)Y^{-1}(j)Y(i) + \mathcal{N}_1(i)\mathcal{N}^{-1}(i)\mathcal{N}_1^\top(i) < 0,$$

and

$$X(i)A(i) + \mathcal{K}_B(i)C_y(i) + A^\top(i)X(i) + C_y^\top(i)\mathcal{K}_B^\top(i)$$

$$+\sum_{j=1}^N \lambda_{ij}X(j) + C_z^\top(i)C_z(i) + \mathcal{N}_2(i)\mathcal{N}^{-1}(i)\mathcal{N}_2^\top(i) < 0.$$

Notice that

$$\sum_{j=1}^N \lambda_{ij}Y^\top(i)Y^{-1}(j)Y(i) = \lambda_{ii}Y(i) + \mathcal{S}_i(Y)\mathcal{Y}_i^{-1}(Y)\mathcal{S}_i^\top(Y),$$

with $\mathcal{S}_i(Y)$ and $\mathcal{Y}_i(Y)$ as defined before.

Using this, the previous stability conditions become

$$\left[\begin{array}{cccc} \begin{array}{c} \mathcal{K}_C^\top(i)D_z^\top(i)C_z(i)Y(i) \\ +Y^\top(i)C_z^\top(i)D_z(i)\mathcal{K}_C(i) \\ +A(i)Y(i)+Y^\top(i)A^\top(i) \\ +B(i)\mathcal{K}_C(i) \\ +\mathcal{K}_C^\top(i)B^\top(i)+\lambda_{ii}Y(i) \end{array} & \mathcal{N}_1(i) & Y^\top(i)C_z^\top(i) & \mathcal{S}_i(Y) \\ \mathcal{N}_1^\top(i) & -\mathcal{N}(i) & 0 & 0 \\ C_z(i)Y(i) & 0 & -\mathbb{I} & 0 \\ \mathcal{S}_i^\top(Y) & 0 & 0 & -\mathcal{Y}_i(Y) \end{array}\right] < 0,$$

$$\left[\begin{array}{cc} \begin{array}{c} X(i)A(i)+\mathcal{K}_B(i)C_y(i) \\ +A^\top(i)X(i)+C_z^\top(i)C_z(i) \\ +C_y^\top(i)\mathcal{K}_B^\top(i)+\sum_{j=1}^N \lambda_{ij}X(j) \end{array} & \mathcal{N}_2(i) \\ \mathcal{N}_2^\top(i) & -\mathcal{N}(i) \end{array}\right] < 0.$$

Notice that if

$$\left[\begin{array}{ccc} \mathcal{K}_C^\top(i)D_z^\top(i)C_z(i)Y(i) & 0 & 0 \\ 0 & 0 & 0 \\ 0 & 0 & 0 \end{array}\right] + \left[\begin{array}{ccc} Y^\top(i)C_z^\top(i)D_z(i)\mathcal{K}_C(i) & 0 & 0 \\ 0 & 0 & 0 \\ 0 & 0 & 0 \end{array}\right]$$

$$\leq \left[\begin{array}{ccc} \mathcal{K}_C^\top(i)D_z^\top(i)D_z(i)\mathcal{K}_C(i) & 0 & 0 \\ 0 & 0 & 0 \\ 0 & 0 & 0 \end{array}\right] + \left[\begin{array}{ccc} Y^\top(i)C_z^\top(i)C_z(i)Y(i) & 0 & 0 \\ 0 & 0 & 0 \\ 0 & 0 & 0 \end{array}\right],$$

we get

$$\left[\begin{array}{cccccc} J_X(i) & Y^\top(i)C_z^\top(i) & \mathcal{K}_C^\top(i)D_z^\top(i) & Y^\top(i)C_z^\top(i) & \mathcal{N}_1(i) & \mathcal{S}_i(Y) \\ C_z(i)Y(i) & -\mathbb{I} & 0 & 0 & 0 & 0 \\ D_z(i)\mathcal{K}_C(i) & 0 & -\mathbb{I} & 0 & 0 & 0 \\ C_z(i)Y(i) & 0 & 0 & -\mathbb{I} & 0 & 0 \\ \mathcal{N}_1^\top(i) & 0 & 0 & 0 & -\mathcal{N}(i) & 0 \\ \mathcal{S}_i(Y) & 0 & 0 & 0 & 0 & -\mathcal{Y}_i(Y) \end{array}\right] < 0,$$

$$\left[\begin{array}{cc} \begin{array}{c} X(i)A(i)+\mathcal{K}_B(i)C_y(i) \\ +A^\top(i)X(i)+C_z^\top(i)C_z(i) \\ +C_y^\top(i)\mathcal{K}_B^\top(i)+\sum_{j=1}^N \lambda_{ij}X(j) \end{array} & \mathcal{N}_2(i) \\ \mathcal{N}_2^\top(i) & -\mathcal{N}(i) \end{array}\right] < 0,$$

with $J_X(i) = A(i)Y(i)+Y^\top(i)A^\top(i)+B(i)\mathcal{K}_C(i)+\mathcal{K}_C^\top(i)B^\top(i)+\lambda_{ii}Y(i)$.

Finally notice that

$$U^\top(i)V^\top(i)P(i)V(i)U(i) = \begin{bmatrix} Y(i) & \mathbb{I} \\ \mathbb{I} & X(i) \end{bmatrix}.$$

The results of the previous development are summarized by the following theorem.

Theorem 54. *System (4.1) is stable if and only if for every $i \in \mathscr{S}$ the following LMIs are feasible for some symmetric and positive-definite matrices $X = (X(1), \cdots, X(N)) > 0$, and $Y = (Y(1), \cdots, Y(N)) > 0$, and matrices $\mathcal{K}_A = (\mathcal{K}_A(1), \cdots, \mathcal{K}_A(N))$, $\mathcal{K}_B = (\mathcal{K}_B(1), \cdots, \mathcal{K}_B(N))$, and $\mathcal{K}_C = (\mathcal{K}_C(1), \cdots, \mathcal{K}_C(N))$:*

$$\begin{bmatrix} J_X(i) & Y^\top(i)C_z^\top(i) & \mathcal{K}_C^\top(i)D_z^\top(i) & Y^\top(i)C_z^\top(i) \\ C_z(i)Y(i) & -\mathbb{I} & 0 & 0 \\ D_z(i)\mathcal{K}_C(i) & 0 & -\mathbb{I} & 0 \\ C_z(i)Y(i) & 0 & 0 & -\mathbb{I} \\ \mathcal{N}_1^\top(i) & 0 & 0 & 0 \\ \mathcal{S}_i(Y) & 0 & 0 & 0 \end{bmatrix}$$

$$\begin{matrix} \mathcal{N}_1(i) & \mathcal{S}_i(Y) \\ 0 & 0 \\ 0 & 0 \\ 0 & 0 \\ -\mathcal{N}(i) & 0 \\ 0 & -\mathcal{Y}_i(Y) \end{matrix} \Bigg] < 0, \qquad (4.39)$$

$$\left[\begin{bmatrix} X(i)A(i) + \mathcal{K}_B(i)C_y(i) \\ +A^\top(i)X(i) + C_z^\top(i)C_z(i) \\ +C_y^\top(i)\mathcal{K}_B^\top(i) + \sum_{j=1}^N \lambda_{ij}X(j) \end{bmatrix} \quad \mathcal{N}_2(i) \\ \mathcal{N}_2^\top(i) \qquad\qquad -\mathcal{N}(i) \right] < 0, \qquad (4.40)$$

$$\begin{bmatrix} Y(i) & \mathbb{I} \\ I & X(i) \end{bmatrix} > 0, \qquad (4.41)$$

with

$$J_X(i) = A(i)Y(i) + Y^\top(i)A^\top(i) + B(i)\mathcal{K}_C(i) + \mathcal{K}_C^\top(i)B^\top(i) + \lambda_{ii}Y(i)$$
$$\mathcal{S}_i(Y) = \left[\sqrt{\lambda_{i1}}Y(i), \cdots, \sqrt{\lambda_{ii-1}}Y(i), \sqrt{\lambda_{ii+1}}Y(i), \right.$$
$$\left. \cdots, \sqrt{\lambda_{iN}}Y(i) \right],$$
$$\mathcal{Y}_i(Y) = \mathrm{diag}\left[Y(1), \cdots, Y(i-1), Y(i+1), \cdots, Y(N) \right].$$

Furthermore the dynamic output feedback controller is given by

$$\begin{cases} K_A(i) = \left[X(i) - Y^{-1}(i)\right]^{-1}\left[A^\top(i) + X(i)A(i)Y(i)\right. \\ \quad +X(i)B(i)\mathcal{K}_C(i) + \mathcal{K}_B(i)C_y(i)Y(i) \\ \quad +\mathcal{N}_2(i)\mathcal{N}^{-1}(i)\mathcal{N}_1^\top(i) \\ \quad \left. + \sum_{j=1}^N \lambda_{ij}Y^{-1}(j)Y(i)\right]Y^{-1}(i), \\ K_B(i) = \left[Y^{-1}(i) - X(i)\right]^{-1}\mathcal{K}_B(i), \\ K_C(i) = \mathcal{K}_C(i)Y^{-1}(i). \end{cases} \qquad (4.42)$$

Example 57. To illustrate the results of this theorem, let us consider the two-mode system of Example 45 with the following extra data:

- mode #1:

$$A(1) = \begin{bmatrix} -0.5 & 1.0 \\ 0.3 & -2.5 \end{bmatrix}, \quad B(1) = \begin{bmatrix} 1.0 & 0.0 \\ 0.0 & 1.0 \end{bmatrix},$$

$$B_w(1) = \begin{bmatrix} 1.0 & 0.0 \\ 0.0 & 1.0 \end{bmatrix}, \quad C_z(1) = \begin{bmatrix} 1.0 & 0.0 \\ 0.0 & 1.0 \end{bmatrix},$$

$$B_z(1) = \begin{bmatrix} 1.0 & 0.0 \\ 0.0 & 1.0 \end{bmatrix}, \quad D_z(1) = \begin{bmatrix} 1.0 & 0.0 \\ 0.0 & 1.0 \end{bmatrix},$$

$$C_y(1) = \begin{bmatrix} 1.0 & 0.0 \\ 0.0 & 1.0 \end{bmatrix},$$

- mode #2:

$$A(2) = \begin{bmatrix} -1.0 & 0.1 \\ 0.2 & -2.0 \end{bmatrix}, \quad B(2) = \begin{bmatrix} 1.0 & 0.0 \\ 0.0 & 1.0 \end{bmatrix},$$

$$B_w(2) = \begin{bmatrix} 1.0 & 0.0 \\ 0.0 & 1.0 \end{bmatrix}, \quad C_z(2) = \begin{bmatrix} 1.0 & 0.0 \\ 0.0 & 1.0 \end{bmatrix},$$

$$B_z(2) = \begin{bmatrix} 1.0 & 0.0 \\ 0.0 & 1.0 \end{bmatrix}, \quad D_z(2) = \begin{bmatrix} 1.0 & 0.0 \\ 0.0 & 1.0 \end{bmatrix},$$

$$C_y(2) = \begin{bmatrix} 1.0 & 0.0 \\ 0.0 & 1.0 \end{bmatrix}.$$

Solving LMIs (4.39)–(4.41), we get

$$X(1) = \begin{bmatrix} 131.7769 & 0.4868 \\ 0.4868 & 131.1781 \end{bmatrix}, \quad X(2) = \begin{bmatrix} 131.7924 & 0.3609 \\ 0.3609 & 131.1062 \end{bmatrix},$$

$$Y(1) = \begin{bmatrix} 0.4843 & -1.1468 \\ -1.1468 & 2.7880 \end{bmatrix}, \quad Y(2) = \begin{bmatrix} 0.2461 & -0.6813 \\ -0.6813 & 2.0074 \end{bmatrix},$$

$$\mathcal{K}_B(1) = 10^3 \cdot \begin{bmatrix} -8.4260 & -0.1195 \\ -0.1195 & -8.1164 \end{bmatrix},$$

$$\mathcal{K}_B(2) = 10^3 \cdot \begin{bmatrix} -8.3574 & -0.0445 \\ -0.0445 & -8.1763 \end{bmatrix},$$

$$\mathcal{K}_C(1) = \begin{bmatrix} -0.5271 & -1.1465 \\ -1.1468 & 1.7772 \end{bmatrix}, \qquad \mathcal{K}_C(2) = \begin{bmatrix} -0.7652 & -0.6807 \\ -0.6808 & 0.9941 \end{bmatrix}.$$

The controller gains are given by

$$K_A(1) = 10^4 \cdot \begin{bmatrix} -1.2755 & -0.3518 \\ -0.4526 & -0.5948 \end{bmatrix}, \quad K_B(1) = \begin{bmatrix} 192.1449 & 52.9780 \\ 53.3990 & 83.5483 \end{bmatrix},$$

$$K_C(1) = \begin{bmatrix} -78.9255 & -32.8749 \\ -32.8544 & -12.8762 \end{bmatrix}, \quad K_A(2) = 10^3 \cdot \begin{bmatrix} -9.2179 & -1.6985 \\ -2.2843 & -4.9327 \end{bmatrix},$$

$$K_B(2) = \begin{bmatrix} 138.2775 & 25.4226 \\ 25.6205 & 71.1886 \end{bmatrix}, \quad K_C(2) = 10^4 \cdot \begin{bmatrix} -66.9426 & -23.0582 \\ -23.0770 & -7.3367 \end{bmatrix}.$$

Based on the results of this theorem, we conclude that the system is stochastically stable under the designed controller and guarantees the disturbance rejection with a level equal to $\gamma = 1.7354$.

Let us now consider the case when the system is subject to norm-bounded uncertainties and see how to design an output feedback controller that robustly stochastically stabilizes the system and simultaneously guarantees the disturbance rejection of level γ. Let us plug the controller expression in the system dynamics and combining the two dynamics, we get the following:

$$\begin{bmatrix} \dot{x}(t) \\ \dot{x}_c(t) \end{bmatrix} = \begin{bmatrix} A(r(t),t) & B(r(t),t)K_C(r(t)) \\ K_B(r(t))C_y(r(t),t) & K_A(r(t)) \end{bmatrix} \begin{bmatrix} x(t) \\ x_c(t) \end{bmatrix}$$
$$+ \begin{bmatrix} B_w(r(t)) \\ 0 \end{bmatrix} w(t)$$
$$= \tilde{A}(r(t),t)\eta(t) + \tilde{B}_w(r(t))\omega(t),$$

with

$$\eta(t) = \begin{bmatrix} x(t) \\ x_c(t) \end{bmatrix},$$

$$\tilde{A}(r(t),t) = \begin{bmatrix} A(r(t),t) & B(r(t),t)K_C(r(t)) \\ K_B(r(t))C_y(r(t),t) & K_A(r(t)) \end{bmatrix},$$

$$\tilde{B}_w(r(t)) = \begin{bmatrix} B_w(r(t)) \\ 0 \end{bmatrix}.$$

For the controlled output, we have

$$z(t) = C_z(r(t),t)x(t) + D_z(r(t),t)K_C(r(t))x_c(t) + B_z(r(t))\omega(t)$$
$$= \left[C_z(r(t),t) \; D_z(r(t),t)K_C(r(t)) \right] \eta(t) + B_z(r(t))\omega(t)$$
$$= \tilde{C}_z(r(t),t)\eta(t) + \tilde{B}_z(r(t))\omega(t),$$

with

$$\tilde{C}_z(r(t),t) = \left[C_z(r(t),t) \; D_z(r(t),t)K_C(r(t)) \right],$$

$$\tilde{B}_z(r(t)) = B_z(r(t)).$$

Notice that the matrices $\tilde{A}(r(t), t)$ and $\tilde{C}_z(r(t), t)$ can be rewritten as follows:

$$\tilde{A}(r(t), t) = \tilde{A}(r(t)) + \widetilde{\Delta A}(r(t), t) + \widetilde{\Delta B}(r(t), t) + \widetilde{\Delta C_y}(r(t), t),$$
$$\tilde{C}_z(r(t), t) = \tilde{C}_z(r(t)) + \widetilde{\Delta C_{C_z}}(r(t), t) + \widetilde{\Delta D_{D_z}}(r(t), t),$$

with

$$\tilde{A}(r(t)) = \begin{bmatrix} A(r(t)) & B(r(t))K_C(r(t)) \\ K_B(r(t))C_y(r(t)) & K_A(r(t)) \end{bmatrix},$$

$$\tilde{C}_z(r(t)) = \begin{bmatrix} C_z(r(t)) & D_z(r(t))K_C(r(t)) \end{bmatrix},$$

$$\widetilde{\Delta A}(r(t), t) = \begin{bmatrix} D_A(r(t))F_A(r(t), t)E_A(r(t)) & 0 \\ 0 & 0 \end{bmatrix}$$
$$= \begin{bmatrix} D_A(r(t)) & 0 \\ 0 & 0 \end{bmatrix} \begin{bmatrix} F_A(r(t), t) & 0 \\ 0 & 0 \end{bmatrix} \begin{bmatrix} E_A(r(t)) & 0 \\ 0 & 0 \end{bmatrix}$$
$$= \tilde{D}_A(r(t))\tilde{F}_A(r(t), t)\tilde{E}_A(r(t)),$$

$$\widetilde{\Delta B}(r(t), t) = \begin{bmatrix} 0 & D_B(r(t))F_B(r(t), t)E_B(r(t))K_C(r(t)) \\ 0 & \end{bmatrix}$$
$$= \begin{bmatrix} 0 & D_B(r(t)) \\ 0 & 0 \end{bmatrix} \begin{bmatrix} 0 & 0 \\ 0 & F_B(r(t), t) \end{bmatrix} \begin{bmatrix} 0 & 0 \\ 0 & E_B(r(t))K_C(r(t)) \end{bmatrix}$$
$$= \tilde{D}_B(r(t))\tilde{F}_B(r(t), t)\tilde{E}_B(r(t)),$$

$$\widetilde{\Delta C_y}(r(t), t) = \begin{bmatrix} 0 & 0 \\ K_B(r(t))D_{C_y}(r(t))F_{C_y}(r(t), t)E_{C_y}(r(t)) & 0 \end{bmatrix}$$
$$= \begin{bmatrix} 0 & 0 \\ 0 & K_B(r(t))D_{C_y}(r(t)) \end{bmatrix} \begin{bmatrix} 0 & 0 \\ 0 & F_{C_y}(r(t), t) \end{bmatrix} \begin{bmatrix} 0 & 0 \\ 0 & E_{C_y}(r(t)) \end{bmatrix}$$
$$= \tilde{D}_{C_y}(r(t))\tilde{F}_{C_y}(r(t), t)\tilde{E}_{C_y}(r(t)),$$

$$\widetilde{\Delta C_{C_z}}(r(t), t) = \begin{bmatrix} D_{C_z}(r(t))F_{C_z}(r(t), t)E_{C_z}(r(t)) & 0 \end{bmatrix}$$
$$= \begin{bmatrix} D_{C_z}(r(t)) & 0 \end{bmatrix} \begin{bmatrix} F_{C_z}(r(t), t) & 0 \\ 0 & 0 \end{bmatrix} \begin{bmatrix} E_{C_z}(r(t)) & 0 \\ 0 & 0 \end{bmatrix}$$
$$= \tilde{D}_{C_z}(r(t))\tilde{F}_{C_z}(r(t), t)\tilde{E}_{C_z}(r(t)),$$

$$\widetilde{\Delta D_{D_z}}(r(t), t) = \begin{bmatrix} 0 & D_{D_z}(r(t))F_{D_z}(r(t), t)E_{D_z}(r(t))K_C(r(t)) \end{bmatrix}$$
$$= \begin{bmatrix} 0 & D_{D_z}(r(t)) \end{bmatrix} \begin{bmatrix} 0 & 0 \\ 0 & F_{D_z}(r(t), t) \end{bmatrix} \begin{bmatrix} 0 & 0 \\ 0 & E_{D_z}(r(t))K_C(r(t)) \end{bmatrix}$$
$$= \tilde{D}_{D_z}(r(t))\tilde{F}_{D_z}(r(t), t)\tilde{E}_{D_z}(r(t)).$$

Using the previous results on stochastic stability and disturbance rejection, the closed-loop system will be robustly stochastically stable and guarantee the disturbance rejection of level γ if the following holds:

$$\begin{bmatrix} \tilde{J}(i,t) & P(i)\tilde{B}_w(i) & \tilde{C}_z^\top(i,t) \\ \tilde{B}_w^\top(i)P(i) & -\gamma^2\mathbb{I} & \tilde{B}_z^\top(i) \\ \tilde{C}_z(i,t) & \tilde{B}_z(i) & -\mathbb{I} \end{bmatrix} < 0,$$

with $\tilde{J}(i,t) = \tilde{A}^\top(i,t)P(i) + P(i)\tilde{A}(i,t) + \sum_{j=1}^N \lambda_{ij}P(j)$.

Using the expression of $\tilde{A}(i,t)$ and $\tilde{C}_z(i,t)$, we rewrite this inequality as follows:

$$\begin{bmatrix} \tilde{J}(i) & P(i)\tilde{B}_w(i) & \tilde{C}_z^\top(i) \\ \tilde{B}_w^\top(i)P(i) & -\gamma^2\mathbb{I} & \tilde{B}_z^\top(i) \\ \tilde{C}_z(i) & \tilde{B}_z(i) & -\mathbb{I} \end{bmatrix}$$

$$+ \begin{bmatrix} P(i)\tilde{D}_A(i)\tilde{F}_A(i,t)\tilde{E}_A(i) + \tilde{E}_A^\top(i)\tilde{F}_A^\top(i,t)\tilde{D}_A^\top(i)P(i) & 0 & 0 \\ 0 & 0 & 0 \\ 0 & 0 & 0 \end{bmatrix}$$

$$+ \begin{bmatrix} P(i)\tilde{D}_B(i)\tilde{F}_B(i,t)\tilde{E}_B(i) + \tilde{E}_B^\top(i)\tilde{F}_B^\top(i,t)\tilde{D}_B^\top(i)P(i) & 0 & 0 \\ 0 & 0 & 0 \\ 0 & 0 & 0 \end{bmatrix}$$

$$+ \begin{bmatrix} \begin{bmatrix} P(i)\tilde{D}_{C_y}(i)\tilde{F}_{C_y}(i,t)\tilde{E}_{C_y}(i) \\ +\tilde{E}_{C_y}^\top(i)\tilde{F}_{C_y}^\top(i,t)\tilde{D}_{C_y}^\top(i)P(i) \end{bmatrix} & 0 & 0 \\ 0 & 0 & 0 \\ 0 & 0 & 0 \end{bmatrix}$$

$$+ \begin{bmatrix} 0 & 0 & \tilde{E}_{C_z}^\top(i)\tilde{F}_{C_z}^\top(i,t)\tilde{D}_{C_z}^\top(i) \\ 0 & 0 & 0 \\ \tilde{D}_{C_z}(i)\tilde{F}_{C_z}(i,t)\tilde{E}_{C_z}(i) & 0 & 0 \end{bmatrix}$$

$$+ \begin{bmatrix} 0 & 0 & \tilde{E}_{D_z}^\top(i)\tilde{F}_{D_z}^\top(i,t)\tilde{D}_{D_z}^\top(i) \\ 0 & 0 & 0 \\ \tilde{D}_{D_z}(i)\tilde{F}_{D_z}(i,t)\tilde{E}_{D_z}(i) & 0 & 0 \end{bmatrix} < 0,$$

with $\tilde{J}(i) = \tilde{A}^\top(i)P(i) + P(i)\tilde{A}(i) + \sum_{j=1}^N \lambda_{ij}P(j)$.

Notice that if

$$\begin{bmatrix} P(i)\tilde{D}_A(i)\tilde{F}_A(i,t)\tilde{E}_A(i) + \tilde{E}_A^\top(i)\tilde{F}_A^\top(i,t)\tilde{D}_A^\top(i)P(i) & 0 & 0 \\ 0 & 0 & 0 \\ 0 & 0 & 0 \end{bmatrix}$$

$$= \begin{bmatrix} P(i)\tilde{D}_A(i) & 0 & 0 \\ 0 & 0 & 0 \\ 0 & 0 & 0 \end{bmatrix} \begin{bmatrix} \tilde{F}_A(i,t) & 0 & 0 \\ 0 & 0 & 0 \\ 0 & 0 & 0 \end{bmatrix} \begin{bmatrix} \tilde{E}_A(i) & 0 & 0 \\ 0 & 0 & 0 \\ 0 & 0 & 0 \end{bmatrix}$$

$$+ \begin{bmatrix} \tilde{E}_A^\top(i,t) & 0 & 0 \\ 0 & 0 & 0 \\ 0 & 0 & 0 \end{bmatrix} \begin{bmatrix} \tilde{F}_A^\top(i) & 0 & 0 \\ 0 & 0 & 0 \\ 0 & 0 & 0 \end{bmatrix} \begin{bmatrix} \tilde{D}_A^\top(i)P(i) & 0 & 0 \\ 0 & 0 & 0 \\ 0 & 0 & 0 \end{bmatrix}$$

$$\leq \varepsilon_A^{-1}(i) \begin{bmatrix} P(i)\tilde{D}_A(i)\tilde{D}_A^\top(i)P(i) & 0 & 0 \\ 0 & 0 & 0 \\ 0 & 0 & 0 \end{bmatrix}$$

$$+\varepsilon_A(i)\begin{bmatrix} \tilde{E}_A^\top(i)\tilde{E}_A(i) & 0 & 0 \\ 0 & 0 & 0 \\ 0 & 0 & 0 \end{bmatrix}$$

$$\leq \begin{bmatrix} \varepsilon_A^{-1}(i)P(i)\tilde{D}_A(i)\tilde{D}_A^\top(i)P(i) + \varepsilon_A(i)\tilde{E}_A^\top(i)\tilde{E}_A(i) & 0 & 0 \\ 0 & 0 & 0 \\ 0 & 0 & 0 \end{bmatrix},$$

$$\begin{bmatrix} P(i)\tilde{D}_B(i)\tilde{F}_B(i,t)\tilde{E}_B(i) + \tilde{E}_B^\top(i)\tilde{F}_B^\top(i,t)\tilde{D}_B^\top(i)P(i) & 0 & 0 \\ 0 & 0 & 0 \\ 0 & 0 & 0 \end{bmatrix}$$

$$= \begin{bmatrix} P(i)\tilde{D}_B(i) & 0 & 0 \\ 0 & 0 & 0 \\ 0 & 0 & 0 \end{bmatrix}\begin{bmatrix} \tilde{F}_B(i,t) & 0 & 0 \\ 0 & 0 & 0 \\ 0 & 0 & 0 \end{bmatrix}\begin{bmatrix} \tilde{E}_B(i) & 0 & 0 \\ 0 & 0 & 0 \\ 0 & 0 & 0 \end{bmatrix}$$

$$+ \begin{bmatrix} \tilde{E}_B^\top(i) & 0 & 0 \\ 0 & 0 & 0 \\ 0 & 0 & 0 \end{bmatrix}\begin{bmatrix} \tilde{F}_B^\top(i,t) & 0 & 0 \\ 0 & 0 & 0 \\ 0 & 0 & 0 \end{bmatrix}\begin{bmatrix} \tilde{D}_B^\top(i)P(i) & 0 & 0 \\ 0 & 0 & 0 \\ 0 & 0 & 0 \end{bmatrix}$$

$$\leq \varepsilon_B^{-1}(i)\begin{bmatrix} P(i)\tilde{D}_B(i)\tilde{D}_B^\top(i)P(i) & 0 & 0 \\ 0 & 0 & 0 \\ 0 & 0 & 0 \end{bmatrix}$$

$$+\varepsilon_B(i)\begin{bmatrix} \tilde{E}_B^\top(i)\tilde{E}_B(i) & 0 & 0 \\ 0 & 0 & 0 \\ 0 & 0 & 0 \end{bmatrix}$$

$$\leq \begin{bmatrix} \varepsilon_B^{-1}(i)P(i)\tilde{D}_B(i)\tilde{D}_B^\top(i)P(i) + \varepsilon_B(i)\tilde{E}_B^\top(i)\tilde{E}_B(i) & 0 & 0 \\ 0 & 0 & 0 \\ 0 & 0 & 0 \end{bmatrix},$$

$$\begin{bmatrix} P(i)\tilde{D}_{C_y}(i)\tilde{F}_{C_y}(i,t)\tilde{E}_{C_y}(i) + \tilde{E}_{C_y}^\top(i)\tilde{F}_{C_y}^\top(i,t)\tilde{D}_{C_y}^\top(i)P(i) & 0 & 0 \\ 0 & 0 & 0 \\ 0 & 0 & 0 \end{bmatrix}$$

$$= \begin{bmatrix} P(i)\tilde{D}_{C_y}(i) & 0 & 0 \\ 0 & 0 & 0 \\ 0 & 0 & 0 \end{bmatrix}\begin{bmatrix} \tilde{F}_{C_y}(i,t) & 0 & 0 \\ 0 & 0 & 0 \\ 0 & 0 & 0 \end{bmatrix}\begin{bmatrix} \tilde{E}_{C_y}(i) & 0 & 0 \\ 0 & 0 & 0 \\ 0 & 0 & 0 \end{bmatrix}$$

$$+ \begin{bmatrix} \tilde{E}_{C_y}^\top(i,t) & 0 & 0 \\ 0 & 0 & 0 \\ 0 & 0 & 0 \end{bmatrix}\begin{bmatrix} \tilde{F}_{C_y}^\top(i) & 0 & 0 \\ 0 & 0 & 0 \\ 0 & 0 & 0 \end{bmatrix}\begin{bmatrix} \tilde{D}_{C_y}^\top(i)P(i) & 0 & 0 \\ 0 & 0 & 0 \\ 0 & 0 & 0 \end{bmatrix}$$

$$\leq \varepsilon_{C_y}^{-1}(i)\begin{bmatrix} P(i)\tilde{D}_{C_y}(i)\tilde{D}_{C_y}^\top(i)P(i) & 0 & 0 \\ 0 & 0 & 0 \\ 0 & 0 & 0 \end{bmatrix}$$

$$+\varepsilon_{C_y}(i)\begin{bmatrix} \tilde{E}_{C_y}^\top(i)\tilde{E}_{C_y}(i) & 0 & 0 \\ 0 & 0 & 0 \\ 0 & 0 & 0 \end{bmatrix}$$

$$\leq \begin{bmatrix} \varepsilon_{C_y}^{-1}(i)P(i)\tilde{D}_{C_y}(i)\tilde{D}_{C_y}^\top(i)P(i) + \varepsilon_{C_y}(i)\tilde{E}_{C_y}^\top(i)\tilde{E}_{C_y}(i) & 0 & 0 \\ 0 & 0 & 0 \\ 0 & 0 & 0 \end{bmatrix},$$

$$\begin{bmatrix} 0 & 0 & \tilde{E}_{C_z}^\top(i)\tilde{F}_{C_z}^\top(i,t)\tilde{D}_{C_z}^\top(i) \\ 0 & 0 & 0 \\ \tilde{D}_{C_z}(i)\tilde{F}_{C_z}(i,t)\tilde{E}_{C_z}(i) & 0 & 0 \end{bmatrix}$$

$$= \begin{bmatrix} 0 & 0 & \tilde{E}_{C_z}^\top(i) \\ 0 & 0 & 0 \\ 0 & 0 & 0 \end{bmatrix}\begin{bmatrix} 0 & 0 & 0 \\ 0 & 0 & 0 \\ 0 & 0 & \tilde{F}_{C_z}^\top(i,t) \end{bmatrix}\begin{bmatrix} 0 & 0 & 0 \\ 0 & 0 & 0 \\ 0 & 0 & \tilde{D}_{C_z}^\top(i) \end{bmatrix}$$

$$+ \begin{bmatrix} 0 & 0 & 0 \\ 0 & 0 & 0 \\ 0 & 0 & \tilde{D}_{C_z}(i) \end{bmatrix}\begin{bmatrix} 0 & 0 & 0 \\ 0 & 0 & 0 \\ 0 & 0 & \tilde{F}_{C_z}(i,t) \end{bmatrix}\begin{bmatrix} 0 & 0 & 0 \\ 0 & 0 & 0 \\ \tilde{E}_{C_z}(i) & 0 & 0 \end{bmatrix}$$

$$= \varepsilon_{C_z}^{-1}(i)\begin{bmatrix} \tilde{E}_{C_z}^\top(i)\tilde{E}_{C_z}(i) & 0 & 0 \\ 0 & 0 & 0 \\ 0 & 0 & 0 \end{bmatrix} + \varepsilon_{C_z}(i)\begin{bmatrix} 0 & 0 & 0 \\ 0 & 0 & 0 \\ 0 & 0 & \tilde{D}_{C_z}(i)\tilde{D}_{C_z}^\top(i) \end{bmatrix}$$

$$= \begin{bmatrix} \varepsilon_{C_z}^{-1}(i)\tilde{E}_{C_z}^\top(i)\tilde{E}_{C_z}(i) & 0 & 0 \\ 0 & 0 & 0 \\ 0 & 0 & \varepsilon_{C_z}(i)\tilde{D}_{C_z}(i)\tilde{D}_{C_z}^\top(i) \end{bmatrix},$$

and

$$\begin{bmatrix} 0 & 0 & \tilde{E}_{D_z}^\top(i)\tilde{F}_{D_z}^\top(i,t)\tilde{D}_{D_z}^\top(i) \\ 0 & 0 & 0 \\ \tilde{D}_{D_z}(i)\tilde{F}_{D_z}(i,t)\tilde{E}_{D_z}(i) & 0 & 0 \end{bmatrix}$$

$$= \begin{bmatrix} 0 & 0 & \tilde{E}_{D_z}^\top(i) \\ 0 & 0 & 0 \\ 0 & 0 & 0 \end{bmatrix}\begin{bmatrix} 0 & 0 & 0 \\ 0 & 0 & 0 \\ 0 & 0 & \tilde{F}_{D_z}^\top(i,t) \end{bmatrix}\begin{bmatrix} 0 & 0 & 0 \\ 0 & 0 & 0 \\ 0 & 0 & \tilde{D}_{D_z}^\top(i) \end{bmatrix}$$

$$+ \begin{bmatrix} 0 & 0 & 0 \\ 0 & 0 & 0 \\ 0 & 0 & \tilde{D}_{D_z}(i) \end{bmatrix}\begin{bmatrix} 0 & 0 & 0 \\ 0 & 0 & 0 \\ 0 & 0 & \tilde{F}_{D_z}(i,t) \end{bmatrix}\begin{bmatrix} 0 & 0 & 0 \\ 0 & 0 & 0 \\ \tilde{E}_{D_z}(i) & 0 & 0 \end{bmatrix}$$

$$= \varepsilon_{D_z}^{-1}(i)\begin{bmatrix} \tilde{E}_{D_z}^\top(i)\tilde{E}_{D_z}(i) & 0 & 0 \\ 0 & 0 & 0 \\ 0 & 0 & 0 \end{bmatrix} + \varepsilon_{D_z}(i)\begin{bmatrix} 0 & 0 & 0 \\ 0 & 0 & 0 \\ 0 & 0 & \tilde{D}_{D_z}(i)\tilde{D}_{D_z}^\top(i) \end{bmatrix}$$

$$= \begin{bmatrix} \varepsilon_{D_z}^{-1}(i)\tilde{E}_{D_z}^\top(i)\tilde{E}_{D_z}(i) & 0 & 0 \\ 0 & 0 & 0 \\ 0 & 0 & \varepsilon_{D_z}(i)\tilde{D}_{D_z}(i)\tilde{D}_{D_z}^\top(i) \end{bmatrix},$$

the previous conditions for robust stochastic stability will be satisfied if the following holds for each $i \in \mathscr{S}$:

$$
\begin{bmatrix}
\tilde{A}^\top(i)P(i) + P(i)\tilde{A}(i) + \sum_{j=1}^N \lambda_{ij}P(j) & P(i)\tilde{B}_w(i) & \tilde{C}_z^\top(i) \\
\tilde{B}_w^\top(i)P(i) & -\gamma^2\mathbb{I} & \tilde{B}_z^\top(i) \\
\tilde{C}_z(i) & \tilde{B}_z(i) & -\mathbb{I}
\end{bmatrix}
$$

$$
+ \begin{bmatrix}
\varepsilon_A^{-1}(i)P(i)\tilde{D}_A(i)\tilde{D}_A^\top(i)P(i) + \varepsilon_A(i)\tilde{E}_A^\top(i)\tilde{E}_A(i) & 0 & 0 \\
0 & 0 & 0 \\
0 & 0 & 0
\end{bmatrix}
$$

$$
+ \begin{bmatrix}
\varepsilon_B^{-1}(i)P(i)\tilde{D}_B(i)\tilde{D}_B^\top(i)P(i) + \varepsilon_B(i)\tilde{E}_B^\top(i)\tilde{E}_B(i) & 0 & 0 \\
0 & 0 & 0 \\
0 & 0 & 0
\end{bmatrix}
$$

$$
+ \begin{bmatrix}
\varepsilon_{C_y}^{-1}(i)P(i)\tilde{D}_{C_y}(i)\tilde{D}_{C_y}^\top(i)P(i) + \varepsilon_{C_y}(i)\tilde{E}_{C_y}^\top(i)\tilde{E}_{C_y}(i) & 0 & 0 \\
0 & 0 & 0 \\
0 & 0 & 0
\end{bmatrix}
$$

$$
+ \begin{bmatrix}
\varepsilon_{C_z}^{-1}(i)\tilde{E}_{C_z}^\top(i)\tilde{E}_{C_z}(i) & 0 & 0 \\
0 & 0 & 0 \\
0 & 0 & \varepsilon_{C_z}(i)\tilde{D}_{C_z}(i)\tilde{D}_{C_z}^\top(i)
\end{bmatrix}
$$

$$
+ \begin{bmatrix}
\varepsilon_{D_z}^{-1}(i)\tilde{E}_{D_z}^\top(i)\tilde{E}_{D_z}(i) & 0 & 0 \\
0 & 0 & 0 \\
0 & 0 & \varepsilon_{D_z}(i)\tilde{D}_{D_z}(i)\tilde{D}_{D_z}^\top(i)
\end{bmatrix} < 0,
$$

which can be rewritten as follows:

$$
\begin{bmatrix}
\tilde{J}(i) & P(i)\tilde{B}_w(i) & \tilde{C}_z^\top(i) \\
\tilde{B}_w^\top(i)P(i) & -\gamma^2\mathbb{I} & \tilde{B}_z^\top(i) \\
\tilde{C}_z(i) & \tilde{B}_z(i) & \begin{bmatrix} -\mathbb{I} \\ +\varepsilon_{C_z}(i)\tilde{D}_{C_z}(i)\tilde{D}_{C_z}^\top(i) \\ +\varepsilon_{D_z}(i)\tilde{D}_{D_z}(i)\tilde{D}_{D_z}^\top(i) \end{bmatrix}
\end{bmatrix} < 0,
$$

where

$$
\begin{aligned}
\tilde{J}(i) =\ & \tilde{A}^\top(i)P(i) + P(i)\tilde{A}(i) + \sum_{j=1}^N \lambda_{ij}P(j) \\
& + \varepsilon_A^{-1}(i)P(i)\tilde{D}_A(i)\tilde{D}_A^\top(i)P(i) + \varepsilon_A(i)\tilde{E}_A^\top(i)\tilde{E}_A(i) \\
& + \varepsilon_B^{-1}(i)P(i)\tilde{D}_B(i)\tilde{D}_B^\top(i)P(i) + \varepsilon_B(i)\tilde{E}_B^\top(i)\tilde{E}_B(i) \\
& + \varepsilon_{C_y}^{-1}(i)P(i)\tilde{D}_{C_y}(i)\tilde{D}_{C_y}^\top(i)P(i) + \varepsilon_{C_y}(i)\tilde{E}_{C_y}^\top(i)\tilde{E}_{C_y}(i) \\
& + \varepsilon_{C_z}^{-1}(i)\tilde{E}_{C_z}^\top(i)\tilde{E}_{C_z}(i) + \varepsilon_{D_z}^{-1}(i)\tilde{E}_{D_z}^\top(i)\tilde{E}_{D_z}(i).
\end{aligned}
$$

If we define $\Upsilon(i)$ and $\Psi(i)$ as follows:

$$
\Upsilon(i) = \begin{bmatrix}
\varepsilon_A(i)\mathbb{I} & 0 & 0 \\
0 & \varepsilon_B(i)\mathbb{I} & 0 \\
0 & 0 & \varepsilon_{C_y}(i)\mathbb{I}
\end{bmatrix},
$$

$$\Psi(i) = \begin{bmatrix} \varepsilon_{C_z}(i)\mathbb{I} & 0 \\ 0 & \varepsilon_{D_z}(i)\mathbb{I} \end{bmatrix},$$

$\tilde{J}(i)$ can be rewritten as

$$\tilde{J}(i) = \tilde{A}^\top(i)P(i) + P(i)\tilde{A}(i) + \sum_{j=1}^N \lambda_{ij}P(j)$$

$$+ P(i)\begin{bmatrix} \tilde{D}_A(i) & \tilde{D}_B(i) & \tilde{D}_{C_y}(i) \end{bmatrix} \Upsilon^{-1}(i) \begin{bmatrix} \tilde{D}_A^\top(i) \\ \tilde{D}_B^\top(i) \\ \tilde{D}_{C_y}^\top(i) \end{bmatrix} P(i)$$

$$+ \begin{bmatrix} \tilde{E}_A^\top(i) & \tilde{E}_B^\top(i) & \tilde{E}_{C_y}^\top(i) \end{bmatrix} \Upsilon(i) \begin{bmatrix} \tilde{E}_A(i) \\ \tilde{E}_B(i) \\ \tilde{E}_{C_y}(i) \end{bmatrix}$$

$$+ \begin{bmatrix} \tilde{E}_{C_z}^\top(i) & \tilde{E}_{D_z}^\top(i) \end{bmatrix} \Psi^{-1}(i) \begin{bmatrix} \tilde{E}_{C_z}(i) \\ \tilde{E}_{D_z}(i) \end{bmatrix}.$$

If we define $\mathscr{R}(i)$ as follows:

$$\mathscr{R}(i) = \mathbb{I} - \varepsilon_{C_z}(i)\tilde{D}_{C_z}(i)\tilde{D}_{C_z}^\top(i) - \varepsilon_{D_z}(i)\tilde{D}_{D_z}(i)\tilde{D}_{D_z}^\top(i)$$
$$= \mathbb{I} - \varepsilon_{C_z}(i)D_{C_z}(i)D_{C_z}^\top(i) - \varepsilon_{D_z}(i)D_{D_z}(i)D_{D_z}^\top(i),$$

which should be positive-definite, the previous LMI becomes

$$\begin{bmatrix} \tilde{J}(i) & P(i)\tilde{B}_w(i) \\ \tilde{B}_w^\top(i)P(i) & -\gamma^2\mathbb{I} \end{bmatrix}$$
$$+ \begin{bmatrix} \tilde{C}_z^\top(i) \\ \tilde{B}_z^\top(i) \end{bmatrix} \mathscr{R}^{-1}(i) \begin{bmatrix} \tilde{C}_z(i) & \tilde{B}_z(i) \end{bmatrix} < 0,$$

which gives in turn the following:

$$\begin{bmatrix} \tilde{J}(i) + \tilde{C}_z^\top(i)\mathscr{R}^{-1}(i)\tilde{C}_z(i) & \begin{bmatrix} P(i)\tilde{B}_w(i) \\ +\tilde{C}_z^\top(i)\mathscr{R}^{-1}(i)\tilde{B}_z(i) \end{bmatrix} \\ \begin{bmatrix} \tilde{B}_w^\top(i)P(i) \\ +\tilde{B}_z^\top(i)\mathscr{R}^{-1}(i)\tilde{C}_z(i) \end{bmatrix} & -\gamma^2\mathbb{I} + \tilde{B}_z^\top(i)\mathscr{R}^{-1}(i)\tilde{B}_z(i) \end{bmatrix} < 0.$$

Using the Schur complement again, we get

$$\tilde{J}(i) + \tilde{C}_z^\top(i)\mathscr{R}^{-1}(i)\tilde{C}_z(i) + \left[P(i)\tilde{B}_w(i) + \tilde{C}_z^\top(i)\mathscr{R}^{-1}(i)\tilde{B}_z(i) \right]$$
$$\times \left[\gamma^2\mathbb{I} - \tilde{B}_z^\top(i)\mathscr{R}^{-1}(i)\tilde{B}_z(i) \right]^{-1} \left[\tilde{B}_w^\top(i)P(i) + \tilde{B}_z^\top(i)\mathscr{R}^{-1}(i)\tilde{C}_z(i) \right] < 0,$$

which gives the following after using the expression of $\tilde{J}(i)$:

$$\tilde{A}^\top(i)P(i) + P(i)\tilde{A}(i) + \sum_{j=1}^N \lambda_{ij}P(j)$$

$$+P(i)\left[\tilde{D}_A(i)\ \tilde{D}_B(i)\ \tilde{D}_{C_y}(i)\right]\varUpsilon^{-1}(i)\begin{bmatrix}\tilde{D}_A^\top(i)\\\tilde{D}_B^\top(i)\\\tilde{D}_{C_y}^\top(i)\end{bmatrix}P(i)$$

$$+\left[\tilde{E}_A^\top(i)\ \tilde{E}_B^\top(i)\ \tilde{E}_{C_y}^\top(i)\right]\varUpsilon(i)\begin{bmatrix}\tilde{E}_A(i)\\\tilde{E}_B(i)\\\tilde{E}_{C_y}(i)\end{bmatrix}$$

$$+\left[\tilde{E}_{C_z}^\top(i)\ \tilde{E}_{D_z}^\top(i)\right]\varPsi^{-1}(i)\begin{bmatrix}\tilde{E}_{C_z}(i)\\\tilde{E}_{D_z}(i)\end{bmatrix}$$

$$+\tilde{C}_z^\top(i)\mathscr{R}^{-1}(i)\tilde{C}_z(i)+\left[P(i)\tilde{B}_w(i)+\tilde{C}_z^\top(i)\mathscr{R}^{-1}(i)\tilde{B}_z(i)\right]$$

$$\times\left[\gamma^2\mathbb{I}-\tilde{B}_z^\top(i)\mathscr{R}^{-1}(i)\tilde{B}_z(i)\right]^{-1}$$

$$\times\left[\tilde{B}_w^\top(i)P(i)+\tilde{B}_z^\top(i)\mathscr{R}^{-1}(i)\tilde{C}_z(i)\right]<0.\qquad(4.43)$$

Let $P(i)$ have the following form:

$$P(i)=\begin{bmatrix}P_1(i)\ P_2(i)\\P_2^\top(i)\ P_3(i)\end{bmatrix}.\qquad(4.44)$$

As before, we pre- and post-multiply the right-hand side of this inequality by, respectively, $U^\top(i)V^\top(i)$ and $V(i)U(i)$ to make it useful.

Notice that some of the terms have already been computed and we do not need to recompute them. For the other terms we have

$$U^\top(i)V^\top(i)P(i)\left[\tilde{D}_A(i)\ \tilde{D}_B(i)\ \tilde{D}_{C_y}(i)\right]$$

$$=\begin{bmatrix}\begin{bmatrix}W^\top(i)P_1(i)\\-W^\top(i)P_2(i)\\\times P_3^{-1}(i)P_2^\top(i)\\P_1(i)\end{bmatrix}\begin{bmatrix}W^\top(i)P_2(i)\\-W^\top(i)P_2(i)\\\times P_3^{-1}(i)P_3(i)\\P_2(i)\end{bmatrix}\end{bmatrix}$$

$$\times\left[\tilde{D}_A(i)\ \tilde{D}_B(i)\ \tilde{D}_{C_y}(i)\right]$$

$$=\begin{bmatrix}\mathbb{I}&0\\P_1(i)&P_2(i)\end{bmatrix}\begin{bmatrix}D_A(i)\ 0\ 0\ D_B(i)\ 0&0\\0&0\ 0&0&0\ K_B(i)D_{C_y}(i)\end{bmatrix}$$

$$=\begin{bmatrix}D_A(i)&0\ 0&D_B(i)&0&0\\P_1(i)D_A(i)\ 0\ 0\ P_1(i)D_B(i)\ 0\ P_2(i)K_B(i)D_{C_y}(i)\end{bmatrix},$$

$$U^\top(i)V^\top(i)\left[\tilde{E}_A^\top(i)\ \tilde{E}_B^\top(i)\ \tilde{E}_{C_y}^\top(i)\right]$$

$$=\begin{bmatrix}W^\top(i)\ -W^\top(i)P_2(i)P_3^{-1}(i)\\\mathbb{I}&0\end{bmatrix}\begin{bmatrix}E_A^\top(i)\ 0\ 0&0\\0&0\ 0\ K_C^\top(i)E_B^\top(i)\\&0&0\\&0\ E_{C_y}^\top(i)\end{bmatrix}$$

$$= \begin{bmatrix} W^\top(i)E_A^\top(i) \ 0 \ 0 \ -W^\top(i)P_2(i)P_3^{-1}(i)K_C^\top(i)E_B^\top(i) \\ E_A^\top(i) \qquad 0 \ 0 \qquad\qquad 0 \end{bmatrix}$$
$$\left. \begin{matrix} 0 \ -W^\top(i)P_2(i)P_3^{-1}(i)E_{C_y}^\top(i) \\ 0 \qquad\qquad 0 \end{matrix} \right],$$

$$U^\top(i)V^\top(i)\left[\, \tilde{E}_{C_z}^\top(i) \ \tilde{E}_{D_z}^\top(i) \,\right]$$
$$= \begin{bmatrix} W^\top(i) \ -W^\top(i)P_2(i)P_3^{-1}(i) \\ \mathbb{I} \qquad\qquad 0 \end{bmatrix} \begin{bmatrix} E_{C_z}^\top(i) \ 0 \\ 0 \qquad 0 \end{bmatrix}$$
$$\begin{bmatrix} 0 \qquad 0 \\ 0 \ K_C^\top(i)E_{D_z}^\top(i) \end{bmatrix}$$
$$= \begin{bmatrix} W^\top(i)E_{C_z}^\top(i) \ 0 \ 0 \begin{bmatrix} -W^\top(i)P_2(i)P_3^{-1}(i) \\ \times K_C^\top(i)E_{D_z}^\top(i) \end{bmatrix} \\ E_{C_z}^\top(i) \qquad 0 \ 0 \qquad\qquad 0 \end{bmatrix},$$

$$U^\top(i)V^\top(i)\tilde{C}_z^\top(i)$$
$$= \begin{bmatrix} W^\top(i) \ -W^\top(i)P_2(i)P_3^{-1}(i) \\ \mathbb{I} \qquad\qquad 0 \end{bmatrix} \begin{bmatrix} C_z^\top(i) \\ K_C^\top(i)D_z^\top(i) \end{bmatrix}$$
$$= \begin{bmatrix} W^\top(i)C_z^\top(i) - W^\top(i)P_2(i)P_3^{-1}(i)K_C^\top(i)D_z^\top(i) \\ C_z^\top(i) \end{bmatrix},$$

$$U^\top(i)V^\top(i)P(i)\tilde{B}_w(i) = \begin{bmatrix} \mathbb{I} \ 0 \\ P_1(i) \ P_2(i) \end{bmatrix} \begin{bmatrix} B_w(i) \\ 0 \end{bmatrix}$$
$$= \begin{bmatrix} B_w(i) \\ P_1(i)B_w(i) \end{bmatrix},$$

and

$$U^\top(i)V^\top(i)\tilde{C}_z^\top(i)\mathscr{R}^{-1}(i)\tilde{B}_z(i)$$
$$= \left[\begin{bmatrix} W^\top(i)C_z^\top(i)\mathscr{R}^{-1}(i)B_z(i) \\ -W^\top(i)P_2(i)P_3^{-1}(i)K_C^\top(i)D_z^\top(i)\mathscr{R}^{-1}(i)B_z(i) \\ C_z^\top(i)\mathscr{R}^{-1}(i)B_z(i) \end{bmatrix} \right].$$

If we define the following variables:

$$\mathscr{N}_1(i) = C_z(i)W(i) - D_z(i)K_C(i)P_3^{-1}(i)P_2^\top(i)W(i),$$
$$\mathscr{N}_2(i) = C_z(i),$$
$$\mathscr{N}_3(i) = \mathbb{I} - \varepsilon_{C_z}(i)D_{C_z}(i)D_{C_z}^\top(i) - \varepsilon_{D_z}(i)D_{D_z}(i)D_{D_z}^\top(i) = \mathscr{R}(i),$$
$$\mathscr{N}_4(i) = W^\top(i)C_z^\top(i)\mathscr{N}_3^{-1}(i)B_z(i)$$
$$\qquad\qquad - W^\top(i)P_2(i)P_3^{-1}(i)K_C^\top(i)D_z^\top(i)\mathscr{N}_3^{-1}(i)B_z(i)$$

$$
\begin{aligned}
&\quad\quad +B_w(i),\\
\mathscr{N}_5(i) =\ & P_1(i)B_w(i) + C_z^\top(i)\mathscr{N}_3^{-1}(i)B_z(i),\\
\mathscr{N}_6(i) =\ & \gamma^2\mathbb{I} - B_z^\top(i)\mathscr{N}_3^{-1}(i)B_z(i),\\
\mathscr{N}_7(i) =\ & \begin{bmatrix} D_A(i)\ 0\ 0\ D_B(i)\ 0\ 0 \end{bmatrix},\\
\mathscr{N}_8(i) =\ & \begin{bmatrix} P_1(i)D_A(i)\ 0\ 0\ P_1(i)D_B(i)\ 0\ P_2(i)K_B(i)D_{C_y}(i) \end{bmatrix},\\
\mathscr{N}_9^\top(i) =\ & \left[\ W^\top(i)E_A^\top(i)\ 0\ 0\ \begin{bmatrix} -W^\top(i)P_2(i)P_3^{-1}(i) \\ \times K_C^\top(i)E_B^\top(i) \end{bmatrix}\right.\\
& \quad\ \left. 0 \begin{bmatrix} -W^\top(i)P_2(i) \\ \times P_3^{-1}(i)E_{C_y}^\top(i) \end{bmatrix}\right],\\
\mathscr{N}_{10}^\top(i) =\ & \begin{bmatrix} E_A^\top(i)\ 0\ 0\ 0\ 0\ 0 \end{bmatrix},\\
\mathscr{N}_{11}^\top(i) =\ & \begin{bmatrix} W^\top(i)E_{C_z}^\top(i)\ 0\ 0\ -W^\top(i)P_2(i)P_3^{-1}(i)K_C^\top(i)E_{D_z}^\top(i) \end{bmatrix},\\
\mathscr{N}_{12}^\top(i) =\ & \begin{bmatrix} E_{C_z}^\top(i)\ 0\ 0\ 0 \end{bmatrix},
\end{aligned}
$$

and using all the previous computations, (4.43) becomes

$$
\mathscr{Z}_1(i) + \mathscr{Z}_2(i) + \mathscr{Z}_3(i) + \mathscr{Z}_4(i) + \mathscr{Z}_5(i) + \mathscr{Z}_6(i) < 0, \tag{4.45}
$$

with

$$
\mathscr{Z}_1(i) = \tilde{A}^\top(i)P(i) + P(i)\tilde{A}(i) + \sum_{j=1}^N \lambda_{ij}P(j) =
$$
$$
\begin{bmatrix}
\begin{bmatrix}
A(i)W(i) \\
+W^\top(i)A^\top(i) \\
+\sum_{j=1}^N \lambda_{ij}W^\top(i)W^{-1}(j)W(i) \\
-B(i)K_C(i)P_3^{-1}(i)P_2^\top(i)W(i) \\
-W^\top(i)P_2(i)P_3^{-1}(i)K_C^\top(i)B^\top(i) \\
+\sum_{j=1}^N \lambda_{ij}W^\top(i)\left[P_2(i)P_3^{-1}(i)P_3(j) - P_2^\top(j)\right] \\
\times P_3^{-1}(j)\left[P_2(i)P_3^{-1}(i)P_3(j) - P_2^\top(j)\right]^\top W(i)
\end{bmatrix} \\
\begin{bmatrix}
A^\top(i) + P_1(i)A(i)W(i) \\
+P_2(i)K_B(i)C_y(i)W(i) \\
-P_1(i)B(i)K_C(i)P_3^{-1}(i)P_2^\top(i)W(i) \\
-P_2(i)K_A(i)P_3^{-1}(i)P_2^\top(i)W(i) \\
+\sum_{j=1}^N \left[P_1(j) - P_2(j)P_3^{-1}(i)P_2^\top(i)\right]W(i)
\end{bmatrix}
\end{bmatrix}
$$

$$
\left[
\begin{array}{c}
\left[
\begin{array}{c}
A(i) + W^\top(i)A^\top(i)P_1(i) \\
+ W^\top(i)C_y^\top(i)K_B^\top(i)P_2^\top(i) \\
- W^\top(i)P_2(i)P_3^{-1}(i)K_C^\top(i)B^\top(i)P_1(i) \\
- W^\top(i)P_2(i)P_3^{-1}(i)K_A^\top(i)P_2^\top(i) \\
+ \sum_{j=1}^{N} \lambda_{ij} W^\top(i)\left[P_1(j) - P_2(i)P_3^{-1}(i)P_2^\top(j) \right]
\end{array}
\right] \\[4ex]
\\
\left[
\begin{array}{c}
P_1(i)A(i) + P_2(i)K_B(i)C_y(i) \\
+ A^\top(i)P_1(i) + C_y^\top(i)K_B^\top(i)P_2^\top(i) \\
+ \sum_{j=1}^{N} \lambda_{ij} P_1(j)
\end{array}
\right]
\end{array}
\right]
$$

$$
= \begin{bmatrix} \mathcal{M}_1(i) & \mathcal{M}_2(i) \\ \mathcal{M}_2^\top(i) & \mathcal{M}_3(i) \end{bmatrix},
$$

$$
\begin{aligned}
\mathcal{L}_2(i) &= U^\top(i)V^\top(i)P(i)\left[\tilde{D}_A(i)\ \tilde{D}_B(i)\ \tilde{D}_{C_y}(i) \right] \\
&\quad \times \Upsilon^{-1}(i) \begin{bmatrix} \tilde{D}_A^\top(i) \\ \tilde{D}_B^\top(i) \\ \tilde{D}_{C_y}^\top(i) \end{bmatrix} P(i)V(i)U(i) \\
&= \begin{bmatrix} D_A(i) & 0\ 0 & D_B(i) & 0 & 0 \\ P_1(i)D_A(i) & 0\ 0 & P_1(i)D_B(i) & 0 & P_2(i)K_B(i)D_{C_y}(i) \end{bmatrix} \\
&\quad \times \Upsilon^{-1}(i) \begin{bmatrix} D_A^\top(i) & D_A^\top(i)P_1(i) \\ 0 & 0 \\ 0 & 0 \\ D_B^\top(i) & D_B^\top(i)P_1(i) \\ 0 & 0 \\ 0 & D_{C_y}^\top(i)K_B^\top(i)P_2(i) \end{bmatrix} \\
&= \begin{bmatrix} \mathcal{N}_7(i)\Upsilon^{-1}(i)\mathcal{N}_7^\top(i) & \mathcal{N}_7(i)\Upsilon^{-1}(i)\mathcal{N}_8^\top(i) \\ \mathcal{N}_8(i)\Upsilon^{-1}(i)\mathcal{N}_7^\top(i) & \mathcal{N}_8(i)\Upsilon^{-1}(i)\mathcal{N}_8^\top(i) \end{bmatrix},
\end{aligned}
$$

$$
\begin{aligned}
\mathcal{L}_3(i) &= U^\top(i)V^\top(i)\left[\tilde{E}_A^\top(i)\ \tilde{E}_B^\top(i)\ \tilde{E}_{C_y}^\top(i) \right] \Upsilon(i) \begin{bmatrix} \tilde{E}_A(i) \\ \tilde{E}_B(i) \\ \tilde{E}_{C_y}(i) \end{bmatrix} \\
&= \begin{bmatrix} W^\top(i)E_A^\top(i) & 0\ 0 & \begin{matrix} -W^\top(i)P_2(i)P_3^{-1}(i) \\ \times K_C^\top(i)E_B^\top(i) \end{matrix} & 0 & \begin{matrix} -W^\top(i)P_2(i) \\ \times P_3^{-1}(i)E_{C_y}^\top(i) \end{matrix} \\ E_A^\top(i) & 0\ 0 & 0 & 0 & 0 \end{bmatrix}
\end{aligned}
$$

$$\times \Upsilon(i) \begin{bmatrix} E_A(i)W(i) & E_A(i) \\ 0 & 0 \\ 0 & 0 \\ -E_B(i)K_C(i)P_3^{-1}(i)P_2^\top(i)W(i) & 0 \\ 0 & 0 \\ -E_C(i)P_3^{-1}(i)P^\top(i)W(i) & 0 \end{bmatrix}$$

$$= \begin{bmatrix} \mathscr{N}_9^\top(i)\Upsilon(i)\mathscr{N}_9(i) & \mathscr{N}_9^\top(i)\Upsilon(i)\mathscr{N}_{10}(i) \\ \mathscr{N}_{10}^\top(i)\Upsilon(i)\mathscr{N}_9(i) & \mathscr{N}_{10}^\top(i)\Upsilon(i)\mathscr{N}_{10}(i) \end{bmatrix},$$

$$\begin{aligned}
\mathscr{Z}_4(i) &= U^\top(i)V^\top(i)\left[\tilde{E}_{C_z}^\top(i)\ \tilde{E}_{D_z}^\top(i)\right]\Psi^{-1}(i)\begin{bmatrix} \tilde{E}_{C_z}(i) \\ \tilde{E}_{D_z}(i) \end{bmatrix}V(i)U(i) \\
&= \begin{bmatrix} W^\top(i)E_{C_z}^\top(i)\ 0\ 0\ -W^\top(i)P_2(i)P_3^{-1}(i)K_C^\top(i)E_{D_z}^\top(i) \\ E_{C_z}^\top(i)\quad 0\ 0\quad\quad\quad 0 \end{bmatrix} \\
&\quad \times \Psi^{-1}(i)\begin{bmatrix} E_{C_z}(i)W(i) & E_{C_z}(i) \\ 0 & 0 \\ 0 & 0 \\ -E_{D_z}(i)K_C(i)P_3^{-1}(i)P_2^\top(i)W(i) & 0 \end{bmatrix} \\
&= \begin{bmatrix} \mathscr{N}_{11}^\top(i)\Psi^{-1}(i)\mathscr{N}_{11}(i) & \mathscr{N}_{11}^\top(i)\Psi^{-1}(i)\mathscr{N}_{12}(i) \\ \mathscr{N}_{12}^\top(i)\Psi^{-1}(i)\mathscr{N}_{11}(i) & \mathscr{N}_{12}^\top(i)\Psi^{-1}(i)\mathscr{N}_{12}(i) \end{bmatrix},
\end{aligned}$$

$$\begin{aligned}
\mathscr{Z}_5(i) &= U^\top(i)V^\top(i)\tilde{C}_z(i)\mathscr{R}^{-1}(i)\tilde{C}_z(i)V(i)U(i) \\
&= \begin{bmatrix} W^\top(i)C_z^\top(i) - W^\top(i)P_2(i)P_3^{-1}(i)K_C^\top(i)D_z^\top(i) \\ C_z^\top(i) \end{bmatrix} \\
&\quad \times \mathscr{R}^{-1}(i)\left[\begin{bmatrix} C_z(i)W(i) \\ -D_z(i)K_C(i)P_3^{-1}(i)P_2^\top(i)W(i) \end{bmatrix}\ C_z(i)\right] \\
&= \begin{bmatrix} \mathscr{N}_1^\top(i)\mathscr{N}_3^{-1}(i)\mathscr{N}_1(i) & \mathscr{N}_1^\top(i)\mathscr{N}_3^{-1}(i)\mathscr{N}_2(i) \\ \mathscr{N}_2^\top(i)\mathscr{N}_3^{-1}(i)\mathscr{N}_1(i) & \mathscr{N}_2^\top(i)\mathscr{N}_3^{-1}(i)\mathscr{N}_2(i) \end{bmatrix},
\end{aligned}$$

$$\begin{aligned}
\mathscr{Z}_6(i) &= U^\top(i)V^\top(i)\left[P(i)\tilde{B}_w(i) + \tilde{C}_z^\top(i)\mathscr{R}^{-1}(i)\tilde{B}_z(i)\right] \\
&\quad \times \left[\gamma^2\mathbb{I} - \tilde{B}_z^\top(i)\mathscr{R}^{-1}(i)\tilde{B}_z(i)\right]^{-1}\left[\tilde{B}_w^\top(i)P(i) + \tilde{B}_z^\top(i)\mathscr{R}^{-1}(i)\tilde{C}_z(i)\right] \\
&= \left[\begin{bmatrix} W^\top(i)C_z^\top(i)\mathscr{N}_3^{-1}(i)B_z(i) \\ -W^\top(i)P_2(i)P_3^{-1}(i)K_C^\top(i)D_z^\top(i) \\ \times\mathscr{N}_3^{-1}(i)B_z(i) + B_w(i) \\ P_1(i)B_w(i) + C_z^\top(i)\mathscr{N}_3^{-1}(i)B_z(i) \end{bmatrix}\right]\begin{bmatrix} \gamma^2\mathbb{I} \\ -B_z^\top(i)\mathscr{R}^{-1}(i)B_z(i) \end{bmatrix}^{-1} \\
&\quad \times \left[\begin{bmatrix} W^\top(i)C_z^\top(i)(i)\mathscr{N}_3^{-1}(i)B_z(i) \\ -W^\top(i)P_2(i)P_3^{-1}(i)K_C^\top(i)D_z^\top(i) \\ \times\mathscr{N}_3^{-1}(i)B_z(i) + B_w(i) \end{bmatrix}\right]^\top\begin{bmatrix} P_1(i)B_w(i) \\ +C_z^\top(i)\mathscr{N}_3^{-1}(i)B_z(i) \end{bmatrix}^\top
\end{aligned}$$

$$= \begin{bmatrix} \mathcal{N}_4(i)\mathcal{N}_6^{-1}(i)\mathcal{N}_4^\top(i) & \mathcal{N}_4(i)\mathcal{N}_6^{-1}(i)\mathcal{N}_5^\top(i) \\ \mathcal{N}_5(i)\mathcal{N}_6^{-1}(i)\mathcal{N}_4^\top(i) & \mathcal{N}_5(i)\mathcal{N}_6^{-1}(i)\mathcal{N}_5^\top(i) \end{bmatrix}.$$

If we use the expression of the controller (4.49), then (4.45) implies

$$\mathcal{M}_2^\top(i) + \mathcal{N}_2^\top(i)\mathcal{N}_3^{-1}(i)\mathcal{N}_1(i) + \mathcal{N}_5(i)\mathcal{N}_6^{-1}(i)\mathcal{N}_4^\top(i)$$
$$+\mathcal{N}_8(i)\Upsilon^{-1}(i)\mathcal{N}_7^\top(i) + \mathcal{N}_{10}^\top(i)\Upsilon(i)\mathcal{N}_9(i)$$
$$+\mathcal{N}_{12}^\top(i)\Psi^{-1}(i)\mathcal{N}_{11}(i) = 0,$$

and using the fact that

$$\sum_{j=1}^{N} \lambda_{ij}W^\top(i)\left[P_2(i)P_3^{-1}(i)P_3(j) - P^\top(j)\right]$$
$$\times P_3^{-1}(i)\left[P_2(i)P_3^{-1}(i)P_3(j) - P^\top(j)\right]^\top W(i) > 0,$$

we get

$$\mathcal{M}_1(i) + \mathcal{N}_1^\top(i)\mathcal{N}_3^{-1}(i)\mathcal{N}_1(i) + \mathcal{N}_4(i)\mathcal{N}_6^{-1}(i)\mathcal{N}_4^\top(i)$$
$$+\mathcal{N}_7(i)\Upsilon^{-1}(i)\mathcal{N}_7^\top(i) + \mathcal{N}_9^\top(i)\Upsilon(i)\mathcal{N}_9(i)$$
$$+\mathcal{N}_{11}^\top(i)\Psi^{-1}(i)\mathcal{N}_{11}(i) < 0,$$
$$\mathcal{M}_3(i) + \mathcal{N}_2^\top(i)\mathcal{N}_3^{-1}(i)\mathcal{N}_2(i) + \mathcal{N}_5(i)\mathcal{N}_6^{-1}(i)\mathcal{N}_5^\top(i)$$
$$+\mathcal{N}_8(i)\Upsilon^{-1}(i)\mathcal{N}_8^\top(i) + \mathcal{N}_{10}^\top(i)\Upsilon(i)\mathcal{N}_{10}(i)$$
$$+\mathcal{N}_{12}^\top(i)\Psi^{-1}(i)\mathcal{N}_{12}(i) < 0,$$

with

$$\begin{aligned}
\mathcal{M}_1(i) =\ & A(i)W(i) + W^\top(i)A^\top(i) \\
& + \sum_{j=1}^{N} \lambda_{ij}W^\top(i)W^{-1}(j)W(i) \\
& -B(i)K_C(i)P_3^{-1}(i)P_2^\top(i)W(i) \\
& -W^\top(i)P_2(i)P_3^{-1}(i)K_C^\top(i)B^\top(i), \\
\mathcal{M}_2(i) =\ & A(i) + W^\top(i)A^\top(i)P_1(i) \\
& +W^\top(i)C_y^\top(i)K_B^\top(i)P_2^\top(i) \\
& -W^\top(i)P_2(i)P_3^{-1}(i)K_C^\top(i)B^\top(i)P_1(i) \\
& -W^\top(i)P_2(i)P_3^{-1}(i)K_A^\top(i)P_2^\top(i) \\
& + \sum_{j=1}^{N} W^\top(i)\left[P_1(j) - P_2(i)P_3^{-1}(i)P_2^\top(j)\right], \\
\mathcal{M}_3(i) =\ & P_1(i)A(i) + A^\top(i)P_1(i) + P_2(i)K_B(i)C_y(i) \\
& +C_y^\top(i)K_B^\top(i)P_2^\top(i) + \sum_{j=1}^{N} \lambda_{ij}P_1(j).
\end{aligned}$$

If we let

$$P(i) = \begin{bmatrix} X(i) & Y^{-1}(i) - X(i) \\ Y^{-1}(i) - X(i) & X(i) - Y^{-1}(i) \end{bmatrix},$$

where $X(i) > 0$ and $Y(i) > 0$ are symmetric and positive-definite matrices, it means that

$$\begin{aligned} P_1(i) &= X(i), \\ P_2(i) &= Y^{-1}(i) - X(i), \\ P_3(i) &= X(i) - Y^{-1}(i), \end{aligned}$$

which implies that

$$W(i) = \left[P_1(i) - P_2(i)P_3^{-1}(i)P_2^\top(i) \right]^{-1} = Y(i),$$

and $P_3^{-1}(i)P_2^\top(i) = -\mathbb{I}$.

Notice also that

$$\sum_{j=1}^{N} \lambda_{ij} Y^\top(i) Y^{-1}(i) Y(i) = \lambda_{ii} Y(i) + \mathcal{S}_i(Y)\mathcal{Y}_i^{-1}(Y)\mathcal{S}_i^\top(Y),$$

where

$$\begin{aligned} \mathcal{S}_i(Y) &= \left[\sqrt{\lambda_{i1}}Y(i), \cdots, \sqrt{\lambda_{ii-1}}Y(i), \sqrt{\lambda_{ii+1}}Y(i), \cdots, \sqrt{\lambda_{iN}}Y(i) \right], \\ \mathcal{Y}_i(Y) &= \mathrm{diag}\left[Y(1), \cdots, Y(i-1), Y(i+1), \cdots, Y(N) \right]. \end{aligned}$$

If we define $\mathcal{K}_B(i)$ and $\mathcal{K}_C(i)$ by

$$\begin{aligned} \mathcal{K}_B(i) &= P_2(i)K_B(i) = \left[Y^{-1}(i) - X(i) \right] K_B(i), \\ \mathcal{K}_C(i) &= -K_C(i)P_3^{-1}(i)P_2^\top(i)W(i) = K_C(i)Y(i), \end{aligned}$$

the expressions of $\mathcal{N}_1(i), \cdots, \mathcal{N}_{12}(i)$ become

$$\begin{aligned} \mathcal{N}_1(i) &= C_z(i)Y(i) + D_z(i)\mathcal{K}_C(i), \\ \mathcal{N}_2(i) &= C_z(i), \\ \mathcal{N}_3(i) &= \mathbb{I} - \varepsilon_{C_z}(i)D_{C_z}(i)D_{C_z}^\top(i) - \varepsilon_{D_z}(i)D_{D_z}(i)D_{D_z}^\top(i), \\ \mathcal{N}_4(i) &= Y^\top(i)C_z^\top(i)\mathcal{N}_3^{-1}(i)B_z(i) \\ &\quad + \mathcal{K}_C^\top(i)D_z^\top(i)\mathcal{N}_3^{-1}(i)B_z(i) + B_w(i), \\ \mathcal{N}_5(i) &= X(i)B_w(i) + C_z^\top(i)\mathcal{N}_3^{-1}(i)B_z(i), \\ \mathcal{N}_6(i) &= \gamma^2\mathbb{I} - B_z^\top(i)\mathcal{N}_3^{-1}(i)B_z(i), \\ \mathcal{N}_7(i) &= \left[D_A(i) \; 0 \; 0 \; D_B(i) \; 0 \; 0 \right], \\ \mathcal{N}_8(i) &= \left[X(i)D_A(i) \; 0 \; 0 \; X(i)D_B(i) \; 0 \; \mathcal{K}_B(i)D_{C_y}(i) \right], \end{aligned}$$

$$\mathcal{N}_9^\top(i) = \left[Y^\top(i)E_A^\top(i)\ 0\ 0\ \mathcal{K}_C^\top(i)E_B^\top(i)\ 0\ Y^\top(i)E_C^\top(i) \right],$$
$$\mathcal{N}_{10}(i) = \left[E_A(i)\ 0\ 0\ 0\ 0\ E_C(i) \right],$$
$$\mathcal{N}_{11}^\top(i) = \left[Y^\top(i)E_{C_z}^\top(i)\ 0\ 0\ \mathcal{K}_C^\top(i)E_{D_z}^\top(i) \right],$$
$$\mathcal{N}_{12}^\top(i) = \left[E_{C_z}^\top(i)\ 0\ 0\ 0 \right].$$

The following theorem summarizes the results of these developments.

Theorem 55. *Let* $\varepsilon_A = (\varepsilon_A(1), \cdots, \varepsilon_A(N))$, $\varepsilon_B = (\varepsilon_B(1), \cdots, \varepsilon_B(N))$, $\varepsilon_{C_y} = (\varepsilon_{C_y}(1), \cdots, \varepsilon_{C_y}(N))$, $\varepsilon_{C_z} = (\varepsilon_{C_z}(1), \cdots, \varepsilon_{C_z}(N))$, *and* $\varepsilon_{D_z} = (\varepsilon_{D_z}(1), \cdots, \varepsilon_{D_z}(N))$ *be sets of positive scalars. Let* γ *be a given positive number. System (4.1) is stable if and only if for every* $i \in \mathscr{S}$, *the following LMIs are feasible for some set of symmetric and positive-definite matrices* $X = (X(1), \cdots, X(N)) > 0$, *and* $Y = (Y(1), \cdots, Y(N))$, *and matrices* $\mathcal{K}_A = (\mathcal{K}_A(1), \cdots, \mathcal{K}_A(N))$, $\mathcal{K}_B = (\mathcal{K}_B(1), \cdots, \mathcal{K}_B(N))$, *and* $\mathcal{K}_C = (\mathcal{K}_C(1), \cdots, \mathcal{K}_C(N))$:

$$
\left[
\begin{array}{ccccccc}
\mathcal{M}_1(i) & \mathcal{N}_1^\top(i) & \mathcal{N}_4(i) & \mathcal{N}_7(i) & \mathcal{N}_9^\top(i) & \mathcal{N}_{11}^\top(i) & \mathcal{S}_i(Y) \\
\mathcal{N}_1(i) & -\mathcal{N}_3(i) & 0 & 0 & 0 & 0 & 0 \\
\mathcal{N}_4^\top(i) & 0 & -\mathcal{N}_6(i) & 0 & 0 & 0 & 0 \\
\mathcal{N}_7^\top(i) & 0 & 0 & -\Upsilon(i) & -\Upsilon^{-1}(i) & 0 & 0 \\
\mathcal{N}_9(i) & 0 & 0 & 0 & 0 & -\Psi(i) & 0 \\
\mathcal{N}_{11}(i) & 0 & 0 & 0 & 0 & 0 & -\mathcal{Y}_i(Y) \\
\mathcal{S}_i^\top(Y) & 0 & 0 & 0 & & &
\end{array}
\right] < 0, \qquad (4.46)
$$

$$
\left[
\begin{array}{cccccc}
\mathcal{M}_3(i) & \mathcal{N}_2^\top(i) & \mathcal{N}_5(i) & \mathcal{N}_8(i) & \mathcal{N}_{10}^\top(i) & \mathcal{N}_{12}^\top(i) \\
\mathcal{N}_2(i) & -\mathcal{N}_3(i) & 0 & 0 & 0 & 0 \\
\mathcal{N}_5^\top(i) & 0 & -\mathcal{N}_6(i) & 0 & 0 & 0 \\
\mathcal{N}_8^\top(i) & 0 & 0 & -\Upsilon(i) & -\Upsilon^{-1}(i) & 0 \\
\mathcal{N}_{10}(i) & 0 & 0 & 0 & 0 & -\Psi(i) \\
\mathcal{N}_{12}(i) & 0 & 0 & 0 & &
\end{array}
\right] < 0, \qquad (4.47)
$$

$$
\left[
\begin{array}{cc}
Y(i) & \mathbb{I} \\
I & X(i)
\end{array}
\right] > 0, \qquad (4.48)
$$

with

$$
\begin{aligned}
\mathscr{M}_1(i) =\ & A(i)Y(i) + Y^\top(i)A^\top(i) + \lambda_{ii}Y(i) \\
& + B(i)\mathcal{K}_C(i) + \mathcal{K}_C^\top(i)B^\top(i), \\
\mathscr{M}_3(i) =\ & X(i)A(i) + A^\top(i)X(i) + \mathcal{K}_B(i)C_y(i) \\
& + C_y^\top(i)\mathcal{K}_B^\top(i) + \sum_{j=1}^{N}\lambda_{ij}X(j), \\
\mathcal{S}_i(Y) =\ & \Big[\sqrt{\lambda_{i1}}Y(i), \\
& \cdots, \sqrt{\lambda_{ii-1}}Y(i), \sqrt{\lambda_{ii+1}}Y(i), \cdots, \sqrt{\lambda_{iN}}Y(i)\Big], \\
\mathcal{Y}_i(Y) =\ & \operatorname{diag}\left[Y(1), \cdots, Y(i-1), Y(i+1), \cdots, Y(N)\right].
\end{aligned}
$$

Furthermore, the dynamic output feedback controller is given by

$$
\begin{cases}
K_A(i) = \left[X(i) - Y^{-1}(i)\right]^{-1}\Big[A^\top(i) + X(i)A(i)Y(i) \\
\qquad + X(i)B(i)\mathcal{K}_C(i) + \mathcal{K}_B(i)C_y(i)Y(i) \\
\qquad + \mathscr{N}_2^\top(i)\mathscr{N}_3^{-1}(i)\mathscr{N}_1(i) + \mathscr{N}_5(i)\mathscr{N}_6^{-1}(i)\mathscr{N}_4^\top(i) \\
\qquad + \mathscr{N}_8(i)\varUpsilon^{-1}(i)\mathscr{N}_7^\top(i) + \mathscr{N}_{10}^\top(i)\varUpsilon(i)\mathscr{N}_9(i) \\
\qquad + \mathscr{N}_{12}^\top(i)\varPsi^{-1}(i)\mathscr{N}_{11}(i) \\
\qquad + \sum_{j=1}^{N}\lambda_{ij}Y^{-1}(j)Y(i)\Big]Y^{-1}(i), \\
K_B(i) = \left[Y^{-1}(i) - X(i)\right]^{-1}\mathcal{K}_B(i), \\
K_C(i) = \mathcal{K}_C(i)Y^{-1}(i).
\end{cases}
\tag{4.49}
$$

4.4 Observer-Based Output Stabilization

In this section we study a second method to design a stabilizing controller that uses the measurement of the input and output vectors of the system to estimate the system state which can then be used for feedback. This technique is called observer-based output feedback control. We focus on the design of such feedback stabilization to see if we can compute the gains of this controller. The structure is given by the following dynamics:

$$
\begin{cases}
\dot{x}_c(t) = A(r(t))x_c(t) + B(r(t))u(t) + B_w(r(t))\omega_1(t) \\
\qquad + L(r(t))\left[C_y(r(t))x_c(t) - y(t)\right], x_c(0) = 0, \\
u(t) = K(r(t))x_c(t),
\end{cases}
\tag{4.50}
$$

where $x_c(t)$ is the state of the observer and $L(i)$ and $K(i)$ are gains to be determined for each mode $i \in \mathscr{S}$.

In this section we assume that the external disturbances of the state equations are different and are denoted, respectively, as $\omega_1(t)$ and $\omega_2(t)$.

For simplicity in development, we assume in the dynamics (4.1) that the matrices $D_y(i)$, $B_z(i)$, and $D_z(i)$ are always equal to zero for each mode $i \in \mathscr{S}$. Using (4.1) and (4.50), and letting the observer error be defined by

$$e(t) = x(t) - x_c(t), \qquad (4.51)$$

we get

$$\dot{x}(t) = A(r(t))x(t) + B(r(t))K(r(t))x_c(t) + B_w(r(t))\omega_1(t)$$
$$= [A(r(t)) + B(r(t))K(r(t))]\, x(t) - B(r(t))K(r(t))e(t) + B_w(r(t))\omega_1(t),$$

and

$$\dot{x}_c(t) = [A(r(t)) + B(r(t))K(r(t))]\, x_c(t) - L(r(t))C_y(r(t))e(t)$$
$$+ B_w(r(t))\omega_1(t) - L(r(t))B_y(r(t))\omega_2(t)$$
$$= [A(r(t)) + B(r(t))K(r(t))]\, x(t)$$
$$- [A(r(t)) + B(r(t))K(r(t)) + L(r(t))C_y(r(t))]\, e(t)$$
$$B_w(r(t))\omega_1(t) - L(r(t))B_y(r(t))\omega_2(t).$$

Combining these equations with the error equation, we get the following dynamics for error:

$$\dot{e}(t) = [A(r(t)) + L(r(t))C_y(r(t))]\, e(t) + L(r(t))B_y(r(t))\omega_2(t),$$

which gives the following dynamics for the extended system:

$$\dot{\eta}(t) = \begin{bmatrix} A(r(t)) + B(r(t))K(r(t)) & -B(r(t))K(r(t)) \\ 0 & \begin{bmatrix} A(r(t)) \\ +L(r(t))C_y(r(t)) \end{bmatrix} \end{bmatrix} \eta(t)$$
$$+ \begin{bmatrix} B_w(r(t)) & 0 \\ 0 & L(r(t))B_y(r(t)) \end{bmatrix} \begin{bmatrix} \omega_1(t) \\ \omega_2(t) \end{bmatrix}$$
$$= \tilde{A}(r(t))\eta(t) + \tilde{B}_w(r(t))\omega(t),$$

with

$$\eta(t) = \begin{bmatrix} x(t) \\ e(t) \end{bmatrix},$$

$$\tilde{A}(r(t)) = \begin{bmatrix} A(r(t)) + B(r(t))K(r(t)) & -B(r(t))K(r(t)) \\ 0 & \begin{bmatrix} A(r(t)) \\ +L(r(t))C_y(r(t)) \end{bmatrix} \end{bmatrix},$$

$$\tilde{B}_w(r(t)) = \begin{bmatrix} B_w(r(t)) & 0 \\ 0 & L(r(t))B_y(r(t)) \end{bmatrix}.$$

For the controlled output we have

$$z(t) = \begin{bmatrix} C_z(r(t)) & 0 \end{bmatrix} \eta(t)$$

$$= \tilde{C}_z(r(t))\eta(t).$$

Based on the previous results of this chapter, the extended system will be stochastically stable and guarantee the disturbance rejection of level γ if there exists a set of symmetric and positive-definite matrices $\tilde{P} = (\tilde{P}(1), \cdots, \tilde{P}(N)) > 0$ such that the following holds for each $i \in \mathscr{S}$:

$$\tilde{J}(i) + \tilde{C}_z^\top(i)\tilde{C}_z(i) + \gamma^{-2}\tilde{P}(i)\tilde{B}_w(i)\tilde{B}_w^\top(i)\tilde{P}(i) < 0, \qquad (4.52)$$

with $\tilde{J}(i) = \tilde{A}^\top(i)\tilde{P}(i) + \tilde{P}(i)\tilde{A}(i) + \sum_{j=1}^N \lambda_{ij}\tilde{P}(j)$.

Let $\tilde{P}(i)$ be given by

$$\tilde{P}(i) = \begin{bmatrix} P(i) & 0 \\ 0 & Q(i) \end{bmatrix},$$

with $P(i)$ and $Q(i)$ symmetric and positive-definite matrices, and using the expression of the matrices $\tilde{A}(i)$, we get

$$\tilde{A}^\top(i)\tilde{P}(i) = \begin{bmatrix} \begin{bmatrix} A^\top(i)P(i) \\ +K^\top(i)B^\top(i)P(i) \end{bmatrix} & 0 \\ -K^\top(i)B^\top(i)P(i) & \begin{bmatrix} A^\top(i)Q(i) \\ +C_y^\top(i)L^\top(i)Q(i) \end{bmatrix} \end{bmatrix},$$

$$\tilde{P}(i)\tilde{B}_w(i) = \begin{bmatrix} P(i)B_w(i) & 0 \\ 0 & Q(i)L(i)B_y(i) \end{bmatrix},$$

$$\tilde{C}_z^\top(i)\tilde{C}_z(i) = \begin{bmatrix} C_z^\top(i)C_z(i) & 0 \\ 0 & 0 \end{bmatrix}.$$

Based on these computations we get

$$\tilde{J}(i) + \tilde{C}_z^\top(i)\tilde{C}_z(i) = \begin{bmatrix} \begin{bmatrix} A^\top(i)P(i) \\ +P(i)A(i) \\ +P(i)B(i)K(i) \\ +K^\top(i)B^\top(i)P(i) \\ +C_z^\top(i)C_z(i) \\ +\sum_{j=1}^N \lambda_{ij}P(j) \end{bmatrix} & -P(i)B(i)K(i) \\ -K^\top(i)B^\top(i)P(i) & \begin{bmatrix} A^\top(i)Q(i) \\ +Q(i)A(i) \\ +Q(i)L(i)C_y(i) \\ +C_y^\top(i)L^\top(i)Q(i) \\ +\sum_{j=1}^N \lambda_{ij}Q(j) \end{bmatrix} \end{bmatrix}.$$

Notice that

$$\begin{bmatrix} 0 & P(i)B(i)K(i) \\ 0 & 0 \end{bmatrix} = \begin{bmatrix} P(i)B(i)K(i) & 0 \\ 0 & 0 \end{bmatrix} \begin{bmatrix} \mathbb{I} & 0 \\ 0 & 0 \end{bmatrix} \begin{bmatrix} 0 & \mathbb{I} \\ 0 & 0 \end{bmatrix}.$$

Using Lemma 7 in Appendix A, we get

$$-\begin{bmatrix} 0 & P(i)B(i)K(i) \\ 0 & 0 \end{bmatrix} - \begin{bmatrix} 0 & P(i)B(i)K(i) \\ 0 & 0 \end{bmatrix}^{\top}$$

$$\leq \varepsilon_P^{-1}(i)\begin{bmatrix} P(i)B(i)K(i)K^{\top}(i)B^{\top}(i)P(i) & 0 \\ 0 & 0 \end{bmatrix} + \varepsilon_P(i)\begin{bmatrix} 0 & 0 \\ 0 & \mathbb{I} \end{bmatrix}.$$

Based on all these computations, the stochastic stability condition (4.52) will be satisfied if the following holds:

$$\begin{bmatrix} \begin{bmatrix} A^{\top}(i)P(i) \\ +P(i)A(i) \\ +P(i)B(i)K(i) \\ +K^{\top}(i)B^{\top}(i)P(i) \\ +C_z^{\top}(i)C_z(i) \\ +\varepsilon_P^{-1}(i)P(i)B(i)K(i) \\ \times K^{\top}(i)B^{\top}(i)P(i) \\ +\gamma^{-2}P(i)B_w(i) \\ \times B_w^{\top}(i)P(i) \\ +\sum_{j=1}^{N}\lambda_{ij}P(j) \end{bmatrix} & 0 \\[4mm] 0 & \begin{bmatrix} A^{\top}(i)Q(i) \\ +Q(i)A(i) \\ +Q(i)L(i)C_y(i) \\ +C_y^{\top}(i)L^{\top}(i)Q(i) \\ +\varepsilon_P(i)\mathbb{I} \\ +\gamma^{-2}Q(i)L(i)B_y(i) \\ \times B_y^{\top}(i)L^{\top}(i)Q(i) \\ +\sum_{j=1}^{N}\lambda_{ij}Q(j) \end{bmatrix} \end{bmatrix} < 0,$$

which implies in turn that

$$A^{\top}(i)P(i) + P(i)A(i) + P(i)B(i)K(i) + K^{\top}(i)B^{\top}(i)P(i)$$
$$+C_z^{\top}(i)C_z(i) + \varepsilon_P^{-1}(i)P(i)B(i)K(i)K^{\top}(i)B^{\top}(i)P(i)$$
$$+\gamma^{-2}P(i)B_w(i)B_w^{\top}(i)P(i) + \sum_{j=1}^{N}\lambda_{ij}P(j) < 0,$$
$$A^{\top}(i)Q(i) + Q(i)A(i) + Q(i)L(i)C_y(i) + C_y^{\top}(i)L^{\top}(i)Q(i)$$
$$+\varepsilon_P(i)\mathbb{I} + \gamma^{-2}Q(i)L(i)B_y(i)B_y^{\top}(i)L^{\top}(i)Q(i) + \sum_{j=1}^{N}\lambda_{ij}Q(j) < 0.$$

Using the Schur complement these conditions become

$$\begin{bmatrix} J_P(i) & P(i)B_w(i) & P(i)B(i)K(i) \\ B_w^{\top}(i)P(i) & -\gamma^2\mathbb{I} & 0 \\ K^{\top}(i)B^{\top}(i)P(i) & 0 & -\varepsilon_P(i)\mathbb{I} \end{bmatrix} < 0,$$

$$\begin{bmatrix} J_Q(i) & Q(i)L(i)B_y(i) \\ B_y^{\top}(i)L^{\top}(i)Q(i) & -\gamma^2\mathbb{I} \end{bmatrix} < 0,$$

where

$$J_P(i) = A^\top(i)P(i) + P(i)A(i) + P(i)B(i)K(i) + K^\top(i)B^\top(i)P(i)$$
$$+C_z^\top(i)C_z(i) + \sum_{j=1}^{N} \lambda_{ij}P(j),$$
$$J_Q(i) = A^\top(i)Q(i) + Q(i)A(i) + Q(i)L(i)C_y(i) + C_y^\top(i)L^\top(i)Q(i)$$
$$+\varepsilon_P(i)\mathbb{I} + \sum_{j=1}^{N} \lambda_{ij}Q(j).$$

These two inequalities are nonlinear in their design parameters. To put them in the LMI setting let us change some variables. For the first LMI, notice that it implies the following:

$$\begin{bmatrix} J_P(i) & P(i)B_w(i) \\ B_w^\top(i)P(i) & -\gamma^2\mathbb{I} \end{bmatrix} < 0.$$

Let $X(i) = P^{-1}(i)$. Pre- and post-multiply this LMI by $\operatorname{diag}(X(i), \mathbb{I})$ to get

$$\begin{bmatrix} X(i)J_P(i)X(i) & B_w(i) \\ B_w^\top(i) & -\gamma^2\mathbb{I} \end{bmatrix} < 0.$$

Using the fact that

$$X(i)J_P(i)X(i) = X(i)A^\top(i) + A(i)X(i) + B(i)K(i)X(i)$$
$$+X(i)K^\top(i)B^\top(i) + X(i)C_z^\top(i)C_z(i)X(i) + \sum_{j=1}^{N} \lambda_{ij}X(i)X^{-1}(j)X(i),$$

and

$$\sum_{j=1}^{N} \lambda_{ij}X(i)X^{-1}(j)X(i) = \lambda_{ii}X(i) + \mathcal{S}_i(X)\mathcal{X}_i^{-1}(X)\mathcal{S}_i^\top(X),$$

with $\mathcal{S}_i(X)$ and $\mathcal{X}_i(X)$ and letting $Y_c(i) = K(i)X(i)$, we get

$$\begin{bmatrix} \mathscr{J}_P(i) & B_w(i) & X(i)C_z^\top(i) & \mathcal{S}_i(X) \\ B_w^\top(i) & -\gamma^2\mathbb{I} & 0 & 0 \\ C_z(i)X(i) & 0 & -\mathbb{I} & 0 \\ \mathcal{S}_i^\top(X) & 0 & 0 & -\mathcal{X}_i(X) \end{bmatrix} < 0,$$

with

$$\mathscr{J}_P(i) = X(i)A^\top(i) + A(i)X(i) + B(i)Y_c(i) + Y_c^\top(i)B^\top(i) + \lambda_{ii}X(i).$$

For the second nonlinear matrix inequality, letting $Y_o(i) = Q(i)L(i)$ implies the following LMI:

$$\begin{bmatrix} \mathscr{J}_Q(i) & Y_o(i)B_y(i) \\ B_y^\top(i)Y_o^\top(i) & -\gamma^2\mathbb{I} \end{bmatrix} < 0,$$

where

$$\mathscr{J}_Q(i) = A^\top(i)Q(i) + Q(i)A(i) + Y_o(i)C_y(i) + C_y^\top(i)Y_o^\top(i)$$

$$+\varepsilon_P(i)\mathbb{I} + \sum_{j=1}^{N} \lambda_{ij}Q(j).$$

The following theorem summarizes the results of this development.

Theorem 56. *Let γ be a given positive constant. If there exist sets of symmetric and positive-definite matrices $X = (X(1),\cdots,X(N)) > 0$ and $Q = (Q(1),\cdots,Q(N)) > 0$ and sets of matrices $Y_c = (Y_c(1),\cdots,Y_c(N))$ and $Y_o = (Y_o(1),\cdots,Y_o(N))$ such that the following sets of coupled LMIs hold for every $i \in \mathscr{S}$:*

$$\begin{bmatrix} \mathscr{J}_P(i) & B_w(i) & X(i)C_z^\top(i) & S_i(X) \\ B_w^\top(i) & -\gamma^2\mathbb{I} & 0 & 0 \\ C_z(i)X(i) & 0 & -\mathbb{I} & 0 \\ S_i^\top(X) & 0 & 0 & -\mathcal{X}_i(X) \end{bmatrix} < 0, \qquad (4.53)$$

$$\begin{bmatrix} \mathscr{J}_Q(i) & Y_o(i)B_y(i) \\ B_y^\top(i)Y_o^\top(i) & -\gamma^2\mathbb{I} \end{bmatrix} < 0, \qquad (4.54)$$

with

$$\mathscr{J}_P(i) = X(i)A^\top(i) + A(i)X(i) + B(i)Y_c(i) + Y_c^\top(i)B^\top(i) + \lambda_{ii}X(i),$$
$$\mathscr{J}_Q(i) = A^\top(i)Q(i) + Q(i)A(i) + Y_o(i)C_y(i) + C_y^\top(i)Y_o^\top(i)$$

$$+\varepsilon_P(i)\mathbb{I} + \sum_{j=1}^{N} \lambda_{ij}Q(j),$$

then the observer-based output feedback control with the following gains:

$$L(i) = Q^{-1}(i)Y_o(i), \qquad (4.55)$$
$$K(i) = Y_c(i)X^{-1}(i), \qquad (4.56)$$

stochastically stabilizes the class of systems we are studying and at the same time guarantees the disturbance rejection of level γ.

From the practical point of view, the observer-based output feedback control that stochastically stabilizes the system and at the same time guarantees the minimum disturbance rejection is of great interest. This controller can be obtained by solving the following optimization problem:

$$P : \begin{cases} \min\limits_{\substack{\nu>0, \\ X=(X(1),\cdots,X(N))>0, \\ Q=(Q(1),\cdots,Q(N))>0, \\ Y_c=(Y_c(1),\cdots,Y_c(N)), \\ Y_o=(Y_o(1),\cdots,Y_o(N)), }} \nu, \\ s.t. : (4.53) \text{ and } (4.54), \text{ with } \nu = \gamma^2. \end{cases}$$

The following corollary gives the results of the design of the controller that stochastically stabilizes the system (4.1) and simultaneously guarantees the smallest disturbance rejection level.

Corollary 11. *Let* $\nu > 0$, $X = (X(1),\cdots,X(N)) > 0$, $Q = (Q(1),\cdots, Q(N)) > 0$, $Y_c = (Y_c(1),\cdots,Y_c(N))$, *and* $Y_o = (Y_o(1),\cdots,Y_o(N))$ *be the solution of the optimization problem P. Then the controller (4.50) with* $K(i) = Y_c(i)X^{-1}(i)$ *and* $L(i) = Q^{-1}(i)Y_o(i)$ *stochastically stabilizes the class of systems we are considering and, moreover, the closed-loop system satisfies the disturbance rejection of level* $\sqrt{\nu}$.

Example 58. To show the validity of the previous results let us consider a system with two modes with the following data:

- mode #1:

$$A(1) = \begin{bmatrix} 1.0 & 0.0 & 1.0 \\ 0.0 & 1.0 & 0.0 \\ 0.0 & 1.0 & 1.0 \end{bmatrix}, \quad B(1) = \begin{bmatrix} 0.3 & 0.0 & 0.1 \\ 0.0 & 0.3 & 0.1 \\ 0.0 & 0.2 & 1.0 \end{bmatrix},$$

$$B_w(1) = \begin{bmatrix} 0.3 & 0.0 & 0.1 \\ 0.0 & 0.3 & 0.1 \\ 0.0 & 0.2 & 1.0 \end{bmatrix}, \quad C_y(1) = \begin{bmatrix} 0.3 & 0.0 & 0.1 \\ 0.0 & 0.3 & 0.1 \\ 0.0 & 0.2 & 1.0 \end{bmatrix},$$

$$B_y(1) = \begin{bmatrix} 0.3 & 0.0 & 0.1 \\ 0.0 & 0.3 & 0.1 \\ 0.0 & 0.2 & 1.0 \end{bmatrix}, \quad C_z(1) = \begin{bmatrix} 0.3 & 0.0 & 0.1 \\ 0.0 & 0.3 & 0.1 \\ 0.0 & 0.2 & 1.0 \end{bmatrix},$$

- mode #2:

$$A(2) = \begin{bmatrix} 1.0 & 0.0 & 1.0 \\ 0.0 & -1.0 & 0.0 \\ 0.0 & 1.0 & -1.0 \end{bmatrix}, \quad B(2) = \begin{bmatrix} 0.1 & 0.0 & 0.1 \\ 0.2 & 0.0 & 0.1 \\ 0.0 & 0.2 & 0.2 \end{bmatrix},$$

$$B_w(2) = \begin{bmatrix} 0.1 & 0.0 & 0.1 \\ 0.2 & 0.0 & 0.1 \\ 0.0 & 0.2 & 0.2 \end{bmatrix}, \quad C_y(2) = \begin{bmatrix} 0.1 & 0.0 & 0.1 \\ 0.2 & 0.0 & 0.1 \\ 0.0 & 0.2 & 0.2 \end{bmatrix},$$

$$B_y(2) = \begin{bmatrix} 0.1 & 0.0 & 0.1 \\ 0.2 & 0.0 & 0.1 \\ 0.0 & 0.2 & 0.2 \end{bmatrix}, \quad C_z(2) = \begin{bmatrix} 0.1 & 0.0 & 0.1 \\ 0.2 & 0.0 & 0.1 \\ 0.0 & 0.2 & 0.2 \end{bmatrix}.$$

The switching between the two modes is described by the following transition matrix:

$$\Lambda = \begin{bmatrix} -2.0 & 2.0 \\ 1.0 & -1.0 \end{bmatrix}.$$

Letting $\gamma = 1.0001$ and $\varepsilon_P(1) = \varepsilon_P(2) = 0.1$ and solving the LMIs (4.53)–(4.54), we get

$$X(1) = \begin{bmatrix} 0.4359 & 0.0001 & -0.0016 \\ 0.0001 & 0.4342 & -0.0130 \\ -0.0016 & -0.0130 & 0.3980 \end{bmatrix},$$

$$X(2) = \begin{bmatrix} 0.5546 & 0.0009 & -0.0044 \\ 0.0009 & 0.5556 & -0.0127 \\ -0.0044 & -0.0127 & 0.5172 \end{bmatrix},$$

$$Q(1) = \begin{bmatrix} 0.2309 & 0.0524 & -0.1490 \\ 0.0524 & 0.3142 & -0.1873 \\ -0.1490 & -0.1873 & 0.3085 \end{bmatrix},$$

$$Q(2) = \begin{bmatrix} 0.1816 & -0.0238 & 0.0133 \\ -0.0238 & 0.9020 & 0.1864 \\ 0.0133 & 0.1864 & 0.7542 \end{bmatrix},$$

$$Y_c(1) = \begin{bmatrix} -1.2225 & -30.8435 & -10.0310 \\ 16.6117 & -17.9955 & -154.9530 \\ -3.7886 & 46.5175 & 29.7948 \end{bmatrix},$$

$$Y_c(2) = \begin{bmatrix} 1.6423 & -6.7984 & -12.8747 \\ 31.2102 & 11.1533 & 32.5000 \\ -11.4035 & 14.7539 & -32.1622 \end{bmatrix},$$

$$Y_o(1) = \begin{bmatrix} -2.0854 & -0.1062 & 0.2608 \\ 0.0702 & -1.8611 & 0.1915 \\ 0.1362 & 0.4518 & -0.6117 \end{bmatrix},$$

$$Y_o(2) = \begin{bmatrix} 4.5711 & -5.1555 & -0.0411 \\ -4.2065 & 2.0675 & 0.5072 \\ 5.2752 & -2.9807 & -0.5636 \end{bmatrix}.$$

The corresponding gains are given by

$$K(1) = \begin{bmatrix} -2.8798 & -71.8554 & -27.5711 \\ 36.7147 & -53.1929 & -390.9547 \\ -8.4448 & 109.4810 & 78.4204 \end{bmatrix},$$

$$K(2) = \begin{bmatrix} 2.7816 & -12.8149 & -25.1842 \\ 56.7476 & 21.4392 & 63.8493 \\ -21.0923 & 25.1832 & -61.7515 \end{bmatrix},$$

$$L(1) = \begin{bmatrix} -13.1190 & -1.2448 & -0.4532 \\ -1.7251 & -8.1503 & -0.9834 \\ -6.9423 & -4.0854 & -2.7990 \end{bmatrix},$$

$$L(2) = \begin{bmatrix} 23.8431 & -27.7796 & -0.0591 \\ -5.6820 & 2.3959 & 0.7534 \\ 7.9790 & -4.0554 & -0.9324 \end{bmatrix}.$$

Previously, we used the following decomposition:

$$\begin{bmatrix} 0 & P(i)B(i)K(i) \\ 0 & 0 \end{bmatrix} = \begin{bmatrix} P(i)B(i)K(i) & 0 \\ 0 & 0 \end{bmatrix} \begin{bmatrix} 0 & \mathbb{I} \\ 0 & 0 \end{bmatrix},$$

which gives us some terms that we have ignored since they are positive, which may give conservative results. We can use another decomposition and get less conservative results. Notice that

$$\begin{bmatrix} 0 & P(i)B(i)K(i) \\ 0 & 0 \end{bmatrix} = \begin{bmatrix} P(i)B(i) & 0 \\ 0 & 0 \end{bmatrix} \begin{bmatrix} 0 & K(i) \\ 0 & 0 \end{bmatrix},$$

which implies

$$\begin{bmatrix} 0 & P(i)B(i)K(i) \\ 0 & 0 \end{bmatrix} + \begin{bmatrix} 0 & P(i)B(i)K(i) \\ 0 & 0 \end{bmatrix}^\top$$
$$\leq \begin{bmatrix} P(i)B(i)B^\top(i)P(i) & 0 \\ 0 & K^\top(i)K(i) \end{bmatrix}.$$

Following the same reasoning as before, we get

$$A^\top(i)P(i) + P(i)A(i) + P(i)B(i)K(i) + K^\top(i)B^\top(i)P(i) + C_z^\top(i)C_z(i)$$
$$+ P(i)B(i)B^\top(i)P(i) + \gamma^{-2}P(i)B_w(i)B_w^\top(i)P(i) + \sum_{j=1}^{N} \lambda_{ij}P(j) < 0,$$

$$A^\top(i)Q(i) + Q(i)A(i) + Q(i)L(i)C_y(i) + C_y^\top(i)L^\top(i)Q(i) + K^\top(i)K(i)$$
$$+ \gamma^{-2}Q(i)L(i)B_y(i)B_y^\top(i)L^\top(i)Q(i) + \sum_{j=1}^{N} \lambda_{ij}Q(j) < 0.$$

These two matrix inequalities are nonlinear in the design parameters that we should put in the LMI form. Let us transform the first one. Let $X(i) = P^{-1}(i)$ and pre- and post-multiply this inequality by $X(i)$ to give

$$X(i)A^\top(i) + A(i)X(i) + B(i)K(i)X(i) + X(i)K^\top(i)B^\top(i)$$
$$+ X(i)C_z^\top(i)C_z(i)X(i) + B(i)B^\top(i) + \gamma^{-2}B_w(i)B_w^\top(i)$$
$$+ \sum_{j=1}^{N} \lambda_{ij}X(i)X^{-1}(j)X(i) < 0.$$

Let $Y_c(i) = K(i)X(i)$ and use the expression

$$\sum_{j=1}^{N} \lambda_{ij}X(i)X^{-1}(j)X(i) = \lambda_{ii}X(i) + \mathcal{S}_i(X)\mathcal{X}_i^{-1}(X)\mathcal{X}_i^\top(X)$$

to give

$$
\begin{bmatrix}
J_{\mathscr{P}}(i) & X(i)C_z^\top(i) & B_w(i) & \mathcal{S}_i(X) \\
C_z(i)X(i) & -\mathbb{I} & 0 & 0 \\
B_w^\top & 0 & -\gamma^2\mathbb{I} & 0 \\
\mathcal{S}_i^\top(X) & 0 & 0 & -\mathcal{X}_i(X)
\end{bmatrix} < 0,
$$

with $J_{\mathscr{P}}(i) = X(i)A^\top(i) + A(i)X(i) + B(i)Y_c(i) + Y_c^\top(i)B^\top(i) + B(i)B^\top(i) + \lambda_{ii}X(i)$.

For the second inequality, let $Y_o(i) = Q(i)L(i)$ to give

$$
\begin{bmatrix}
J_{\mathscr{Q}}(i) & K^\top(i) & Y_o(i)B_y(i) \\
K(i) & -\mathbb{I} & 0 \\
B_y^\top(i)Y_o^\top(i) & 0 & -\gamma^2\mathbb{I}
\end{bmatrix} < 0,
$$

with $J_{\mathscr{Q}}(i) = A^\top(i)Q(i) + Q(i)A(i) + Y_o(i)C_y(i) + C_y^\top(i)Y_o^\top(i) + \sum_{j=1}^N \lambda_{ij}Q(j)$.
The following theorem summarizes the results of this development.

Theorem 57. *Let γ be a given positive constant. If there exist sets of symmetric and positive-definite matrices $X = (X(1), \cdots, X(N)) > 0$ and $Q = (Q(1), \cdots, Q(N)) > 0$ and sets of matrices $Y_c = (Y_c(1), \cdots, Y_c(N))$ and $Y_o = (Y_o(1), \cdots, Y_o(N))$ such that the following sets of coupled LMIs hold for every $i \in \mathscr{S}$:*

$$
\begin{bmatrix}
\mathscr{J}_{\mathscr{P}}(i) & X(i)C_z^\top(i) & B_w(i) & \mathcal{S}_i(X) \\
C_z(i)X(i) & -\mathbb{I} & 0 & 0 \\
B_w^\top & 0 & -\gamma^2\mathbb{I} & 0 \\
\mathcal{S}_i^\top(X) & 0 & 0 & -\mathcal{X}_i(X)
\end{bmatrix} < 0, \qquad (4.57)
$$

$$
\begin{bmatrix}
\mathscr{J}_{\mathscr{Q}}(i) & K^\top(i) & Y_o(i)B_y(i) \\
K(i) & -\mathbb{I} & 0 \\
B_y^\top(i)Y_o^\top(i) & 0 & -\gamma^2\mathbb{I}
\end{bmatrix} < 0, \qquad (4.58)
$$

with

$$
\begin{aligned}
\mathscr{J}_{\mathscr{P}}(i) &= X(i)A^\top(i) + A(i)X(i) + B(i)Y_c(i) + Y_c^\top(i)B^\top(i) \\
&\quad + B(i)B^\top(i) + \lambda_{ii}X(i), \\
\mathscr{J}_{\mathscr{Q}}(i) &= A^\top(i)Q(i) + Q(i)A(i) + Y_o(i)C_y(i) + C_y^\top(i)Y_o^\top(i) \\
&\quad + \varepsilon_P(i)\mathbb{I} + \sum_{j=1}^N \lambda_{ij}Q(j),
\end{aligned}
$$

then the observer-based output feedback control with the following gains:

$$
L(i) = Q^{-1}(i)Y_o(i), \qquad (4.59)
$$

$$
K(i) = Y_c(i)X^{-1}(i), \qquad (4.60)
$$

stochastically stabilizes the class of systems we are studying and at the same time guarantees the disturbance rejection of level γ.

Remark 16. Notice that the second LMI depends on the solution of the first one. Therefore, to get the solution, we should solve the first set of LMIs to get $K = (K(1), \cdots, K(N))$, which enters in the second set of LMIs.

Let us now consider the uncertainties in the dynamics and see how to modify conditions (4.53) and (4.54) to design a robust observer-based output feedback control that robustly stochastically stabilizes the class of systems under study.

Combining the system dynamics and the controller dynamics and using the same techniques as before for the nominal system, we get the following:

- for the system dynamics:

$$
\begin{aligned}
\dot{x}(t) =\; & [A(r(t)) + \Delta A(r(t), t)]\, x(t) + [B(r(t)) + \Delta B(r(t), t)]\, K(r(t))x_c(t) \\
& + B_w(r(t))\omega_1(t) \\
=\; & [A(r(t)) + \Delta A(r(t), t) + [B(r(t)) + \Delta B(r(t), t)]\, K(r(t))]\, x(t) \\
& - [B(r(t)) + \Delta B(r(t), t)]\, K(r(t))e(t) + B_w(r(t))\omega_1(t),
\end{aligned}
$$

- for the controller dynamics:

$$
\begin{aligned}
\dot{x}_c(t) =\; & A(r(t))x_c(t) + B(r(t))u(t) + L(r(t))\, [C_y(r(t))x_c(t) - y(t)] \\
& + B_w(r(t))\omega_1(t) \\
=\; & [A(r(t)) + B(r(t))K(r(t))]\, x_c(t) + B_w(r(t))\omega_1(t) \\
& + L(r(t))\, [C_y(r(t))x_c(t) - [C_y(r(t)) + \Delta C_y(t)]\, x(t) \\
& - B_y(r(t))\omega_2(t)] \\
=\; & [A(r(t)) + B(r(t))K(r(t)) - L(r(t))\Delta C_y(r(t))]\, x(t) \\
& - [L(r(t))C_y(r(t)) + [A(r(t)) + B(r(t))K(r(t))]]\, e(t) \\
& + B_w(r(t))\omega_1(t) - L(r(t))B_y(r(t))\omega_2(t),
\end{aligned}
$$

- for the error dynamics:

$$
\begin{aligned}
\dot{e}(t) =\; & \dot{x}(t) - \dot{x}_c(t) \\
=\; & [A(r(t)) + \Delta A(r(t), t) + [B(r(t)) + \Delta B(r(t), t)]\, K(r(t))]\, x(t) \\
& - [B(r(t)) + \Delta B(r(t), t)]\, K(r(t))e(t) + B_w(r(t))\omega_1(t) \\
& - [A(r(t)) + B(r(t))K(r(t)) - L(r(t))\Delta C_y(r(t), t)]\, x(t) \\
& + [L(r(t))C_y(r(t)) + [A(r(t)) + B(r(t))K(r(t))]]\, e(t) \\
& + L(r(t))B_y(r(t))\omega_2(t) - B_w(r(t))\omega_1(t) \\
=\; & [\Delta A(r(t), t) + \Delta B(r(t), t)K(r(t)) + L(r(t))\Delta C_y(r(t), t)]\, x(t) \\
& + [A(r(t)) + L(r(t))C_y(r(t)) - \Delta B(r(t), t)K(r(t))]\, e(t) \\
& + L(r(t))B_y(r(t))\omega_2(t),
\end{aligned}
$$

- for the extended dynamics:

$$\dot{\eta}(t) = \left[\tilde{A}(r(t)) + \Delta\tilde{A}(r(t),t)\right]\eta(t) + \tilde{B}_w(r(t))\omega(t)$$
$$= \tilde{A}(r(t),t)\eta(t) + \tilde{B}_w(r(t))\omega(t),$$

where

$$\eta(t) = \begin{bmatrix} x(t) \\ e(t) \end{bmatrix},$$

$$\omega(t) = \begin{bmatrix} \omega_1(t) \\ \omega_2(t) \end{bmatrix},$$

$$\tilde{A}(r(t)) = \begin{bmatrix} A(r(t)) + B(r(t))K(r(t)) & -B(r(t))K(r(t)) \\ 0 & A(r(t)) + L(r(t))C_y(r(t)) \end{bmatrix},$$

$$\tilde{\Delta A}(r(t),t) = \begin{bmatrix} \Delta A(r(t),t) + \Delta B(r(t),t)K(r(t)) & -\Delta B(r(t),t)K(r(t)) \\ \begin{bmatrix} \Delta A(r(t),t) \\ +\Delta B(r(t),t)K(r(t)) \\ +L(r(t))\Delta C_y(r(t),t) \end{bmatrix} & -\Delta B(r(t),t)K(r(t)) \end{bmatrix},$$

$$\tilde{B}_w(r(t)) = \begin{bmatrix} B_w(r(t)) & 0 \\ 0 & L(r(t))B_y(r(t)) \end{bmatrix}.$$

For the controlled output, we have

$$z(t) = [C_z(r(t)) + \Delta C_z(r(t),t)]\,x(t) = \left[\tilde{C}_z(r(t)) + \tilde{\Delta}C_z(r(t),t)\right]\eta(t),$$

with

$$\tilde{C}_z(r(t)) = \left[\, C_z(r(t))\; 0\,\right],$$
$$\Delta\tilde{C}_z(r(t),t) = \left[\,\Delta C_z(r(t),t)\; 0\,\right].$$

The extended uncertain system will be stochastically stable and guarantee the disturbance rejection of level γ, if there exists a set of symmetric and positive-definite matrices $\tilde{P} = (\tilde{P}(1),\cdots,\tilde{P}(N)) > 0$ such that the following holds for each $i \in \mathscr{S}$:

$$\tilde{J}_u(i,t) + \tilde{C}_z^\top(i,t)\tilde{C}_z(i,t) + \gamma^{-2}\tilde{P}(i)\tilde{B}_w(i)\tilde{B}_w^\top(i)\tilde{P}(i) < 0, \qquad (4.61)$$

with $\tilde{J}_u(i,t) = \tilde{A}^\top(i,t)\tilde{P}(i) + \tilde{P}(i)\tilde{A}(i,t) + \sum_{j=1}^N \lambda_{ij}\tilde{P}(j)$.

Using the Schur complement, we obtain

$$\begin{bmatrix} \tilde{J}_u(i,t) & \tilde{P}(i)\tilde{B}_w(i) & \tilde{C}_z^\top(i,t) \\ \tilde{B}_w^\top(i)\tilde{P}(i) & -\gamma^2\mathbb{I} & 0 \\ \tilde{C}_z(i,t) & 0 & -\mathbb{I} \end{bmatrix} < 0.$$

Based on the expressions of $\tilde{A}(i,t)$, $\tilde{C}_z(i,t)$, we get

$$\begin{bmatrix} \tilde{J}(i) & \tilde{P}(i)\tilde{B}_w(i) & \tilde{C}_z^\top(i) \\ \tilde{B}_w^\top(i)\tilde{P}(i) & -\gamma^2\mathbb{I} & 0 \\ \tilde{C}_z(i) & 0 & -\mathbb{I} \end{bmatrix}$$

$$+ \begin{bmatrix} \Delta\tilde{A}^\top(i,t)\tilde{P}(i) + \tilde{P}(i)\Delta\tilde{A}(i,t) & 0 & \Delta\tilde{C}_z^\top(i,t) \\ 0 & 0 & 0 \\ \Delta\tilde{C}_z(i,t) & 0 & 0 \end{bmatrix} < 0,$$

where $\tilde{J}(i) = \tilde{A}^\top(i)P(i) + P(i)\tilde{A}(i) + \sum_{j=1}^N \lambda_{ij}P(j)$.

Let $\tilde{P}(i)$ be given by

$$\tilde{P}(i) = \begin{bmatrix} P(i) & 0 \\ 0 & Q(i) \end{bmatrix},$$

where $P(i)$ and $Q(i)$ are symmetric and positive-definite matrices, and using the expression of the matrices $\tilde{A}(i)$, we get

$$\tilde{A}^\top(i)P(i) = \begin{bmatrix} \begin{bmatrix} A^\top(i)P(i) \\ +K^\top(i)B^\top(i)P(i) \end{bmatrix} & 0 \\ -K^\top(i)B^\top(i)P(i) & \begin{bmatrix} A^\top(i)Q(i) \\ +C_y^\top(i)L^\top(i)Q(i) \end{bmatrix} \end{bmatrix},$$

$$P(i)\tilde{B}_w(i) = \begin{bmatrix} P(i)B_w(i) & 0 \\ 0 & Q(i)L(i)B_y(i) \end{bmatrix},$$

$$\sum_{j=1}^N \lambda_{ij}\tilde{P}(j) = \begin{bmatrix} \sum_{j=1}^N \lambda_{ij}P(j) & 0 \\ 0 & \sum_{j=1}^N \lambda_{ij}Q(j) \end{bmatrix},$$

$$\tilde{P}(i)\begin{bmatrix} \Delta A(i,t) & 0 \\ 0 & 0 \end{bmatrix} = \begin{bmatrix} P(i)D_A(i)F_A(i,t)E_A(i) & 0 \\ 0 & 0 \end{bmatrix},$$

$$\tilde{P}(i)\begin{bmatrix} \Delta B(i,t)K(i) & 0 \\ 0 & 0 \end{bmatrix} = \begin{bmatrix} P(i)D_B(i)F_B(i,t)E_B(i)K(i) & 0 \\ 0 & 0 \end{bmatrix},$$

$$\tilde{P}(i)\begin{bmatrix} 0 & 0 \\ \Delta A(i,t) & 0 \end{bmatrix} = \begin{bmatrix} 0 & 0 \\ Q(i)D_A(i)F_A(i,t)E_A(i) & 0 \end{bmatrix},$$

$$\tilde{P}(i)\begin{bmatrix} 0 & 0 \\ \Delta B(i,t)K(i) & 0 \end{bmatrix} = \begin{bmatrix} 0 & 0 \\ Q(i)D_B(i)F_B(i,t)E_B(i)K(i) & 0 \end{bmatrix},$$

$$\tilde{P}(i)\begin{bmatrix} 0 & 0 \\ L(i)\Delta C_y(i,t) & 0 \end{bmatrix} = \begin{bmatrix} 0 & 0 \\ Q(i)L(i)D_{C_y}(i)F_{C_y}(i,t)E_{C_y}(i) & 0 \end{bmatrix},$$

$$\tilde{P}(i)\begin{bmatrix} 0 & -\Delta B(i,t)K(i) \\ 0 & 0 \end{bmatrix} = \begin{bmatrix} 0 & -P(i)D_B(i)F_B(i,t)E_B(i)K(i) \\ 0 & 0 \end{bmatrix},$$

$$\tilde{P}(i)\begin{bmatrix} 0 & 0 \\ 0 & -\Delta B(i,t)K(i) \end{bmatrix} = \begin{bmatrix} 0 & 0 \\ 0 & -Q(i)D_B(i)F_B(i,t)E_B(i)K(i) \end{bmatrix}.$$

Based on Lemma 7 in Appendix A, we get

$$\begin{bmatrix} P(i)D_A(i)F_A(i,t)E_A(i) & 0 \\ 0 & 0 \end{bmatrix} + \begin{bmatrix} P(i)D_A(i)F_A(i,t)E_A(i) & 0 \\ 0 & 0 \end{bmatrix}^\top$$

$$\leq \varepsilon_A(i)\begin{bmatrix} P(i)D_A(i)D_A^\top(i)P(i) & 0 \\ 0 & 0 \end{bmatrix} + \varepsilon_A^{-1}(i)\begin{bmatrix} E_A^\top(i)E_A(i) & 0 \\ 0 & 0 \end{bmatrix},$$

$$\begin{bmatrix} P(i)D_B(i)F_B(i,t)E_B(i)K(i) & 0 \\ 0 & 0 \end{bmatrix} + \begin{bmatrix} P(i)D_B(i)F_B(i,t)E_B(i)K(i) & 0 \\ 0 & 0 \end{bmatrix}^\top$$
$$\leq \varepsilon_B(i) \begin{bmatrix} P(i)D_B(i)D_B^\top(i)P(i) & 0 \\ 0 & 0 \end{bmatrix} + \varepsilon_B^{-1}(i) \begin{bmatrix} K^\top(i)E_B^\top(i)E_B(i)K(i) & 0 \\ 0 & 0 \end{bmatrix},$$

$$\begin{bmatrix} 0 & 0 \\ Q(i)D_A(i)F_A(i,t)E_A(i) & 0 \end{bmatrix} + \begin{bmatrix} 0 & 0 \\ Q(i)D_A(i)F_A(i,t)E_A(i) & 0 \end{bmatrix}^\top$$
$$\leq \varepsilon_C^{-1}(i) \begin{bmatrix} 0 & 0 \\ 0 & Q(i)D_A(i)D_A^\top(i)Q(i) \end{bmatrix} + \varepsilon_C(i) \begin{bmatrix} E_A^\top(i)E_A(i) & 0 \\ 0 & 0 \end{bmatrix},$$

$$\begin{bmatrix} 0 & 0 \\ Q(i)D_B(i)F_B(i,t)E_B(i)K(i) & 0 \end{bmatrix} + \begin{bmatrix} 0 & 0 \\ Q(i)D_B(i)F_B(i,t)E_B(i)K(i) & 0 \end{bmatrix}^\top$$
$$\leq \varepsilon_D^{-1}(i) \begin{bmatrix} 0 & 0 \\ 0 & Q(i)D_B(i)D_B^\top(i)Q(i) \end{bmatrix} + \varepsilon_D(i) \begin{bmatrix} K^\top(i)E_B^\top(i)E_B(i)K(i) & 0 \\ 0 & 0 \end{bmatrix},$$

$$\begin{bmatrix} 0 & 0 \\ Q(i)L(i)D_{C_y}(i)F_{C_y}(i,t)E_{C_y}(i) & 0 \end{bmatrix} + \begin{bmatrix} 0 & 0 \\ Q(i)L(i)D_{C_y}(i)F_{C_y}(i,t)E_{C_y}(i) & 0 \end{bmatrix}^\top$$
$$\leq \varepsilon_{C_y}^{-1}(i) \begin{bmatrix} 0 & 0 \\ 0 & Q(i)L(i)D_{C_y}(i)D_{C_y}^\top(i)L^\top(i)Q(i) \end{bmatrix} + \varepsilon_{C_y}(i) \begin{bmatrix} E_{C_y}^\top(i)E_{C_y}(i) & 0 \\ 0 & 0 \end{bmatrix},$$

$$\begin{bmatrix} 0 & -P(i)D_B(i)F_B(i,t)E_B(i)K(i) \\ 0 & 0 \end{bmatrix} + \begin{bmatrix} 0 & -P(i)D_B(i)F_B(i,t)E_B(i)K(i) \\ 0 & 0 \end{bmatrix}^\top$$
$$\leq \varepsilon_E(i) \begin{bmatrix} P(i)D_B(i)D_B^\top(i)P(i) & 0 \\ 0 & 0 \end{bmatrix} + \varepsilon_E^{-1}(i) \begin{bmatrix} 0 & 0 \\ 0 & K^\top(i)E_B^\top(i)E_B(i)K(i) \end{bmatrix},$$

$$\begin{bmatrix} 0 & 0 \\ 0 & -Q(i)D_B(i)F_B(i,t)E_B(i)K(i) \end{bmatrix} + \begin{bmatrix} 0 & 0 \\ 0 & -Q(i)D_B(i)F_B(i,t)E_B(i)K(i) \end{bmatrix}^\top$$
$$\leq \varepsilon_F^{-1}(i) \begin{bmatrix} 0 & 0 \\ 0 & Q(i)D_B(i)D_B^\top(i)Q(i) \end{bmatrix} + \varepsilon_F(i) \begin{bmatrix} 0 & 0 \\ 0 & K^\top(i)E_B^\top(i)E_B(i)K(i) \end{bmatrix},$$

$$\begin{bmatrix} 0 & 0 & 0 & 0 \\ 0 & 0 & 0 & 0 \\ 0 & 0 & 0 & 0 \\ D_{C_z}(i)F_{C_z}(i,t)E_{C_z}(i) & 0 & 0 & 0 \end{bmatrix} + \begin{bmatrix} 0 & 0 & 0 & 0 \\ 0 & 0 & 0 & 0 \\ 0 & 0 & 0 & 0 \\ D_{C_z}(i)F_{C_z}(i,t)E_{C_z}(i) & 0 & 0 & 0 \end{bmatrix}^\top$$

$$\leq \varepsilon_G(i) \begin{bmatrix} 0 & 0 & 0 & 0 \\ 0 & 0 & 0 & 0 \\ 0 & 0 & 0 & 0 \\ 0 & 0 & 0 & D_{C_z}(i)D_{C_z}^\top(i) \end{bmatrix} + \varepsilon_G^{-1}(i) \begin{bmatrix} E_{C_z}^\top(i)E_{C_z}(i) & 0 & 0 & 0 \\ 0 & 0 & 0 & 0 \\ 0 & 0 & 0 & 0 \\ 0 & 0 & 0 & 0 \end{bmatrix}.$$

Based on these computations and the ones we did for the nominal dynamics, we get

$$\begin{bmatrix} \begin{bmatrix} A^\top(i)P(i) \\ +P(i)A(i) \\ +P(i)B(i)K(i) \\ +K^\top(i)B^\top(i)P(i) \\ +C_z^\top(i)\left[\mathbb{I} - \varepsilon_G(i)D_{C_z}(i) \right. \\ \left. \times D_{C_z}^\top(i)\right]^{-1} C_z(i) \\ +\varepsilon_P^{-1}(i)P(i)B(i)K(i) \\ \times K^\top(i)B^\top(i)P(i) \\ +\gamma^{-2}P(i)B_w(i) \\ \times B_w^\top(i)P(i) \\ +\varepsilon_A(i)P(i)D_A(i)D_A^\top(i)P(i) \\ +\varepsilon_A^{-1}(i)E_A^\top(i)E_A(i) \\ +\varepsilon_B(i)P(i)D_B(i)D_B^\top(i)P(i) \\ +\varepsilon_B^{-1}(i)K^\top(i)E_B^\top(i)E_B(i)K(i) \\ +\varepsilon_C(i)E_A^\top(i)E_A(i) \\ +\varepsilon_D(i)K^\top(i)E_B^\top(i)E_B(i)K(i) \\ +\varepsilon_{C_y}(i)E_{C_y}^\top(i)E_{C_y}(i) \\ +\varepsilon_E(i)P(i)D_B(i)D_B^\top(i)P(i) \\ +\varepsilon_G^{-1}(i)E_{C_z}^\top(i)E_{C_z}(i) \\ +\sum_{j=1}^N \lambda_{ij}P(j) \end{bmatrix} & 0 \\[2em] 0 & \begin{bmatrix} A^\top(i)Q(i) \\ +Q(i)A(i) \\ +Q(i)L(i)C_y(i) \\ +C_y^\top(i)L^\top(i)Q(i) \\ +\varepsilon_P(i)\mathbb{I} \\ +\gamma^{-2}Q(i)L(i)B_y(i) \\ \times B_y^\top(i)L^\top(i)Q(i) \\ +\varepsilon_C^{-1}(i)Q(i)D_A(i)D_A^\top(i)Q(i) \\ +\varepsilon_D^{-1}(i)Q(i)D_B(i)D_B^\top(i)Q(i) \\ +\varepsilon_{C_y}^{-1}(i)Q(i)L(i)D_{C_y}(i) \\ \times D_{C_y}^\top(i)L^\top(i)Q(i) \\ +\varepsilon_E^{-1}(i)K^\top(i)E_B^\top(i)E_B(i)K(i) \\ +\varepsilon_F^{-1}(i)Q(i)D_B(i)D_B^\top(i)Q(i) \\ +\varepsilon_F(i)K^\top(i)E_B^\top(i)E_B(i)K(i) \\ +\varepsilon_G(i)D_{C_z}(i)D_{C_z}^\top(i) \\ +\sum_{j=1}^N \lambda_{ij}Q(j) \end{bmatrix} \end{bmatrix} < 0,$$

which implies in turn that

$$A^\top(i)P(i) + P(i)A(i) + P(i)B(i)K(i) + K^\top(i)B^\top(i)P(i)$$
$$+C_z^\top(i)\left[\mathbb{I} - \varepsilon_G(i)D_{C_z}(i)D_{C_z}^\top(i)\right]^{-1}C_z(i)$$
$$+\varepsilon_P^{-1}(i)P(i)B(i)K(i)K^\top(i)B^\top(i)P(i) + \gamma^{-2}P(i)B_w(i)B_w^\top(i)P(i)$$
$$+\varepsilon_A(i)P(i)D_A(i)D_A^\top(i)P(i) + \varepsilon_A^{-1}(i)E_A^\top(i)E_A(i)$$
$$+\varepsilon_B(i)P(i)D_B(i)D_B^\top(i)P(i) + \varepsilon_B^{-1}(i)K^\top(i)E_B^\top(i)E_B(i)K(i)$$
$$+\varepsilon_C(i)E_A^\top(i)E_A(i) + \varepsilon_D(i)K^\top(i)E_B^\top(i)E_B(i)K(i)$$
$$+\varepsilon_{C_y}(i)E_{C_y}^\top(i)E_{C_y}(i) + \varepsilon_E(i)P(i)D_B(i)D_B^\top(i)P(i)$$
$$+\varepsilon_G^{-1}(i)E_{C_z}^\top(i)E_{C_z}(i) + \sum_{j=1}^N \lambda_{ij}P(j) < 0,$$
$$A^\top(i)Q(i) + Q(i)A(i) + Q(i)L(i)C_y(i) + C_y^\top(i)L^\top(i)Q(i)$$
$$+\varepsilon_P(i)\mathbb{I} + \gamma^{-2}Q(i)L(i)B_y(i)B_y^\top(i)L^\top(i)Q(i)$$
$$+\varepsilon_C^{-1}(i)Q(i)D_A(i)D_A^\top(i)Q(i) + \varepsilon_D^{-1}(i)Q(i)D_B(i)D_B^\top(i)Q(i)$$
$$+\varepsilon_{C_y}^{-1}(i)Q(i)L(i)D_{C_y}(i)D_{C_y}^\top(i)L^\top(i)Q(i) + \varepsilon_E^{-1}(i)K^\top(i)E_B^\top(i)E_B(i)K(i)$$
$$+\varepsilon_F^{-1}(i)Q(i)D_B(i)D_B^\top(i)Q(i) + \varepsilon_F(i)K^\top(i)E_B^\top(i)E_B(i)K(i)$$
$$+\varepsilon_G(i)D_{C_z}(i)D_{C_z}^\top(i) + \sum_{j=1}^N \lambda_{ij}Q(j) < 0.$$

These two conditions are nonlinear in the design parameters $P(i)$, $Q(i)$, $K(i)$, and $L(i)$. To put them in the LMI framework let us proceed as before. For this purpose, let us transform the first condition. Let $X(i) = P^{-1}(i)$, and pre- and post-multiply the first condition by $X(i)$ to yield

$$X(i)A^\top(i) + A(i)X(i) + B(i)K(i)X(i) + X(i)K^\top(i)B^\top(i)$$
$$+X(i)C_z^\top(i)\left[\mathbb{I} - \varepsilon_G(i)D_{C_z}(i)D_{C_z}^\top(i)\right]^{-1}C_z(i)X(i)$$
$$+\varepsilon_P^{-1}(i)B(i)K(i)K^\top(i)B^\top(i) + \gamma^{-2}B_w(i)B_w^\top(i)$$
$$+\varepsilon_A(i)D_A(i)D_A^\top(i) + \varepsilon_A^{-1}(i)X(i)E_A^\top(i)E_A(i)X(i)$$
$$+\varepsilon_B(i)D_B(i)D_B^\top(i) + \varepsilon_B^{-1}(i)X(i)K^\top(i)E_B^\top(i)E_B(i)K(i)X(i)$$
$$+\varepsilon_C(i)X(i)E_A^\top(i)E_A(i)X(i) + \varepsilon_D(i)X(i)K^\top(i)E_B^\top(i)E_B(i)K(i)X(i)$$
$$+\varepsilon_{C_y}(i)X(i)E_{C_y}^\top(i)E_{C_y}(i)X(i) + \varepsilon_E(i)D_B(i)D_B^\top(i)$$
$$+\varepsilon_G^{-1}(i)X(i)E_{C_z}^\top(i)E_{C_z}(i)X(i) + \sum_{j=1}^N \lambda_{ij}X(i)X^{-1}(j)X(i) < 0.$$

Letting $Y_c(i) = K(i)X(i)$ and using the previous expression for $\sum_{j=1}^N \lambda_{ij}$ $X(i)X^{-1}(j)X(i)$, and noticing that $\varepsilon_P^{-1}(i)B(i)K(i)K^\top(i)B^\top(i) > 0$, we get the following LMI:

$$
\left[\begin{array}{cccccccccc}
\mathscr{J}_X(i) & X(i)C_z^\top(i) & B_w(i) & X(i)E_A^\top(i) & Y_c^\top(i)E_B^\top(i) & Y_c^\top(i)E_B^\top(i) & X(i)E_A^\top(i) & X(i)E_{C_y}^\top(i) & X(i)E_{C_z}^\top(i) & \mathcal{S}_i(X) \\
C_z(i)X(i) & -\mathbb{I}+\varepsilon_G(i)D_{C_z}(i)D_{C_z}^\top(i) & 0 & 0 & 0 & 0 & 0 & 0 & 0 & 0 \\
B_w^\top(i) & 0 & -\gamma^2\mathbb{I} & 0 & 0 & 0 & 0 & 0 & 0 & 0 \\
E_A(i)X(i) & 0 & 0 & -\varepsilon_A(i)\mathbb{I} & 0 & 0 & 0 & 0 & 0 & 0 \\
E_B(i)Y_c(i) & 0 & 0 & 0 & -\varepsilon_D^{-1}(i)\mathbb{I} & 0 & 0 & 0 & 0 & 0 \\
E_B(i)Y_c(i) & 0 & 0 & 0 & 0 & -\varepsilon_B(i)\mathbb{I} & 0 & 0 & 0 & 0 \\
E_A(i)X(i) & 0 & 0 & 0 & 0 & 0 & -\varepsilon_C^{-1}(i)\mathbb{I} & 0 & 0 & 0 \\
E_{C_y}(i)X(i) & 0 & 0 & 0 & 0 & 0 & 0 & -\varepsilon_{C_y}^{-1}(i)\mathbb{I} & 0 & 0 \\
E_{C_z}(i)X(i) & 0 & 0 & 0 & 0 & 0 & 0 & 0 & -\varepsilon_G(i)\mathbb{I} & 0 \\
\mathcal{S}_i^\top(X) & 0 & 0 & 0 & 0 & 0 & 0 & 0 & 0 & -\mathcal{X}_i(X)
\end{array}\right] < 0,
$$

where

$$
\mathscr{J}_X(i) = X(i)A^\top(i) + A(i)X(i) + B(i)Y_c(i) + Y_c^\top(i)B^\top(i) + \lambda_{ii}X(i)
$$
$$
+\varepsilon_A(i)D_A(i)D_A^\top(i) + \varepsilon_B(i)D_B(i)D_B^\top(i) + \varepsilon_E(i)D_B(i)D_B^\top(i).
$$

For the second condition, letting $Y_o(i) = Q(i)L(i)$ yields

$$
\left[\begin{array}{cccccccc}
\mathscr{J}_Q(i) & Y_o(i)B_y(i) & Q(i)D_A(i) & Q(i)D_B(i) & Y_o(i)D_{C_y}(i) & Q(i)D_B(i) & K^\top(i)E_B^\top(i) & K^\top(i)E_B^\top(i) \\
B_y^\top(i)Y_o^\top(i) & -\gamma^2\mathbb{I} & 0 & 0 & 0 & 0 & 0 & 0 \\
D_A^\top(i)Q(i) & 0 & -\varepsilon_C(i)\mathbb{I} & 0 & 0 & 0 & 0 & 0 \\
D_B^\top(i)Q(i) & 0 & 0 & -\varepsilon_D(i)\mathbb{I} & 0 & 0 & 0 & 0 \\
D_{C_y}^\top(i)Y_o^\top(i) & 0 & 0 & 0 & -\varepsilon_{C_y}(i)\mathbb{I} & 0 & 0 & 0 \\
D_B^\top(i)Q(i) & 0 & 0 & 0 & 0 & -\varepsilon_F(i)\mathbb{I} & 0 & 0 \\
E_B(i)K(i) & 0 & 0 & 0 & 0 & 0 & -\varepsilon_E(i)\mathbb{I} & 0 \\
E_B(i)K(i) & 0 & 0 & 0 & 0 & 0 & 0 & -\varepsilon_F^{-1}(i)\mathbb{I}
\end{array}\right] < 0,
$$

where

$$\mathscr{J}_Q(i) = A^\top(i)Q(i) + Q(i)A(i) + Y_o(i)C_y(i) + C_y^\top(i)Y_o^\top(i)$$

$$+\varepsilon_P(i)\mathbb{I} + \varepsilon_G(i)D_{C_z}(i)D_{C_z}^\top(i) + \sum_{j=1}^{N} \lambda_{ij}Q(j).$$

The following theorem summarizes the results of this development.

Theorem 58. *Let* $\varepsilon_A = (\varepsilon_A(1), \cdots, \varepsilon_A(N)) > 0,$ $\varepsilon_B = (\varepsilon_B(1), \cdots, \varepsilon_B(N)) > 0,$ $\varepsilon_C = (\varepsilon_C(1), \cdots, \varepsilon_C(N)) > 0,$ $\varepsilon_{C_y} = (\varepsilon_{C_y}(1), \cdots, \varepsilon_{C_y}(N)) > 0,$ $\varepsilon_D = (\varepsilon_D(1), \cdots, \varepsilon_D(N)) > 0,$ $\varepsilon_E = (\varepsilon_E(1), \cdots, \varepsilon_E(N)) > 0,$ $\varepsilon_F = (\varepsilon_F(1), \cdots, \varepsilon_F(N)) > 0,$ $\varepsilon_G = (\varepsilon_G(1), \cdots, \varepsilon_G(N)) > 0,$ $\varepsilon_P = (\varepsilon_P(1), \cdots, \varepsilon_P(N)) > 0$ *be sets of positive scalars. Let* γ *be a given positive constant. If there exist sets of symmetric and positive-definite matrices* $X = (X(1), \cdots, X(N)) > 0$ *and* $Q = (Q(1), \cdots, Q(N)) > 0$ *and sets of matrices* $Y_c = (Y_c(1), \cdots, Y_c(N))$ *and* $Y_o = (Y_o(1), \cdots, Y_o(N))$ *such that the following LMIs hold for every* $i \in \mathscr{S}$:

$$\begin{bmatrix} \mathscr{J}_X(i) & X(i)C_z^\top(i) & B_w(i) & X(i)E_A^\top(i) & Y_c^\top(i)E_B^\top(i) \\ C_z(i)X(i) & -\mathbb{I} + \varepsilon_G(i)D_{C_z}(i)D_{C_z}^\top(i) & 0 & 0 & 0 \\ B_w^\top(i) & 0 & -\gamma^2\mathbb{I} & 0 & 0 \\ E_A(i)X(i) & 0 & 0 & -\varepsilon_A(i)\mathbb{I} & 0 \\ E_B(i)Y_c(i) & 0 & 0 & 0 & -\varepsilon_D^{-1}(i)\mathbb{I} \\ E_B(i)Y_c(i) & 0 & 0 & 0 & 0 \\ E_A(i)X(i) & 0 & 0 & 0 & 0 \\ E_{C_y}(i)X(i) & 0 & 0 & 0 & 0 \\ E_{C_z}(i)X(i) & 0 & 0 & 0 & 0 \\ \mathcal{S}_i^\top(X) & 0 & 0 & 0 & 0 \end{bmatrix}$$

$$\begin{bmatrix} Y_c^\top(i)E_B^\top(i) & X(i)E_A^\top(i) & X(i)E_{C_y}^\top(i) & X(i)E_{C_z}^\top(i) & \mathcal{S}_i(X) \\ 0 & 0 & 0 & 0 & 0 \\ 0 & 0 & 0 & 0 & 0 \\ 0 & 0 & 0 & 0 & 0 \\ 0 & 0 & 0 & 0 & 0 \\ -\varepsilon_B(i)\mathbb{I} & 0 & 0 & 0 & 0 \\ 0 & -\varepsilon_C^{-1}(i)\mathbb{I} & 0 & 0 & 0 \\ 0 & 0 & -\varepsilon_{C_y}^{-1}(i)\mathbb{I} & 0 & 0 \\ 0 & 0 & 0 & -\varepsilon_G(i)\mathbb{I} & 0 \\ 0 & 0 & 0 & 0 & -\mathcal{X}_i(X) \end{bmatrix} < 0, \quad (4.62)$$

$$\begin{bmatrix} \mathscr{J}_Q(i) & Y_o(i)B_y(i) & Q(i)D_A(i) & Q(i)D_B(i) \\ B_y^\top(i)Y_o^\top(i) & -\gamma^2\mathbb{I} & 0 & 0 \\ D_A^\top(i)Q(i) & 0 & -\varepsilon_C(i)\mathbb{I} & 0 \\ D_B^\top(i)Q(i) & 0 & 0 & -\varepsilon_D(i)\mathbb{I} \\ D_{C_y}^\top(i)Y_o^\top(i) & 0 & 0 & 0 \\ D_B^\top(i)Q(i) & 0 & 0 & 0 \\ E_B(i)K(i) & 0 & 0 & 0 \\ E_B(i)K(i) & 0 & 0 & 0 \end{bmatrix}$$

$$\begin{bmatrix} Y_o(i)D_{C_y}(i) & Q(i)D_B(i) & K^\top(i)E_B^\top(i) & K^\top(i)E_B^\top(i) \\ 0 & 0 & 0 & 0 \\ 0 & 0 & 0 & 0 \\ 0 & 0 & 0 & 0 \\ -\varepsilon_{C_y}(i)\mathbb{I} & 0 & 0 & 0 \\ 0 & -\varepsilon_F(i)\mathbb{I} & 0 & 0 \\ 0 & 0 & -\varepsilon_E(i)\mathbb{I} & 0 \\ 0 & 0 & 0 & -\varepsilon_F^{-1}(i)\mathbb{I} \end{bmatrix} < 0, \qquad (4.63)$$

then the observer-based control with the following gains:

$$L(i) = Q^{-1}(i)Y_o(i), \qquad (4.64)$$

$$K(i) = Y_c(i)X^{-1}(i), \qquad (4.65)$$

robustly stochastically stabilizes the class of systems we are studying and at the same time guarantees the disturbance rejection of level γ.

From the practical point of view, the observer-based control that stochastically robustly stabilizes the system and at the same time guarantees the minimum disturbance rejection is of great interest. This controller can be obtained by solving the following optimization problem:

$$\mathrm{P}: \begin{cases} \min\limits_{\substack{\nu>0, \\ X=(X(1),\cdots,X(N))>0, \\ Q=(Q(1),\cdots,Q(N))>0, \\ Y_c=(Y_c(1),\cdots,Y_c(N)), \\ Y_o=(Y_o(1),\cdots,Y_o(N)),}} \nu, \\ s.t.: (4.62) \text{ and } (4.63) \text{ with } \nu = \gamma^2. \end{cases}$$

The following corollary gives the results of the design of the controller that stochastically stabilizes the system (4.1) and simultaneously guarantees the smallest disturbance rejection level.

Corollary 12. *Let $\nu > 0$, $X = (X(1),\cdots,X(N)) > 0$, $Q = (Q(1),\cdots, Q(N)) > 0$, $Y_c = (Y_c(1),\cdots,Y_c(N))$, and $Y_o = (Y_o(1),\cdots,Y_o(N))$ be the solution of the optimization problem P. Then the controller (4.50) with $K(i) = Y_c(i)X^{-1}(i)$ and $L(i) = Q^{-1}(i)Y_o(i)$ stochastically stabilizes the class of systems we are considering and, moreover, the closed-loop system satisfies the disturbance rejection of level $\sqrt{\nu}$.*

Example 59. To illustrate the results of the previous theorem, let us consider the two-mode system with state space in $\mathbb{R}^3$ of the previous example with the following extra data:

- mode #1:

$$D_A(1) = \begin{bmatrix} 0.1 \\ 0.2 \\ 0.2 \end{bmatrix}, \qquad E_A(1) = \begin{bmatrix} 0.2 & 0.1 & 0.1 \end{bmatrix},$$

$$D_B(1) = \begin{bmatrix} 0.1 \\ 0.2 \\ 0.2 \end{bmatrix}, \quad E_B(1) = \begin{bmatrix} 0.2 \ 0.1 \ 0.1 \end{bmatrix},$$

$$D_{C_y}(1) = \begin{bmatrix} 0.1 \\ 0.2 \\ 0.2 \end{bmatrix}, \quad E_{C_y}(1) = \begin{bmatrix} 0.2 \ 0.1 \ 0.1 \end{bmatrix},$$

- mode #2:

$$D_A(2) = \begin{bmatrix} 0.13 \\ 0.1 \\ 0.1 \end{bmatrix}, \quad E_A(2) = \begin{bmatrix} 0.1 \ 0.2 \ 0.2 \end{bmatrix},$$

$$D_B(2) = \begin{bmatrix} 0.13 \\ 0.1 \\ 0.1 \end{bmatrix}, \quad E_B(2) = \begin{bmatrix} 0.1 \ 0.2 \ 0.2 \end{bmatrix},$$

$$D_{C_y}(2) = \begin{bmatrix} 0.13 \\ 0.1 \\ 0.1 \end{bmatrix}, \quad E_{C_y}(2) = \begin{bmatrix} 0.1 \ 0.2 \ 0.2 \end{bmatrix}.$$

Let

$$\varepsilon_A(1) = \varepsilon_A(2) = 0.5, \varepsilon_B(1) = \varepsilon_B(2) = 0.1, \varepsilon_C(1) = \varepsilon_C(2) = 0.1,$$
$$\varepsilon_{C_y}(1) = \varepsilon_{C_y}(2) = 0.1, \varepsilon_D(1) = \varepsilon_D(2) = 0.1, \varepsilon_E(1) = \varepsilon_E(2) = 0.1,$$
$$\varepsilon_F(1) = \varepsilon_F(2) = 0.1, \varepsilon_G(1) = \varepsilon_G(2) = 0.1, \varepsilon_P(1) = \varepsilon_P(2) = 0.1.$$

Solving the optimization problem iteratively we get $\gamma = 1.0001$ and the corresponding matrices are

$$X(1) = \begin{bmatrix} 1.1891 & -1.2714 & -0.3465 \\ -1.2714 & 4.7803 & -1.8519 \\ -0.3465 & -1.8519 & 4.7348 \end{bmatrix},$$

$$X(2) = \begin{bmatrix} 2.5471 & 1.1317 & -0.6506 \\ 1.1317 & 5.9863 & -5.0774 \\ -0.6506 & -5.0774 & 8.4276 \end{bmatrix},$$

$$Y_c(1) = \begin{bmatrix} -11.6591 & 23.8088 & 6.5857 \\ 27.5583 & -58.2554 & 9.1138 \\ -5.2481 & 9.8404 & -23.0591 \end{bmatrix},$$

$$Y_c(2) = \begin{bmatrix} 35.2732 & -57.7214 & 17.8767 \\ 164.8969 & -19.2224 & 147.5903 \\ -182.4561 & 47.9403 & -156.7615 \end{bmatrix},$$

$$Q(1) = \begin{bmatrix} 0.2121 & -0.0225 & -0.1250 \\ -0.0225 & 0.1953 & -0.1768 \\ -0.1250 & -0.1768 & 0.3743 \end{bmatrix},$$

$$Q(2) = \begin{bmatrix} 0.2453 & -0.2391 & -0.1802 \\ -0.2391 & 7.3322 & -3.5416 \\ -0.1802 & -3.5416 & 4.0056 \end{bmatrix},$$

$$Y_o(1) = \begin{bmatrix} -2.6968 & 0.8976 & 0.2141 \\ 1.0410 & -1.0789 & 0.0700 \\ 0.1182 & 0.9326 & -1.0586 \end{bmatrix},$$

$$Y_o(2) = \begin{bmatrix} 5.4077 & -7.4053 & 0.4305 \\ 1.9615 & -1.2240 & -1.5041 \\ 3.9213 & -2.0942 & -3.0715 \end{bmatrix},$$

$$L(1) = \begin{bmatrix} -16.7451 & 6.7031 & -4.0813 \\ -2.3959 & -0.8230 & -6.8220 \\ -6.4068 & 4.3406 & -7.4131 \end{bmatrix},$$

$$L(2) = \begin{bmatrix} 30.5904 & -39.0788 & -0.5410 \\ 4.1933 & -4.4382 & -1.0558 \\ 6.0624 & -6.2045 & -1.7246 \end{bmatrix},$$

$$K(1) = \begin{bmatrix} -2.9970 & 5.4655 & 3.3093 \\ 13.6955 & -8.7333 & -0.4888 \\ -9.8416 & -3.2112 & -6.8462 \end{bmatrix},$$

$$K(2) = \begin{bmatrix} 21.0037 & -21.3461 & -9.1176 \\ 68.6938 & 6.4495 & 26.7015 \\ -79.1390 & 4.1125 & -22.2328 \end{bmatrix}.$$

As we did for the nominal system, we can establish other results that do not neglect anything resulting from the transformation of the terms:

$$\begin{bmatrix} 0 & P(i)B(i)K(i) \\ 0 & 0 \end{bmatrix} + \begin{bmatrix} 0 & P(i)B(i)K(i) \\ 0 & 0 \end{bmatrix}^\top.$$

We can use another decomposition and get less conservative results. In fact, notice that

$$\begin{bmatrix} 0 & P(i)B(i)K(i) \\ 0 & 0 \end{bmatrix} = \begin{bmatrix} P(i)B(i) & 0 \\ 0 & 0 \end{bmatrix} \begin{bmatrix} 0 & K(i) \\ 0 & 0 \end{bmatrix},$$

which implies

$$\begin{bmatrix} 0 & P(i)B(i)K(i) \\ 0 & 0 \end{bmatrix} + \begin{bmatrix} 0 & P(i)B(i)K(i) \\ 0 & 0 \end{bmatrix}^\top$$
$$\leq \begin{bmatrix} P(i)B(i)B^\top(i)P(i) & 0 \\ 0 & K^\top(i)K(i) \end{bmatrix}.$$

Following the same reasoning as before, we get

$$A^\top(i)P(i) + P(i)A(i) + P(i)B(i)K(i) + K^\top(i)B^\top(i)P(i)$$

$$+C_z^\top(i)\left[\mathbb{I}-\varepsilon_G(i)D_{C_z}(i)D_{C_z}^\top(i)\right]^{-1}C_z(i)$$
$$+P(i)B(i)B^\top(i)P(i)+\gamma^{-2}P(i)B_w(i)B_w^\top(i)P(i)$$
$$+\varepsilon_A(i)P(i)D_A(i)D_A^\top(i)P(i)+\varepsilon_A^{-1}(i)E_A^\top(i)E_A(i)$$
$$+\varepsilon_B(i)P(i)D_B(i)D_B^\top(i)P(i)+\varepsilon_B^{-1}(i)K^\top(i)E_B^\top(i)E_B(i)K(i)$$
$$+\varepsilon_C(i)E_A^\top(i)E_A(i)+\varepsilon_D(i)K^\top(i)E_B^\top(i)E_B(i)K(i)$$
$$+\varepsilon_{C_y}(i)E_{C_y}^\top(i)E_{C_y}(i)+\varepsilon_E^{-1}(i)P(i)D_B(i)D_B^\top(i)P(i)$$
$$+\varepsilon_G^{-1}(i)E_{C_z}^\top(i)E_{C_z}(i)+\sum_{j=1}^N\lambda_{ij}P(j)<0,$$
$$A^\top(i)Q(i)+Q(i)A(i)+Q(i)L(i)C_y(i)+C_y^\top(i)L^\top(i)Q(i)$$
$$+K^\top(i)K(i)+\gamma^{-2}Q(i)L(i)B_y(i)B_y^\top(i)L^\top(i)Q(i)$$
$$+\varepsilon_C^{-1}(i)Q(i)D_A(i)D_A^\top(i)Q(i)+\varepsilon_D^{-1}(i)Q(i)D_B(i)D_B^\top(i)Q(i)$$
$$+\varepsilon_{C_y}^{-1}(i)Q(i)L(i)D_{C_y}(i)D_{C_y}^\top(i)L^\top(i)Q(i)+\varepsilon_E^{-1}(i)K^\top(i)E_B^\top(i)E_B(i)K(i)$$
$$+\varepsilon_F^{-1}(i)Q(i)D_B(i)D_B^\top(i)Q(i)+\varepsilon_F(i)K^\top(i)E_B^\top(i)E_B(i)K(i)$$
$$+\varepsilon_G(i)D_{C_z}(i)D_{C_z}^\top(i)+\sum_{j=1}^N\lambda_{ij}Q(j)<0.$$

These two conditions are nonlinear in design parameters $P(i)$, $Q(i)$, $K(i)$, and $L(i)$. To put them in the LMI framework let us proceed as before. For this purpose, let us transform the first condition. Let $X(i)=P^{-1}(i)$ and pre- and post-multiply the first condition by $X(i)$ to yield

$$X(i)A^\top(i)+A(i)X(i)+B(i)K(i)X(i)+X(i)K^\top(i)B^\top(i)$$
$$+X(i)C_z^\top(i)\left[\mathbb{I}-\varepsilon_G(i)D_{C_z}(i)D_{C_z}^\top(i)\right]^{-1}C_z(i)X(i)$$
$$+B(i)B^\top(i)+\gamma^{-2}B_w(i)B_w^\top(i)$$
$$+\varepsilon_A(i)D_A(i)D_A^\top(i)+\varepsilon_A^{-1}(i)X(i)E_A^\top(i)E_A(i)X(i)$$
$$+\varepsilon_B(i)D_B(i)D_B^\top(i)+\varepsilon_B^{-1}(i)X(i)K^\top(i)E_B^\top(i)E_B(i)K(i)X(i)$$
$$+\varepsilon_C(i)X(i)E_A^\top(i)E_A(i)X(i)+\varepsilon_D(i)X(i)K^\top(i)E_B^\top(i)E_B(i)K(i)X(i)$$
$$+\varepsilon_{C_y}(i)X(i)E_{C_y}^\top(i)E_{C_y}(i)X(i)+\varepsilon_E^{-1}(i)D_B(i)D_B^\top(i)$$
$$+\varepsilon_G^{-1}(i)X(i)E_{C_z}^\top(i)E_{C_z}(i)X(i)+\sum_{j=1}^N\lambda_{ij}X(i)X^{-1}(j)X(i)<0.$$

Letting $Y_c(i)=K(i)X(i)$ and using the previous expression for

$$\sum_{j=1}^N\lambda_{ij}X(i)X^{-1}(j)X(i),$$

and noticing that $\varepsilon_P^{-1}(i)B(i)K(i)K^\top(i)B^\top(i)>0$, we get the following LMI:

$$
\left[
\begin{array}{ccccc}
\mathscr{J}_X(i) & X(i)C_z^\top(i) & B_w(i) & X(i)E_A^\top(i) & Y_c^\top(i)E_B^\top(i) \\
C_z(i)X(i) & -\mathbb{I}+\varepsilon_G(i)D_{C_z}(i)D_{C_z}^\top(i) & 0 & 0 & 0 \\
B_w^\top(i) & 0 & -\gamma^2\mathbb{I} & 0 & 0 \\
E_A(i)X(i) & 0 & 0 & -\varepsilon_A(i)\mathbb{I} & 0 \\
E_B(i)Y_c(i) & 0 & 0 & 0 & -\varepsilon_D^{-1}(i)\mathbb{I} \\
E_B(i)Y_c(i) & 0 & 0 & 0 & 0 \\
E_A(i)X(i) & 0 & 0 & 0 & 0 \\
E_{C_y}(i)X(i) & 0 & 0 & 0 & 0 \\
E_{C_z}(i)X(i) & 0 & 0 & 0 & 0 \\
\mathcal{S}_i^\top(X) & 0 & 0 & 0 & 0
\end{array}
\right.
$$

$$
\left.
\begin{array}{ccccc}
Y_c^\top(i)E_B^\top(i) & X(i)E_A^\top(i) & X(i)E_{C_y}^\top(i) & X(i)E_{C_z}^\top(i) & \mathcal{S}_i(X) \\
0 & 0 & 0 & 0 & 0 \\
0 & 0 & 0 & 0 & 0 \\
0 & 0 & 0 & 0 & 0 \\
0 & 0 & 0 & 0 & 0 \\
-\varepsilon_B(i)\mathbb{I} & 0 & 0 & 0 & 0 \\
0 & -\varepsilon_C^{-1}(i)\mathbb{I} & 0 & 0 & 0 \\
0 & 0 & -\varepsilon_{C_y}^{-1}(i)\mathbb{I} & 0 & 0 \\
0 & 0 & 0 & -\varepsilon_G(i)\mathbb{I} & 0 \\
0 & 0 & 0 & 0 & -\mathcal{X}_i(X)
\end{array}
\right] < 0, \quad (4.66)
$$

where

$$
\begin{aligned}
\mathscr{J}_X(i) &= X(i)A^\top(i) + A(i)X(i) + B(i)Y_c(i) + Y_c^\top(i)B^\top(i) + B(i)B^\top(i) \\
&\quad + \lambda_{ii}X(i) + \varepsilon_A(i)D_A(i)D_A^\top(i) + \varepsilon_B(i)D_B(i)D_B^\top(i) \\
&\quad + \varepsilon_E^{-1}(i)D_B(i)D_B^\top(i).
\end{aligned}
$$

For the second condition, letting $Y_o(i) = Q(i)L(i)$ yields

$$
\left[
\begin{array}{ccccc}
\mathscr{J}_Q(i) & Y_o(i)B_y(i) & Q(i)D_A(i) & Q(i)D_B(i) & Y_o(i)D_{C_y}(i) \\
B_y^\top(i)Y_o^\top(i) & -\gamma^2\mathbb{I} & 0 & 0 & 0 \\
D_A^\top(i)Q(i) & 0 & -\varepsilon_C(i)\mathbb{I} & 0 & 0 \\
D_B^\top(i)Q(i) & 0 & 0 & -\varepsilon_D(i)\mathbb{I} & 0 \\
D_{C_y}^\top(i)Y_o^\top(i) & 0 & 0 & 0 & -\varepsilon_{C_y}(i)\mathbb{I} \\
D_B^\top(i)Q(i) & 0 & 0 & 0 & 0 \\
E_B(i)K(i) & 0 & 0 & 0 & 0 \\
E_B(i)K(i) & 0 & 0 & 0 & 0 \\
K(i) & 0 & 0 & 0 & 0
\end{array}
\right.
$$

$$\begin{bmatrix} Q(i)D_B(i) & K^\top(i)E_B^\top(i) & K^\top(i)E_B^\top(i) & K^\top(i) \\ 0 & 0 & 0 & 0 \\ 0 & 0 & 0 & 0 \\ 0 & 0 & 0 & 0 \\ 0 & 0 & 0 & 0 \\ -\varepsilon_F(i)\mathbb{I} & 0 & 0 & 0 \\ 0 & -\varepsilon_E(i)\mathbb{I} & 0 & 0 \\ 0 & 0 & -\varepsilon_F^{-1}(i)\mathbb{I} & 0 \\ 0 & 0 & 0 & -\mathbb{I} \end{bmatrix} < 0, \qquad (4.67)$$

where

$$\mathscr{J}_Q(i) = A^\top(i)Q(i) + Q(i)A(i) + Y_o(i)C_y(i) + C_y^\top(i)Y_o^\top(i)$$

$$+\varepsilon_G(i)D_{C_z}(i)D_{C_z}^\top(i) + \sum_{j=1}^{N}\lambda_{ij}Q(j).$$

The following theorem summarizes the results of this development.

Theorem 59. *Let* $\varepsilon_A = (\varepsilon_A(1), \cdots, \varepsilon_A(N)) > 0$, $\varepsilon_B = (\varepsilon_B(1), \cdots, \varepsilon_B(N)) > 0$, $\varepsilon_C = (\varepsilon_C(1), \cdots, \varepsilon_C(N)) > 0$, $\varepsilon_{C_y} = (\varepsilon_{C_y}(1), \cdots, \varepsilon_{C_y}(N)) > 0$, $\varepsilon_D = (\varepsilon_D(1), \cdots, \varepsilon_D(N)) > 0$, $\varepsilon_E = (\varepsilon_E(1), \cdots, \varepsilon_E(N)) > 0$, $\varepsilon_F = (\varepsilon_F(1), \cdots, \varepsilon_F(N)) > 0$, $\varepsilon_G = (\varepsilon_G(1), \cdots, \varepsilon_G(N)) > 0$, $\varepsilon_P = (\varepsilon_P(1), \cdots, \varepsilon_P(N)) > 0$ *be sets of positive scalars. Let* γ *be a given positive constant. If there exist sets of symmetric and positive-definite matrices* $X = (X(1), \cdots, X(N)) > 0$ *and* $Q = (Q(1), \cdots, Q(N)) > 0$ *and sets of matrices* $Y_c = (Y_c(1), \cdots, Y_c(N))$ *and* $Y_o = (Y_o(1), \cdots, Y_o(N))$ *such that the following LMIs (4.66) and (4.67) hold for every* $i \in \mathscr{S}$, *then the observer-based control with the following gains:*

$$L(i) = Q^{-1}(i)Y_o(i), \qquad (4.68)$$
$$K(i) = Y_c(i)X^{-1}(i), \qquad (4.69)$$

robustly stochastically stabilizes the class of systems we are studying and at the same time guarantees the disturbance rejection of level γ.

4.5 Stochastic Systems with Multiplicative Noise

Let us consider a dynamical system defined in a probability space $(\Omega, \mathscr{F}, \mathscr{P})$ and assume that its dynamics are described by the following differential equations:

$$\begin{cases} dx(t) = A(r(t), t)x(t)dt + B(r(t), t)u(t)dt + B_w(r(t))\omega(t)dt \\ \qquad + \mathbb{W}(r(t))x(t)dw(t), x(0) = x_0, \\ y(t) = C_y(r(t), t)x(t) + D_y(r(t), t)u(t) + B_y(r(t))\omega(t), \\ z(t) = C_z(r(t), t)x(t) + D_z(r(t), t)u(t) + B_z(r(t))\omega(t), \end{cases} \qquad (4.70)$$

where $x(t) \in \mathbb{R}^n$ is the state vector; $x_0 \in \mathbb{R}^n$ is the initial state; $y(t) \in \mathbb{R}^{n_y}$ is the measured output; $z(t) \in \mathbb{R}^{n_z}$ is the controlled output; $u(t) \in \mathbb{R}^m$ is the control input; and $\omega(t) \in \mathbb{R}^l$ is the system external disturbance. $w(t) \in \mathbb{R}$ is a standard Wiener process that is assumed to be independent of $\{r(t), t \geq 0\}$, which is a continuous-time Markov process taking values in a finite space $\mathscr{S} = \{1, \cdots, N\}$ and describing the evolution of the mode at time t, when $r(t) = i$. The matrices $A(r(t), t)$, $B(r(t), t)$, $C_y(r(t), t)$, $D_y(r(t), t)$, $C_z(r(t), t)$, and $D_z(r(t), t)$ are given by

$$\begin{cases} A(i,t) = A(i) + D_A(i)F_A(i,t)E_A(i), \\ B(i,t) = B(i) + D_B(i)F_B(i,t)E_B(i), \\ C_y(i,t) = C_y(i) + D_{C_y}(i)F_{C_y}(i,t)E_{C_y}(i), \\ C_z(i,t) = C_z(i) + D_{C_z}(i)F_{C_z}(i,t)E_{C_z}(i), \\ D_y(i,t) = C_y(i) + D_{D_y}(i)F_{D_y}(i,t)E_{D_y}(i), \\ D_z(i,t) = D_z(i) + D_{D_z}(i)F_{D_z}(i,t)E_{D_z}(i), \end{cases}$$

where the matrices $A(i)$, $B(i)$, $B_w(i)$, $\mathbb{W}(i)$, $C_y(i)$, $D_y(i)$, $B_y(i)$, $C_z(i)$, $D_z(i)$, and $B_z(i)$ are given matrices with appropriate dimensions.

Let us drop the uncertainties and see how we can design a state feedback controller that stochastically stabilizes the nominal system. Before giving the results that determine such a controller, let us prove the following theorem.

Theorem 60. *If system (4.70) with $u(t) \equiv 0$ is internally MSQS, then it is stochastically stable.*

Proof: To prove this theorem, let us consider a candidate Lyapunov function defined as follows:

$$V(x(t), r(t)) = x^\top(t)P(r(t))x(t),$$

where $P(i) > 0$ is symmetric and positive-definite matrix for every $i \in \mathscr{S}$.

The infinitesimal operator $\mathscr{L}$ of the Markov process $\{(x(t), r(t)), t \geq 0\}$ acting on $V(.)$ and emanating from the point (x, i) at time t, when at time t, $x(t) = x$ and $r(t) = i$ for $i \in \mathscr{S}$, is given by:

$$\begin{aligned} \mathscr{L}V(x(t), i) = {} & \dot{x}^\top(t)P(i)x(t) + x^\top(t)P(i)\dot{x}(t) \\ & + x^\top(t)\mathbb{W}^\top(i)P(i)\mathbb{W}(i)x(t) + \sum_{j=1}^N \lambda_{ij}x^\top(t)P(j)x(t) \\ = {} & x^\top(t)\Big[A^\top(i)P(i) + P(i)A(i) + \mathbb{W}^\top(i)P(i)\mathbb{W}(i) \\ & + \sum_{j=1}^N \lambda_{ij}P(j)\Big]x(t) \\ & + 2x^\top(t)P(i)B_\omega(i)\omega(t). \end{aligned}$$

Using Lemma 7 from Appendix A, we get the following for any $\varepsilon_w(i) > 0$:

$$2x^\top(t)P(i)B_\omega(i)\omega(t) \leq \varepsilon_w(i)x^\top(t)P(i)B_\omega(i)B_\omega^\top(i)P(i)x(t)$$
$$+\varepsilon_w^{-1}(i)\omega^\top(t)\omega(t).$$

Combining this with the expression $\mathscr{L}V(x(t), i)$ yields

$$\mathscr{L}V(x(t), i) \leq x^\top(t)\Big[A^\top(i)P(i) + P(i)A(i) + \mathbb{W}^\top(i)P(i)\mathbb{W}(i)$$
$$+ \sum_{j=1}^N \lambda_{ij}P(j)\Big]x(t) + \varepsilon_w(i)x^\top(t)P(i)B_\omega(i)B_\omega^\top(i)P(i)x(t)$$

$$+\varepsilon_w^{-1}(i)\omega^\top(t)\omega(t)$$
$$= x^\top(t)\Big[A^\top(i)P(i) + P(i)A(i) + \mathbb{W}^\top(i)P(i)\mathbb{W}(i)$$
$$+ \sum_{j=1}^N \lambda_{ij}P(j)\Big]x(t) + x^\top(t)\left[\varepsilon_w(i)P(i)B_\omega(i)B_\omega^\top(i)P(i)\right]x(t)$$

$$+\varepsilon_w^{-1}(i)\omega^\top(t)\omega(t),$$
$$= x^\top(t)\Xi(i)x(t) + \varepsilon^{-1}(i)\omega^\top(t)\omega(t), \quad (4.71)$$

with

$$\Xi(i) = A^\top(i)P(i) + P(i)A(i) + \mathbb{W}^\top(i)P(i)\mathbb{W}(i) + \sum_{j=1}^N \lambda_{ij}P(j)$$
$$+\varepsilon_w(i)P(i)B_\omega(i)B_\omega^\top(i)P(i).$$

Based on Dynkin's formula, we get the following:

$$\mathbb{E}\left[V(x(t), i)\right] - V(x_0, r_0) = \mathbb{E}\left[\int_0^t \mathscr{L}V(x(s), r(s))ds|x_0, r_0\right],$$

which combined with (4.71) yields

$$\mathbb{E}[V(x(t), i) - V(x_0, r_0)] \leq \mathbb{E}\left[\int_0^t x^\top(s)\Xi(r(s))x(s)ds|x_0, r_0\right]$$
$$+\varepsilon_w^{-1}(i)\int_0^t \omega^\top(s)\omega(s)ds. \quad (4.72)$$

Since $V(x(t), i)$ is nonnegative, (4.72) implies

$$\mathbb{E}[V(x(t), i)] + \mathbb{E}\left[\int_0^t x^\top(s)[-\Xi(r(s))]x(s)ds|x_0, r_0\right]$$
$$\leq V(x_0, r_0)] + \varepsilon_w^{-1}(i)\int_0^t \omega^\top(s)\omega(s)ds,$$

which yields

$$\min_{i\in\mathscr{S}}\{\lambda_{min}(-\Xi(i))\}\mathbb{E}\left[\int_0^t x^\top(s)x(s)ds\right] \leq \mathbb{E}\left[\int_0^t x^\top(s)[-\Xi(r(s))]x(s)ds\right]$$

$$\leq V(x_0,r_0) + \varepsilon_w^{-1}(i)\int_0^\infty \omega^\top(s)\omega(s)ds.$$

This proves that system (4.70) is stochastically stable.   □

Let us now establish what conditions should we satisfy if we want to get system (4.70), with $u(t) = 0$ for all $t \geq 0$, stochastically stable and has γ-disturbance rejection. The following theorem gives such conditions.

Theorem 61. *Let γ be a given positive constant. If there exists a set of symmetric and positive-definite matrices $P = (P(1), \cdots , P(N)) > 0$ such that the following LMI holds for every $i \in \mathscr{S}$*

$$\begin{bmatrix} J_0(i) & \begin{bmatrix} C_z^\top(i)B_z(i) \\ +P(i)B_\omega(i) \end{bmatrix} \\ \begin{bmatrix} B_z^\top(i)C_z(i) \\ +B_\omega^\top(i)P(i) \end{bmatrix} & B_z^\top(i)B_z(i) - \gamma^2\mathbb{I} \end{bmatrix} < 0, \qquad (4.73)$$

where $J_0(i) = A^\top(i)P(i) + P(i)A(i) + \mathbb{W}^\top(i)P(i)\mathbb{W}(i) + \sum_{j=1}^N \lambda_{ij}P(j) + C_z^\top(i)C_z(i)$, then system (4.70) with $u(t) \equiv 0$ is stochastically stable and satisfies the following:

$$\|z\|_2 \leq \left[\gamma^2\|w\|_2^2 + x_0^\top P(r_0)x_0\right]^{\frac{1}{2}}, \qquad (4.74)$$

which means that the system with $u(t) = 0$ for all $t \geq 0$ is stochastically stable with γ-disturbance attenuation.

Proof: From (4.73) and using the Schur complement, we get the following inequality:

$$A^\top(i)P(i) + P(i)A(i) + \mathbb{W}^\top(i)P(i)\mathbb{W}(i) + \sum_{j=1}^N \lambda_{ij}P(j) + C_z^\top(i)C_z(i) < 0,$$

which implies the following since $C_z^\top(i)C_z(i) \geq 0$:

$$A^\top(i)P(i) + P(i)A(i) + \mathbb{W}^\top(i)P(i)\mathbb{W}(i) + \sum_{j=1}^N \lambda_{ij}P(j) < 0.$$

Based on Definition 6, this proves that the system under study is internally MSQS. Using Theorem 60, we conclude that system (4.70) with $u(t) \equiv 0$ is stochastically stable.

Let us now prove that (4.74) is satisfied. To this end, define the following performance function:

$$J_T = \mathbb{E}\left[\int_0^T [z^\top(t)z(t) - \gamma^2\omega^\top(t)\omega(t)]dt\right].$$

To prove (4.74), it suffices to establish that J_∞ is bounded, that is,

$$J_\infty \leq V(x_0, r_0) = x_0^\top P(r_0)x_0.$$

Notice that for $V(x(t), i) = x^\top(t)P(i)x(t)$ we have

$$\mathscr{L}V(x(t), i) = x^\top(t)\left[A^\top(i)P(i) + P(i)A(i) + \mathbb{W}^\top(i)P(i)\mathbb{W}(i)\right.$$

$$\left. + \sum_{j=1}^N \lambda_{ij}P(j)\right]x(t) + x^\top(t)P(i)B_\omega(i)\omega(t) + \omega^\top(t)B_\omega^\top(i)P(i)x(t),$$

and

$$z^\top(t)z(t) - \gamma^2\omega^\top(t)\omega(t)$$

$$= [C_z(i)x(t) + B_z(i)\omega(t)]^\top [C_z(i)x(t) + B_z(i)\omega(t)] - \gamma^2\omega^\top(t)\omega(t)$$

$$= x^\top(t)C_z^\top(i)C_z(i)x(t) + x^\top(t)C_z^\top(i)B_z(i)\omega(t)$$

$$+ \omega^\top(t)B_z^\top(i)C_z(i)x(t) + \omega^\top(t)B_z^\top(i)B_z(i)\omega(t) - \gamma^2\omega^\top(t)\omega(t),$$

which implies the following equality:

$$z^\top(t)z(t) - \gamma^2\omega^\top(t)\omega(t) + \mathscr{L}V(x(t), i) = \eta^\top(t)\Theta(i)\eta(t),$$

with

$$\Theta(i) = \begin{bmatrix} J_0(i) & \begin{bmatrix} C_z^\top(i)B_z(i) \\ +P(i)B_\omega(i) \end{bmatrix} \\ \begin{bmatrix} B_z^\top(i)C_z(i) \\ +B_\omega^\top(i)P(i) \end{bmatrix} & B_z^\top(i)B_z(i) - \gamma^2\mathbb{I} \end{bmatrix},$$

$$\eta^\top(t) = \begin{bmatrix} x^\top(t) & \omega^\top(t) \end{bmatrix}.$$

Therefore,

$$J_T = \mathbb{E}\left[\int_0^T \left[z^\top(t)z(t) - \gamma^2\omega^\top(t)\omega(t) + \mathscr{L}V(x(t), r(t))\right]dt\right]$$

$$- \mathbb{E}\left[\int_0^T \mathscr{L}V(x(t), r(t))dt\right].$$

Using Dynkin's formula,

$$\mathbb{E}\left[\int_0^T \mathscr{L}V(x(t), r(t))]dt|x_0, r_0\right] = \mathbb{E}[V(x(T), r(T))] - V(x_0, r_0),$$

we get

$$J_T = \mathbb{E}\left[\int_0^T \eta^\top(t)\Theta(r(t))\eta(t)dt\right] - \mathbb{E}[V(x(T), r(T))] + V(x_0, r_0).$$

Since $\Theta(i) < 0$ and $\mathbb{E}[V(x(T), r(T))] \geq 0$, this implies the following:

$$J_T \leq V(x_0, r_0),$$

which yields $J_\infty \leq V(x_0, r_0)$, i.e., $\|z\|_2^2 - \gamma^2\|\omega\|_2^2 \leq x_0^\top P(r_0)x_0$.
This gives the desired results:

$$\|z\|_2 \leq \left[\gamma^2\|\omega\|_2^2 + x_0^\top P(r_0)x_0\right]^{\frac{1}{2}}.$$

This ends the proof of the theorem. □

Let us see how we can design a controller of the form (4.17). Plugging the expression of the controller in the dynamics (4.70), we get

$$\begin{cases} dx(t) = \bar{A}(i)x(t)dt + B_w(i)w(t)dt + \mathbb{W}(i)x(t)d\omega(t), \\ z(t) = \bar{C}_z(i)x(t) + B_z(i)w(t), \end{cases} \quad (4.75)$$

where $\bar{A}(i) = A(i) + B(i)K(i)$ and $\bar{C}_z(i) = C_z(i) + D_z(i)K(i)$.

Using the results of Theorem 61, we get the following for the stochastic stability and the disturbance rejection of level $\gamma > 0$ for the closed-loop dynamics.

Theorem 62. *Let γ be a given positive constant and $K = (K(1), \cdots, K(N))$ be a set of given gains. If there exists a set of symmetric and positive-definite matrices $P = (P(1), \cdots, P(N)) > 0$ such that the following LMI holds for every $i \in \mathscr{S}$:*

$$\begin{bmatrix} \bar{J}_0(i) & \begin{bmatrix} \bar{C}_z^\top(i)B_z(i) \\ +P(i)B_\omega(i) \end{bmatrix} \\ \begin{bmatrix} B_z^\top(i)\bar{C}_z(i) \\ +B_\omega^\top(i)P(i) \end{bmatrix} & B_z^\top(i)B_z(i) - \gamma^2\mathbb{I} \end{bmatrix} < 0, \quad (4.76)$$

with $\bar{J}_0(i) = \bar{A}^\top(i)P(i) + P(i)\bar{A}(i) + \mathbb{W}(i)P(i)\mathbb{W}(i) + \sum_{j=1}^N \lambda_{ij}P(j) + \bar{C}_z^\top(i)\bar{C}_z(i)$, then system (4.70) is stochastically stable under the controller (4.17) and satisfies the following:

$$\|z\|_2 \leq \left[\gamma^2\|w\|_2^2 + x_0^\top P(r_0)x_0\right]^{\frac{1}{2}}, \quad (4.77)$$

which means that the system is stochastically stable with γ-disturbance attenuation.

To synthesize the controller gain, let us transform the LMI (4.76) into a form that can be used easily to compute the gain for every mode $i \in \mathscr{S}$. Notice that

$$
\begin{bmatrix} \bar{J}_0(i) & \begin{bmatrix} \bar{C}_z^\top(i)B_z(i) \\ +P(i)B_\omega(i) \end{bmatrix} \\ \begin{bmatrix} B_z^\top(i)\bar{C}_z(i) \\ +B_\omega^\top(i)P(i) \end{bmatrix} & B_z^\top(i)B_z(i) - \gamma^2\mathbb{I} \end{bmatrix} =
$$

$$
\begin{bmatrix} \bar{J}_1(i) & P(i)B_\omega(i) \\ B_\omega^\top(i)P(i) & -\gamma^2\mathbb{I} \end{bmatrix}
$$

$$
+ \begin{bmatrix} \bar{C}_z^\top(i) \\ B_z^\top(i) \end{bmatrix} \begin{bmatrix} \bar{C}_z(i) & B_z(i) \end{bmatrix},
$$

with $\bar{J}_1(i) = \bar{A}^\top(i)P(i) + P(i)\bar{A}(i) + \mathbb{W}(i)P(i)\mathbb{W}(i) + \sum_{j=1}^N \lambda_{ij}P(j)$.

Using the Schur complement we show that (4.76) is equivalent to the following inequality:

$$
\begin{bmatrix} \bar{J}_1(i) & P(i)B_\omega(i) & \bar{C}_z^\top(i) \\ B_\omega^\top(i)P(i) & -\gamma^2\mathbb{I} & B_z^\top(i) \\ \bar{C}_z(i) & B_z(i) & -\mathbb{I} \end{bmatrix} < 0.
$$

Since $\bar{A}(i)$ is nonlinear in $K(i)$ and $P(i)$, the previous inequality is nonlinear and therefore cannot be solved using existing linear algorithms. To transform it to an LMI, let $X(i) = P^{-1}(i)$. Pre- and post-multiply this inequality by $\mathrm{diag}[X(i), \mathbb{I}, \mathbb{I}]$ to give

$$
\begin{bmatrix} \bar{J}_X(i) & B_\omega(i) & X(i)\bar{C}_z^\top(i) \\ B_\omega^\top(i) & -\gamma^2\mathbb{I} & B_z^\top(i) \\ \bar{C}_z(i)X(i) & B_z(i) & -\mathbb{I} \end{bmatrix} < 0,
$$

with

$$
\bar{J}_X(i) = X(i)\bar{A}^\top(i) + \bar{A}(i)X(i) + X(i)\mathbb{W}(i)X^{-1}(i)\mathbb{W}(i)X(i)
$$

$$
+ \sum_{j=1}^N \lambda_{ij}X(i)X^{-1}(j)X(i).
$$

Notice that

$$
X(i)\bar{A}^\top(i) + \bar{A}(i)X(i) = X(i)A^\top(i) + A(i)X(i) + Y^\top(i)B^\top(i)
$$

$$
+ B(i)Y(i),
$$

$$
\sum_{j=1}^N \lambda_{ij}X(i)X^{-1}(j)X(i) = \lambda_{ii}X(i) + \mathcal{S}_i(X)\mathcal{X}_i^{-1}(X)\mathcal{S}_i^\top(X),
$$

$$
X(i)\left[C_z(i) + D_z(i)K(i)\right]^\top = X(i)C_z^\top(i) + Y^\top(i)D_z^\top(i),
$$

where $Y(i) = K(i)X(i)$, and $\mathcal{S}_i(X)$ and $\mathcal{X}_i(X)$ are defined as before.

Using the Schur complement again implies that the previous inequality is equivalent to the following:

$$\begin{bmatrix} J(i) & B_\omega(i) & \begin{bmatrix} X(i)C_z^\top(i) \\ +Y^\top(i)D_z^\top(i) \end{bmatrix} & X(i)\mathbb{W}^\top(i) & \mathcal{S}_i(X) \\ B_\omega^\top(i) & -\gamma^2\mathbb{I} & B_z^\top(i) & 0 & 0 \\ \begin{bmatrix} C_z(i)X(i) \\ +D_z(i)Y(i) \end{bmatrix} & B_z(i) & -\mathbb{I} & 0 & 0 \\ \mathbb{W}(i)X(i) & 0 & 0 & -X(i) & 0 \\ \mathcal{S}_i^\top(X) & 0 & 0 & 0 & -\mathcal{X}_i(X) \end{bmatrix} < 0,$$

with $J(i) = X(i)A^\top(i) + A(i)X(i) + Y^\top(i)B^\top(i) + B(i)Y(i) + \lambda_{ii}X(i)$.

From this discussion we get the following theorem.

Theorem 63. *Let γ be a positive constant. If there exist a set of symmetric and positive-definite matrices $X = (X(1), \cdots, X(N)) > 0$ and a set of matrices $Y = (Y(1), \cdots, Y(N))$ such that the following LMI holds for every $i \in \mathscr{S}$:*

$$\begin{bmatrix} J(i) & B_\omega(i) & \begin{bmatrix} X(i)C_z^\top(i) \\ +Y^\top(i)D_z^\top(i) \end{bmatrix} & X(i)\mathbb{W}^\top(i) & \mathcal{S}_i(X) \\ B_\omega^\top(i) & -\gamma^2\mathbb{I} & B_z^\top(i) & 0 & 0 \\ \begin{bmatrix} C_z(i)X(i) \\ +D_z(i)Y(i) \end{bmatrix} & B_z(i) & -\mathbb{I} & 0 & 0 \\ \mathbb{W}(i)X(i) & 0 & 0 & -X(i) & 0 \\ \mathcal{S}_i^\top(X) & 0 & 0 & 0 & -\mathcal{X}_i(X) \end{bmatrix} < 0, \quad (4.78)$$

with $J(i) = X(i)A^\top(i) + A(i)X(i) + Y^\top(i)B^\top(i) + B(i)Y(i) + \lambda_{ii}X(i)$, then the system (4.70) under the controller (4.17) with $K(i) = Y(i)X^{-1}(i)$ is stochastically stable and, moreover, the closed-loop system satisfies the disturbance rejection of level γ.

From the practical point of view, the controller that stochastically stabilizes the class of systems and simultaneously guarantees the minimum disturbance rejection is of great interest. This controller can be obtained by solving the following optimization problem:

$$P : \begin{cases} \min\limits_{\substack{\nu>0, \\ X=(X(1),\cdots,X(N))>0, \\ Y=(Y(1),\cdots,Y(N)),}} \nu, \\ s.t. : (4.78) \text{ with } \nu = \gamma^2. \end{cases}$$

The following corollary gives the results of the design of the controller that stochastically stabilizes the system (4.70) and simultaneously guarantees the smallest disturbance rejection level.

Corollary 13. *Let $\nu > 0$, $X = (X(1), \cdots, X(N)) > 0$, and $Y = (Y(1), \cdots, Y(N))$ be the solution of the optimization problem P. Then the controller*

(4.17) with $K(i) = Y(i)X^{-1}(i)$ stochastically stabilizes the class of systems we are considering and, moreover, the closed-loop system satisfies the disturbance rejection of level $\sqrt{\nu}$.

Previously we developed results that determine the state feedback controller that stochastically stabilizes the class of systems we are treating in this chapter and at the same time rejects the disturbance $w(t)$ with the desired level $\gamma > 0$. The conditions we developed are in the LMI form, which makes their resolution easy. In the rest of this section we give some numerical examples to show the effectiveness of our results. Two numerical examples are presented.

Example 60. Let us consider a system with two modes with the following data:

* transition probability rate matrix:

$$A = \begin{bmatrix} -2.0 & 2.0 \\ 3.0 & -3.0 \end{bmatrix},$$

* mode #1:

$$A(1) = \begin{bmatrix} 1.0 & -0.5 \\ 0.1 & 1.0 \end{bmatrix}, \quad B(1) = \begin{bmatrix} 1.0 & 0.0 \\ 0.0 & 1.0 \end{bmatrix}, \quad B_w(1) = \begin{bmatrix} 1.0 & 0.0 \\ 0.0 & 1.0 \end{bmatrix},$$

$$B_z(1) = \begin{bmatrix} 1.0 & 0.0 \\ 0.0 & 1.0 \end{bmatrix}, \quad W(1) = \begin{bmatrix} 0.1 & 0.0 \\ 0.0 & 0.1 \end{bmatrix}, \quad C_z(1) = \begin{bmatrix} 1.0 & 0.0 \\ 0.0 & 1.0 \end{bmatrix},$$

$$D_z(1) = \begin{bmatrix} 1.0 & 0.0 \\ 0.0 & 1.0 \end{bmatrix},$$

* mode #2:

$$A(2) = \begin{bmatrix} -0.2 & -0.5 \\ 0.5 & -0.25 \end{bmatrix}, \quad B(2) = \begin{bmatrix} 1.0 & 0.0 \\ 0.0 & 1.0 \end{bmatrix}, \quad B_w(2) = \begin{bmatrix} 1.0 & 0.0 \\ 0.0 & 1.0 \end{bmatrix},$$

$$B_z(2) = \begin{bmatrix} 1.0 & 0.0 \\ 0.0 & 1.0 \end{bmatrix}, \quad W(2) = \begin{bmatrix} 0.2 & 0.0 \\ 0.0 & 0.2 \end{bmatrix}, \quad C_z(2) = \begin{bmatrix} 1.0 & 0.0 \\ 0.0 & 1.0 \end{bmatrix},$$

$$D_z(2) = \begin{bmatrix} 1.0 & 0.0 \\ 0.0 & 1.0 \end{bmatrix}.$$

Notice that the system is instable in mode 1 and is stochastically instable. Letting $\gamma = 10$ and solving the LMI (4.78), we get

$$X(1) = \begin{bmatrix} 35.2579 & 2.3259 \\ 2.3259 & 29.7626 \end{bmatrix}, \quad Y(1) = \begin{bmatrix} -37.3452 & -2.3342 \\ -2.3117 & -31.8415 \end{bmatrix},$$

$$X(2) = \begin{bmatrix} 44.3439 & -1.2064 \\ -1.2064 & 39.3224 \end{bmatrix}, \quad Y(2) = \begin{bmatrix} -45.5124 & 1.4861 \\ 1.4905 & -40.7028 \end{bmatrix},$$

which gives the following gains:

$$K(1) = \begin{bmatrix} -1.0595 & 0.0044 \\ 0.0050 & -1.0702 \end{bmatrix}, \quad K(2) = \begin{bmatrix} -1.0262 & 0.0063 \\ 0.0055 & -1.0349 \end{bmatrix}.$$

All the conditions in Theorem 63 are satisfied and therefore the closed-loop system is stochastically stable under the state feedback controller designed for this system. The system also assures the disturbance rejection of level 10.

Example 61. To design a stabilizing controller that assures the minimum disturbance rejection, let us reconsider the system with two modes from the previous example and solve the optimization problem P. The resolution of such a system gives

$$X(1) = \begin{bmatrix} 1.9289 & -0.0471 \\ -0.0471 & 1.8714 \end{bmatrix}, \quad Y(1) = \begin{bmatrix} -2.9289 & 0.0471 \\ 0.0471 & -2.8714 \end{bmatrix},$$

$$X(2) = \begin{bmatrix} 2.0797 & -0.1916 \\ -0.1916 & 2.1665 \end{bmatrix}, \quad Y(2) = \begin{bmatrix} -3.0797 & 0.1916 \\ 0.1916 & -3.1665 \end{bmatrix},$$

which gives the following gains:

$$K(1) = \begin{bmatrix} -1.5188 & -0.0131 \\ -0.0131 & -1.5347 \end{bmatrix}, \quad K(2) = \begin{bmatrix} -1.4848 & -0.0429 \\ -0.0429 & -1.4654 \end{bmatrix}.$$

Using the results of Corollary 13, the system of this example is stochastically stable under the state feedback controller with the computed constant gain and ensures the disturbance rejection of level $\gamma = 1.0$.

Let us now consider the effect of the uncertainties and see how to design the state feedback controller that robustly stochastically stabilizes the class of systems we are considering in this section. Before doing this let us give the following definition.

Definition 13. *System (4.70) with $u(t) \equiv 0$ and $w(t) \equiv 0$ for all $t \geq 0$ is said to be internally mean square quadratically stable (MSQS) if there exists a set of symmetric and positive-definite matrices $P = (P(1), \cdots, P(N)) > 0$ satisfying the following for every $i \in \mathscr{S}$:*

$$A^\top(i)P(i) + P(i)A(i) + \mathbb{W}(i)P(i)\mathbb{W}(i) + \sum_{j=1}^{N} \lambda_{ij} P(j) < 0. \tag{4.79}$$

Based on the previous definition, the uncertain system with $u(t), \forall t \geq 0$, will be stochastically stable if the following holds for every $i \in \mathscr{S}$ and for all admissible uncertainties:

$$A^\top(i,t)P(i) + P(i)A(i,t) + \mathbb{W}^\top(i)P(i)\mathbb{W}(i) + \sum_{j=1}^{N} \lambda_{ij} P(j) < 0.$$

This condition is useless since it contains the uncertainties $F_A(i,t)$. Let us now transform it into a useful condition that can be used to check the robust stability.

If we use the expression $A(i, t)$, we get

$$A^\top(i)P(i) + P(i)A(i) + \mathbb{W}^\top(i)P(i)\mathbb{W}(i) + \sum_{j=1}^{N} \lambda_{ij} P(j)$$
$$+ P(i)D_A(i)F_A(i, t)E_A(i) + E_A^\top(i)F_A^\top(i, t)D_A^\top(i)P(i) < 0.$$

Using Lemma 7 of Appendix A, the previous inequality will be satisfied if the following holds:

$$A^\top(i)P(i) + P(i)A(i) + \mathbb{W}^\top(i)P(i)\mathbb{W}(i) + \sum_{j=1}^{N} \lambda_{ij} P(j) + \varepsilon_A(i)E_A^\top(i)E_A(i)$$
$$+ \varepsilon_A^{-1}(i)P(i)D_A(i)D_A^\top(i)P(i) < 0,$$

with $\varepsilon_A(i) > 0$ for all $i \in \mathscr{S}$.

Using the Schur complement we get the desired condition:

$$\begin{bmatrix} J_0(i) & P(i)D_A(i) \\ D_A^\top(i)P(i) & -\varepsilon_A(i)\mathbb{I} \end{bmatrix} < 0, \tag{4.80}$$

with

$$J_0(i) = A^\top(i)P(i) + P(i)A(i) + \mathbb{W}^\top(i)P(i)\mathbb{W}(i) + \sum_{j=1}^{N} \lambda_{ij} P(j)$$
$$+ \varepsilon_A(i)E_A^\top(i)E_A(i).$$

The results of this development are summarized by the following theorem.

Theorem 64. *If there exist a set of symmetric and positive-definite matrices* $P = (P(1), \cdots, P(N)) > 0$ *and a set of positive scalars* $\varepsilon_A = (\varepsilon_A(1), \cdots, \varepsilon_A(N))$ *such that the following LMI (4.80) holds for every* $i \in \mathscr{S}$ *and for all admissible uncertainties, then system (4.70) with* $u(t) = 0$ *and* $w(t) = 0$ *for all* $t \geq 0$ *is internally mean square quadratically stable.*

Theorem 65. *If system (4.70) with* $u(t) \equiv 0$ *is internally mean square stochastically stable for all admissible uncertainties, then it is also robust stochastically stable.*

Proof: To prove this theorem, consider a candidate Lyapunov function to be defined as follows:

$$V(x(t), r(t)) = x^\top(t)P(r(t))x(t),$$

where $P(i) > 0$ is a symmetric and positive-definite matrix for every $i \in \mathscr{S}$.

The infinitesimal operator $\mathscr{L}$ of the Markov process $\{(x(t), r(t)), t \geq 0\}$ acting on $V(.)$ and emanating from the point (x, i) at time t, where $x(t) = x$ and $r(t) = i$ for $i \in \mathscr{S}$, is given by

$$
\begin{aligned}
\mathscr{L}V(x(t), i) = {}& \dot{x}^\top(t)P(i)x(t) + x^\top(t)P(i)\dot{x}(t) \\
& + x^\top(t)\mathbb{W}^\top(i)P(i)\mathbb{W}(i)x(t) + \sum_{j=1}^{N}\lambda_{ij}x^\top(t)P(j)x(t) \\
= {}& x^\top(t)\Big[A^\top(i)P(i) + P(i)A(i) + \mathbb{W}^\top(i)P(i)\mathbb{W}(i) \\
& + \sum_{j=1}^{N}\lambda_{ij}P(j)\Big]x(t) + 2x^\top(t)P(i)D_A(i)F_A(i,t)E_A(i)x(t) \\
& + 2x^\top(t)P(i)B_\omega(i)\omega(t).
\end{aligned}
$$

Using Lemma 7 in Appendix A, we get the following:

$$
\begin{aligned}
2x^\top(t)P(i)D_A(i)F_A(i,t)E_A(i)x(t) &\leq \varepsilon_A(i)x^\top(t)E_A^\top(i)E_A(i)x(t) \\
&\quad + \varepsilon_A^{-1}(i)x^\top(t)P(i)D_A(i)D_A^\top(i)P(i)x(t) \\
2x^\top(t)P(i)B_\omega(i)\omega(t) &\leq \varepsilon_w^{-1}(i)x^\top(t)P(i)B_\omega(i)B_\omega^\top(i)P(i)x(t) \\
&\quad + \varepsilon_w(i)\omega^\top(t)\omega(t).
\end{aligned}
$$

Combining this with the expression $\mathscr{L}V(x(t), i)$ yields

$$
\begin{aligned}
\mathscr{L}V(x(t), i) \leq {}& x^\top(t)\Big[A^\top(i)P(i) + P(i)A(i) + \mathbb{W}^\top(i)P(i)\mathbb{W}(i) \\
& + \sum_{j=1}^{N}\lambda_{ij}P(j)\Big]x(t) + \varepsilon_A(i)x^\top(t)E_A^\top(i)E_A(i)x(t) \\
& + \varepsilon_A^{-1}(i)x^\top(t)P(i)D_A(i)D_A^\top(i)P(i)x(t) \\
& + \varepsilon_w^{-1}(i)x^\top(t)P(i)B_\omega(i)B_\omega^\top(i)P(i)x(t) + \varepsilon_w(i)\omega^\top(t)\omega(t) \\
= {}& x^\top(t)\left[A^\top(i)P(i) + P(i)A(i) + \mathbb{W}^\top(i)P(i)\mathbb{W}(i) + \sum_{j=1}^{N}\lambda_{ij}P(j)\right]x(t) \\
& + \varepsilon_A(i)x^\top(t)E_A^\top(i)E_A(i)x(t) + \varepsilon_A^{-1}(i)x^\top(t)P(i)D_A(i)D_A^\top(i)P(i)x(t) \\
& + x^\top(t)\left[\varepsilon_w^{-1}(i)P(i)B_\omega(i)B_\omega^\top(i)P(i)\right]x(t) + \varepsilon_w(i)\omega^\top(t)\omega(t), \\
= {}& x^\top(t)\Upsilon(i)x(t) + \varepsilon(i)\omega^\top(t)\omega(t), \qquad\qquad (4.81)
\end{aligned}
$$

with

$$
\begin{aligned}
\Upsilon(i) = {}& A^\top(i)P(i) + P(i)A(i) + \mathbb{W}^\top(i)P(i)\mathbb{W}(i) + \sum_{j=1}^{N}\lambda_{ij}P(j) \\
& + \varepsilon_A(i)E_A^\top(i)E_A(i) + \varepsilon_A^{-1}(i)P(i)D_A(i)D_A^\top(i)P(i)
\end{aligned}
$$

$$+\varepsilon_w^{-1}(i)P(i)B_w(i)B_w^\top(i)P(i).$$

If $\Upsilon(i) < 0$ for each $i \in \mathscr{S}$, we get the following equivalent inequality matrix:

$$\Xi(i) = \begin{bmatrix} J_w(i) & P(i)B_w(i) & P(i)D_A(i) \\ B_w^\top(i)P(i) & -\varepsilon_w(i)\mathbb{I} & 0 \\ D_A^\top(i)P(i) & 0 & -\varepsilon_A(i)\mathbb{I} \end{bmatrix} < 0,$$

with $J_w(i) = A^\top(i)P(i) + P(i)A(i) + \mathbb{W}^\top(i)P(i)\mathbb{W}(i) + \sum_{j=1}^{N} \lambda_{ij}P(j) + \varepsilon_A(i)E_A^\top(i)E_A(i)$.

Based on Dynkin's formula, we get the following:

$$\mathbb{E}\left[V(x(t),i)\right] - V(x_0,r_0) = \mathbb{E}\left[\int_0^t \mathscr{L}V(x(s),r(s))ds | x_0, r_0\right],$$

which combined with (4.81) yields

$$\mathbb{E}\left[V(x(t),i)\right] - V(x_0,r_0) \leq \mathbb{E}\left[\int_0^t x^\top(s)\Xi(r(s))x(s)ds | x_0, r_0\right]$$

$$+\varepsilon_w(i)\int_0^t \omega^\top(s)\omega(s)ds. \quad (4.82)$$

Since $V(x(t),i)$ is nonnegative, (4.82) implies

$$\mathbb{E}\left[V(x(t),i)\right] + \mathbb{E}\left[\int_0^t x^\top(s)\left[-\Xi(r(s))\right]x(s)ds | x_0, r_0\right]$$

$$\leq V(x_0,r_0) + \varepsilon_w(i)\int_0^t \omega^\top(s)\omega(s)ds,$$

which yields

$$\min_{i \in \mathscr{S}}\{\lambda_{min}(-\Xi(i))\}\mathbb{E}\left[\int_0^t x^\top(s)x(s)ds\right] \leq \mathbb{E}\left[\int_0^t x^\top(s)[-\Xi(r(s))]x(s)ds\right]$$

$$\leq V(x_0,r_0) + \varepsilon_w(i)\int_0^\infty \omega^\top(s)\omega(s)ds.$$

This proves that system (4.70) is stochastically stable. □

Let us now establish what conditions we should satisfy if we want system (4.70), with $u(t) = 0$ for all $t \geq 0$, to be stochastically stable with γ-disturbance rejection. The following theorem gives such conditions.

Theorem 66. *Let γ be a given positive constant. If there exists a set of symmetric and positive-definite matrices $P = (P(1), \cdots, P(N)) > 0$ such that the following LMI holds for every $i \in \mathscr{S}$:*

$$\begin{bmatrix} J_u(i) & \begin{bmatrix} C_z^\top(i,t)B_z(i) \\ +P(i)B_\omega(i) \end{bmatrix} \\ \begin{bmatrix} B_z^\top(i)C_z(i,t) \\ +B_\omega^\top(i)P(i) \end{bmatrix} & B_z^\top(i)B_z(i) - \gamma^2 \mathbb{I} \end{bmatrix} < 0, \qquad (4.83)$$

where $J_u(i) = A^\top(i,t)P(i) + P(i)A(i,t) + \mathbb{W}^\top(i)P(i)\mathbb{W}(i) + \sum_{j=1}^N \lambda_{ij}P(j) + C_z^\top(i,t)C_z(i,t)$, then system (4.70) with $u(t) \equiv 0$ is robustly stochastically stable and satisfies the following:

$$\|z\|_2 \le \left[\gamma^2 \|w\|_2^2 + x_0^\top P(r_0)x_0 \right]^{\frac{1}{2}}, \qquad (4.84)$$

which means that the system with $u(t) = 0$ for all $t \ge 0$ is stochastically stable with γ-disturbance attenuation.

Proof: From (4.83) and using the Schur complement, we get the following inequality:

$$A^\top(i,t)P(i) + P(i)A(i,t) + \mathbb{W}^\top(i)P(i)\mathbb{W}(i) + \sum_{j=1}^N \lambda_{ij}P(j)$$
$$+ C_z^\top(i,t)C_z(i,t) < 0,$$

which implies the following since $C_z^\top(i,t)C_z(i,t) \ge 0$:

$$A^\top(i,t)P(i) + P(i)A(i,t) + \mathbb{W}^\top(i)P(i)\mathbb{W}(i) + \sum_{j=1}^N \lambda_{ij}P(j) < 0.$$

Based on Definition 13, this proves that the system under study is internally MSQS. Using Theorem 65, we conclude that system (4.70) with $u(t) \equiv 0$ is robust stochastically stable.

Let us now prove that (4.84) is satisfied. To this end, let us define the following performance function:

$$J_T = \mathbb{E}\left[\int_0^T [z^\top(t)z(t) - \gamma^2 \omega^\top(t)\omega(t)]dt \right].$$

To prove (4.84), it suffices to establish that J_∞ is bounded, that is:

$$J_\infty \le V(x_0, r_0) = x_0^\top P(r_0)x_0.$$

Notice that for $V(x(t), i) = x^\top(t)P(i)x(t)$, we have

$$\mathscr{L}V(x(t), i) = x^\top(t)\left[A^\top(i,t)P(i) + P(i)A(i,t) + \mathbb{W}^\top(i)P(i)\mathbb{W}(i) \right.$$
$$\left. + \sum_{j=1}^N \lambda_{ij}P(j) \right]x(t) + x^\top(t)P(i)B_\omega(i)\omega(t)$$

$$+\omega^{\top}(t)B_{\omega}^{\top}(i)P(i)x(t),$$

and

$$z^{\top}(t)z(t) - \gamma^2\omega^{\top}(t)\omega(t)$$
$$= [C_z(i,t)x(t) + B_z(i)\omega(t)]^{\top}[C_z(i,t)x(t) + B_z(i)\omega(t)] - \gamma^2\omega^{\top}(t)\omega(t)$$
$$= x^{\top}(t)C_z^{\top}(i,t)C_z(i,t)x(t) + x^{\top}(t)C_z^{\top}(i,t)B_z(i)\omega(t)$$
$$+\omega^{\top}(t)B_z^{\top}(i)C_z(i,t)x(t) + \omega^{\top}(t)B_z^{\top}(i)B_z(i)\omega(t) - \gamma^2\omega^{\top}(t)\omega(t),$$

which implies the following equality:

$$z^{\top}(t)z(t) - \gamma^2\omega^{\top}(t)\omega(t) + \mathscr{L}V(x(t),i) = \eta^{\top}(t)\Theta_u(i)\eta(t),$$

with

$$\Theta_u(i) = \begin{bmatrix} J_u(i) & \begin{bmatrix} C_z^{\top}(i,t)B_z(i) \\ +P(i)B_\omega(i) \end{bmatrix} \\ \begin{bmatrix} B_z^{\top}(i)C_z(i,t) \\ +B_\omega^{\top}(i)P(i) \end{bmatrix} & B_z^{\top}(i)B_z(i) - \gamma^2\mathbb{I} \end{bmatrix},$$
$$\eta^{\top}(t) = \begin{bmatrix} x^{\top}(t) & \omega^{\top}(t) \end{bmatrix}.$$

Therefore,

$$J_T = \mathbb{E}\left[\int_0^T [z^{\top}(t)z(t) - \gamma^2\omega^{\top}(t)\omega(t) + \mathscr{L}V(x(t),r(t))]dt\right]$$
$$-\mathbb{E}\left[\int_0^T \mathscr{L}V(x(t),r(t))]dt\right].$$

Using Dynkin's formula, that is,

$$\mathbb{E}\left[\int_0^T \mathscr{L}V(x(t),r(t))dt|x_0,r_0\right] = \mathbb{E}[V(x(T),r(T))] - V(x_0,r_0),$$

we get

$$J_T = \mathbb{E}\left[\int_0^T \eta^{\top}(t)\Theta_u(r(t))\eta(t)dt\right] - \mathbb{E}[V(x(T),r(T))] + V(x_0,r_0). \quad (4.85)$$

Since $\Theta_u(i) < 0$ and $\mathbb{E}[V(x(T),r(T))] \geq 0$, (4.85) implies the following:

$$J_T \leq V(x_0,r_0),$$

which yields $J_\infty \leq V(x_0,r_0)$, i.e., $\|z\|_2^2 - \gamma^2\|\omega\|_2^2 \leq x_0^{\top}P(r_0)x_0$.
This gives the desired results:

$$\|z\|_2 \leq [\gamma^2\|\omega\|_2^2 + x_0^{\top}P(r_0)x_0]^{\frac{1}{2}}.$$

This ends the proof of the theorem. $\square$

Let us see how we can design a controller of the form (4.17). Plugging the expression of the controller in the dynamics (4.70), we get

$$\begin{cases} dx(t) = \bar{A}(i,t)x(t)dt + B_w(i)w(t)dt + \mathbb{W}(i)x(t)d\omega(t), \\ z(t) = \bar{C}_z(i,t)x(t) + B_z(i)w(t), \end{cases} \tag{4.86}$$

where $\bar{A}(i,t) = A(i,t) + B(i,t)K(i)$ and $\bar{C}_z(i,t) = C_z(i,t) + D_z(i,t)K(i)$.

Using the results of Theorem 66, we get the following for the stochastic stability and the disturbance rejection of level $\gamma > 0$ for the closed-loop dynamics.

Theorem 67. *Let γ be a given positive constant and $K = (K(1), \cdots, K(N))$ be a set of given gains. If there exists a set of symmetric and positive-definite matrices $P = (P(1), \cdots, P(N)) > 0$ such that the following LMI holds for every $i \in \mathscr{S}$:*

$$\begin{bmatrix} \bar{J}_0(i,t) & \begin{bmatrix} \bar{C}_z^\top(i,t)B_z(i) \\ +P(i)B_\omega(i) \end{bmatrix} \\ \begin{bmatrix} B_z^\top(i)\bar{C}_z(i,t) \\ +B_\omega^\top(i)P(i) \end{bmatrix} & B_z^\top(i)B_z(i) - \gamma^2\mathbb{I} \end{bmatrix} < 0, \tag{4.87}$$

with $\bar{J}_0(i,t) = \bar{A}^\top(i,t)P(i) + P(i)\bar{A}(i,t) + \mathbb{W}^\top(i)P(i)\mathbb{W}(i) + \sum_{j=1}^N \lambda_{ij}P(j) + \bar{C}_z^\top(i,t)\bar{C}_z(i,t)$, then system (4.70) is stochastically stable under the controller (4.17) and satisfies the following:

$$\|z\|_2 \leq \left[\gamma^2\|w\|_2^2 + x_0^\top P(r_0)x_0\right]^{\frac{1}{2}}, \tag{4.88}$$

which means that the system is stochastically stable with γ-disturbance attenuation.

To synthesize the controller gain, let us transform the LMI (4.87) into a form that can be used easily to compute the gain for every mode $i \in \mathscr{S}$. Notice that

$$\begin{bmatrix} \bar{J}_0(i,t) & \begin{bmatrix} \bar{C}_z^\top(i,t)B_z(i) \\ +P(i)B_\omega(i) \end{bmatrix} \\ \begin{bmatrix} B_z^\top(i)\bar{C}_z(i,t) \\ +B_\omega^\top(i)P(i) \end{bmatrix} & B_z^\top(i)B_z(i) - \gamma^2\mathbb{I} \end{bmatrix} = $$
$$\begin{bmatrix} \bar{J}_1(i,t) & P(i)B_\omega(i) \\ B_\omega^\top(i)P(i) & -\gamma^2\mathbb{I} \end{bmatrix}$$
$$+ \begin{bmatrix} \bar{C}_z^\top(i,t) \\ B_z^\top(i) \end{bmatrix} \begin{bmatrix} \bar{C}_z(i,t) & B_z(i) \end{bmatrix},$$

with $\bar{J}_1(i,t) = \bar{A}^\top(i,t)P(i) + P(i)\bar{A}(i,t) + \mathbb{W}^\top(i)P(i)\mathbb{W}(i) + \sum_{j=1}^N \lambda_{ij}P(j)$.

Using the Schur complement we show that (4.87) is equivalent to the following inequality:

$$
\begin{bmatrix}
\bar{J}_1(i,t) & P(i)B_\omega(i) & \bar{C}_z^\top(i,t) \\
B_\omega^\top(i)P(i) & -\gamma^2\mathbb{I} & B_z^\top(i) \\
\bar{C}_z(i,t) & B_z(i) & -\mathbb{I}
\end{bmatrix} < 0.
$$

Using the expressions of $\bar{A}(i,t)$ and $\bar{C}_z(i,t)$ and their components, we obtain the following inequality:

$$
\begin{bmatrix}
J_1(i) & P(i)B_\omega(i) & \begin{bmatrix} C_z^\top(i) \\ +K^\top(i)D_z^\top(i) \end{bmatrix} \\
B_\omega^\top(i)P(i) & -\gamma^2\mathbb{I} & B_z^\top(i) \\
\begin{bmatrix} D_z(i)K(i) \\ +C_z(i) \end{bmatrix} & B_z(i) & -\mathbb{I}
\end{bmatrix}
$$

$$
+ \begin{bmatrix}
E_A^\top(i)F_A^\top(i,t)D_A^\top(i)P(i) & 0 & 0 \\
0 & 0 & 0 \\
0 & 0 & 0
\end{bmatrix}
$$

$$
+ \begin{bmatrix}
P(i)D_A(i)F_A(i,t)E_A(i) & 0 & 0 \\
0 & 0 & 0 \\
0 & 0 & 0
\end{bmatrix}
$$

$$
+ \begin{bmatrix}
P(i)D_B(i)F_B(i,t)E_B(i)K(i) & 0 & 0 \\
0 & 0 & 0 \\
0 & 0 & 0
\end{bmatrix}
$$

$$
+ \begin{bmatrix}
K^\top(i)E_B^\top(i)F_B^\top(i,t)D_B^\top(i)P(i) & 0 & 0 \\
0 & 0 & 0 \\
0 & 0 & 0
\end{bmatrix}
$$

$$
+ \begin{bmatrix}
0 & 0 & K^\top(i)E_{D_z}^\top(i)F_{D_z}^\top(i,t)D_{D_z}^\top(i) \\
0 & 0 & 0 \\
0 & 0 & 0
\end{bmatrix}
$$

$$
+ \begin{bmatrix}
0 & 0 & 0 \\
0 & 0 & 0 \\
D_{D_z}(i)F_{D_z}(i,t)E_{D_z}(i)K(i) & 0 & 0
\end{bmatrix}
$$

$$
+ \begin{bmatrix}
0 & 0 & E_{C_z}^\top(i)F_{C_z}^\top(i,t)D_{C_z}^\top(i) \\
0 & 0 & 0 \\
0 & 0 & 0
\end{bmatrix}
$$

$$
+ \begin{bmatrix}
0 & 0 & 0 \\
0 & 0 & 0 \\
D_{C_z}(i)F_{C_z}(i,t)E_{C_z}(i) & 0 & 0
\end{bmatrix} < 0,
$$

with

$$
J_1(i) = A^\top(i)P(i) + P(i)A(i) + K^\top(i)B^\top(i)P(i) + \mathbb{W}^\top(i)P(i)\mathbb{W}(i)
$$

$$+ P(i)B(i)K(i) + \sum_{j=1}^{N} \lambda_{ij} P(j).$$

Based on Lemma 7 in Appendix A, we get

$$\begin{bmatrix} J_1(i) & P(i)B_\omega(i) & C_z^\top(i) + K^\top(i)D_z^\top(i) \\ B_\omega^\top(i)P(i) & -\gamma^2 \mathbb{I} & B_z^\top(i) \\ D_z(i)K(i) + C_z(i) & B_z(i) & -\mathbb{I} \end{bmatrix}$$

$$+ \begin{bmatrix} \varepsilon_A(i)P(i)D_A(i)D_A^\top(i)P(i) + \varepsilon_A^{-1}(i)E_A^\top(i)E_A(i) & 0 & 0 \\ 0 & 0 & 0 \\ 0 & 0 & 0 \end{bmatrix}$$

$$+ \begin{bmatrix} \begin{bmatrix} \varepsilon_B(i)P(i)D_B(i)D_B^\top(i)P(i) \\ +\varepsilon_B^{-1}K^\top(i)E_B^\top(i)E_B(i)K(i) \end{bmatrix} & 0 & 0 \\ 0 & 0 & 0 \\ 0 & 0 & 0 \end{bmatrix}$$

$$+ \begin{bmatrix} \varepsilon_{D_z}^{-1}(i)K^\top(i)E_{D_z}^\top(i)E_{D_z}(i)K(i) & 0 & 0 \\ 0 & 0 & 0 \\ 0 & 0 & \varepsilon_{D_z}(i)D_{D_z}(i)D_{D_z}^\top(i) \end{bmatrix}$$

$$+ \begin{bmatrix} \varepsilon_{C_z}^{-1}(i)E_{C_z}^\top(i)E_{C_z}(i,t) & 0 & 0 \\ 0 & 0 & 0 \\ 0 & 0 & \varepsilon_{C_z}(i)D_{C_z}(i)D_{C_z}^\top(i) \end{bmatrix} < 0,$$

with

$$J_1(i) = A^\top(i)P(i) + P(i)A(i) + K^\top(i)B^\top(i)P(i)$$

$$+ P(i)B(i)K(i) + \mathbb{W}^\top(i)P(i)\mathbb{W}(i) + \sum_{j=1}^{N} \lambda_{ij} P(j).$$

Let $J_2(i)$, $\mathcal{W}(i)$, and $\mathcal{T}(i)$ be defined as

$$J_2(i) = J_1(i) + \varepsilon_A^{-1}(i)E^\top(i)E_A(i) + \varepsilon_B^{-1}(i)K^\top(i)E_B^\top(i)E_B(i)K(i),$$
$$\mathcal{W}(i) = \mathrm{diag}[\varepsilon_A^{-1}(i)\mathbb{I}, \varepsilon_B^{-1}(i)\mathbb{I}, \varepsilon_{C_z}(i)\mathbb{I}, \varepsilon_{D_z}(i)\mathbb{I}],$$
$$\mathcal{T}(i) = \left(P(i)D_A(i), P(i)D_B(i), E_{C_z}^\top(i), K^\top(i)E_{D_z}^\top(i) \right),$$

and using the Schur complement we get the equivalent inequality:

$$\begin{bmatrix} J_2(i) & P(i)B_\omega(i) & \begin{bmatrix} C_z^\top(i) \\ +K^\top(i)D_z^\top(i) \end{bmatrix} & \mathcal{T}(i) \\ B_\omega^\top(i)P(i) & -\gamma^2 \mathbb{I} & B_z^\top(i) & 0 \\ \begin{bmatrix} D_z(i)K(i) \\ +C_z(i) \end{bmatrix} & B_z(i) & -\mathcal{U}(i) & 0 \\ \mathcal{T}^\top(i) & 0 & 0 & -\mathcal{W}(i) \end{bmatrix} < 0,$$

with $\mathcal{U}(i) = \mathbb{I} - \varepsilon_{D_z}(i)D_{D_z}(i)D_{D_z}^\top(i) - \varepsilon_{C_z}(i)D_{C_z}(i)D_{C_z}^\top(i).$

This inequality is nonlinear in $K(i)$ and $P(i)$ and therefore it cannot solved using existing linear algorithms. To transform it to an LMI, let $X(i) = P^{-1}(i)$. Pre- and post-multiply this inequality by $\mathrm{diag}[X(i), \mathbb{I}, \mathbb{I}, \mathbb{I}]$ to give

$$
\begin{bmatrix}
J_3(i) & B_\omega(i) & \begin{bmatrix} X(i)C_z^\top(i) \\ +X(i)K^\top(i)D_z^\top(i) \end{bmatrix} & X(i)T(i) \\
B_\omega^\top(i) & -\gamma^2\mathbb{I} & B_z^\top(i) & 0 \\
\begin{bmatrix} D_z(i)K(i)X(i) \\ +C_z(i)X(i) \end{bmatrix} & B_z(i) & -\mathcal{U}(i) & 0 \\
T^\top(i)X(i) & 0 & 0 & -\mathcal{W}(i)
\end{bmatrix} < 0,
$$

with

$$
\begin{aligned}
J_3(i) = {} & X(i)A^\top(i) + A(i)X(i) + X(i)K^\top(i)B^\top(i) \\
& + X(i)\mathbb{W}^\top(i)X^{-1}(i)\mathbb{W}(i)X(i) \\
& + B(i)K(i)X(i) + \varepsilon_A^{-1}(i)X(i)E_A^\top(i)E_A(i)X(i) \\
& + \varepsilon_B^{-1}(i)X(i)K^\top(i)E_B^\top(i)E_B(i)K(i)X(i) \\
& + \sum_{j=1}^N \lambda_{ij}X(i)X^{-1}(j)X(i).
\end{aligned}
$$

Notice that

$$
X(i)T(i) = \left(D_A(i), D_B(i), X(i)E_{C_z}^\top(i), X(i)K^\top(i)E_{D_z}^\top(i) \right),
$$

and

$$
\sum_{j=1}^N \lambda_{ij}X(i)X^{-1}(j)X(i) = \lambda_{ii}X(i) + \mathcal{S}_i(X)\mathcal{X}^{-1}(X)\mathcal{S}_i^\top(X).
$$

Letting $Y(i) = K(i)X(i)$ and using the Schur complement we obtain

$$
\begin{bmatrix}
\tilde{J}(i) & B_\omega(i) & \begin{bmatrix} X(i)C_z^\top(i) \\ +Y^\top(i)D_z^\top(i) \end{bmatrix} & X(i)\mathbb{W}^\top(i) & \mathcal{Z}(i) & \mathcal{S}_i(X) \\
B_\omega^\top(i) & -\gamma^2\mathbb{I} & B_z^\top(i) & 0 & 0 & 0 \\
\begin{bmatrix} D_z(i)Y(i) \\ +C_z(i)X(i) \end{bmatrix} & B_z(i) & -\mathcal{U}(i) & 0 & 0 & 0 \\
X(i)\mathbb{W}(i) & 0 & 0 & -X(i) & 0 & 0 \\
\mathcal{Z}^\top(i) & 0 & 0 & 0 & -\mathcal{V}(i) & 0 \\
\mathcal{S}_i^\top(X) & 0 & 0 & 0 & 0 & -\mathcal{X}_i(X)
\end{bmatrix} < 0, \qquad (4.89)
$$

with

$$\tilde{J}(i) = X(i)A^\top(i) + A(i)X(i) + Y^\top(i)B^\top(i) + B(i)Y(i)$$
$$+ \varepsilon_A(i)D_A(i)D_A^\top(i) + \varepsilon_B(i)D_B(i)D_B^\top(i) + \lambda_{ii}X(i),$$
$$\mathcal{U}(i) = \mathbb{I} - \varepsilon_{D_z}(i)D_{D_z}(i)D_{D_z}^\top(i) - \varepsilon_{C_z}(i)D_{C_z}(i)D_{C_z}^\top(i),$$
$$\mathcal{Z}(i) = \left(X(i)E_A^\top(i), Y^\top(i)E_B^\top(i), X^\top(i)E_{C_z}^\top(i), Y^\top(i)E_{D_z}^\top(i)\right),$$
$$\mathcal{V}(i) = \mathrm{diag}[\varepsilon_A(i)\mathbb{I}, \varepsilon_B(i)\mathbb{I}, \varepsilon_{C_z}(i)\mathbb{I}, \varepsilon_{D_z}(i)\mathbb{I}].$$

The following theorem summarizes the results of this development.

Theorem 68. *Let γ be a positive constant. If there exist a set of symmetric and positive-definite matrices $X = (X(1), \cdots, X(N)) > 0$ and a set of matrices $Y = (Y(1), \cdots, Y(N))$ and sets of positive scalars $\varepsilon_A = (\varepsilon_A(1), \cdots, \varepsilon_A(N))$, $\varepsilon_B = (\varepsilon_B(1), \cdots, \varepsilon_B(N))$, $\varepsilon_{C_z} = (\varepsilon_{C_z}(1), \cdots, \varepsilon_{C_z}(N))$, and $\varepsilon_{D_z} = (\varepsilon_{D_z}(1), \cdots, \varepsilon_{D_z}(N))$ such that the following LMI (4.89) holds for every $i \in \mathscr{S}$ and for all admissible uncertainties, then the system (4.70) under the controller (4.17) with $K(i) = Y(i)X^{-1}(i)$ is stochastically stable and, moreover, the closed-loop system satisfies the disturbance rejection of level γ.*

From the practical point of view, the controller that stochastically stabilizes the class of systems and simultaneously guarantees the minimum disturbance rejection is of great interest. This controller can be obtained by solving the following optimization problem:

$$\mathrm{Pu}: \begin{cases} \min\limits_{\substack{\nu > 0, \\ \varepsilon_A = (\varepsilon_A(1), \cdots, \varepsilon_A(N)) > 0, \\ \varepsilon_B = (\varepsilon_B(1), \cdots, \varepsilon_B(N)) > 0, \\ \varepsilon_{C_z} = (\varepsilon_{D_z}(1), \cdots, \varepsilon_{C_z}(N)) > 0, \\ \varepsilon_{D_z} = (\varepsilon_{D_z}(1), \cdots, \varepsilon_{D_z}(N)) > 0, \\ X = (X(1), \cdots, X(N)) > 0, \\ Y = (Y(1), \cdots, Y(N))}} \nu, \\ s.t. : (4.89) \text{ with } \nu = \gamma^2. \end{cases}$$

The following corollary gives the results of the design of the controller that stochastically stabilizes the system (4.70) and simultaneously guarantees the smallest disturbance rejection level.

Corollary 14. *Let $\nu > 0$, $\varepsilon_A = (\varepsilon_A(1), \cdots, \varepsilon_A(N)) > 0$, $\varepsilon_B = (\varepsilon_B(1), \cdots, \varepsilon_B(N)) > 0$, $\varepsilon_{C_z} = (\varepsilon_{C_z}(1), \cdots, \varepsilon_{C_z}(N)) > 0$, $\varepsilon_{D_z} = (\varepsilon_{D_z}(1), \cdots, \varepsilon_{D_z}(N)) > 0$, $X = (X(1), \cdots, X(N)) > 0$, and $Y = (Y(1), \cdots, Y(N))$ be the solution of the optimization problem Pu. Then the controller (4.17) with $K(i) = Y(i)X^{-1}(i)$ stochastically stabilizes the class of systems we are considering and, moreover, the closed-loop system satisfies the disturbance rejection of level $\sqrt{\nu}$.*

We developed results that determine the state feedback controller that stochastically stabilizes the class of systems treated in this chapter and at the same time rejects the disturbance $w(t)$ with the desired level $\gamma > 0$. The conditions we developed are in the LMI form, which makes their resolution easy. In the rest of this section we will give a numerical example to show the effectiveness of our results.

Example 62. Let us consider the two-mode system of Example 49 with the following extra data:

- mode #1:

$$\mathbb{W}(1) = \begin{bmatrix} 0.1 & 0.0 \\ 0.0 & 0.1 \end{bmatrix},$$

- mode #2:

$$\mathbb{W}(2) = \begin{bmatrix} 0.2 & 0.0 \\ 0.0 & 0.2 \end{bmatrix}.$$

The required positive scalars are fixed to the following values:

$$\varepsilon_A(1) = \varepsilon_A(2) = 0.50,$$
$$\varepsilon_B(1) = \varepsilon_B(2) = \varepsilon_{C_z}(1) = \varepsilon_{C_z}(2) = \varepsilon_{D_z}(1) = \varepsilon_{D_z}(2) = 0.10.$$

Solving the problem Pu gives

$$X(1) = \begin{bmatrix} 0.0769 & -0.1335 \\ -0.1335 & 0.5794 \end{bmatrix}, \quad Y(1) = \begin{bmatrix} -1.0610 & 0.1437 \\ 0.1392 & -1.5717 \end{bmatrix},$$
$$X(2) = \begin{bmatrix} 0.1433 & -0.1433 \\ -0.1433 & 0.5533 \end{bmatrix}, \quad Y(2) = \begin{bmatrix} -1.1449 & 0.1518 \\ 0.1478 & -1.5304 \end{bmatrix},$$

which gives the following gains:

$$K(1) = \begin{bmatrix} -22.2878 & -4.8864 \\ -4.8327 & -3.8260 \end{bmatrix}, \quad K(2) = \begin{bmatrix} -10.4069 & -2.4204 \\ -2.3394 & -3.3718 \end{bmatrix}.$$

Using the results of Corollary 14, the system of this example is stochastically stable under the state feedback controller with the computed constant gain and assures the disturbance rejection of level $\gamma = 1.01$.

The goal of this section is to design an observer-based output feedback control that stochastically stabilizes the class of stochastic switching systems with Wiener process we are considering in this paper and at the same time rejects the effect of the external disturbance $w(t)$ with a desired level $\gamma > 0$. The structure of the controller we use here is given by the following expression:

$$\begin{cases} dx_c(t) = A(r(t))x_c(t)dt + B(r(t))u(t)dt + B_w(r(t))w(t)dt \\ \quad + L(r(t))\left[C_y(r(t))x_c(t) - y(t)\right]dt + \mathbb{W}(r(t))x_c(t)d\omega(t), x_c(0) = 0, \quad (4.90) \\ u(t) = K(r(t))x_c(t), \end{cases}$$

where $x_c(t)$ is the state of the controller and $L(i)$ and $K(i)$ are gains to be determined for each mode $i \in \mathscr{S}$.

We are mainly concerned with the design of such a controller. LMI-based conditions are searched since the design becomes easier and the gain can be obtained by solving the appropriate LMIs using the developed algorithms. In the rest of this section, we assume the complete access to the mode and the state vector at time t.

For simplicity, we will assume in the dynamics (4.70) that the matrices $D_y(i)$ and $B_z(i)$ are always equal to zero for each mode $i \in \mathscr{S}$. Using (4.70) and (4.90) and letting the observer error be defined by $e(t) = x(t) - x_c(t)$, we get

$$
\begin{aligned}
dx(t) &= A(r(t))x(t)dt + B(r(t))K(r(t))x_c(t)dt + B_w(r(t))w(t)dt \\
&\quad + \mathbb{W}(r(t))x(t)d\omega(t) \\
&= [A(r(t)) + B(r(t))K(r(t))]\, x(t)dt - B(r(t))K(r(t))e(t)dt + B_w(r(t))w(t)dt \\
&\quad + \mathbb{W}(r(t))x(t)d\omega(t),
\end{aligned}
$$

and

$$
\begin{aligned}
dx_c(t) &= [A(r(t)) + B(r(t))K(r(t))]\, x_c(t)dt - L(r(t))C_y(r(t))e(t)dt \\
&\quad [B_w(r(t)) - L(r(t))B_y(r(t))]\, w(t)dt + \mathbb{W}(r(t))x_c(t)d\omega(t).
\end{aligned}
$$

Combining these equations with the error expression, we get the following dynamics for error:

$$
\begin{aligned}
de(t) &= [A(r(t)) + L(r(t))C_y(r(t))]\, e(t)dt \\
&\quad + L(r(t))B_y(r(t))w(t)dt + \mathbb{W}(r(t))e(t)dt,
\end{aligned}
$$

which gives the following dynamics for the extended system:

$$
\begin{aligned}
\dot{\eta}(t) &= \begin{bmatrix} A(r(t)) + B(r(t))K(r(t)) & -B(r(t))K(r(t)) \\ 0 & \begin{bmatrix} A(r(t)) \\ +L(r(t))C_y(r(t)) \end{bmatrix} \end{bmatrix} \eta(t) \\
&\quad + \begin{bmatrix} B_w(r(t)) \\ L(r(t))B_y(r(t)) \end{bmatrix} w(t)dt + \begin{bmatrix} \mathbb{W}(r(t)) & 0 \\ 0 & \mathbb{W}(r(t)) \end{bmatrix} \eta(t)d\omega(t) \\
&= \tilde{A}(r(t))\eta(t)dt + \tilde{B}_w(r(t))w(t)dt + \tilde{\mathbb{W}}(r(t))\eta(t)d\omega(t),
\end{aligned}
$$

with

$$
\eta(t) = \begin{bmatrix} x(t) \\ e(t) \end{bmatrix},
$$

$$
\tilde{A}(r(t)) = \begin{bmatrix} A(r(t)) + B(r(t))K(r(t)) & -B(r(t))K(r(t)) \\ 0 & \begin{bmatrix} A(r(t)) \\ +L(r(t))C_y(r(t)) \end{bmatrix} \end{bmatrix},
$$

$$\tilde{B}_w(r(t)) = \begin{bmatrix} B_w(r(t)) \\ L(r(t))B_y(r(t)) \end{bmatrix}, \tilde{\mathbb{W}}(r(t)) = \begin{bmatrix} \mathbb{W}(r(t)) & 0 \\ 0 & \mathbb{W}(r(t)) \end{bmatrix}.$$

For the controlled output, we have

$$\begin{aligned}
z(t) &= C_z(r(t)) + D_z(r(t))K(r(t))x_c(t) + D_z(r(t))K(r(t))x(t) \\
&\quad -D_z(r(t))K(r(t))x(t) \\
&= \left[C_z(r(t)) + D_z(r(t))K(r(t)) \ -D_z(r(t))K(r(t)) \right] \eta(t) \\
&= \tilde{C}_z(r(t))\eta(t).
\end{aligned}$$

Based on the previous results, the extended system will be stochastically stable and guarantee the disturbance rejection of level γ if there exists a set of symmetric and positive-definite matrices $\tilde{P} = (\tilde{P}(1), \cdots, \tilde{P}(N)) > 0$ such that the following holds for each $i \in \mathscr{S}$:

$$\tilde{J}(i) + \tilde{C}_z^\top(i)\tilde{C}_z(i) + \gamma^{-2}\tilde{P}(i)\tilde{B}_w(i)\tilde{B}_w^\top(i)\tilde{P}(i) < 0, \tag{4.91}$$

with $\tilde{J}(i) = \tilde{A}^\top(i)\tilde{P}(i) + \tilde{P}(i)\tilde{A}(i) + \sum_{j=1}^N \lambda_{ij}\tilde{P}(j) + \tilde{\mathbb{W}}(i)\tilde{P}(i)\mathbb{W}(i)$.
Let $\tilde{P}(i)$ be given by

$$\tilde{P}(i) = \begin{bmatrix} P(i) & 0 \\ 0 & Q(i) \end{bmatrix},$$

with $P(i)$ and $Q(i)$ symmetric and positive-definite matrices, and using the expression of the matrices $\tilde{A}(i)$, we get

$$\tilde{A}^\top(i)\tilde{P}(i) = \begin{bmatrix} \begin{bmatrix} A^\top(i)P(i) \\ +K^\top(i)B^\top(i)P(i) \end{bmatrix} & 0 \\ -K^\top(i)B^\top(i)P(i) & \begin{bmatrix} A^\top(i)Q(i) \\ +C_y^\top(i)L^\top(i)Q(i) \end{bmatrix} \end{bmatrix},$$

$$\tilde{P}(i)\tilde{B}_w(i) = \begin{bmatrix} P(i)B_w(i) \\ Q(i)L(i)B_y(i) \end{bmatrix},$$

$$\tilde{P}(i)\tilde{B}_w(i)\tilde{B}_w^\top(i)\tilde{P}(i) = \begin{bmatrix} \begin{bmatrix} P(i)B_w(i) \\ \times B_w^\top(i)P(i) \end{bmatrix} & \begin{bmatrix} P(i)B_w(i) \\ \times B_y^\top(i)L^\top(i)Q(i) \end{bmatrix} \\ \begin{bmatrix} Q(i)L(i) \\ \times B_y(i)B_w^\top(i)P(i) \end{bmatrix} & \begin{bmatrix} Q(i)L(i) \\ \times B_y(i)B_y^\top(i)L^\top(i)Q(i) \end{bmatrix} \end{bmatrix},$$

$$\tilde{C}_z^\top(i)\tilde{C}_z(i) = \begin{bmatrix} \begin{bmatrix} [C_z^\top(i) + K^\top(i)D_z^\top(i)] \\ \times [C_z(i) + D_z(i)K(i)] \\ -K^\top(i)D_z^\top(i) \\ \times [C_z(i) + D_z(i)K(i)] \end{bmatrix} & \begin{bmatrix} -[C_z^\top(i) + K^\top(i)D_z^\top(i)] \\ \times D_z(i)K(i) \\ K^\top(i)D_z^\top(i) \\ \times D_z(i)K(i) \end{bmatrix} \end{bmatrix},$$

$$\tilde{\mathbb{W}}(i)\tilde{P}(i)\mathbb{W}(i) = \begin{bmatrix} \mathbb{W}^\top(i)P(i)\mathbb{W}(i) & 0 \\ 0 & \mathbb{W}^\top(i)Q(i)\mathbb{W}(i) \end{bmatrix},$$

$$\sum_{j=1}^{N} \lambda_{ij} \tilde{P}(i) = \begin{bmatrix} \sum_{j=1}^{N} \lambda_{ij} P(j) & 0 \\ 0 & \sum_{j=1}^{N} \lambda_{ij} Q(j) \end{bmatrix}.$$

Based on these computations we get

$$\tilde{J}(i) + \tilde{C}_z^\top(i)\tilde{C}_z(i) + \gamma^{-2} P(i)B_w(i)B_w^\top(i)P(i) =$$

$$\begin{bmatrix} \begin{bmatrix} A^\top(i)P(i) + P(i)A(i) \\ +P(i)B(i)K(i) \\ +K^\top(i)B^\top(i)P(i) \\ +C_z^\top(i)C_z(i) \\ +C_z^\top(i)D_z(i)K(i) \\ +K^\top(i)D_z^\top(i)C_z(i) \\ +K^\top(i)D_z^\top(i)D_z(i)K(i) \\ +\gamma^{-2}P(i)B_w(i)B_w^\top(i)P(i) \\ +\sum_{j=1}^{N}\lambda_{ij}P(j) \\ +\mathbb{W}^\top(i)P(i)\mathbb{W}(i) \end{bmatrix} & \begin{bmatrix} -P(i)B(i)K(i) \\ +\gamma^{-2}P(i)B_w(i)B_y^\top(i)L^\top(i)Q(i) \\ -C_z^\top(i)D_z(i)K(i) \\ -K^\top(i)D_z^\top(i)D_z(i)K(i) \end{bmatrix} \\ \\ \begin{bmatrix} -K^\top(i)B^\top(i)P(i) \\ +\gamma^{-2}Q(i)L(i)B_y(i)B_w^\top(i)P(i) \\ -K^\top(i)D_z^\top(i)C_z(i) \\ -K^\top(i)D_z^\top(i)D_z(i)K(i) \end{bmatrix} & \begin{bmatrix} A^\top(i)Q(i) + Q(i)A(i) \\ +Q(i)L(i)C_y(i) \\ +C_y^\top(i)L^\top(i)Q(i) \\ +K^\top(i)D_z^\top(i)D_z(i)K(i) \\ +\gamma^{-2}Q(i)L(i)B_y(i)B_y^\top(i)L^\top(i)Q(i) \\ +\sum_{j=1}^{N}\lambda_{ij}Q(j) \\ +\mathbb{W}^\top(i)Q(i)\mathbb{W}(i) \end{bmatrix} \end{bmatrix}.$$

Notice that

$$\begin{bmatrix} 0 & P(i)B(i)K(i) \\ 0 & 0 \end{bmatrix} = \begin{bmatrix} P(i)B(i)K(i) & 0 \\ 0 & 0 \end{bmatrix} \begin{bmatrix} 0 & \mathbb{I} \\ 0 & 0 \end{bmatrix},$$

$$\begin{bmatrix} 0 & P(i)B_w(i)B_y^\top(i)L^\top(i)Q(i) \\ 0 & 0 \end{bmatrix} = \begin{bmatrix} P(i)B_w(i) & 0 \\ 0 & 0 \end{bmatrix} \begin{bmatrix} 0 & B_y^\top(i)L^\top(i)Q(i) \\ 0 & 0 \end{bmatrix},$$

$$\begin{bmatrix} 0 & C_z^\top(i)D_z(i)K(i) \\ 0 & 0 \end{bmatrix} = \begin{bmatrix} 0 & C_z^\top(i) \\ 0 & 0 \end{bmatrix} \begin{bmatrix} 0 & D_z(i)K(i) \\ 0 & 0 \end{bmatrix},$$

$$\begin{bmatrix} 0 & K^\top(i)D_z^\top(i)D_z(i)K(i) \\ 0 & 0 \end{bmatrix} = \begin{bmatrix} K^\top(i)D_z^\top(i) & 0 \\ 0 & 0 \end{bmatrix} \begin{bmatrix} 0 & D_z(i)K(i) \\ 0 & 0 \end{bmatrix}.$$

Using Lemma 7 in Appendix A, we get

$$-\begin{bmatrix} 0 & P(i)B(i)K(i) \\ 0 & 0 \end{bmatrix} - \begin{bmatrix} 0 & P(i)B(i)K(i) \\ 0 & 0 \end{bmatrix}^\top$$

$$\leq \begin{bmatrix} P(i)B(i)K(i)K^\top(i)B^\top(i)P(i) & 0 \\ 0 & 0 \end{bmatrix} + \begin{bmatrix} 0 & 0 \\ 0 & \mathbb{I} \end{bmatrix},$$

$$\begin{bmatrix} 0 & P(i)B_w(i)B_y^\top(i)L^\top(i)Q(i) \\ 0 & 0 \end{bmatrix} + \begin{bmatrix} 0 & P(i)B_w(i)B_y^\top(i)L^\top(i)Q(i) \\ 0 & 0 \end{bmatrix}^\top$$
$$\leq \begin{bmatrix} P(i)B_w(i)B_w^\top(i)P(i) & \\ 0 & 0 \end{bmatrix} + \begin{bmatrix} 0 & 0 \\ 0 & Q(i)L(i)B_y(i)B_y^\top(i)L^\top(i)Q(i) \end{bmatrix},$$

$$-\begin{bmatrix} 0 & C_z^\top(i)D_z(i)K(i) \\ 0 & 0 \end{bmatrix} - \begin{bmatrix} 0 & C_z^\top(i)D_z(i)K(i) \\ 0 & 0 \end{bmatrix}^\top$$
$$\leq \begin{bmatrix} C_z^\top(i)C_z(i) & 0 \\ 0 & 0 \end{bmatrix} + \begin{bmatrix} 0 & 0 \\ 0 & K^\top(i)D_z^\top(i)D_z(i)K(i) \end{bmatrix},$$

$$-\begin{bmatrix} 0 & K^\top(i)D_z^\top(i)D_z(i)K(i) \\ 0 & 0 \end{bmatrix} - \begin{bmatrix} 0 & K^\top(i)D_z^\top(i)D_z(i)K(i) \\ 0 & 0 \end{bmatrix}^\top$$
$$\leq \begin{bmatrix} K^\top(i)D_z^\top(i)D_z(i)K(i) & 0 \\ 0 & 0 \end{bmatrix} + \begin{bmatrix} 0 & 0 \\ 0 & K^\top(i)D_z^\top(i)D_z(i)K(i) \end{bmatrix}.$$

Based on all these computations, the stochastic stability condition (4.91) will be satisfied if the following holds:

$$\begin{bmatrix} \begin{bmatrix} \begin{matrix} A^\top(i)P(i) + P(i)A(i) \\ +P(i)B(i)K(i) \\ +K^\top(i)B^\top(i)P(i) \\ +C_z^\top(i)C_z(i) \\ +C_z^\top(i)D_z(i)K(i) \\ +K^\top(i)D_z^\top(i)C_z(i) \\ +K^\top(i)D_z^\top(i)D_z(i)K(i) \\ +\gamma^{-2}P(i)B_w(i)B^\top(i)P(i) \\ +P(i)B(i)K(i)K^\top(i)B^\top(i)P(i) \\ +\gamma^{-2}P(i)B_w(i)B_w^\top(i)P(i) \\ +C_z^\top(i)C_z(i) \\ +K^\top(i)D_z^\top(i)D_z(i)K(i) \\ +\sum_{j=1}^{N}\lambda_{ij}P(j) \\ +\mathbb{W}^\top(i)P(i)\mathbb{W}(i) \end{matrix} & 0 \\[2em] 0 & \begin{bmatrix} \begin{matrix} A^\top(i)Q(i) + Q(i)A(i) \\ +Q(i)L(i)C_y(i) \\ +C_y^\top(i)L^\top(i)Q(i) \\ +K^\top(i)D_z^\top(i)D_z(i)K(i) \\ +\gamma^{-2}Q(i)L(i)B_y(i)B_y^\top(i)L^\top(i)Q(i) \\ +\mathbb{I} \\ +\gamma^{-2}Q(i)L(i)B_y(i)B_y^\top(i)L^\top(i)Q(i) \\ +K^\top(i)D_z^\top(i)D_z(i)K(i) \\ +K^\top(i)D_z^\top(i)D_z(i)K(i) \\ +\sum_{j=1}^{N}\lambda_{ij}Q(j) \\ +\mathbb{W}^\top(i)Q(i)\mathbb{W}(i) \end{matrix} \end{bmatrix} \end{bmatrix} \end{bmatrix} < 0,$$

which implies in turn that

$$A^\top(i)P(i) + P(i)A(i) + P(i)B(i)K(i) + K^\top(i)B^\top(i)P(i) + 2C_z^\top(i)C_z(i)$$
$$+C_z^\top(i)D_z(i)K(i) + K^\top(i)D_z^\top(i)C_z(i) + 2K^\top(i)D_z^\top(i)D_z(i)K(i)$$
$$+2\gamma^{-2}P(i)B_w(i)B_w^\top(i)P(i) + P(i)B(i)K(i)K^\top(i)B^\top(i)P(i)$$
$$+\sum_{j=1}^N \lambda_{ij}P(j) + \mathbb{W}^\top(i)P(i)\mathbb{W}(i) < 0$$

$$A^\top(i)Q(i) + Q(i)A(i) + Q(i)L(i)C_y(i) + C_y^\top(i)L^\top(i)Q(i)$$
$$+3K^\top(i)D_z^\top(i)D_z(i)K(i) + 2\gamma^{-2}Q(i)L(i)B_y(i)B_y^\top(i)L^\top(i)Q(i)$$

$$+\mathbb{I} + \sum_{j=1}^N \lambda_{ij}Q(j) + \mathbb{W}^\top(i)Q(i)\mathbb{W}(i) < 0.$$

Noticing from Lemma 7 that

$$C_z^\top(i)D_z(i)K(i) + D_z^\top(i)K^\top(i)C_z(i) \le C_z^\top(i)C_z(i) + K^\top(i)D_z^\top(i)D_z(i)K(i),$$

and using the Schur complement, these conditions become

$$\begin{bmatrix} J_P(i) & \sqrt{2}P(i)B_w(i) & \sqrt{3}K^\top(i)D_z^\top(i) & P(i)B(i)K(i) \\ \sqrt{2}B_w^\top(i)P(i) & -\gamma^2\mathbb{I} & 0 & 0 \\ \sqrt{3}D_z(i)K(i) & 0 & -\mathbb{I} & 0 \\ K^\top(i)B^\top(i)P(i) & 0 & 0 & -\mathbb{I} \end{bmatrix} < 0,$$

$$\begin{bmatrix} J_Q(i) & \sqrt{2}Q(i)L(i)B_y(i) & \sqrt{3}K^\top(i)D_z^\top(i) \\ \sqrt{2}B_y^\top(i)L^\top(i)Q(i) & -\gamma^2\mathbb{I} & 0 \\ \sqrt{3}D_z(i)K(i) & 0 & -\mathbb{I} \end{bmatrix} < 0,$$

where

$$J_P(i) = A^\top(i)P(i) + P(i)A(i) + P(i)B(i)K(i) + K^\top(i)B^\top(i)P(i)$$
$$+3C_z^\top(i)C_z(i) + \sum_{j=1}^N \lambda_{ij}P(j) + \mathbb{W}^\top(i)P(i)\mathbb{W}(i),$$

$$J_Q(i) = A^\top(i)Q(i) + Q(i)A(i) + Q(i)L(i)C_y(i) + C_y^\top(i)L^\top(i)Q(i)$$
$$+\mathbb{I} + \sum_{j=1}^N \lambda_{ij}Q(j) + \mathbb{W}^\top(i)Q(i)\mathbb{W}(i).$$

These two inequalities are nonlinear in the design parameters. To put them in the LMI setting let us change some variables. Notice that the first LMI implies the following:

$$\begin{bmatrix} J_P(i) & \sqrt{2}P(i)B_w(i) & \sqrt{3}K^\top(i)D_z^\top(i) \\ \sqrt{2}B_w^\top(i)P(i) & -\gamma^2\mathbb{I} & 0 \\ \sqrt{3}D_z(i)K(i) & 0 & -\mathbb{I} \end{bmatrix} < 0.$$

Let $X(i) = P^{-1}(i)$. Pre- and post-multiply this LMI by $\mathrm{diag}(X(i), \mathbb{I})$ to get

$$\begin{bmatrix} X(i)J_P(i)X(i) & \sqrt{2}B_w(i) & \sqrt{3}X(i)K^\top(i)D_z^\top(i) \\ \sqrt{2}B_w^\top(i) & -\gamma^2\mathbb{I} & 0 \\ \sqrt{3}D_z(i)K(i)X(i) & 0 & -\mathbb{I} \end{bmatrix} < 0.$$

Using the fact that

$$X(i)J_P(i)X(i) = X(i)A^\top(i) + A(i)X(i) + B(i)K(i)X(i)$$

$$+X(i)K^\top(i)B^\top(i) + 3X(i)C_z^\top(i)C_z(i)X(i) + \sum_{j=1}^{N}\lambda_{ij}X(i)X^{-1}(j)X(i)$$

$$+X(i)\mathbb{W}^\top(i)X^{-1}(i)\mathbb{W}(i)X(i),$$

$$\sum_{j=1}^{N}\lambda_{ij}X(i)X^{-1}(j)X(i) = \lambda_{ii}X(i) + \mathcal{S}_i(X)\mathcal{X}_i^{-1}(X)\mathcal{S}_i^\top(X),$$

and after letting $Y_c(i) = K(i)X(i)$, we get

$$\begin{bmatrix} X(i)J_P(i)X(i) & \sqrt{2}B_w(i) & \sqrt{3}Y_o^\top(i)D_z^\top(i) \\ \sqrt{2}B_w^\top(i) & -\gamma^2\mathbb{I} & 0 \\ \sqrt{3}D_z(i)Y_o(i) & 0 & -\mathbb{I} \\ \sqrt{3}C_z(i)X(i) & 0 & 0 \\ \mathbb{W}(i)X(i) & 0 & 0 \\ \mathcal{S}_i^\top(X) & 0 & 0 \end{bmatrix}$$

$$\left.\begin{matrix} \sqrt{3}X(i)C_z^\top(i) & X(i)\mathbb{W}^\top(i) & \mathcal{S}_i(X) \\ 0 & 0 & 0 \\ 0 & 0 & 0 \\ -\mathbb{I} & 0 & 0 \\ 0 & -X(i) & 0 \\ 0 & 0 & -\mathcal{X}_i(X) \end{matrix}\right] < 0, \qquad (4.92)$$

with

$$\mathscr{J}_P(i) = X(i)A^\top(i) + A(i)X(i) + B(i)Y_c(i) + Y_c^\top(i)B^\top(i) + \lambda_{ii}X(i).$$

For the second nonlinear matrix inequality, letting $Y_o(i) = Q(i)L(i)$ implies the following LMI:

$$\begin{bmatrix} J_Q(i) & \sqrt{2}Y_o(i)B_y(i) & \sqrt{3}K^\top(i)D_z^\top(i) \\ \sqrt{2}B_y^\top(i)Y_o^\top(i) & -\gamma^2\mathbb{I} & 0 \\ \sqrt{3}D_z(i)K(i) & 0 & -\mathbb{I} \end{bmatrix} < 0, \qquad (4.93)$$

where

$$\mathscr{J}_Q(i) = A^\top(i)Q(i) + Q(i)A(i) + Y_o(i)C_y(i) + C_y^\top(i)Y_o^\top(i)$$

$$+\mathbb{I} + \mathbb{W}^\top(i)Q(i)\mathbb{W}(i) + \sum_{j=1}^{N} \lambda_{ij}Q(j).$$

The following theorem summarizes the results of this development.

Theorem 69. *Let γ be a given positive constant. If there exist sets of symmetric and positive-definite matrices $X = (X(1), \cdots, X(N)) > 0$ and $Q = (Q(1), \cdots, Q(N)) > 0$ and sets of matrices $Y_c = (Y_c(1), \cdots, Y_c(N))$ and $Y_o = (Y_o(1), \cdots, Y_o(N))$ such that the following LMIs (4.92)–(4.93) hold for every $i \in \mathscr{S}$, then the observer-based output feedback control with the following gains:*

$$L(i) = Q^{-1}(i)Y_o(i), \tag{4.94}$$

$$K(i) = Y_c(i)X^{-1}(i), \tag{4.95}$$

stochastically stabilizes the class of systems we are studying and at the same time guarantees the disturbance rejection of level γ.

Remark 17. Notice that the second LMI depends on the solution of the first. Therefore, to solve these two LMIs, we should solve the first one to get the gains $K = (K(1), \cdots, K(N))$ that enter in the second LMI.

From the practical point of view, the observer-based output feedback control that stochastically stabilizes the system and at the same time guarantees the minimum disturbance rejection is of great interest. This controller can be obtained by solving the following optimization problem:

$$P : \begin{cases} \min\limits_{\substack{\nu>0, \\ X=(X(1),\cdots,X(N))>0, \\ Q=(Q(1),\cdots,Q(N))>0, \\ Y_c=(Y_c(1),\cdots,Y_c(N)), \\ Y_o=(Y_o(1),\cdots,Y_o(N)),}} \nu, \\ s.t. : (4.92) \text{ and } (4.93) \text{ with } \nu = \gamma^2. \end{cases}$$

Remark 18. To solve this optimization, we can proceed by the search method. We fix γ and solve the first LMI that gives the gain $K(i)$ that enters in the second LMI. Then decrease the disturbance rejection parameter γ until an infeasible solution for one of the two LMIs is obtained. The previous solution gives the optimal solution.

The following corollary gives the results on the design of the controller that stochastically stabilizes the system (4.70) and simultaneously guarantees the smallest disturbance rejection level.

Corollary 15. *Let $\nu > 0$, $X = (X(1), \cdots, X(N)) > 0$, $X = (X(1), \cdots, X(N)) > 0$, $Y_c = (Y_c(1), \cdots, Y_c(N))$, and $Y_o = (Y_o(1), \cdots, Y_o(N))$ be the solution of the optimization problem P. Then the controller (4.90) with $K(i) = Y_c(i)X^{-1}(i)$ and $L(i) = Q^{-1}(i)Y_o(i)$ stochastically stabilizes the class of systems we are considering and, moreover, the closed-loop system satisfies the disturbance rejection of level $\sqrt{\nu}$.*

Example 63. To show the validity of the previous results let us consider a system with two modes with the following data:

- mode #1:

$$A(1) = \begin{bmatrix} 1.0\ 0.0\ 1.0 \\ 0.0\ 1.0\ 0.0 \\ 0.0\ 1.0\ 1.0 \end{bmatrix}, \quad B(1) = \begin{bmatrix} 0.3\ 0.0\ 0.1 \\ 0.0\ 0.3\ 0.1 \\ 0.0\ 0.2\ 1.0 \end{bmatrix},$$

$$B_w(1) = \begin{bmatrix} 0.3\ 0.0\ 0.1 \\ 0.0\ 0.3\ 0.1 \\ 0.0\ 0.2\ 1.0 \end{bmatrix}, \quad C_y(1) = \begin{bmatrix} 0.3\ 0.0\ 0.1 \\ 0.0\ 0.3\ 0.1 \\ 0.0\ 0.2\ 1.0 \end{bmatrix},$$

$$B_y(1) = \begin{bmatrix} 0.3\ 0.0\ 0.1 \\ 0.0\ 0.3\ 0.1 \\ 0.0\ 0.2\ 1.0 \end{bmatrix}, \quad C_z(1) = \begin{bmatrix} 0.3\ 0.0\ 0.1 \\ 0.0\ 0.3\ 0.1 \\ 0.0\ 0.2\ 1.0 \end{bmatrix},$$

$$D_z(1) = \begin{bmatrix} 0.3\ 0.0\ 0.1 \\ 0.0\ 0.3\ 0.1 \\ 0.0\ 0.2\ 1.0 \end{bmatrix}, \quad \mathbb{W}(1) = \begin{bmatrix} 0.3\ 0.0\ 0.1 \\ 0.0\ 0.3\ 0.1 \\ 0.0\ 0.2\ 1.0 \end{bmatrix},$$

- mode #2:

$$A(2) = \begin{bmatrix} 1.0\ \ 0.0\ \ \ 1.0 \\ 0.0\ -1.0\ \ 0.0 \\ 0.0\ \ 1.0\ -1.0 \end{bmatrix}, \quad B(2) = \begin{bmatrix} 0.1\ 0.0\ 0.1 \\ 0.2\ 0.0\ 0.1 \\ 0.0\ 0.2\ 0.2 \end{bmatrix},$$

$$B_w(2) = \begin{bmatrix} 0.1\ 0.0\ 0.1 \\ 0.2\ 0.0\ 0.1 \\ 0.0\ 0.2\ 0.2 \end{bmatrix}, \quad C_y(2) = \begin{bmatrix} 0.1\ 0.0\ 0.1 \\ 0.2\ 0.0\ 0.1 \\ 0.0\ 0.2\ 0.2 \end{bmatrix},$$

$$B_y(2) = \begin{bmatrix} 0.1\ 0.0\ 0.1 \\ 0.2\ 0.0\ 0.1 \\ 0.0\ 0.2\ 0.2 \end{bmatrix}, \quad C_z(2) = \begin{bmatrix} 0.1\ 0.0\ 0.1 \\ 0.2\ 0.0\ 0.1 \\ 0.0\ 0.2\ 0.2 \end{bmatrix},$$

$$D_z(2) = \begin{bmatrix} 0.1\ 0.0\ 0.1 \\ 0.2\ 0.0\ 0.1 \\ 0.0\ 0.2\ 0.2 \end{bmatrix}, \quad \mathbb{W}(2) = \begin{bmatrix} 0.1\ 0.0\ 0.1 \\ 0.2\ 0.0\ 0.1 \\ 0.0\ 0.2\ 0.2 \end{bmatrix}.$$

The switching between the two modes is described by the following transition matrix:

$$\Lambda = \begin{bmatrix} -2.0\ \ 2.0 \\ 1.0\ -1.0 \end{bmatrix}.$$

Letting $\gamma = 3.34$ and solving the LMIs (4.92)–(4.93), we get:

$$X(1) = \begin{bmatrix} 0.0785\ \ \ 0.0187\ -0.0323 \\ 0.0187\ \ \ 0.1166\ -0.0576 \\ -0.0323\ -0.0576\ \ 0.0829 \end{bmatrix},$$

$$X(2) = \begin{bmatrix} 0.1202\ \ \ 0.0501\ -0.0828 \\ 0.0501\ \ \ 0.2936\ -0.1708 \\ -0.0828\ -0.1708\ \ 0.2657 \end{bmatrix},$$

$$Q(1) = \begin{bmatrix} 2.3823 & -0.0324 & -0.6153 \\ -0.0324 & 6.7195 & -3.1886 \\ -0.6153 & -3.1886 & 3.2197 \end{bmatrix},$$

$$Q(2) = \begin{bmatrix} 1.1971 & -0.0835 & -0.0826 \\ -0.0835 & 1.7468 & -0.2846 \\ -0.0826 & -0.2846 & 2.0237 \end{bmatrix},$$

$$Y_c(1) = \begin{bmatrix} -1.1284 & -0.0805 & 0.1328 \\ 0.0159 & -1.1255 & 0.1105 \\ 0.0047 & 0.2289 & -0.3606 \end{bmatrix},$$

$$Y_c(2) = \begin{bmatrix} 3.3509 & -3.5017 & -0.3577 \\ 6.6375 & -3.6129 & -2.2431 \\ -6.5908 & 3.5848 & 0.3956 \end{bmatrix},$$

$$Y_o(1) = \begin{bmatrix} -9.6863 & -6.8668 & -0.7993 \\ -7.3390 & -5.8202 & -2.9695 \\ -7.9262 & -5.7653 & -2.7436 \end{bmatrix},$$

$$Y_o(2) = \begin{bmatrix} -10.4511 & -11.0495 & -6.4701 \\ -7.0889 & -4.1117 & -10.1378 \\ -8.7223 & -8.9604 & -8.7097 \end{bmatrix}.$$

The corresponding gains are given by

$$K(1) = \begin{bmatrix} -16.3710 & -0.6398 & -5.2117 \\ 0.1224 & -13.6946 & -8.1380 \\ -2.0789 & -0.3847 & -5.4259 \end{bmatrix},$$

$$K(2) = \begin{bmatrix} 33.6556 & -19.7178 & -3.5320 \\ 62.0216 & -26.4275 & -6.1011 \\ -67.8515 & 19.7155 & -6.9842 \end{bmatrix},$$

$$L(1) = \begin{bmatrix} -6.4854 & -4.6746 & -1.0927 \\ -5.4330 & -4.0794 & -1.7935 \\ -9.0818 & -6.7240 & -2.8371 \end{bmatrix},$$

$$L(2) = \begin{bmatrix} -9.4840 & -9.8576 & -6.2766 \\ -5.4007 & -3.6969 & -7.0073 \\ -5.4569 & -5.3503 & -5.5456 \end{bmatrix}.$$

4.6 Case Study

To end this chapter let us return to the VTOL helicopter example of Chapter 1 and apply some of the developed results to this system. We will mainly restrict ourselves to the nominal system. Using the data of Chapter 1, let us first design a state feedback controller that stochastically stabilizes the helicopter and at the same time rejects the disturbance with a given level γ. Letting $\gamma = 0.5$ and solving the appropriate set of LMIs we get

$$X(1) = \begin{bmatrix} 0.2948 & -0.0967 & -0.0509 & 0.2359 \\ -0.0967 & 0.4735 & 0.0806 & -0.0692 \\ -0.0509 & 0.0806 & 0.5861 & -0.1301 \\ 0.2359 & -0.0692 & -0.1301 & 0.3039 \end{bmatrix},$$

$$X(2) = \begin{bmatrix} 0.5358 & -0.1826 & -0.1263 & 0.4308 \\ -0.1826 & 0.7168 & 0.1990 & -0.1771 \\ -0.1263 & 0.1990 & 0.9789 & -0.2827 \\ 0.4308 & -0.1771 & -0.2827 & 0.5069 \end{bmatrix},$$

$$X(3) = \begin{bmatrix} 0.5136 & -0.3434 & -0.2898 & 0.4981 \\ -0.3434 & 0.8304 & 0.2714 & -0.3010 \\ -0.2898 & 0.2714 & 1.0378 & -0.3592 \\ 0.4981 & -0.3010 & -0.3592 & 0.6783 \end{bmatrix},$$

$$Y(1) = \begin{bmatrix} 0.1380 & -0.5555 & 0.3817 & 0.3061 \\ -0.0559 & -0.2238 & 0.4803 & 0.0584 \end{bmatrix},$$

$$Y(2) = \begin{bmatrix} 0.0208 & 0.2869 & 0.0214 & 0.1261 \\ -0.1425 & 0.1284 & -0.0317 & -0.1510 \end{bmatrix},$$

$$Y(3) = \begin{bmatrix} 0.3216 & 0.0463 & -0.0813 & 0.7911 \\ 0.0347 & 0.1631 & 0.0593 & 0.3081 \end{bmatrix},$$

which gives the following gains:

$$K(1) = \begin{bmatrix} -2.0082 & -1.3939 & 1.2906 & 2.8017 \\ -1.7305 & -0.7550 & 1.1888 & 1.8727 \end{bmatrix},$$

$$K(2) = \begin{bmatrix} -0.5063 & 0.4613 & 0.1257 & 0.9104 \\ 0.0552 & 0.1437 & -0.1664 & -0.3874 \end{bmatrix},$$

$$K(3) = \begin{bmatrix} -1.4854 & 0.2581 & 0.3186 & 2.5403 \\ -1.0905 & 0.2063 & 0.2018 & 1.4535 \end{bmatrix}.$$

We can also design a constant gain state feedback controller that guarantees the same objectives. Letting $\gamma = 0.5$ and solving the appropriate set of LMIs, we get

$$X = \begin{bmatrix} 0.6490 & -0.5358 & -0.2832 & 0.6274 \\ -0.5358 & 0.8143 & 0.2957 & -0.5365 \\ -0.2832 & 0.2957 & 0.9708 & -0.4392 \\ 0.6274 & -0.5365 & -0.4392 & 0.6911 \end{bmatrix},$$

$$Y = \begin{bmatrix} 0.1268 & 0.1184 & 0.2606 & 0.2574 \\ -0.1550 & 0.3269 & 0.3146 & -0.1189 \end{bmatrix},$$

which gives the gain:

$$\mathcal{K} = \begin{bmatrix} -2.5407 & 0.6112 & 1.0778 & 3.8384 \\ -1.0915 & 0.5199 & 0.5620 & 1.5796 \end{bmatrix}.$$

If we assume that we do not have access to the state vector, we can use either output feedback controller or observer-based output feedback controller. Letting again $\gamma = 0.5$ and solving the appropriate set of LMIs, we get the following matrices for the output feedback controller:

$$X(1) = \begin{bmatrix} 18.3522 & -4.8565 & -6.6867 & -5.0359 \\ -4.8565 & 8.2721 & 7.4568 & 9.1974 \\ -6.6867 & 7.4568 & 16.9620 & 3.6472 \\ -5.0359 & 9.1974 & 3.6472 & 21.9054 \end{bmatrix},$$

$$X(2) = \begin{bmatrix} 14.3232 & -0.5857 & -1.8638 & -1.3489 \\ -0.5857 & 7.4600 & 5.7177 & 8.6247 \\ -1.8638 & 5.7177 & 15.6892 & 4.1220 \\ -1.3489 & 8.6247 & 4.1220 & 18.1248 \end{bmatrix},$$

$$X(3) = \begin{bmatrix} 25.6622 & -16.2530 & -8.8191 & -7.0241 \\ -16.2530 & 26.1734 & 12.4299 & 15.0785 \\ -8.8191 & 12.4299 & 17.5084 & -2.9124 \\ -7.0241 & 15.0785 & -2.9124 & 22.9078 \end{bmatrix},$$

$$Y(1) = \begin{bmatrix} 2.4550 & -2.2490 & -2.0781 & 2.6473 \\ -2.2490 & 4.7090 & 1.7095 & -2.2865 \\ -2.0781 & 1.7095 & 5.0399 & -1.9624 \\ 2.6473 & -2.2865 & -1.9624 & 3.1506 \end{bmatrix},$$

$$Y(2) = \begin{bmatrix} 4.4938 & -4.1125 & -3.6648 & 4.7086 \\ -4.1125 & 6.9810 & 3.4775 & -4.2270 \\ -3.6648 & 3.4775 & 7.5519 & -3.7221 \\ 4.7086 & -4.2270 & -3.7221 & 5.1686 \end{bmatrix},$$

$$Y(3) = \begin{bmatrix} 4.7736 & -5.2514 & -4.9484 & 5.3736 \\ -5.2514 & 8.5485 & 4.1335 & -5.9808 \\ -4.9484 & 4.1335 & 10.5045 & -4.2725 \\ 5.3736 & -5.9808 & -4.2725 & 6.8252 \end{bmatrix},$$

$$\mathscr{K}_B(1) = \begin{bmatrix} -8.6649 & 21.3022 \\ -13.1845 & -10.6297 \\ 0.3871 & -0.0000 \\ -17.8135 & 13.2153 \end{bmatrix}, \qquad \mathscr{K}_B(2) = \begin{bmatrix} -6.8412 & 5.9010 \\ -7.3190 & -0.4590 \\ -2.1510 & -0.0000 \\ -2.1010 & 19.4061 \end{bmatrix},$$

$$\mathscr{K}_B(3) = \begin{bmatrix} -13.6704 & 27.2062 \\ -12.9241 & -6.7154 \\ 8.7135 & -0.0000 \\ -29.4058 & 38.3058 \end{bmatrix},$$

$$\mathscr{K}_C(1) = \begin{bmatrix} 2.4658 & 0.9919 & 2.1873 & 4.5482 \\ 0.7630 & 2.4123 & 1.2451 & 1.8877 \end{bmatrix},$$

$$\mathscr{K}_C(2) = \begin{bmatrix} 1.4441 & 2.5580 & 3.4823 & 2.5940 \\ -0.9652 & 3.5037 & 0.9561 & -0.7786 \end{bmatrix},$$

$$\mathscr{K}_C(3) = \begin{bmatrix} 4.3388 & 4.5172 & 5.4320 & 12.6411 \\ 1.6440 & 6.8877 & 2.4956 & 6.6399 \end{bmatrix},$$

which gives the following set of gains:

$$K_A(1) = \begin{bmatrix} -12.2726 & -10.2856 & -13.3531 & -10.7779 \\ -29.7243 & -46.3163 & -37.9057 & -52.8599 \\ 14.4585 & 8.1245 & 2.6309 & -4.6477 \\ 12.9983 & 19.6996 & 18.1768 & 22.4359 \end{bmatrix},$$

$$K_A(2) = \begin{bmatrix} -25.0359 & -36.2574 & -36.8525 & -46.5552 \\ -66.5947 & -111.8581 & -100.9157 & -140.8424 \\ 18.3847 & 11.7572 & 7.5583 & 0.2403 \\ 44.9711 & 73.1096 & 70.7291 & 93.6719 \end{bmatrix},$$

$$K_A(3) = 10^3 \cdot \begin{bmatrix} -0.0782 & -1.6168 & -0.4124 & -1.8744 \\ -0.1571 & -3.3930 & -0.8562 & -3.9326 \\ 0.0961 & 1.7956 & 0.4522 & 2.0679 \\ 0.1091 & 2.4361 & 0.6161 & 2.8263 \end{bmatrix},$$

$$K_B(1) = \begin{bmatrix} 11.2210 & 10.8168 \\ 31.4836 & 40.0747 \\ -5.7709 & -8.4136 \\ -13.8496 & -19.4622 \end{bmatrix},$$

$$K_B(2) = \begin{bmatrix} 28.7384 & 40.1719 \\ 79.2098 & 114.4381 \\ -10.4118 & -15.0794 \\ -53.9396 & -79.2146 \end{bmatrix}, \qquad K_B(3) = 10^3 \cdot \begin{bmatrix} 0.2614 & 1.8556 \\ 0.5431 & 3.8799 \\ -0.2881 & -2.0635 \\ -0.3886 & -2.7910 \end{bmatrix},$$

$$K_C(1) = \begin{bmatrix} -2.6646 & 1.0334 & 0.9382 & 5.0169 \\ -1.2987 & 1.0624 & 0.4087 & 2.7160 \end{bmatrix},$$

$$K_C(2) = \begin{bmatrix} -1.9223 & 1.0750 & 0.8945 & 3.7764 \\ -0.2365 & 0.7963 & -0.0030 & 0.7139 \end{bmatrix},$$

$$K_C(3) = \begin{bmatrix} -5.1469 & 3.8145 & 0.4729 & 9.5429 \\ -3.1381 & 3.3207 & 0.0492 & 6.3842 \end{bmatrix}.$$

Letting $\gamma = 0.5$ and solving the appropriate set of LMIs for the design of the observer-based output feedback control, we get

$$X(1) = \begin{bmatrix} 0.7041 & -0.0223 & 0.0102 & 0.2514 \\ -0.0223 & 0.7798 & 0.0491 & -0.0623 \\ 0.0102 & 0.0491 & 0.8747 & -0.1517 \\ 0.2514 & -0.0623 & -0.1517 & 0.5995 \end{bmatrix},$$

$$X(2) = \begin{bmatrix} 1.7575 & 0.0620 & 0.4759 & 0.4574 \\ 0.0620 & 1.3634 & 0.1096 & -0.0410 \\ 0.4759 & 0.1096 & 1.9321 & -0.3803 \\ 0.4574 & -0.0410 & -0.3803 & 1.1229 \end{bmatrix},$$

$$X(3) = \begin{bmatrix} 1.7911 & 0.0498 & 0.2979 & 0.4661 \\ 0.0498 & 1.4406 & 0.0857 & 0.0003 \\ 0.2979 & 0.0857 & 1.8743 & -0.3977 \\ 0.4661 & 0.0003 & -0.3977 & 1.5035 \end{bmatrix},$$

$$Y_c(1) = \begin{bmatrix} 0.1261 & 0.1834 & -0.0142 & 0.3168 \\ -0.0360 & 0.1073 & -0.0000 & -0.0880 \end{bmatrix},$$

$$Y_c(2) = \begin{bmatrix} -0.1719 & 0.2195 & -0.0986 & -0.0392 \\ -0.2433 & -0.0331 & -0.0000 & -0.5692 \end{bmatrix},$$

$$Y_c(3) = \begin{bmatrix} 0.0191 & 0.0968 & -0.2871 & 0.9983 \\ -0.2292 & -0.0383 & 0.0000 & -0.1035 \end{bmatrix},$$

$$Q(1) = \begin{bmatrix} 2.1557 & -0.6623 & -0.9628 & -0.6915 \\ -0.6623 & 1.0232 & 0.9562 & 1.0889 \\ -0.9628 & 0.9562 & 2.5423 & 1.4246 \\ -0.6915 & 1.0889 & 1.4246 & 4.7794 \end{bmatrix},$$

$$Q(2) = \begin{bmatrix} 1.2563 & -0.0912 & -0.2356 & 0.3869 \\ -0.0912 & 0.7063 & 0.6606 & 1.1761 \\ -0.2356 & 0.6606 & 2.7263 & 0.9792 \\ 0.3869 & 1.1761 & 0.9792 & 3.1284 \end{bmatrix},$$

$$Q(3) = \begin{bmatrix} 2.8558 & -1.8796 & -1.0045 & -1.0200 \\ -1.8796 & 2.1295 & 1.3480 & 1.6847 \\ -1.0045 & 1.3480 & 1.9694 & 0.0808 \\ -1.0200 & 1.6847 & 0.0808 & 2.7941 \end{bmatrix},$$

$$Y_o(1) = \begin{bmatrix} -0.7389 & -0.1460 \\ 0.2689 & -0.6476 \\ -0.5037 & -0.7826 \\ -1.1943 & 2.6974 \end{bmatrix}, \qquad Y_o(2) = \begin{bmatrix} -0.9843 & 0.0656 \\ -0.1650 & -0.2153 \\ -0.0724 & -0.1663 \\ 0.0406 & 3.5329 \end{bmatrix},$$

$$Y_o(3) = \begin{bmatrix} -0.7060 & -1.6128 \\ -0.0168 & 0.6717 \\ 0.3243 & -0.4543 \\ -2.0357 & 4.6174 \end{bmatrix},$$

which gives the following set of gains:

$$K(1) = \begin{bmatrix} -0.0217 & 0.2769 & 0.0698 & 0.5840 \\ 0.0047 & 0.1283 & -0.0321 & -0.1435 \end{bmatrix},$$

$$K(2) = \begin{bmatrix} -0.0921 & 0.1682 & -0.0387 & -0.0044 \\ 0.0426 & -0.0336 & -0.1200 & -0.5661 \end{bmatrix},$$

$$K(3) = \begin{bmatrix} -0.1856 & 0.0719 & 0.0277 & 0.7288 \\ -0.1228 & -0.0232 & 0.0149 & -0.0268 \end{bmatrix},$$

$$L(1) = \begin{bmatrix} -0.4152 & -0.4223 \\ 0.7996 & -1.5455 \\ -0.4566 & -0.4392 \\ -0.3560 & 0.9863 \end{bmatrix}, \qquad L(2) = \begin{bmatrix} -1.1547 & -1.7478 \\ -1.7195 & -7.5510 \\ 0.0024 & 0.1293 \\ 0.8014 & 4.1436 \end{bmatrix},$$

$$L(3) = \begin{bmatrix} -1.1547 & -1.7478 \\ -1.7195 & -7.5510 \\ 0.0024 & 0.1293 \\ 0.8014 & 4.1436 \end{bmatrix}.$$

4.7 Notes

This chapter dealt with the $\mathscr{H}_\infty$ control problem of the class of piecewise deterministic systems. Different types of controllers such as state feedback, output feedback, and observer-based output feedback were discussed and a design approach of each controller was developed in LMI form. Robust and nonfragile controllers were designed. The results of this chapter are based mainly on the work of the author and his coauthors. All of our results have been tested on the practical example presented in Chapter 1, which is borrowed from aerospace industry. The corresponding model we presented was developed by Narendra and Tripathi [53] and was used recently by de Farias et al. [32]. This system was also used by Kose et al. [47] in the deterministic framework.

5

Filtering Problem

In the previous chapters we assumed complete access to the state vector to compute state feedback control. But as is well known, sometimes this access is not possible for physical or cost reasons. In order to continue to use the state feedback controller, we can estimate the state vector. This technique has been used for many years and continues in many industrial applications ranging from aerospace to economics, including engineering, biology, geoscience, and management. In real-time applications, care should be taken to guarantee that the estimation dynamics be faster than those of the closed loop of the considered systems.

The estimation problem is fundamental in control theory. It consists of estimating the unavailable state variables of a studied system. This problem has been extensively investigated, and nowadays many techniques, such as Kalman filtering and $\mathcal{H}_\infty$ filtering, can be used to solve practical estimation problems.

Kalman filtering estimates the state vector of a given system perturbed by a Wiener process (Gaussian). The design of the desired filter is based on the critical assumption that the model of the system is well known and that the external Wiener process disturbance has some known statistical properties that are difficult to satisfy in practice. All the models used in practice for systems to be controlled have uncertainties that result from different causes, such as wearing and neglected dynamics.

To overcome the difficulties of the Kalman filtering problem, an alternative approach, called the $\mathcal{H}_\infty$ filtering problem, has been developed to deal with the issues of external disturbances and measurement noises. Contrary to Kalman filtering in this estimation approach, the external disturbance is an arbitrary signal with norm-bounded energy or bounded average power rather than Gaussian. In the $\mathcal{H}_\infty$ setting the filter is designed to minimize the worst case induced $\mathcal{L}_2$ gain from system disturbance error to estimation error. Compared to Kalman filtering, this approach presents many advantages, among them that this technique does not require the statistical assumptions of the

Kalman filter. Also, the filter designed by this approach is more robust when the system has uncertainties.

In the rest of this chapter, we cover the techniques of Kalman filtering and $\mathscr{H}_\infty$ filtering and their robustness. The rest of this chapter is organized as follows. In Section 5.1, the filtering problem is stated and some necessary definitions are given. Section 5.2 treats Kalman filtering and its robustness. In Section 5.3, the $\mathscr{H}_\infty$ filtering problem is developed and its robustness is also treated. Some numerical examples are provided to show the validity of the developed results.

5.1 Problem Statement

Let us consider a dynamical system defined in a probability space $(\Omega, \mathcal{F}, \mathbb{P})$ and assume that its dynamics are described by the following differential equations:

$$\begin{cases} \dot{x}(t) = A(r(t), t)x(t) + B(r(t), t)w(t), x(0) = x_0, \\ y(t) = C_y(r(t), t)x(t) + D_y(r(t))v(t), \\ z(t) = C_z(r(t), t)x(t) + D_z(r(t))v(t), \end{cases} \tag{5.1}$$

where $x(t) \in \mathbb{R}^n$ is the state vector; $w(t) \in \mathbb{R}^q$ and $v(t)$ are the noise signals; $y(t) \in \mathbb{R}^m$ is the measurement; and $z(t) \in \mathbb{R}^p$ is the signal to be estimated. The matrices $A(r(t), t)$, $B(r(t), t)$, $C_y(r(t), t)$, and $C_z(r(t), t)$ are given by the following expressions:

$$\begin{aligned} A(r(t), t) &= A(r(t)) + D_A(r(t))F_A(r(t), t)E_A(r(t)), \\ B(r(t), t) &= B(r(t)) + D_B(r(t))F_B(r(t), t)E_B(r(t)), \\ C_y(r(t), t) &= C_y(r(t)) + D_{C_y}(r(t))F_{C_y}(r(t), t)E_{C_y}(r(t)), \\ C_z(r(t), t) &= C_z(r(t)) + D_{C_z}(r(t))F_{C_z}(r(t), t)E_{C_z}(r(t)), \end{aligned}$$

where $A(r(t))$, $B(r(t))$, $C_y(r(t))$, $D_y(r(t))$, $C_z(r(t))$, $D_z(r(t))$, $D_A(r(t))$, $E_A(r(t))$, $D_B(r(t))$, $E_B(r(t))$, $D_{C_y}(r(t))$, $E_{C_y}(r(t))$, $D_{C_z}(r(t))$, and $E_{C_z}(r(t))$ are known real matrices with appropriate dimensions and $F_A(r(t), t)$, $F_B(r(t), t)$, $F_{C_y}(r(t), t)$, and $F_{C_z}(r(t), t)$ are unknown matrices representing parameter uncertainties and are assumed to satisfy the following assumption.

Assumption 5.1.1 *Let the uncertainties $F_A(r(t), t)$, $F_B(r(t), t)$, $F_{C_y}(r(t), t)$, and $F_{C_z}(r(t), t)$ satisfy the following for every $r(t) = i \in \mathscr{S}$ and $t \geq 0$:*

$$\begin{cases} F_A^\top(i, t)F_A(i, t) \leq \mathbb{I}, \\ F_B^\top(i, t)F_B(i, t) \leq \mathbb{I}, \\ F_{C_y}^\top(i, t)F_{C_y}(i, t) \leq \mathbb{I}, \\ F_{C_z}^\top(i, t)F_{C_z}(i, t) \leq \mathbb{I}. \end{cases} \tag{5.2}$$

The Markov process $\{r(t), t \geq 0\}$, besides taking values in the finite set $\mathscr{S}$, represents the switching between the different modes. Its dynamics are described by the following probability transitions:

$$\mathbb{P}\left[r(t+h) = j | r(t) = i\right]$$
$$= \begin{cases} \lambda_{ij}h + o(h) & \text{when } r(t) \text{ jumps from } i \text{ to } j, \\ 1 + \lambda_{ii}h + o(h) & \text{otherwise,} \end{cases} \tag{5.3}$$

where λ_{ij} is the transition rate from mode i to mode j with $\lambda_{ij} \geq 0$ when $i \neq j$ and $\lambda_{ii} = -\sum_{j=1, j\neq i}^{N} \lambda_{ij}$ and $o(h)$ is such that $\lim_{h \to 0} \frac{o(h)}{h} = 0$.

Remark 19. The uncertainties that satisfy this assumption will be referred to as admissible. In our case we are considering uncertainties that depend on mode $r(t)$ and time t only. This is not a restriction of our results, which will remain valid even if the uncertainties are chosen to depend on the mode $r(t)$, state $x(t)$, and time t.

The filtering problem consists of computing an estimate $\hat{z}(t)$ of the signal $z(t)$ via a causal Markovian jump linear filter, which provides a uniformly small estimation error $z(t) - \hat{z}(t)$ for all $w(t)$ satisfying some properties irrespective of the admissible uncertainties when there exist.

The following definitions will be used in this chapter. For more details we refer the reader to Chapter 2.

Definition 14. *System (5.1), with $w(t) = 0$ for all $t \geq 0$, is said to be stochastically quadratically stable if there exists a set of symmetric and positive-definite matrices $P = (P(1), \cdots, P(N)) > 0$ such that the following holds for each $i \in \mathscr{S}$:*

$$A(i)P(i) + P(i)A^\top(i) + \sum_{i=1}^{N} \lambda_{ij} P(j) < 0. \tag{5.4}$$

For the uncertain case, we have the following definition.

Definition 15. *System (5.1), with $w(t) = 0$ for all $t \geq 0$, is said to be robustly stochastically quadratically stable if there exists a set of symmetric and positive-definite matrices $P = (P(1), \cdots, P(N)) > 0$ such that the following holds for each $i \in \mathscr{S}$ and for all admissible uncertainties:*

$$[A(i) + \Delta A(i,t)] P(i) + P(i) [A(i) + \Delta A(i,t)]^\top + \sum_{i=1}^{N} \lambda_{ij} P(j) < 0. \tag{5.5}$$

Remark 20. Notice that if we transpose the previous inequalities we get equivalent conditions for stochastic quadratic stability and robust stochastic quadratic stability, respectively, that is,

$$A^\top(i)P(i) + P(i)A(i) + \sum_{i=1}^{N} \lambda_{ij} P(j) < 0,$$

$$[A(i) + \Delta A(i,t)]^\top P(i) + P(i)\left[A(i) + \tilde{A}(i,t)\right] + \sum_{i=1}^{N} \lambda_{ij} P(j) < 0.$$

In this chapter we deal with Kalman filtering and $\mathscr{H}_\infty$ filtering techniques. In both cases, we cover the determination of the filter when the uncertainties are equal to zero and when they are present in the dynamics.

5.2 Kalman Filtering

Kalman filtering is one of the most popular estimation techniques and has been applied with success to different fields such as, for instance, engineering, biology, geoscience, economics, and management.

In the rest of this section we concentrate on the design of a stochastic stable quadratic state estimator that guarantees that the estimator error covariance has a guaranteed bound for all admissible uncertainties when they are acting on the system dynamics. The noise signal will be assumed to be a stationary Gaussian noise. Notice that the noise for the measurement can be different from the one in the state equation.

The theory of Kalman filtering requires the following assumptions for the system state and measurement noises:

$$\begin{cases} \mathbb{E}\left[w(t)\right] = 0, \\ \mathbb{E}\left[w(t)w^\top(t)\right] = W\delta(t), W > 0, \\ \mathbb{E}\left[v(t)\right] = 0, \\ \mathbb{E}\left[v(t)v^\top(t)\right] = V\delta(t), V > 0, \\ \mathbb{E}\left[w(t)v^\top(t)\right] = 0, \end{cases} \tag{5.6}$$

where $\delta(.)$ is the Dirac function.

Theorem 70. *System (5.1) with $w(t) = 0$ for all $t \geq 0$ is stochastically stable if it is stochastically quadratically stable.*

Proof: Since system (5.1) is stochastically quadratically stable, there is a set of symmetric and positive-definite matrices $P = (P(1), \cdots, P(N)) > 0$ satisfying for every $i \in \mathscr{S}$:

$$A^\top(i)P(i) + P(i)A(i) + \sum_{j=1}^{N} \lambda_{ij} P(j) < 0.$$

Let us consider the Lyapunov candidate function given by

$$V(x(t), r(t)) = x^\top(t)P(r(t))x(t). \tag{5.7}$$

where $P(i)$ is a symmetric and positive-definite matrix for every $i \in \mathscr{S}$.

The infinitesimal operator $\mathscr{L}$ of the Markov process $\{(x(t), r(t)), t \geq 0\}$ acting on $V(.)$ and emanating from the point (x, i) at time t, where $x(t) = x$ and $r(t) = i$ for $i \in \mathscr{S}$, is given by:

$$\mathscr{L}V(x(t), i) = \dot{x}^\top(t)P(i)x(t) + x^\top(t)P(i)\dot{x}(t) + \sum_{j=1}^{N} \lambda_{ij} x^\top(t)P(j)x(t).$$

Using the expression of the dynamics (5.1), we get:

$$\mathscr{L}V(x(t), i) = x^\top(t)\left[A^\top(i)P(i) + P(i)A(i) + \sum_{j=1}^{N} \lambda_{ij}P(j)\right]x(t)$$

$$= x^\top(t)\Theta(i)x(t),$$

where $\Theta(i) = A^\top(i)P(i) + P(i)A(i) + \sum_{j=1}^{N} \lambda_{ij}P(j)$.

Since $\Theta(i) < 0$ for all $i \in \mathscr{S}$, it can be easily shown that the following holds:

$$\mathscr{L}V(x(t), i) \leq -\alpha\|x(t)\|^2,$$

where $\alpha = \min_{i \in \mathscr{S}}\{\lambda_{min}(-\Theta(i))\} > 0$.

Using Dynkin's formula, we obtain

$$\mathbb{E}[V(x(t), i)] - \mathbb{E}[V(x_0, r_0)] = \mathbb{E}\left[\int_0^t [\mathscr{L}V(x_s, r(s))]\, ds | x_0, r_0\right]$$

$$\leq -\alpha\mathbb{E}\left[\int_0^t \|x_s\|^2 ds | x_0, r_0\right].$$

Since $\mathbb{E}[V(x(t), i)] \geq 0$, the last equation implies

$$\alpha\mathbb{E}\left[\int_0^t \|x_s\|^2 ds | x_0, r_0\right] \leq \mathbb{E}[V(x(t), i)] + \alpha\mathbb{E}\left[\int_0^t \|x_s\|^2 ds | x_0, r_0\right]$$

$$\leq \mathbb{E}[V(x_0, r_0)], \forall t > 0.$$

This holds for all $t \geq 0$, which proves that the system under study is stochastically stable. This completes the proof of Theorem 70. □

Theorem 71. *System (5.1) with $w(t) = 0$ for all $t \geq 0$ is robustly stochastically stable for all admissible uncertainties if it is stochastically quadratically stable.*

Proof: Since system (5.1) is stochastically quadratically stable, then there is a set of symmetric and positive-definite matrices $P = (P(1), \cdots, P(N)) > 0$ satisfying for every $i \in \mathscr{S}$:

$$A^\top(i,t)P(i) + P(i)A(i,t) + \sum_{j=1}^N \lambda_{ij}P(j) < 0,$$

for all admissible uncertainties. Let the Lyapunov candidate function be given by

$$V(x(t), r(t)) = x^\top(t)P(r(t))x(t).$$

The infinitesimal operator $\mathscr{L}$ of the Markov process $\{(x(t), r(t)), t \geq 0\}$ acting on $V(.)$ and emanating from the point (x, i) at time t, where $x(t) = x$ and $r(t) = i$ for $i \in \mathscr{S}$, is given by

$$\mathscr{L}V(x(t), i) = \dot{x}^\top(t)P(i)x(t) + x^\top(t)P(i)\dot{x}(t) + \sum_{j=1}^N \lambda_{ij}x^\top(t)P(j)x(t).$$

Using the expression of the dynamics (5.1), we get

$$\mathscr{L}V(x(t), i) = x^\top(t)\left[A^\top(i,t)P(i) + P(i)A(i,t) + \sum_{j=1}^N \lambda_{ij}P(j)\right]x(t)$$

$$= x^\top(t)\left[A^\top(i)P(i) + P(i)A(i) + E_A^\top(i)F_A^\top(i,t)D_A^\top(i)P(i)\right.$$

$$\left. + P(i)D_A(i)F_A(i,t)E_A(i) + \sum_{j=1}^N \lambda_{ij}P(j)\right]x(t)$$

$$= x^\top(t)\Theta_0(i)x(t),$$

where

$$\Theta_0(i) = A^\top(i)P(i) + P(i)A(i) + E_A^\top(i)F_A^\top(i,t)D_A^\top(i)P(i)$$

$$+ P(i)D_A(i)F_A(i,t)E_A(i) + \sum_{j=1}^N \lambda_{ij}P(j).$$

Using Lemma 7 in Appendix A, we can show that there exists a set of positive scalars $\varepsilon_A = (\varepsilon_A(1), \cdots, \varepsilon_A(N)) > 0$ such that the following holds for every $i \in \mathscr{S}$:

$$\mathscr{L}V(x(t), i) \leq x^\top(t)\Theta(i)x(t),$$

where

$$\Theta(i) = A^\top(i)P(i) + P(i)A(i) + \varepsilon_A(i)P(i)D_A(i)D_A^\top(i)P(i)$$

$$+ \varepsilon_A^{-1}(i)E_A^\top(i)E_A(i) + \sum_{j=1}^N \lambda_{ij}P(j).$$

Since $\Theta(i) < 0$ for all $i \in \mathscr{S}$, it can be easily shown that the following holds:

$$\mathscr{L}V(x(t), i) \leq -\alpha \|x(t)\|^2,$$

where $\alpha = \min_{i \in \mathscr{S}}\{\lambda_{min}(-\Theta(i))\} > 0$.

Using Dynkin's formula, we obtain

$$\mathbb{E}[V(x(t), i)] - \mathbb{E}[V(x_0, r_0)] = \mathbb{E}\left[\int_0^t [\mathscr{L}V(x_s, r(s))] \, ds | x_0, r_0\right]$$

$$\leq -\alpha\mathbb{E}\left[\int_0^t \|x_s\|^2 ds | x_0, r_0\right].$$

Since $\mathbb{E}[V(x(t), i)] \geq 0$, the last equation implies

$$\alpha\mathbb{E}\left[\int_0^t \|x_s\|^2 ds | x_0, r_0\right] \leq \mathbb{E}[V(x(t), i)] + \alpha\mathbb{E}\left[\int_0^t \|x_s\|^2 ds | x_0, r_0\right]$$

$$\leq \mathbb{E}[V(x_0, r_0)], \forall t > 0.$$

This holds for all $t \geq 0$, which proves that the system under study is stochastically stable. This completes the proof of Theorem 71. □

Theorem 72. *System (5.1) with $w(t) = 0$ for all $t \geq 0$ is robustly stochastically quadratically stable if there exist a sequence of positive numbers $\varepsilon_A = (\varepsilon_A(1), \cdots, \varepsilon_A(N))$ and a set of symmetric and positive-definite matrices $P = (P(1), \cdots, P(N)) > 0$ satisfying the following for each $i \in \mathscr{S}$ and for all admissible uncertainties:*

$$\left[\begin{array}{cc} \left[\begin{array}{c} A^\top(i)P(i) \\ +P(i)A(i) \\ +\sum_{j=1}^N \lambda_{ij}P(j) \\ +\varepsilon_A(i)E_A^\top(i)E_A(i) \end{array}\right] & P(i)D_A(i) \\ D_A^\top(i)P(i) & -\varepsilon_A(i)\mathbb{I} \end{array}\right] < 0. \tag{5.8}$$

Proof: By virtue of the previous result, system (5.1) with $w(t) = 0$ for all $t \geq 0$ is robustly stochastically quadratically stable if the following holds for each $i \in \mathscr{S}$ and all the admissible uncertainties:

$$A^\top(i, t)P(i) + P(i)A(i, t) + \sum_{j=1}^N \lambda_{ij}P(j) < 0.$$

Since $A(i, t) = A(i) + D_A(i)F_A(i, t)E_A(i)$, this inequality becomes

$$A^\top(i)P(i) + P(i)A(i) + \sum_{j=1}^N \lambda_{ij}P(j) + P(i)D_A(i)F_A(i, t)E_A(i)$$

$$+E_A^\top(i)F_A^\top(i,t)D_A^\top(i)P(i) < 0.$$

Using Lemma 7 in Appendix A, we have

$$P(i)D_A(i)F_A(i,t)E_A(i) + E_A^\top(i)F_A^\top(i,t)D_A^\top(i)P(i)$$
$$\leq \varepsilon_A^{-1}(i)P(i)D_A(i)D_A^\top(i)P(i) + \varepsilon_A(i)E_A^\top(i)E_A(i).$$

Taking into account this inequality and using the Schur complement, we obtain the desired results and this ends the proof.   □

Let us now return to the goal of this section, which is designing a stochastically stable estimator such that the error covariance of the state vector $x(t)$ is bounded even in the presence of all admissible uncertainties in the system dynamics.

The state estimator we use in this section is given by:

$$\begin{cases} \dot{\hat{x}}(t) = K_A(r(t))\hat{x}(t) + K_B(r(t))y(t), \\ \hat{x}(0) = x_0, \end{cases} \tag{5.9}$$

where $K_A(r(t))$ and $K_B(r(t))$ are the filter gains to be determined for all the modes.

The estimator error is defined as follows:

$$e(t) = x(t) - \hat{x}(t).$$

From this, (5.1), and the filter expression (5.9), we get

$$\begin{aligned} \dot{e}(t) =\ & \dot{x}(t) - \dot{\hat{x}}(t) \\ =\ & A(r(t),t)x(t) + B(r(t),t)w(t) - K_A(r(t))\hat{x}(t) \\ & -K_B(r(t))C_y(r(t),t)x(t) - K_B(r(t))D_y(r(t))w(t) \\ =\ & [A(r(t),t) - K_A(r(t)) - K_B(r(t))C_y(r(t),t)]\,x(t) \\ & +K_A(r(t))\,[x(t) - \hat{x}(t)] + B(r(t),t)w(t) - K_B(r(t))D_y(r(t))w(t) \\ =\ & [A(r(t),t) - K_A(r(t)) - K_B(r(t))C_y(r(t),t)]\,x(t) + K_A(r(t))e(t) \\ & + [B(r(t),t) - K_B(r(t))D_y(r(t))]\,w(t). \end{aligned}$$

Using this expression, the augmented dynamics system becomes

$$\dot{\tilde{x}}(t) = \tilde{A}(r(t),t)\tilde{x}(t) + \tilde{B}(r(t),t)w(t), \tilde{x}(0) = \tilde{x}_0,$$

where

$$\tilde{x}(t) = \begin{bmatrix} x(t) \\ x(t) - \hat{x}(t) \end{bmatrix},$$

$$\begin{aligned} \tilde{A}(r(t),t) &= \begin{bmatrix} A(r(t),t) & 0 \\ A(r(t),t) - K_A(r(t)) - K_B(r(t))C_y(r(t),t) & K_A(r(t)) \end{bmatrix} \\ &= \begin{bmatrix} A(r(t)) & 0 \\ A(r(t)) - K_A(r(t)) - K_B(r(t))C_y(r(t),t) & K_A(r(t)) \end{bmatrix} \end{aligned}$$

$$+ \begin{bmatrix} \Delta A(r(t),t) & 0 \\ \Delta A(r(t),t) - K_B(r(t))\Delta C_y(r(t),t) & 0 \end{bmatrix},$$

$$\tilde{B}(r(t),t) = \begin{bmatrix} B(r(t),t) \\ B(r(t),t) - K_B(r(t))D_y(r(t)) \end{bmatrix}$$

$$= \begin{bmatrix} B(r(t)) \\ B(r(t)) - K_B(r(t))D_y(r(t)) \end{bmatrix} + \begin{bmatrix} \Delta B(r(t),t) \\ \Delta B(r(t),t) \end{bmatrix}.$$

Notice that

$$\Delta \tilde{A}(r(t),t) = \begin{bmatrix} \Delta A(r(t),t) & 0 \\ \Delta A(r(t),t) - K_B(r(t))\Delta C_y(r(t),t) & 0 \end{bmatrix}$$

$$= \begin{bmatrix} \Delta A(r(t),t) & 0 \\ \Delta A(r(t),t) & 0 \end{bmatrix} + \begin{bmatrix} 0 & 0 \\ -K_B(r(t))\Delta C_y(r(t),t) & 0 \end{bmatrix}$$

$$= \begin{bmatrix} D_A(r(t))F_A(r(t),t)E_A(r(t)) & 0 \\ D_A(r(t))F_A(r(t),t)E_A(r(t)) & 0 \end{bmatrix}$$

$$+ \begin{bmatrix} 0 & 0 \\ -K_B(r(t))D_{C_y}(r(t))F_{C_y}(r(t),t)E_{C_y}(r(t)) & 0 \end{bmatrix}$$

$$= \begin{bmatrix} D_A(r(t)) \\ D_A(r(t)) \end{bmatrix} F_A(r(t),t) \begin{bmatrix} E_A(r(t)) & 0 \end{bmatrix}$$

$$+ \begin{bmatrix} 0 \\ -K_B(r(t))D_{C_y}(r(t)) \end{bmatrix} F_{C_y}(r(t),t) \begin{bmatrix} E_{C_y}(r(t)) & 0 \end{bmatrix}$$

$$= \tilde{D}_A(r(t))F_A(r(t),t)\tilde{E}_A(r(t)) + \tilde{D}_{C_y}(r(t))F_{C_y}(r(t),t)\tilde{E}_{C_y}(r(t)),$$

with

$$\tilde{D}_A(r(t)) = \begin{bmatrix} D_A(r(t)) \\ D_A(r(t)) \end{bmatrix},$$

$$\tilde{E}_A(r(t)) = \begin{bmatrix} E_A(r(t)) & 0 \end{bmatrix},$$

$$\tilde{D}_{C_y}(r(t)) = \begin{bmatrix} 0 \\ -K_B(r(t))D_{C_y}(r(t)) \end{bmatrix},$$

$$\tilde{E}_{C_y}(r(t)) = \begin{bmatrix} E_{C_y}(r(t)) & 0 \end{bmatrix},$$

and

$$\Delta \tilde{B}(r(t),t) = \begin{bmatrix} \Delta B(r(t),t) \\ \Delta B(r(t),t) \end{bmatrix} = \begin{bmatrix} D_B(r(t))F_B(r(t),t)E_B(r(t)) \\ D_B(r(t))F_B(r(t),t)E_B(r(t)) \end{bmatrix}$$

$$= \begin{bmatrix} D_B(r(t)) \\ D_B(r(t)) \end{bmatrix} F_B(r(t),t)E_B(r(t))$$

$$= \tilde{D}_B(r(t))F_B(r(t),t)\tilde{E}_B(r(t)),$$

with

$$\tilde{D}_B(r(t)) = \begin{bmatrix} D_B(r(t)) \\ D_B(r(t)) \end{bmatrix},$$

$$\tilde{E}_B(r(t)) = E_B(r(t)).$$

Definition 16. *Given the dynamics (5.1), the state estimator (5.9) is said to be a stochastically stable estimator if there exists a set of symmetric and positive-definite matrices* $P = (P(1), \cdots, P(N)) > 0$ *such that the following holds for each* $i \in \mathscr{S}$:

$$\tilde{A}(i)P(i) + P(i)\tilde{A}^\top(i) + \sum_{i=1}^{N} \lambda_{ij}P(j) + \tilde{B}(i)\tilde{B}^\top(i) < 0. \qquad (5.10)$$

For the robust case, we have the following definition.

Definition 17. *Given the dynamics (5.1), the state estimator (5.9) is said to be a robustly stochastically stable estimator if there exists a set of symmetric and positive-definite matrices* $P = (P(1), \cdots, P(N)) > 0$ *such that the following holds for each* $i \in \mathscr{S}$ *and for all admissible uncertainties:*

$$\left[\tilde{A}(i) + \Delta\tilde{A}(i,t)\right] P(i) + P(i) \left[\tilde{A}(i) + \Delta\tilde{A}(i,t)\right]^\top + \sum_{i=1}^{N} \lambda_{ij}P(j)$$

$$+ \left[\tilde{B}(i) + \Delta\tilde{B}(i,t)\right] \left[\tilde{B}(i) + \Delta\tilde{B}(i,t)\right]^\top < 0. \quad (5.11)$$

Definition 18. *The state estimator (5.9) is said to be a guaranteed cost state estimator for the class of systems we are considering if there exists a symmetric and positive-definite matrix* R *such that*

$$\mathbb{E}\{(x - \hat{x})(x - \hat{x})^\top\} \leq R \qquad (5.12)$$

or

$$\mathbb{E}\{(x - \hat{x})(x - \hat{x})^\top\} \leq tr(R) \qquad (5.13)$$

holds for all admissible uncertainties.

Remark 21. In the rest of this chapter we refer to the state estimator (5.9) as an estimator that provides a guaranteed cost matrix R.

Let us now drop the uncertainties from the dynamics (5.1) and see how we can design the gain filter parameters $K_A(i)$ and $K_B(i)$ for all the modes. Based on the previous definition, the filter (5.9) is quadratically stochastically stable if there exists a set of symmetric and positive-definite matrices $P = (P(1), \cdots, P(N)) > 0$ such that the following holds for every $i \in \mathscr{S}$:

$$\tilde{A}(i)P(i) + P(i)\tilde{A}^\top(i) + \sum_{i=1}^{N} \lambda_{ij}P(j) + \tilde{B}(i)\tilde{B}^\top(i) < 0.$$

Now let $P(i) = \text{diag}[P_1(i), P_2(i)]$, where $P_1(i)$ and $P_2(i)$ are symmetric and positive-definite matrices. Let $X(i) = P^{-1}(i)$. Pre- and post-multiply the previous inequality by $X(i)$ to get

$$X(i)\tilde{A}(i) + \tilde{A}^\top(i)X(i) + \sum_{i=1}^{N}\lambda_{ij}X(i)X^{-1}(j)X(i)$$
$$+X(i)\tilde{B}(i)\tilde{B}^\top(i)X(i) < 0.$$

Using the fact that

$$\sum_{i=1}^{N}\lambda_{ij}X(i)X^{-1}(j)X(i) = \lambda_{ii}X(i) + \mathcal{S}_i(X)\mathcal{X}_i^{-1}(X)\mathcal{S}_i^\top(X),$$

with

$$\mathcal{S}_i(X) = \begin{bmatrix} \mathcal{S}_{1i}(X) & 0 \\ 0 & \mathcal{S}_{2i}(X) \end{bmatrix},$$

$$\mathcal{X}_i(X) = \begin{bmatrix} \mathcal{X}_{1i}(X) & 0 \\ 0 & \mathcal{X}_{2i}(X) \end{bmatrix},$$

$$\mathcal{S}_{1i}(X) = \left[\sqrt{\lambda_{i1}}X_1(i), \cdots, \sqrt{\lambda_{ii-1}}X_1(i), \sqrt{\lambda_{ii+1}}X_1(i), \right.$$
$$\left. \cdots, \sqrt{\lambda_{iN}}X_1(i) \right],$$

$$\mathcal{S}_{2i}(X) = \left[\sqrt{\lambda_{i1}}X_2(i), \cdots, \sqrt{\lambda_{ii-1}}X_2(i), \sqrt{\lambda_{ii+1}}X_2(i), \right.$$
$$\left. \cdots, \sqrt{\lambda_{iN}}X_2(i) \right],$$

$$\mathcal{X}_{1i}(X) = \text{diag}\left[X_1(1), \cdots, X_1(i-1), X_1(i+1), \cdots, X_1(N)\right],$$
$$\mathcal{X}_{2i}(X) = \text{diag}\left[X_2(1), \cdots, X_2(i-1), X_2(i+1), \cdots, X_2(N)\right].$$

Using the Schur complement, we get

$$\begin{bmatrix} X(i)\tilde{A}(i) + \tilde{A}^\top(i)X(i) + \lambda_{ii}X(i) & X(i)\tilde{B}(i) & \mathcal{S}_i(X) \\ \tilde{B}^\top(i)X(i) & -\mathbb{I} & 0 \\ \mathcal{S}_i^\top(X) & 0 & -\mathcal{X}_i(X) \end{bmatrix} < 0.$$

Based on the expressions of $\tilde{A}(i)$, $\tilde{B}(i)$, and $X(i)$, we get

$$X(i)\tilde{A}(i) = \begin{bmatrix} X_1(i)A(i) & 0 \\ X_2(i)A(i) - X_2(i)K_A(i) - X_2(i)K_B(i)C_y(i) & X_2(i)K_A(i) \end{bmatrix},$$

$$X(i)\tilde{B}(i) = \begin{bmatrix} X_1(i)B(i) \\ X_2(i)B(i) - X_2(i)K_B(i)D_y(i) \end{bmatrix},$$

$$\lambda_{ii}X(i) = \begin{bmatrix} \lambda_{ii}X_1(i) & 0 \\ 0 & \lambda_{ii}X_2(i) \end{bmatrix},$$

which imply

$$
\left[\begin{array}{cc}
\begin{bmatrix} A^\top(i)X_1(i) \\ +X_1(i)A(i) \\ +\lambda_{ii}X_1(i) \end{bmatrix} & \begin{bmatrix} A^\top(i)X_2(i) \\ -K_A^\top(i)X_2(i) \\ -C_y^\top(i)K_B^\top(i)X_2(i) \end{bmatrix} \\[1.5em]
\begin{bmatrix} X_2(i)A(i) \\ -X_2(i)K_A(i) \\ -X_2(i)K_B(i)C_y(i) \end{bmatrix} & \begin{bmatrix} K_A^\top(i)X_2(i) \\ +X_2(i)K_A(i) \\ +\lambda_{ii}X_2(i) \end{bmatrix} \\[1.5em]
B^\top(i)X_1(i) & \begin{bmatrix} B^\top(i)X_2(i) \\ -D_y^\top(i)K_B^\top(i)X_2(i) \end{bmatrix} \\[1em]
\mathcal{S}_{1i}^\top(X) & 0 \\
0 & \mathcal{S}_{2i}^\top(X)
\end{array}\right.
$$

$$
\left.\begin{array}{ccc}
\begin{bmatrix} X_1(i)B(i) \\ X_2(i)B(i) \\ -X_2(i)K_B(i)D_y(i) \end{bmatrix} & \mathcal{S}_{1i}(X) & 0 \\[1.5em]
-\mathbb{I} & 0 & \mathcal{S}_{2i}(X) \\
0 & 0 & 0 \\
0 & -\mathcal{X}_{1i}(X) & 0 \\
0 & 0 & -\mathcal{X}_{2i}(X)
\end{array}\right] < 0.
$$

Letting $Y(i) = X_2(i)K_A(i)$ and $Z(i) = X_2(i)K_B(i)$, we get the following LMI:

$$
\left[\begin{array}{cc}
\begin{bmatrix} A^\top(i)X_1(i) + X_1(i)A(i) \\ +\lambda_{ii}X_1(i) \end{bmatrix} & \begin{bmatrix} A^\top(i)X_2(i) - Y^\top(i) \\ -C_y^\top(i)Z^\top(i) \end{bmatrix} \\[1.5em]
\begin{bmatrix} X_2(i)A(i) - Y(i) \\ -Z(i)C_y(i) \end{bmatrix} & \begin{bmatrix} Y^\top(i) + Y(i) \\ +\lambda_{ii}X_2(i) \end{bmatrix} \\[1.5em]
B^\top(i)X_1(i) & \begin{bmatrix} B^\top(i)X_2(i) \\ -D_y^\top(i)Z^\top(i) \end{bmatrix} \\[1em]
\mathcal{S}_{1i}^\top(X) & 0 \\
0 & \mathcal{S}_{2i}^\top(X)
\end{array}\right.
$$

$$
\left.\begin{array}{ccc}
\begin{bmatrix} X_1(i)B(i) \\ X_2(i)B(i) \\ -Z(i)D_y(i) \end{bmatrix} & \mathcal{S}_{1i}(X) & 0 \\[1.5em]
-\mathbb{I} & 0 & \mathcal{S}_{2i}(X) \\
0 & 0 & 0 \\
0 & -\mathcal{X}_{1i}(X) & 0 \\
0 & 0 & -\mathcal{X}_{2i}(X)
\end{array}\right] < 0. \qquad (5.14)
$$

The results of this development are summarized by the following theorem.

Theorem 73. *If there exist sets of symmetric and positive-definite matrices $X_1 = (X_1(1), \cdots, X_1(N)) > 0$ and $X_2 = (X_2(1), \cdots, X_2(N)) > 0$ and sets of matrices $Y = (Y(1), \cdots, Y(N))$ and $Z = (Z(1), \cdots, Z(N))$ satisfying the following set of LMIs (5.14) for every $i \in \mathscr{S}$, then the filter (5.9) is a guaranteed cost state estimator and the gains are given by*

$$
\begin{cases} K_A(i) = X_2^{-1}(i)Y(i), \\ K_B(i) = X_2^{-1}(i)Z(i). \end{cases} \qquad (5.15)
$$

Example 64. To illustrate the results of this theorem, let us consider a two-mode system with the following data:

- mode #1:

$$A(1) = \begin{bmatrix} -3.0 & 1.0 & 0.0 \\ 0.3 & -2.5 & 1.0 \\ -0.1 & 0.3 & -3.8 \end{bmatrix}, \quad B(1) = \begin{bmatrix} 1.0 \\ 0.0 \\ 1.0 \end{bmatrix},$$

$$C_y(1) = \begin{bmatrix} 0.1 & 1.0 & 0.0 \end{bmatrix}, \quad D_y(1) = \begin{bmatrix} 0.2 \end{bmatrix},$$

- mode #2:

$$A(2) = \begin{bmatrix} -4.0 & 1.0 & 0.0 \\ 0.3 & -3.0 & 1 \\ -0.1 & 0.3 & -4.8 \end{bmatrix}, \quad B(2) = \begin{bmatrix} 1.0 \\ 0.0 \\ 2.0 \end{bmatrix},$$

$$C_y(2) = \begin{bmatrix} 0.8 & 0.4 & 0.0 \end{bmatrix}, \quad D_y(2) = \begin{bmatrix} 0.1 \end{bmatrix}.$$

The switching between the two modes is described by the following transition rate matrix:

$$\Lambda = \begin{bmatrix} -2.0 & 2.0 \\ 1.0 & -1.0 \end{bmatrix}.$$

Solving the LMI of the previous theorem, we get

$$X_1(1) = \begin{bmatrix} 0.3251 & 0.0421 & -0.0137 \\ 0.0421 & 0.4056 & 0.0347 \\ -0.0137 & 0.0347 & 0.2965 \end{bmatrix},$$

$$X_1(2) = \begin{bmatrix} 0.3846 & 0.0433 & -0.0297 \\ 0.0433 & 0.4565 & 0.0291 \\ -0.0297 & 0.0291 & 0.3044 \end{bmatrix},$$

$$X_2(1) = \begin{bmatrix} 0.4124 & 0.1013 & 0.0755 \\ 0.1013 & 0.4537 & 0.1262 \\ 0.0755 & 0.1262 & 0.3977 \end{bmatrix},$$

$$X_2(2) = \begin{bmatrix} 0.4517 & 0.0913 & 0.0555 \\ 0.0913 & 0.5413 & 0.1197 \\ 0.0555 & 0.1197 & 0.4215 \end{bmatrix},$$

$$Y(1) = \begin{bmatrix} -1.0652 & -0.6438 & -0.6657 \\ -0.5275 & -0.8897 & -0.5892 \\ -0.5667 & -0.5879 & -1.1556 \end{bmatrix},$$

$$Y(2) = \begin{bmatrix} -1.2414 & -0.6077 & -0.6817 \\ -0.5667 & -1.1770 & -0.6455 \\ -0.5416 & -0.6030 & -1.2854 \end{bmatrix},$$

$$Z(1) = \begin{bmatrix} 0.1278 \\ -0.6279 \\ -0.5518 \end{bmatrix}, \qquad Z(2) = \begin{bmatrix} -0.1660 \\ -0.6951 \\ -0.5598 \end{bmatrix},$$

which give the following gains:

$$K_A(1) = \begin{bmatrix} -2.3272 & -1.0449 & -1.0551 \\ -0.4056 & -1.5047 & -0.3409 \\ -0.8543 & -0.8023 & -2.5969 \end{bmatrix},$$

$$K_A(2) = \begin{bmatrix} 2.5593 & -0.8739 & -1.0864 \\ -0.4328 & -1.8525 & -0.3910 \\ -0.8249 & -0.7895 & -2.7957 \end{bmatrix},$$

$$K_B(1) = \begin{bmatrix} 0.8269 \\ -1.2492 \\ -1.1480 \end{bmatrix}, \qquad\qquad K_B(2) = \begin{bmatrix} 0.0286 \\ -1.0525 \\ -1.0256 \end{bmatrix}.$$

Let us now consider the effect of the system uncertainties on the design of the guaranteed cost state estimator. For this purpose we should satisfy the following:

$$\left[\tilde{A}(i) + \Delta\tilde{A}(i,t)\right] P(i) + P(i)\left[\tilde{A}(i) + \Delta\tilde{A}(i,t)\right]^\top$$
$$+ \sum_{j=1}^{N} \lambda_{ij} P(j) + \left[\tilde{B}(i) + \Delta\tilde{B}(i,t)\right]\left[\tilde{B}(i) + \Delta\tilde{B}(i,t)\right]^\top < 0.$$

Using the expressions of $\Delta\tilde{A}(i,t)$ and $\Delta\tilde{B}(i,t)$ and Lemma 7 and Lemma 8 in Appendix A, respectively, we get

$$\tilde{D}_A(i)F_A(i,t)\tilde{E}_A(i)P(i) + P(i)\tilde{E}_A^\top(i)F_A^\top(i,t)\tilde{D}_A^\top(i) \leq \varepsilon_A^{-1}(i)\tilde{D}_A(i)\tilde{D}_A^\top(i)$$
$$+ \varepsilon_A(i)P(i)\tilde{E}_A^\top(i)\tilde{E}_A(i)P(i),$$
$$\tilde{D}_{C_y}(i)F_{C_y}(i,t)\tilde{E}_{C_y}(i)P(i) + P(i)\tilde{E}_{C_y}^\top(i)F_{C_y}^\top(i,t)\tilde{D}_{C_y}^\top(i)$$
$$\leq \varepsilon_{C_y}^{-1}(i)\tilde{D}_{C_y}(i)\tilde{D}_{C_y}^\top(i) + \varepsilon_{C_y}(i)P(i)\tilde{E}_{C_y}^\top(i)\tilde{E}_{C_y}(i)P(i),$$
$$\left[\tilde{B}(i) + \Delta\tilde{B}(i,t)\right]\left[\tilde{B}(i) + \Delta\tilde{B}(i,t)\right]^\top \leq \tilde{B}(i)\tilde{B}^\top(i) + \tilde{B}(i)\tilde{E}_B^\top(i)$$
$$\times \left[\varepsilon_B(i)\mathbb{I} - \tilde{E}_B(i)\tilde{E}_B^\top(i)\right]^{-1}\tilde{E}_B(i)\tilde{B}^\top(i) + \varepsilon_B(i)\tilde{D}_B(i)\tilde{D}_B^\top(i).$$

Based on these inequalities, the previous condition becomes

$$\tilde{A}(i)P(i) + P(i)\tilde{A}^\top(i) + \sum_{j=1}^{N} \lambda_{ij}P(j) + \varepsilon_A^{-1}(i)\tilde{D}_A(i)\tilde{D}_A^\top(i)$$
$$+ \varepsilon_A(i)P(i)\tilde{E}_A^\top(i)\tilde{E}_A(i)P(i) + \tilde{B}(i)\tilde{B}^\top(i)$$
$$+ \varepsilon_{C_y}^{-1}(i)\tilde{D}_{C_y}(i)\tilde{D}_{C_y}^\top(i) + \varepsilon_{C_y}(i)P(i)\tilde{E}_{C_y}^\top(i)\tilde{E}_{C_y}(i)P(i)$$
$$+ \tilde{B}(i)\tilde{E}_B^\top(i)\left[\varepsilon_B(i)\mathbb{I} - \tilde{E}_B(i)\tilde{E}_B^\top(i)\right]^{-1}\tilde{E}_B(i)\tilde{B}^\top(i)$$
$$+ \varepsilon_B(i)\tilde{D}_B(i)\tilde{D}_B^\top(i) < 0.$$

Let $P(i) = \text{diag}[P_1(i), P_2(i)]$ with $P_1(i)$ and $P_2(i)$ symmetric and positive-definite matrices. Let $X(i) = P^{-1}(i)$. Pre- and post-multiply the previous inequality by $X(i)$ to get

$$X(i)\tilde{A}(i) + \tilde{A}^\top(i)X(i) + \sum_{j=1}^{N} \lambda_{ij} X(i)X^{-1}(j)X(i) + \varepsilon_A(i)\tilde{E}_A^\top(i)\tilde{E}_A(i)$$

$$+\varepsilon_A^{-1}(i)X(i)\tilde{D}_A(i)\tilde{D}_A^\top(i)X(i) + X(i)\tilde{B}(i)\tilde{B}^\top(i)X(i)$$

$$+\varepsilon_{C_y}^{-1}(i)X(i)\tilde{D}_{C_y}(i)\tilde{D}_{C_y}^\top(i)X(i) + \varepsilon_{C_y}(i)\tilde{E}_{C_y}^\top(i)\tilde{E}_{C_y}(i)$$

$$+X(i)\tilde{B}(i)\tilde{E}_B^\top(i)\left[\varepsilon_B(i)\mathbb{I} - \tilde{E}_B(i)\tilde{E}_B^\top(i)\right]^{-1}\tilde{E}_B(i)\tilde{B}^\top(i)X(i)$$

$$+\varepsilon_B(i)X(i)\tilde{D}_B(i)\tilde{D}_B^\top(i)X(i) < 0.$$

Using the Schur complement, we get

$$\left[\begin{array}{cccc}
\begin{bmatrix} X(i)\tilde{A}(i) \\ +\tilde{A}^\top(i)X(i) \\ +\lambda_{ii}X(i) \\ +\varepsilon_A(i)\tilde{E}_A^\top(i)\tilde{E}_A(i) \\ +\varepsilon_{C_y}(i)\tilde{E}_{C_y}^\top(i)\tilde{E}_{C_y}(i) \end{bmatrix} & X(i)\tilde{B}(i) & \mathcal{S}_i(X) & X(i)\tilde{D}_A(i) \\
\tilde{B}^\top(i)X(i) & -\mathbb{I} & 0 & 0 \\
\mathcal{S}_i^\top(X) & 0 & -\mathscr{X}_i(X) & 0 \\
\tilde{D}_A^\top(i)X(i) & 0 & 0 & -\varepsilon_A(i)\mathbb{I} \\
\tilde{E}_B(i)\tilde{B}^\top(i)X(i) & 0 & 0 & 0 \\
\tilde{D}_B^\top(i)X(i) & 0 & 0 & 0 \\
\tilde{D}_{C_y}^\top(i)X(i) & 0 & 0 & 0
\end{array}\right.$$

$$\left.\begin{array}{ccc}
X(i)\tilde{B}(i)\tilde{E}_B^\top(i) & X(i)\tilde{D}_B(i) & X(i)\tilde{D}_{C_y}(i) \\
0 & 0 & 0 \\
0 & 0 & 0 \\
0 & 0 & 0 \\
-\left[\varepsilon_B(i)\mathbb{I} - \tilde{E}_B(i)\tilde{E}_B^\top(i)\right] & 0 & 0 \\
0 & -\varepsilon_B^{-1}(i)\mathbb{I} & 0 \\
0 & 0 & -\varepsilon_{C_y}(i)\mathbb{I}
\end{array}\right] < 0.$$

Noticing that

$$\tilde{E}_A^\top(i)\tilde{E}_A(i) = \begin{bmatrix} E_A^\top(i)E_A(i) & 0 \\ 0 & 0 \end{bmatrix},$$

$$\tilde{E}_{C_y}^\top(i)\tilde{E}_{C_y}(i) = \begin{bmatrix} E_{C_y}^\top(i)E_{C_y}(i) & 0 \\ 0 & 0 \end{bmatrix},$$

$$X(i)\tilde{D}_A(i) = \begin{bmatrix} X_1(i)D_A(i) \\ X_2(i)D_A(i) \end{bmatrix},$$

$$X(i)\tilde{B}(i)\tilde{E}_B^\top(i) = \begin{bmatrix} X_1(i)B(i)E_B^\top(i) \\ X_2(i)B(i)E_B^\top(i) - X_2(i)K_B(i)D_y(i)E_B^\top(i) \end{bmatrix},$$

$$X(i)\tilde{D}_B(i) = \begin{bmatrix} X_1(i)D_B(i) \\ X_2(i)D_B(i) \end{bmatrix},$$

$$X(i)\tilde{D}_{C_y}(i) = \begin{bmatrix} 0 \\ -X_2(i)K_B(i)D_{C_y}(i) \end{bmatrix},$$

$$\tilde{E}_B(i)\tilde{E}_B^\top(i) = E_B(i)E_B^\top(i),$$

and using the previous expression for $X(i)\tilde{A}(i)$, $\tilde{A}^\top(i)X(i)$, $X(i)\tilde{B}(i)$, $\tilde{B}^\top(i)X(i)$, $\lambda_{ii}X(i)$, $\mathcal{S}_i(X)$, and $\mathcal{X}_i(X)$, we get the following equivalent LMI after letting $Y(i) = X_2(i)K_A(i)$ and $Z(i) = X_2(i)K_B(i)$:

$$
\left[\begin{array}{c}
\begin{bmatrix} A^\top(i)X_1(i) + X_1(i)A(i) \\ +\lambda_{ii}X_1(i) \\ +\varepsilon_A(i)E_A^\top(i)E_A(i) \\ +\varepsilon_{C_y}(i)E_{C_y}^\top(i)E_{C_y}(i) \end{bmatrix} \\
\begin{bmatrix} X_2(i)A(i) - Y(i) \\ -Z(i)C_y(i) \end{bmatrix} \\
B^\top(i)X_1(i) \\
\mathcal{S}_{1i}^\top(X) \\
0 \\
D_A^\top(i)X_1(i) \\
E_B(i)B^\top(i)X_1(i) \\
D_B^\top(i)X_1(i) \\
0
\end{array} \right.
\left. \begin{array}{c}
\begin{bmatrix} A^\top(i)X_2(i) - Y^\top(i) \\ -C_y^\top(i)Z^\top(i) \end{bmatrix} \\
\begin{bmatrix} Y^\top(i) + Y(i) \\ +\lambda_{ii}X_2(i) \end{bmatrix} \\
\begin{bmatrix} B^\top(i)X_2(i) \\ -D_y^\top(i)Z^\top(i) \end{bmatrix} \\
0 \\
\mathcal{S}_{2i}^\top(X) \\
D_A^\top(i)X_2(i) \\
\begin{bmatrix} E_B(i)B^\top(i)X_2(i) \\ -E_B(i)D_y^\top(i)Z^\top(i) \end{bmatrix} \\
D_B^\top(i)X_2(i) \\
-D_{C_y}^\top(i)Z^\top(i)
\end{array} \right.
$$

$$
\begin{array}{ccccc}
X_1(i)B(i) & \mathcal{S}_{1i}(X) & 0 & X_1(i)D_A(i) \\
\begin{bmatrix} X_2(i)B(i) \\ -Z(i)D_y(i) \end{bmatrix} & 0 & \mathcal{S}_{2i}(X) & X_2(i)D_A(i) \\
-\mathbb{I} & 0 & 0 & 0 \\
0 & -\mathcal{X}_{1i}(X) & 0 & 0 \\
0 & 0 & -\mathcal{X}_{2i}(X) & 0 \\
0 & 0 & 0 & -\varepsilon_A(i)\mathbb{I} \\
0 & 0 & 0 & 0 \\
0 & 0 & 0 & 0 \\
0 & 0 & 0 & 0
\end{array}
$$

$$
\left. \begin{array}{ccc}
X_1(i)B(i)E_B^\top(i) & X_1(i)D_B(i) & 0 \\
\begin{bmatrix} X_2(i)B(i)E_B^\top(i) \\ -Z(i)D_y(i)E_B^\top(i) \end{bmatrix} & X_2(i)D_B(i) & -Z(i)D_{C_y}(i) \\
0 & 0 & 0 \\
0 & 0 & 0 \\
0 & 0 & 0 \\
0 & 0 & 0 \\
-\left[\varepsilon_B(i)\mathbb{I} - E_B(i)E_B^\top(i)\right] & 0 & 0 \\
0 & -\varepsilon_B^{-1}(i)\mathbb{I} & 0 \\
0 & 0 & -\varepsilon_{C_y}(i)\mathbb{I}
\end{array} \right] < 0. \qquad (5.16)
$$

The results of this development are summarized by the following theorem.

Theorem 74. *Let $\varepsilon_B = (\varepsilon_B(1), \cdots, \varepsilon_B(N))$ be a given set of positive scalars. If there exist sets of symmetric and positive-definite matrices $X_1 = (X_1(1), \cdots, X_1(N)) > 0$ and $X_2 = (X_2(1), \cdots, X_2(N)) > 0$, sets of matrices $Y = (Y(1), \cdots, Y(N))$ and $Z = (Z(1), \cdots, Z(N))$, and sets of positive scalars $\varepsilon_A = (\varepsilon_A(1), \cdots, \varepsilon_A(N))$ and $\varepsilon_{C_y} = (\varepsilon_{C_y}(1), \cdots, \varepsilon_{C_y}(N))$ satisfying the following the set of LMIs (5.16) for every $i \in \mathscr{S}$, then the filter (5.9) is a guaranteed cost state estimator and the gains are given by*

$$\begin{cases} K_A(i) = X_2^{-1}(i)Y(i), \\ K_B(i) = X_2^{-1}(i)Z(i). \end{cases} \quad (5.17)$$

Example 65. To illustrate the results of the previous theorem, let us consider the system of the previous example with the following extra data:

- mode #1:

$$D_A(1) = \begin{bmatrix} 0.1 \\ 0.01 \\ 0.0 \end{bmatrix}, \quad E_A(1) = \begin{bmatrix} 0.1 & 0.2 & 0.0 \end{bmatrix},$$

$$D_B(1) = \begin{bmatrix} 0.1 \\ 0.1 \\ 0.0 \end{bmatrix}, \quad E_B(1) = \begin{bmatrix} 0.1 \end{bmatrix},$$

$$D_{C_y}(1) = \begin{bmatrix} 0.2 \end{bmatrix}, \quad E_{C_y}(1) = \begin{bmatrix} 0.1 \end{bmatrix},$$

- mode #2:

$$D_A(2) = \begin{bmatrix} 0.2 \\ 0.1 \\ 0.0 \end{bmatrix}, \quad E_A(2) = \begin{bmatrix} 0.1 & 1.0 & 0.02 \end{bmatrix},$$

$$D_B(2) = \begin{bmatrix} 0.1 & 1.0 & 0.0 \end{bmatrix}, \quad E_B(2) = \begin{bmatrix} 0.02 \end{bmatrix},$$

$$D_{C_y}(2) = \begin{bmatrix} 0.1 \end{bmatrix}, \quad E_{C_y}(2) = \begin{bmatrix} 0.01 \end{bmatrix}.$$

Solving the LMI of the previous theorem, we get

$$X_1(1) = 10^{-3} \cdot \begin{bmatrix} 0.3901 & 0.4781 & 0.4119 \\ 0.4781 & 0.6968 & 0.5443 \\ 0.4119 & 0.5443 & 0.5219 \end{bmatrix},$$

$$X_1(2) = \begin{bmatrix} 0.0001 & 0.0001 & 0.0001 \\ 0.0001 & 0.0014 & 0.0002 \\ 0.0001 & 0.0002 & 0.0002 \end{bmatrix},$$

$$X_2(1) = 10^{-3} \cdot \begin{bmatrix} 0.0907 & 0.2818 & 0.3070 \\ 0.2818 & 0.4971 & 0.5069 \\ 0.3070 & 0.5069 & 0.3563 \end{bmatrix},$$

$$X_2(2) = 10^{-3} \cdot \begin{bmatrix} -0.4562 & -0.1674 & 0.6966 \\ -0.1674 & 0.2448 & 0.3879 \\ 0.6966 & 0.3879 & 0.1134 \end{bmatrix},$$

$$Y(1) = \begin{bmatrix} -0.0001 & -0.0012 & -0.0009 \\ -0.0007 & -0.0000 & -0.0012 \\ -0.0008 & -0.0009 & -0.0001 \end{bmatrix},$$

$$Y(2) = \begin{bmatrix} -0.2736 & -0.1927 & -0.0025 \\ 0.0011 & -0.0002 & -0.0015 \\ -0.0018 & -0.0005 & -0.0000 \end{bmatrix},$$

$$Z(1) = 10^{-3} \cdot \begin{bmatrix} 0.5416 \\ 0.0362 \\ 0.0269 \end{bmatrix}, \qquad Z(2) = \begin{bmatrix} 0.9537 \\ -0.0007 \\ -0.0002 \end{bmatrix},$$

which give the following gains:

$$K_A(1) = \begin{bmatrix} -4.9348 & 19.2437 & 0.9510 \\ 1.2661 & -18.8446 & 4.3377 \\ 0.0685 & 7.7659 & -7.2276 \end{bmatrix},$$

$$K_A(2) = \begin{bmatrix} -219.3289 & -151.5680 & 0.7011 \\ 510.1277 & 355.7673 & -0.4170 \\ -413.8264 & -290.4277 & -3.2239 \end{bmatrix},$$

$$K_B(1) = \begin{bmatrix} -7.7939 \\ 5.4013 \\ -0.8926 \end{bmatrix}, \qquad K_B(2) = 10^3 \cdot \begin{bmatrix} 0.7490 \\ -1.7651 \\ 1.4356 \end{bmatrix}.$$

This section covered the filtering problem under a severe assumption on the external disturbance that is hard to justify in practice. The next section will relax this assumption and replace it with another one that requires only that the external disturbance have finite energy or finite power.

5.3 $\mathcal{H}_\infty$ Filtering

To overcome the limitation of Kalman filtering, the $\mathcal{H}_\infty$ approach has been developed. Its advantage is that does not require knowledge of the external disturbance statistics. It also tolerates the existence of uncertainties in the system dynamics. This approach is based on the assumption that external disturbances can be chosen arbitrarily with the only restriction to have bounded energy or bounded average power. The approach of $\mathcal{H}_\infty$ filtering provides an estimator that guarantees that the $\mathcal{L}_2$-gain from the noise signals to the estimation error is bounded by a prescribed level γ (given positive constant). In the rest of this section, we use this technique to estimate the state vector of the class of systems we are considering.

In order to put the $\mathcal{H}_\infty$ filtering problem of the class of systems (5.1) in the stochastic setting, let us introduce the space $\mathscr{L}_2\left[\Omega, \mathcal{F}, \mathbb{P}\right]$ of $\mathcal{F}$-measurable processes $z(t) - \hat{z}(t)$ for which the following holds:

$$\|z - \hat{z}\|_2 \triangleq \left\{ \mathbb{E}\left[\int_0^\infty \left[z(t) - \hat{z}(t)\right]^\top \left[z(t) - \hat{z}(t)\right] dt\right]\right\}^{\frac{1}{2}} < \infty. \qquad (5.18)$$

The goal of this section is to design a linear n-order filter of the following form:

$$\begin{cases} \dot{\hat{x}}(t) = K_A(r(t))\hat{x}(t) + K_B(r(t))y(t), \hat{x}(0) = 0, \\ \hat{z}(t) = K_C(r(t))\hat{x}(t), \end{cases} \qquad (5.19)$$

which gives an estimate of the state vector $\hat{x}(t)$ at time t and can ensure that the extended system $(x(t), x(t) - \hat{x}(t))$ is stochastically stable and the estimation error $z(t) - \hat{z}(t)$ is bounded for all noises $w(t) \in \mathscr{L}_2[0, \infty)$. The matrices $K_A(r(t))$, $K_B(r(t))$, and $K_C(r(t))$ are design parameters that should be determined in order to estimate the state vector properly.

Let us now drop the uncertainties from the dynamics and see how we can design the $\mathcal{H}_\infty$ filter with the structure defined by (5.19) for the nominal system. We consider that $v(t) = w(t)$ for the rest of this chapter.

If we combine the dynamical system dynamics (5.1) with the filter dynamics (5.19), we get the following:

$$\dot{\tilde{x}}(t) = \tilde{A}(r(t))\tilde{x}(t) + \tilde{B}(r(t))w(t), \tilde{x}(0) = (x_0^\top, x_0^\top)^\top, \qquad (5.20)$$

where

$$\tilde{x}(t) = \begin{bmatrix} x(t) \\ x(t) - \hat{x}(t) \end{bmatrix},$$

$$\tilde{A}(r(t)) = \begin{bmatrix} A(r(t)) & 0 \\ A(r(t)) - K_B(r(t))C_y(r(t)) - K_A(r(t)) & K_A(r(t)) \end{bmatrix},$$

$$\tilde{B}(r(t)) = \begin{bmatrix} B(r(t)) \\ B(r(t)) - K_B(r(t))D_y(r(t)) \end{bmatrix}.$$

The estimation error $e(t) = z(t) - \hat{z}(t)$ satisfies the following:

$$e(t) = \tilde{C}(r(t))\tilde{x}(t) + \tilde{D}(r(t))w(t), \qquad (5.21)$$

with

$$\tilde{C}(r(t)) = \begin{bmatrix} C_z(r(t)) - K_C(r(t)) & K_C(r(t)) \end{bmatrix},$$
$$\tilde{D}(t) = D_z(r(t)).$$

Remark 22. To get the extended dynamics we computed

$$\dot{x}(t) - \dot{\hat{x}}(t) = A(r(t))x(t) + B(r(t))w(t) - K_A(r(t))\hat{x}(t) - K_B(r(t))y(t)$$

$$= [A(r(t)) - K_B(r(t))C_y(r(t)) - K_A(r(t))] x(t)$$
$$+ K_A(r(t)) (x(t) - \hat{x}(t)) + [B(r(t)) - K_B(r(t))D_y(r(t))] w(t),$$

for the second component of the state vector $\tilde{x}(t)$ and

$$\begin{aligned} e(t) &= z(t) - \hat{z}(t) \\ &= C_z(r(t))x(t) + D_z(r(t))w(t) - K_C(r(t))\hat{x}(t) \\ &= \left[C_z(r(t)) - K_C(r(t)) \; K_C(r(t)) \right] \begin{bmatrix} x(t) \\ x(t) - \hat{x}(t) \end{bmatrix} + D_z(r(t))w(t) \\ &= \tilde{C}(r(t))\tilde{x}(t) + \tilde{D}(r(t))w(t) \end{aligned}$$

for the estimation error equation.

The next theorem states that when the filter (5.19) exists (that is, we can get a set of gains $K_A = (K_A(1), \cdots, K_A(N))$, $K_B = (K_B(1), \cdots, K_B(N))$, and $K_C = (K_C(1), \cdots, K_C(N))$), the extended dynamics are stochastically stable and the estimation error $z(t) - \hat{z}(t)$ is bounded for all signals $w(t) \in \mathscr{L}_2[0, \infty)$ if some given conditions are satisfied.

Theorem 75. *Let γ be a given positive constant and R a given symmetric positive-definite matrix representing the weighting of the initial conditions. Let $K_A = (K_A(1), \cdots, K_A(N))$, $K_B = (K_B(1), \cdots, K_B(N))$, and $K_C = (K_C(1), \cdots, K_C(N))$ be given gains. If there exist symmetric and positive-definite matrices $P = (P(1), \cdots, P(N)) > 0$ such that the following hold for every $i \in \mathscr{S}$:*

$$\begin{bmatrix} \tilde{J}_1(i) & P(i)\tilde{B}(i) & \tilde{C}^\top(i) \\ \tilde{B}^\top(i)P(i) & -\gamma^2\mathbb{I} & \tilde{D}^\top(i) \\ \tilde{C}(i) & \tilde{D}(i) & -\mathbb{I} \end{bmatrix} < 0, \tag{5.22}$$

$$\begin{bmatrix} \mathbb{I} & \mathbb{I} \end{bmatrix} P(r_0) \begin{bmatrix} \mathbb{I} \\ \mathbb{I} \end{bmatrix} \leq \gamma^2 R, \tag{5.23}$$

where $\tilde{J}_1(i) = \tilde{A}^\top(i)P(i) + P(i)\tilde{A}(i) + \sum_{j=1}^N \lambda_{ij}P(j)$, then the extended system is stochastically stable and, moreover, the estimation error satisfies the following:

$$\|z(t) - \hat{z}\|_2 \leq \gamma \left[\|w\|_2^2 + x_0^\top R x_0 \right]^{\frac{1}{2}}. \tag{5.24}$$

Proof: Let us first set $r(t) = i \in \mathscr{S}$ and prove the stochastic stability of the extended system. For this purpose, from (5.22) we get

$$\tilde{A}^\top(i)P(i) + P(i)\tilde{A}(i) + \sum_{j=1}^N \lambda_{ij}P(j) < 0,$$

which implies that the extended system is stochastically stable.

Let us now prove the second part of the theorem, which indicates that the estimation error is bounded for all signals $w(t) \in \mathscr{L}_2[0, \infty)$. To this end, let us define the following $\mathscr{H}_\infty$ performance:

$$J_T = \mathbb{E}\left[\int_0^T [e^\top(t)e(t) - \gamma^2 w^\top(t)w(t)]\, dt\right], \forall T > 0, \qquad (5.25)$$

and let $\mathscr{L}$ be the infinitesimal generator of the Markov process $\{(x(t), r(t)), t \geq 0\}$.

Consider the following Lyapunov function candidate as follows:

$$V(\tilde{x}(t), r(t)) = \tilde{x}^\top(t)P(r(t))\tilde{x}(t). \qquad (5.26)$$

The infinitesimal operator $\mathscr{L}$ of the Markov process $\{(\tilde{x}(t), r(t)), t \geq 0\}$ acting on $V(.)$ and emanating from the point $(\tilde{x}, i)$ at time t, where $\tilde{x}(t) = \tilde{x}$ and $r(t) = i$ for $i \in \mathscr{S}$, is given by:

$$\mathscr{L}V(\tilde{x}_t, i) = \dot{\tilde{x}}^\top(t)P(i)\tilde{x}(t) + \tilde{x}^\top(t)P(i)\dot{\tilde{x}}(t) + \sum_{j=1}^N \lambda_{ij}\tilde{x}^\top(t)P(j)\tilde{x}(t)$$

$$= \left[\tilde{A}(i)\tilde{x}(t) + \tilde{B}(i)w(t)\right]^\top P(i)\tilde{x}(t) + \tilde{x}^\top(t)P(i)\left[\tilde{A}(i)\tilde{x}(t) + \tilde{B}(i)w(t)\right]$$

$$+ \sum_{j=1}^N \lambda_{ij}\tilde{x}^\top(t)P(j)\tilde{x}(t)$$

$$= \tilde{x}^\top(t)\left[\tilde{A}^\top(i)P(i) + P(i)\tilde{A}(i) + \sum_{j=1}^N \lambda_{ij}P(j)\right]\tilde{x}(t)$$

$$+ \tilde{x}^\top(t)P(i)\tilde{B}(i)w(t) + w^\top(t)\tilde{B}^\top(i)P(i)\tilde{x}(t),$$

and

$$e^\top(t)e(t) - \gamma^2 w^\top(t)w(t)$$

$$= \left[\tilde{C}(i)\tilde{x}(t) + \tilde{D}(i)w(t)\right]^\top \left[\tilde{C}(i)\tilde{x}(t) + \tilde{D}(i)w(t)\right] - \gamma^2 w^\top(t)w(t)$$

$$= \tilde{x}^\top(t)\tilde{C}^\top(i)\tilde{C}(i)\tilde{x}(t) + \tilde{x}^\top(t)\tilde{C}^\top(i)\tilde{D}(i)w(t)$$

$$+ w^\top(t)\tilde{D}^\top(i)\tilde{C}(i)\tilde{x}(t) + w^\top(t)\tilde{D}^\top(i)\tilde{D}(i)w(t) - \gamma^2 w^\top(t)w(t).$$

Combining these two relations, we get

$$e^\top(t)e(t) - \gamma^2 w^\top(t)w(t) + \mathscr{L}V(x_t, i)$$

$$= \tilde{x}^\top(t)\left[\tilde{A}^\top(i)P(i) + P(i)\tilde{A}(i)\right.$$

$$+ \sum_{j=1}^N \lambda_{ij}P(j) + \tilde{C}^\top(i)\tilde{C}(i)\Big]\tilde{x}(t)$$

$$+ \tilde{x}^\top(t) \left[P(i)\tilde{B}(i) + \tilde{C}^\top(i)\tilde{D}(i) \right] w(t)$$

$$+ w^\top(t) \left[\tilde{B}^\top(i)P(i) + \tilde{D}^\top(i)\tilde{C}(i) \right] \tilde{x}(t)$$

$$+ w^\top(t) \left[\tilde{D}^\top(i)\tilde{D}(i) - \gamma^2\mathbb{I} \right] w(t),$$

which gives in matrix form:

$$
\begin{aligned}
e^\top(t)e(t) - \gamma^2 w^\top(t)w(t) + \mathscr{L}V(x_t, i) \\
= \begin{bmatrix} \tilde{x}^\top(t) \ w^\top(t) \end{bmatrix} \Lambda_n(i) \begin{bmatrix} \tilde{x}(t) \\ w(t) \end{bmatrix} \\
= \tilde{\xi}^\top(t)\Lambda_n(i)\tilde{\xi}(t),
\end{aligned}
\tag{5.27}
$$

with $\tilde{\xi}(t) = \begin{bmatrix} \tilde{x}(t) \\ w(t) \end{bmatrix}$ and $\Lambda_n(i)$ defined by

$$
\Lambda_n(i) = \begin{bmatrix} \begin{bmatrix} \tilde{A}^\top(i)P(i) \\ +P(i)\tilde{A}(i) \\ +\sum_{j=1}^N \lambda_{ij}P(j) \\ +\tilde{C}^\top(i)\tilde{C}(i) \end{bmatrix} & \begin{bmatrix} P(i)\tilde{B}(i) \\ +\tilde{C}^\top(i)\tilde{D}(i) \end{bmatrix} \\ \begin{bmatrix} \tilde{B}^\top(i)P(i) \\ +D^\top(i)\tilde{C}(i) \end{bmatrix} & \begin{bmatrix} \tilde{D}^\top(i)\tilde{D}(i) \\ -\gamma^2\mathbb{I} \end{bmatrix} \end{bmatrix}.
$$

Adding and subtracting $\mathscr{L}V(x_t, i)$ to the $\mathscr{H}_\infty$ performance (5.25), we get the following:

$$
J_T = \mathbb{E}\left[\int_0^T [e^\top(t)e(t) - \gamma^2 w^\top(t)w(t) + \mathscr{L}V(\tilde{x}_t, r(t))]dt \right]
$$
$$
- \mathbb{E}\left[\int_0^T \mathscr{L}V(\tilde{x}_t, r(t))dt \right].
$$

Using Dynkin's formula, we obtain

$$
\mathbb{E}\left[\int_0^T \mathscr{L}V(\tilde{x}_t, r(t))dt \right] = \mathbb{E}[V(\tilde{x}_T, r(T))] - \mathbb{E}[V(\tilde{x}_0, r_0)].
$$

From (5.26), we have

$$
\mathbb{E}[V(\tilde{x}_0, r_0)] = \mathbb{E}[\tilde{x}^\top(0)P(r_0)\tilde{x}(0)].
\tag{5.28}
$$

Note that $\tilde{x}^\top(0) = \left[x^\top(0) \ x^\top(0) - \hat{x}^\top(0) \right]^\top = \left[x_0^\top, x_0^\top \right]^\top$.
In view of (5.23) and (5.28), we have

$$
\mathbb{E}[V(\tilde{x}(0), r(0))] = \left\{ \mathbb{E}\left[x^\top(0) \begin{bmatrix} \mathbb{I} \ \mathbb{I} \end{bmatrix} P(r_0) \begin{bmatrix} \mathbb{I} \\ \mathbb{I} \end{bmatrix} x(0) \right] \right\}
$$

$$\leq \gamma^2 \mathbb{E}\left[x^\top(0)Rx(0)\right].$$

Notice that the $\mathscr{H}_\infty$ performance J_T can be rewritten as follows:

$$J_T = \mathbb{E}\left[\int_0^T \tilde{\xi}_t^\top \Lambda_n(r(t))\tilde{\xi}_t dt\right] + \mathbb{E}[V(\tilde{x}(0), r_0)] - \mathbb{E}[V(\tilde{x}(T), r(T))],$$

which implies

$$J_T \leq \mathbb{E}\left[\int_0^T \tilde{\xi}_t^\top \Lambda_n(r(t))\tilde{\xi}_t dt\right] + \mathbb{E}[V(\tilde{x}(0), r_0)]. \tag{5.29}$$

Combining this with the fact that $\Lambda_n(i) < 0$ for all $i \in \mathscr{S}$, the following holds for all $T > 0$:

$$J_T \leq \mathbb{E}[V(\tilde{x}(0), r_0)] \leq \gamma^2 x^\top(0)Rx(0).$$

Therefore, we get

$$\begin{aligned} J_\infty &= \mathbb{E}\left[\int_0^\infty [e^\top(t)e(t) - \gamma^2 w^\top(t)w(t)]dt\right] \\ &\leq \gamma^2 x^\top(0)Rx(0). \end{aligned}$$

This gives in turn that

$$\|e\|_2^2 \leq \gamma\left[\|w\|_2^2 + x^\top(0)Rx(0)\right]$$

and this ends the proof of Theorem 75. $\qquad\square$

For a given set of gains of the filter of the form (5.19), we can compute the minimum disturbance rejection by solving the following convex optimization problem:

$$\text{Pn:} \begin{cases} \min\limits_{\substack{v>0,\\ P=(P(1),\cdots,P(N))>0}} v, \\ \text{s.t.:} \\ \begin{bmatrix} \tilde{J}_1(i) & P(i)\tilde{B}(i) & \tilde{C}^\top(i) \\ \tilde{B}^\top(i)P(i) & -v\mathbb{I} & \tilde{D}^\top(i) \\ \tilde{C}(i) & \tilde{D}(i) & -\mathbb{I} \end{bmatrix} < 0, \\ \begin{bmatrix} \mathbb{I} & \mathbb{I} \end{bmatrix} P(r_0) \begin{bmatrix} \mathbb{I} \\ \mathbb{I} \end{bmatrix} \leq vR, \end{cases}$$

where $v = \gamma^2$.

But since we have not yet developed a way to choose the filter gains, this optimization problem is useless. The design of the filter gains should be included in an optimization problem similar to this one, which can help us to determine simultaneously the filter gains and the minimum disturbance rejection.

Notice that the condition (5.22) is nonlinear in $P(i)$ and the design filter parameters. To cast the design of the $\mathscr{H}_\infty$ filter in the LMI framework, let us transform this condition in order to compute the gains $K_A(i)$, $K_B(i)$, and $K_C(i)$.

Compute $\tilde{J}_1(i)$, $P(i)\tilde{B}(i)$, $\tilde{C}^\top(i)$, $\tilde{B}^\top(i)P(i)$, and $\tilde{D}^\top(i)$ in function of $A(i)$, $B(i)$, $C_y(i)$, $D_y(i)$, $C_z(i)$, and $D_z(i)$. Using the expression of $\tilde{A}(i)$, $\tilde{B}(i)$, $\tilde{C}(i)$, and $\tilde{D}(i)$, and assuming that $P(i) = \mathrm{diag}[X_1(i), X_2(i)]$ we get

$$
\begin{aligned}
\tilde{J}_1(i) =\ & \tilde{A}^\top(i)P(i) + P(i)\tilde{A}(i) + \sum_{j=1}^{N} \lambda_{ij} P(j) \\
=\ & \begin{bmatrix} A(i) & 0 \\ A(i) - K_B(i)C_y(i) - K_A(i) & K_A(i) \end{bmatrix}^\top \begin{bmatrix} X_1(i) & 0 \\ 0 & X_2(i) \end{bmatrix} \\
& + \begin{bmatrix} X_1(i) & 0 \\ 0 & X_2(i) \end{bmatrix} \begin{bmatrix} A(i) & 0 \\ A(i) - K_B(i)C_y(i) - K_A(i) & K_A(i) \end{bmatrix} \\
& + \sum_{j=1}^{N} \lambda_{ij} \begin{bmatrix} X_1(j) & 0 \\ 0 & X_2(j) \end{bmatrix} \\
=\ & \begin{bmatrix} A^\top(i) & A^\top(i) - C_y^\top(i)K_B^\top(i) - K_A^\top(i) \\ 0 & K_A^\top(i) \end{bmatrix} \begin{bmatrix} X_1(i) & 0 \\ 0 & X_2(i) \end{bmatrix} \\
& + \begin{bmatrix} X_1(i) & 0 \\ 0 & X_2(i) \end{bmatrix} \begin{bmatrix} A(i) & 0 \\ A(i) - K_B(i)C_y(i) - K_A(i) & K_A(i) \end{bmatrix} \\
& + \sum_{j=1}^{N} \lambda_{ij} \begin{bmatrix} X_1(j) & 0 \\ 0 & X_2(j) \end{bmatrix} \\
=\ & \begin{bmatrix} \begin{bmatrix} A^\top(i)X_1(i) \\ +X_1(i)A(i) \\ +\sum_{j=1}^{N}\lambda_{ij}X_1(j) \end{bmatrix} & \begin{bmatrix} A^\top(i)X_2(i) \\ -C_y^\top(i)K_B^\top(i)X_2(i) \\ -K_A^\top(i)X_2(i) \end{bmatrix} \\[2mm] \begin{bmatrix} X_2(i)A(i) \\ -X_2(i)K_B(i)C_y(i) \\ -X_2(i)K_A(i) \end{bmatrix} & \begin{bmatrix} K_A^\top(i)X_2(i) \\ +X_2(i)K_A(i) \\ +\sum_{j=1}^{N}\lambda_{ij}X_2(j) \end{bmatrix} \end{bmatrix},
\end{aligned}
$$

$$
\begin{aligned}
P(i)\tilde{B}(i) =\ & \begin{bmatrix} X_1(i) & 0 \\ 0 & X_2(i) \end{bmatrix} \begin{bmatrix} B(i) \\ B(i) - K_B(i)D_y(i) \end{bmatrix} \\
=\ & \begin{bmatrix} X_1(i)B(i) \\ X_2(i)B(i) - X_2(i)K_B(i)D_y(i) \end{bmatrix}, \\
\tilde{C}(i) =\ & \begin{bmatrix} C_z(i) - K_C(i) & K_C(i) \end{bmatrix}, \\
\tilde{D}(i) =\ & D_z(i).
\end{aligned}
$$

Using these relations, (5.22) becomes

$$
\left[\begin{array}{cc}
\begin{bmatrix} A^\top(i)X_1(i) \\ +X_1(i)A(i) \\ +\sum_{j=1}^N \lambda_{ij}X_1(j) \end{bmatrix} & \begin{bmatrix} A^\top(i)X_2(i) \\ -C_y^\top(i)K_B^\top(i)X_2(i) \\ -K_A^\top(i)X_2(i) \end{bmatrix} \\[4ex]
\begin{bmatrix} X_2(i)A(i) \\ -X_2(i)K_B(i)C_y(i) \\ -X_2(i)K_A(i) \end{bmatrix} & \begin{bmatrix} K_A^\top(i)X_2(i) \\ +X_2(i)K_A(i) \\ +\sum_{j=1}^N \lambda_{ij}X_2(j) \end{bmatrix} \\[4ex]
B^\top(i)X_1(i) & \begin{bmatrix} B^\top(i)X_2(i) \\ -D_y^\top(i)K_B^\top(i)X_2(i) \end{bmatrix} \\[2ex]
C_z(i)-K_C(i) & K_C(i)
\end{array}\right.
$$

$$
\left.\begin{array}{cc}
X_1(i)B(i) & C_z^\top(i)-K_C^\top(i) \\[1ex]
\begin{bmatrix} X_2(i)B(i) \\ -X_2(i)K_B(i)D_y(i) \end{bmatrix} & K_C^\top(i) \\[2ex]
-\gamma^2\mathbb{I} & D_z^\top(i) \\[1ex]
D_z(i) & -\mathbb{I}
\end{array}\right] < 0.
$$

Letting $Y(i) = X_2(i)K_A(i)$, $Z(i) = X_2(i)K_B(i)$, and $W(i) = K_C(i)$, we get

$$
\left[\begin{array}{cc}
\begin{bmatrix} A^\top(i)X_1(i) \\ +X_1(i)A(i) \\ +\sum_{j=1}^N \lambda_{ij}X_1(j) \end{bmatrix} & \begin{bmatrix} A^\top(i)X_2(i) \\ -C_y^\top(i)Z^\top(i) \\ -Y^\top(i) \end{bmatrix} \\[4ex]
\begin{bmatrix} X_2(i)A(i) \\ -Z(i)C_y(i) \\ -Y(i) \end{bmatrix} & \begin{bmatrix} Y^\top(i) \\ +Y(i) \\ +\sum_{j=1}^N \lambda_{ij}X_2(j) \end{bmatrix} \\[4ex]
B^\top(i)X_1(i) & \begin{bmatrix} B^\top(i)X_2(i) \\ -D_y^\top(i)Z^\top(i) \end{bmatrix} \\[2ex]
C_z(i)-W(i) & W(i)
\end{array}\right.
$$

$$
\left.\begin{array}{cc}
X_1(i)B(i) & \begin{bmatrix} C_z^\top(i) \\ -W^\top(i) \end{bmatrix} \\[2ex]
\begin{bmatrix} X_2(i)B(i) \\ -Z(i)D_y(i) \end{bmatrix} & W^\top(i) \\[2ex]
-\gamma^2\mathbb{I} & D_z^\top(i) \\[1ex]
D_z(i) & -\mathbb{I}
\end{array}\right] < 0. \tag{5.30}
$$

For the second relation of the theorem, we have for $r_0 \in \mathscr{S}$:

$$
\begin{bmatrix} \mathbb{I} & \mathbb{I} \end{bmatrix} \begin{bmatrix} X_1(r_0) & 0 \\ 0 & X_2(r_0) \end{bmatrix} \begin{bmatrix} \mathbb{I} \\ \mathbb{I} \end{bmatrix} = X_1(r_0) + X_2(r_0) < \gamma^2 R. \tag{5.31}
$$

The following theorem gives the results for the design of the gains of the $\mathscr{H}_\infty$ filter.

Theorem 76. *Let γ and R be, respectively, a given positive constant and a symmetric and positive-definite matrix representing the weighting of the initial conditions. If there exist sets of symmetric and positive matrices*

$X_1 = (X_1(1), \cdots, X_1(N)) > 0$, $X_2 = (X_2(1), \cdots, X_2(N)) > 0$, and matrices $Y = (Y(1), \cdots, Y(N))$, $Z = (Z(1), \cdots, Z(N))$, and $W = (W(1), \cdots, W(N))$ satisfying the following LMIs (5.30)–(5.31) for every $i \in \mathscr{S}$, then there exists a filter of the form (5.19) such that the estimation error is stochastically stable and bounded by

$$\|z - \hat{z}\|_2 \leq \gamma \left[\|w\|_2^2 + x_0^\top R x_0 \right]^{\frac{1}{2}}. \tag{5.32}$$

The filter's gains are given by

$$\begin{cases} K_A(i) = X_2^{-1}(i)Y(i), \\ K_B(i) = X_2^{-1}(i)Z(i), \\ K_C(i) = W(i). \end{cases} \tag{5.33}$$

Example 66. To show the usefulness of the results of the previous theorem, let us consider a two-mode system with the following data:

- mode #1:

$$A(1) = \begin{bmatrix} -3.0 & 1.0 & 0.0 \\ 0.3 & -2.5 & 1.0 \\ -0.1 & 0.3 & -3.8 \end{bmatrix}, \quad B(1) = \begin{bmatrix} 1.0 \\ 0.0 \\ 1.0 \end{bmatrix},$$

$$C_y(1) = \begin{bmatrix} 0.1 \ 1.0 \ 0.0 \end{bmatrix}, \qquad D_y(1) = \begin{bmatrix} 0.2 \end{bmatrix},$$
$$C_z(1) = \begin{bmatrix} 0.1 \ 1.0 \ 0.0 \end{bmatrix}, \qquad D_z(1) = \begin{bmatrix} 2 \end{bmatrix},$$

- mode #2:

$$A(2) = \begin{bmatrix} -4.0 & 1.0 & 0.0 \\ 0.3 & -3.0 & 1.0 \\ -0.1 & 0.3 & -4.8 \end{bmatrix}, \quad B(2) = \begin{bmatrix} 1.0 \\ 0.0 \\ 2.0 \end{bmatrix},$$

$$C_y(2) = \begin{bmatrix} 0.8 \ 0.4 \ 0.0 \end{bmatrix}, \qquad D_y(2) = \begin{bmatrix} 0.1 \end{bmatrix},$$
$$C_z(2) = \begin{bmatrix} 0.7 \ 0.1 \ 0.0 \end{bmatrix}, \qquad D_z(2) = \begin{bmatrix} 3.0 \end{bmatrix}.$$

The switching between the two modes is described by the following transition rate matrix:

$$\Lambda = \begin{bmatrix} -2.0 & 2.0 \\ 1.0 & -1.0 \end{bmatrix}.$$

Let $\gamma = 3.2$. Solving the LMI of the previous theorem, we get

$$X_1(1) = \begin{bmatrix} 0.4550 & 0.0594 & -0.0937 \\ 0.0594 & 0.8202 & 0.1464 \\ -0.0937 & 0.1464 & 0.3339 \end{bmatrix},$$

$$X_1(2) = \begin{bmatrix} 0.6481 & 0.1897 & -0.0466 \\ 0.1897 & 0.6430 & 0.1188 \\ -0.0466 & 0.1188 & 0.2227 \end{bmatrix},$$

$$X_2(1) = \begin{bmatrix} 0.2468 & 0.0736 & -0.0579 \\ 0.0736 & 0.5412 & 0.0858 \\ -0.0579 & 0.0858 & 0.2679 \end{bmatrix},$$

$$X_2(2) = \begin{bmatrix} 0.6846 & 0.1937 & -0.0594 \\ 0.1937 & 0.4612 & 0.0575 \\ -0.0594 & 0.0575 & 0.1159 \end{bmatrix},$$

$$Y(1) = \begin{bmatrix} -1.9012 & -0.4551 & -0.4489 \\ -0.9434 & -1.1422 & -0.6304 \\ -0.8838 & -0.6897 & -1.3025 \end{bmatrix},$$

$$Y(2) = \begin{bmatrix} -2.6610 & -0.1501 & -0.0875 \\ -1.1614 & -0.9220 & -0.2901 \\ -0.5432 & -0.1691 & -0.3244 \end{bmatrix},$$

$$Z(1) = \begin{bmatrix} 0.6668 \\ -0.6417 \\ -0.6286 \end{bmatrix}, \qquad Z(2) = \begin{bmatrix} 0.9559 \\ -0.2011 \\ -0.0397 \end{bmatrix},$$

$$W(1) = \begin{bmatrix} 0.4269 & -0.1865 & -0.2150 \end{bmatrix},$$
$$W(2) = \begin{bmatrix} 0.4434 & -0.0228 & -0.0067 \end{bmatrix},$$

which gives the following gains for the $\mathscr{H}_\infty$ filter:

$$K_A(1) = \begin{bmatrix} -9.0652 & -2.0157 & -3.1796 \\ 0.3417 & -1.4316 & 0.1559 \\ -5.3691 & -2.5520 & -5.6001 \end{bmatrix},$$

$$K_A(2) = \begin{bmatrix} -4.6004 & 0.3629 & -0.3474 \\ 0.3121 & -2.1243 & -0.1192 \\ -7.2000 & -0.2180 & -2.9176 \end{bmatrix},$$

$$K_B(1) = \begin{bmatrix} 2.8010 \\ -1.3597 \\ -1.3052 \end{bmatrix}, \qquad K_B(2) = \begin{bmatrix} 1.9095 \\ -1.4044 \\ 1.3337 \end{bmatrix},$$

$$K_C(1) = \begin{bmatrix} 0.4269 & -0.1865 & -0.2150 \end{bmatrix},$$
$$K_C(2) = \begin{bmatrix} 0.4434 & -0.0228 & -0.0067 \end{bmatrix}.$$

If the initial conditions are equal to zero, the previous results become easier and are given by the following corollary.

Corollary 16. *Let the initial conditions of system (5.1) be zero. Let γ be a given positive constant. If there exist sets of symmetric and positive matrices $X_1 = (X_1(1), \cdots, X_1(N)) > 0$, $X_2 = (X_2(1), \cdots, X_2(N)) > 0$, and matrices $Y = (Y(1), \cdots, Y(N))$, $Z = (Z(1), \cdots, Z(N))$, and $W = (W(1), \cdots, W(N))$*

satisfying the LMI (5.30) for every $i \in \mathscr{S}$, then there exists a filter of the form (5.19) such that the estimation error is stochastically stable and bounded by

$$\|z - \hat{z}\|_2 \leq \gamma \|w\|_2.$$

The filter's gains are given by (5.33).

The minimal noise attenuation level γ that can be verified by the filter of the form of (5.19) can be obtained by solving the following optimization problem:

$$
\mathcal{P}_0 : \begin{cases}
\min\limits_{\substack{v > 0, \\ X_1 = (X_1(1), \cdots, X_1(N)) > 0, \\ X_2 = (X_2(1), \cdots, X_2(N)) > 0, \\ Y = (Y(1), \cdots, Y(N)), \\ Z = (Z(1), \cdots, Z(N)), \\ W = (W(1), \cdots, W(N)),}} v, \\
s.t. \\
\Theta_v(i) < 0, \\
X_1(r_0) + X_2(r_0) < vR,
\end{cases}
$$

where $\Theta_v(i)$ is obtained from (5.30) by replacing γ^2 by v. Thus, if the convex optimization problem $\mathcal{P}_0$ has a solution v, then by using Theorem 76, the corresponding error of the filter (5.19) is stable with noise attenuation level $\sqrt{v}$.

Let us now return to the dynamics of this chapter as given by (5.1) and consider the effect of the uncertainties. As we did previously, let us assume that there exists a filter of the form (5.19) and see under which conditions the estimation error will be robustly stable in the stochastic sense and bounded for all admissible uncertainties and all signals $w(t) \in \mathscr{L}_2[0, \infty)$.

Let us establish the uncertain extended dynamics in a different form. From the dynamics (5.1) notice that we have

$$\dot{x}(t) = [A(i) + \Delta A(i,t)] x(t) + [B(i) + \Delta B(i,t)] w(t), \tag{5.34}$$

with $\Delta A(i,t) = D_A(i) F_A(i,t) E_A(i)$ and $\Delta B(i,t) = D_B(i) F_B(i,t) E_B(i)$.

For the estimation error dynamics $\tilde{x}(t) = x(t) - \hat{x}(t)$ we have

$$
\begin{aligned}
\dot{x}(t) - \dot{\hat{x}}(t) &= [A(i) + \Delta A(i,t)] x(t) + [B(i) + \Delta B(i,t)] w(t) \\
&\quad - K_A(i)\hat{x}(t) - K_B(i)y(t) \\
&= [A(i) + \Delta A(i,t) - K_B(i)[C_y(i) + \Delta C_y(i,t)] - K_A(i)] x(t) \\
&\quad + K_A(i)[x(t) - \hat{x}(t)] + [B(i) + \Delta B(i,t) - K_B(i)D_y(i)] w(t),
\end{aligned}
$$

with $\Delta C_y(i,t) = D_{C_y}(i) F_{C_y}(i,t) E_{C_y}(i)$.

Based on these calculations, the extended dynamics become

$$\dot{\tilde{x}}(t) = \left[\tilde{A}(i) + \Delta\tilde{A}(i,t) \right] \tilde{x}(t) + \left[\tilde{B}(i) + \Delta\tilde{B}(i,t) \right] w(t), \tag{5.35}$$

where

$$\tilde{x}(t) = \begin{bmatrix} x(t) \\ x(t) - \hat{x}(t) \end{bmatrix},$$

$$\tilde{A}(i) = \begin{bmatrix} A(i) & 0 \\ A(i) - K_B(i)C_y(i) - K_A(i) & K_A(i) \end{bmatrix},$$

$$\Delta\tilde{A}(i,t) = \begin{bmatrix} \Delta A(i,t) & 0 \\ \Delta A(i,t) - K_B(i)\Delta C_y(i,t) & 0 \end{bmatrix},$$

$$\tilde{B}(i) = \begin{bmatrix} B(i) \\ B(i) - K_B(i)D_y(i) \end{bmatrix},$$

$$\Delta\tilde{B}(i,t) = \begin{bmatrix} \Delta B(i,t) \\ \Delta B(i,t) \end{bmatrix}.$$

Notice that

$$\begin{aligned}
\Delta\tilde{A}(i,t) &= \begin{bmatrix} \Delta A(i,t) & 0 \\ \Delta A(i,t) - K_B(i)\Delta C_y(i,t) & 0 \end{bmatrix} \\
&= \begin{bmatrix} \Delta A(i,t) & 0 \\ \Delta A(i,t) & 0 \end{bmatrix} + \begin{bmatrix} 0 & 0 \\ -K_B(i)\Delta C_y(i,t) & 0 \end{bmatrix} \\
&= \begin{bmatrix} D_A(i)F_A(i,t)E_A(i) & 0 \\ D_A(i)F_A(i,t)E_A(i) & 0 \end{bmatrix} \\
&\quad + \begin{bmatrix} 0 & 0 \\ -K_B(i)D_{C_y}(i)F_{C_y}(i,t)E_{C_y}(i) & 0 \end{bmatrix} \\
&= \begin{bmatrix} D_A(i) \\ D_A(i) \end{bmatrix} F_A(i,t) \begin{bmatrix} E_A(i) & 0 \end{bmatrix} \\
&\quad + \begin{bmatrix} 0 \\ -K_B(i)D_{C_y}(i) \end{bmatrix} F_{C_y}(i,t) \begin{bmatrix} E_{C_y}(i) & 0 \end{bmatrix} \\
&= \tilde{D}_A(i)F_A(i,t)\tilde{E}_A(i) + \tilde{D}_{C_y}(i)F_{C_y}(i,t)\tilde{E}_{C_y}(i),
\end{aligned}$$

with

$$\tilde{D}_A(i) = \begin{bmatrix} D_A(i) \\ D_A(i) \end{bmatrix},$$

$$\tilde{E}_A(i) = \begin{bmatrix} E_A(i) & 0 \end{bmatrix},$$

$$\tilde{D}_{C_y}(i) = \begin{bmatrix} 0 \\ -K_B(i)D_{C_y}(i) \end{bmatrix},$$

$$\tilde{E}_{C_y}(i) = \begin{bmatrix} E_{C_y}(i) & 0 \end{bmatrix},$$

and

$$\begin{aligned}
\Delta\tilde{B}(i,t) &= \begin{bmatrix} \Delta B(i,t) \\ \Delta B(i,t) \end{bmatrix} = \begin{bmatrix} D_B(i)F_B(i,t)E_B(i) \\ D_B(i)F_B(i,t)E_B(i) \end{bmatrix} \\
&= \begin{bmatrix} D_B(i) \\ D_B(i) \end{bmatrix} F_B(i,t)E_B(i)
\end{aligned}$$

$$= \tilde{D}_B(i)F_B(i,t)\tilde{E}_B(i),$$

with

$$\tilde{D}_B(i) = \begin{bmatrix} D_B(i) \\ D_B(i) \end{bmatrix},$$

$$\tilde{E}_B(i) = E_B(i).$$

For the estimation error, we have

$$\begin{aligned}
e(t) =\ & z(t) - \hat{z}(t) = C_z(i,t)x(t) + D_z(i)w(t) - K_C(i)\hat{x}(t) \\
=\ & [C_z(i)x(t) + \Delta C_z(i,t)]\, x(t) - K_C(i)x(t) \\
& + K_C(i)\,[x(t) - \hat{x}(t)] + D_z(i)w(t) \\
=\ & \left[\left[C_z(i) - K_C(i)\ K_C(i) \right] + \left[\Delta C_z(i,t)\ 0 \right] \right] \\
& \times \begin{bmatrix} x(t) \\ x(t) - \hat{x}(t) \end{bmatrix} + D_z(i)w(t) \\
=\ & \left[\tilde{C}(i) + \Delta\tilde{C}(i,t) \right]\tilde{x}(t) + \tilde{D}(i)w(t),
\end{aligned}$$

where

$$\begin{aligned}
\tilde{C}(i) =\ & \left[C_z(i) - K_C(i)\ K_C(i) \right], \\
\Delta\tilde{C}_z(i,t) =\ & \left[\Delta C_z(i,t)\ 0 \right] = \left[D_{C_z}(i)F_{C_z}(i,t)E_{C_z}(i)\ 0 \right] \\
=\ & D_{C_z}(i)F_{C_z}(i,t)\left[E_{C_z}(i)\ 0 \right] \\
=\ & \tilde{D}_{C_z}(i)F_{C_z}(i,t)\tilde{E}_{C_z}(i), \\
\tilde{D}(i) =\ & D_z(i).
\end{aligned}$$

Using the first condition of Theorem 75 for the uncertain extended system, we get

$$\begin{bmatrix}
\tilde{J}(i,t) & P(i)\left[\tilde{B}(i) + \Delta\tilde{B}(i,t) \right] \\
\left[\tilde{B}^\top(i) + \Delta\tilde{B}^\top(i,t) \right] P(i) & -\gamma^2\mathbb{I} \\
\left[\tilde{C}(i) + \Delta\tilde{C}(i,t) \right] & \tilde{D}(i) \\
& \begin{matrix} \left[\tilde{C}^\top(i) + \Delta\tilde{C}^\top(i,t) \right] \\ \tilde{D}^\top(i) \\ -\mathbb{I} \end{matrix}
\end{bmatrix} < 0,$$

with

$$\tilde{J}(i,t) = \left[\tilde{A}(i) + \Delta\tilde{A}(i,t) \right]^\top P(i) + P(i)\left[\tilde{A}(i) + \Delta\tilde{A}(i,t) \right] + \sum_{j=1}^{N} \lambda_{ij}P(j).$$

This last relation can be rewritten as follows:

$$\begin{bmatrix} \tilde{J}(i) & P(i)\tilde{B}(i) & \tilde{C}_z^\top(i) \\ \tilde{B}^\top(i)P(i) & -\gamma^2\mathbb{I} & \tilde{D}_z^\top(i) \\ \tilde{C}_z(i) & \tilde{D}_z(i) & -\mathbb{I} \end{bmatrix}$$

$$+ \begin{bmatrix} \Delta\tilde{A}^\top(i,t)P(i) + P(i)\Delta\tilde{A}(i,t) & P(i)\Delta\tilde{B}(i,t) & \Delta\tilde{C}_z^\top(i,t) \\ \Delta\tilde{B}^\top(i,t)P(i) & 0 & 0 \\ \Delta\tilde{C}_z(i,t) & 0 & 0 \end{bmatrix} < 0,$$

with $\tilde{J}(i) = \tilde{A}^\top(i)P(i) + P(i)\tilde{A}(i) + \sum_{j=1}^N \lambda_{ij}P(j)$.

This gives in turn:

$$\begin{bmatrix} \tilde{J}(i) & P(i)\tilde{B}(i) & \tilde{C}_z^\top(i) \\ \tilde{B}^\top(i)P(i) & -\gamma^2\mathbb{I} & \tilde{D}_z^\top(i) \\ \tilde{C}_z(i) & \tilde{D}_z(i) & -\mathbb{I} \end{bmatrix}$$

$$+ \begin{bmatrix} P(i)\Delta\tilde{A}(i,t) & P(i)\Delta\tilde{B}(i,t) & \Delta\tilde{C}_z^\top(i,t) \\ 0 & 0 & 0 \\ 0 & 0 & 0 \end{bmatrix}$$

$$+ \begin{bmatrix} \Delta\tilde{A}^\top(i,t)P(i) & 0 & 0 \\ \Delta\tilde{B}^\top(i,t)P(i) & 0 & 0 \\ \Delta\tilde{C}_z(i,t) & 0 & 0 \end{bmatrix} < 0.$$

Using now the expressions of the uncertainties, we get

$$\begin{bmatrix} \tilde{J}(i) & P(i)\tilde{B}(i) & \tilde{C}_z^\top(i) \\ \tilde{B}^\top(i)P(i) & -\gamma^2\mathbb{I} & \tilde{D}_z^\top(i) \\ \tilde{C}_z(i) & \tilde{D}_z(i) & -\mathbb{I} \end{bmatrix}$$

$$+ \begin{bmatrix} P(i)\tilde{D}_A(i)F_A(i,t)\tilde{E}_A(i) & 0 & 0 \\ 0 & 0 & 0 \\ 0 & 0 & 0 \end{bmatrix}$$

$$+ \begin{bmatrix} \tilde{E}_A^\top(i)F_A^\top(i,t)\tilde{D}_A^\top(i)P(i) & 0 & 0 \\ 0 & 0 & 0 \\ 0 & 0 & 0 \end{bmatrix}$$

$$+ \begin{bmatrix} P(i)\tilde{D}_{C_y}(i)F_{C_y}(i,t)\tilde{E}_{C_y}(i) & 0 & 0 \\ 0 & 0 & 0 \\ 0 & 0 & 0 \end{bmatrix}$$

$$+ \begin{bmatrix} \tilde{E}_{C_y}^\top(i)F_{C_y}^\top(i,t)\tilde{D}_{C_y}^\top(i)P(i) & 0 & 0 \\ 0 & 0 & 0 \\ 0 & 0 & 0 \end{bmatrix}$$

$$+ \begin{bmatrix} 0 & P(i)\tilde{D}_B(i)F_B(i,t)\tilde{E}_B(i) & 0 \\ 0 & 0 & 0 \\ 0 & 0 & 0 \end{bmatrix}$$

$$+ \begin{bmatrix} 0 & 0 & 0 \\ \tilde{E}_B^\top(i)F_B^\top(i,t)\tilde{D}_B^\top(i)P(i) & 0 & 0 \\ 0 & 0 & 0 \end{bmatrix}$$

$$+ \begin{bmatrix} 0 & 0 & \tilde{E}_{C_z}^{\top}(i)F_{C_z}^{\top}(i,t)\tilde{D}_{C_z}^{\top}(i) \\ 0 & 0 & 0 \\ 0 & 0 & 0 \end{bmatrix}$$

$$+ \begin{bmatrix} 0 & & 0 & 0 \\ 0 & & 0 & 0 \\ \tilde{D}_{C_z}(i)F_{C_z}(i,t)\tilde{E}_{C_z}(i) & 0 & 0 \end{bmatrix} < 0.$$

Notice that

$$\begin{bmatrix} P(i)\tilde{D}_A(i)F_A(i,t)\tilde{E}_A(i) & 0 & 0 \\ 0 & 0 & 0 \\ 0 & 0 & 0 \end{bmatrix}$$

$$= \begin{bmatrix} P(i)\tilde{D}_A(i) & 0 & 0 \\ 0 & 0 & 0 \\ 0 & 0 & 0 \end{bmatrix} \begin{bmatrix} F_A(i,t) & 0 & 0 \\ 0 & 0 & 0 \\ 0 & 0 & 0 \end{bmatrix} \begin{bmatrix} \tilde{E}_A(i) & 0 & 0 \\ 0 & 0 & 0 \\ 0 & 0 & 0 \end{bmatrix},$$

$$\begin{bmatrix} P(i)\tilde{D}_{C_y}(i)F_{C_y}(i,t)\tilde{E}_{C_y}(i) & 0 & 0 \\ 0 & 0 & 0 \\ 0 & 0 & 0 \end{bmatrix}$$

$$= \begin{bmatrix} P(i)\tilde{D}_{C_y}(i) & 0 & 0 \\ 0 & 0 & 0 \\ 0 & 0 & 0 \end{bmatrix} \begin{bmatrix} F_{C_y}(i,t) & 0 & 0 \\ 0 & 0 & 0 \\ 0 & 0 & 0 \end{bmatrix} \begin{bmatrix} \tilde{E}_{C_y}(i) & 0 & 0 \\ 0 & 0 & 0 \\ 0 & 0 & 0 \end{bmatrix},$$

$$\begin{bmatrix} 0 & P(i)\tilde{D}_B(i)F_B(i,t)\tilde{E}_B(i) & 0 \\ 0 & 0 & 0 \\ 0 & 0 & 0 \end{bmatrix}$$

$$= \begin{bmatrix} 0 & P(i)\tilde{D}_B(i) & 0 \\ 0 & 0 & 0 \\ 0 & 0 & 0 \end{bmatrix} \begin{bmatrix} 0 & 0 & 0 \\ 0 & F_B(i,t) & 0 \\ 0 & 0 & 0 \end{bmatrix} \begin{bmatrix} 0 & 0 & 0 \\ 0 & \tilde{E}_B(i) & 0 \\ 0 & 0 & 0 \end{bmatrix},$$

$$\begin{bmatrix} 0 & 0 & \tilde{E}_{C_z}^{\top}(i)F_{C_z}^{\top}(i,t)\tilde{D}_{C_z}^{\top}(i) \\ 0 & 0 & 0 \\ 0 & 0 & 0 \end{bmatrix}$$

$$= \begin{bmatrix} 0 & 0 & \tilde{E}_{C_z}^{\top}(i) \\ 0 & 0 & 0 \\ 0 & 0 & 0 \end{bmatrix} \begin{bmatrix} 0 & 0 & 0 \\ 0 & 0 & 0 \\ 0 & 0 & F_{C_z}^{\top}(i,t) \end{bmatrix} \begin{bmatrix} 0 & 0 & 0 \\ 0 & 0 & 0 \\ 0 & 0 & \tilde{D}_{C_z}^{\top}(i) \end{bmatrix},$$

and

$$\begin{bmatrix} \tilde{E}_A^{\top}(i)F^{\top}(i,t)\tilde{D}_A^{\top}(i)P(i) & 0 & 0 \\ 0 & 0 & 0 \\ 0 & 0 & 0 \end{bmatrix}$$

$$= \begin{bmatrix} \tilde{E}_A^\top(i) & 0 & 0 \\ 0 & 0 & 0 \\ 0 & 0 & 0 \end{bmatrix} \begin{bmatrix} F_A^\top(i,t) & 0 & 0 \\ 0 & 0 & 0 \\ 0 & 0 & 0 \end{bmatrix}$$

$$\times \begin{bmatrix} \tilde{D}_A^\top(i)P(i) & 0 & 0 \\ 0 & 0 & 0 \\ 0 & 0 & 0 \end{bmatrix},$$

$$\begin{bmatrix} \tilde{E}_{C_y}^\top(i)F_{C_y}^\top(i,t)\tilde{D}_{C_y}^\top(i)P(i) & 0 & 0 \\ 0 & 0 & 0 \\ 0 & 0 & 0 \end{bmatrix}$$

$$= \begin{bmatrix} \tilde{E}_{C_y}^\top(i) & 0 & 0 \\ 0 & 0 & 0 \\ 0 & 0 & 0 \end{bmatrix} \begin{bmatrix} F_{C_y}^\top(i,t) & 0 & 0 \\ 0 & 0 & 0 \\ 0 & 0 & 0 \end{bmatrix}$$

$$\times \begin{bmatrix} \tilde{D}_{C_y}^\top(i)P(i) & 0 & 0 \\ 0 & 0 & 0 \\ 0 & 0 & 0 \end{bmatrix},$$

$$\begin{bmatrix} 0 & 0 & 0 \\ \tilde{E}_B^\top(i)F_B^\top(i,t)\tilde{D}_B^\top(i)P(i) & 0 & 0 \\ 0 & 0 & 0 \end{bmatrix}$$

$$= \begin{bmatrix} 0 & 0 & 0 \\ 0 & \tilde{E}_B^\top(i) & 0 \\ 0 & 0 & 0 \end{bmatrix} \begin{bmatrix} 0 & 0 & 0 \\ 0 & F_B^\top(i,t) & 0 \\ 0 & 0 & 0 \end{bmatrix}$$

$$\times \begin{bmatrix} 0 & 0 & 0 \\ \tilde{D}_B^\top(i)P(i) & 0 & 0 \\ 0 & 0 & 0 \end{bmatrix},$$

$$\begin{bmatrix} 0 & 0 & 0 \\ 0 & 0 & 0 \\ \tilde{D}_{C_z}(i)F_{C_z}(i,t)\tilde{E}_{C_z}(i) & 0 & 0 \end{bmatrix}$$

$$= \begin{bmatrix} 0 & 0 & 0 \\ 0 & 0 & 0 \\ 0 & 0 & \tilde{D}_{C_z}(i) \end{bmatrix} \begin{bmatrix} 0 & 0 & 0 \\ 0 & 0 & 0 \\ 0 & 0 & F_{C_z}(i,t) \end{bmatrix}$$

$$\times \begin{bmatrix} 0 & 0 & 0 \\ 0 & 0 & 0 \\ \tilde{E}_{C_z}(i) & 0 & 0 \end{bmatrix}.$$

Using Lemma 7 in Appendix A, we get

$$\begin{bmatrix} P(i)\tilde{D}_A(i)F_A(i,t)\tilde{E}_A(i) & 0 & 0 \\ 0 & 0 & 0 \\ 0 & 0 & 0 \end{bmatrix}$$

$$+ \begin{bmatrix} \tilde{E}_A^\top(i)F_A^\top(i,t)\tilde{D}_A^\top(i)P(i) & 0 & 0 \\ 0 & 0 & 0 \\ 0 & 0 & 0 \end{bmatrix}$$

$$\leq \tilde{\varepsilon}_A^{-1}(i)\begin{bmatrix} P(i)\tilde{D}_A(i) & 0 & 0 \\ 0 & 0 & 0 \\ 0 & 0 & 0 \end{bmatrix}\begin{bmatrix} \tilde{D}_A^\top(i)P(i) & 0 & 0 \\ 0 & 0 & 0 \\ 0 & 0 & 0 \end{bmatrix}$$

$$+\tilde{\varepsilon}_A(i)\begin{bmatrix} \tilde{E}_A^\top(i) & 0 & 0 \\ 0 & 0 & 0 \\ 0 & 0 & 0 \end{bmatrix}\begin{bmatrix} \tilde{E}_A(i) & 0 & 0 \\ 0 & 0 & 0 \\ 0 & 0 & 0 \end{bmatrix}$$

$$= \begin{bmatrix} \tilde{\varepsilon}_A^{-1}(i)P(i)\tilde{D}_A(i)\tilde{D}_A^\top(i)P(i) + \tilde{\varepsilon}_A(i)\tilde{E}_A^\top(i)E_A(i) & 0 & 0 \\ 0 & 0 & 0 \\ 0 & 0 & 0 \end{bmatrix},$$

$$\begin{bmatrix} P(i)\tilde{D}_{C_y}(i)F_{C_y}(i,t)\tilde{E}_{C_y}(i) & 0 & 0 \\ 0 & 0 & 0 \\ 0 & 0 & 0 \end{bmatrix}$$

$$+ \begin{bmatrix} \tilde{E}_{C_y}^\top(i)F_{C_y}^\top(i,t)\tilde{D}_{C_y}^\top(i)P(i) & 0 & 0 \\ 0 & 0 & 0 \\ 0 & 0 & 0 \end{bmatrix}$$

$$\leq \tilde{\varepsilon}_{C_y}^{-1}(i)\begin{bmatrix} P(i)\tilde{D}_{C_y}(i) & 0 & 0 \\ 0 & 0 & 0 \\ 0 & 0 & 0 \end{bmatrix}\begin{bmatrix} \tilde{D}_{C_y}^\top(i)P(i) & 0 & 0 \\ 0 & 0 & 0 \\ 0 & 0 & 0 \end{bmatrix}$$

$$+\tilde{\varepsilon}_{C_y}(i)\begin{bmatrix} \tilde{E}_{C_y}^\top(i) & 0 & 0 \\ 0 & 0 & 0 \\ 0 & 0 & 0 \end{bmatrix}\begin{bmatrix} \tilde{E}_{C_y}(i) & 0 & 0 \\ 0 & 0 & 0 \\ 0 & 0 & 0 \end{bmatrix}$$

$$= \begin{bmatrix} \tilde{\varepsilon}_{C_y}^{-1}(i)P(i)\tilde{D}_{C_y}(i)\tilde{D}_{C_y}^\top(i)P(i) + \tilde{\varepsilon}_{C_y}(i)\tilde{E}_{C_y}^\top(i)E_{C_y}(i) & 0 & 0 \\ 0 & 0 & 0 \\ 0 & 0 & 0 \end{bmatrix},$$

$$\begin{bmatrix} 0 & P(i)\tilde{D}_B(i)F_B(i,t)\tilde{E}_B(i) & 0 \\ 0 & 0 & 0 \\ 0 & 0 & 0 \end{bmatrix}$$

$$+ \begin{bmatrix} 0 & 0 & 0 \\ \tilde{E}_B^\top(i)F_B^\top(i,t)\tilde{D}_B^\top(i)P(i) & 0 & 0 \\ 0 & 0 & 0 \end{bmatrix}$$

$$\leq \tilde{\varepsilon}_B^{-1}(i)\begin{bmatrix} 0 & P(i)\tilde{D}_B(i) & 0 \\ 0 & 0 & 0 \\ 0 & 0 & 0 \end{bmatrix}\begin{bmatrix} 0 & 0 & 0 \\ \tilde{D}_B^\top(i)P(i) & 0 & 0 \\ 0 & 0 & 0 \end{bmatrix}$$

$$+\tilde{\varepsilon}_B(i)\begin{bmatrix} 0 & 0 & 0 \\ 0 & \tilde{E}_B^\top(i) & 0 \\ 0 & 0 & 0 \end{bmatrix}\begin{bmatrix} 0 & 0 & 0 \\ 0 & \tilde{E}_B(i) & 0 \\ 0 & 0 & 0 \end{bmatrix}$$

$$=\begin{bmatrix} \tilde{\varepsilon}_B^{-1}(i)P(i)\tilde{D}_B(i)\tilde{D}_B^\top(i)P(i) & 0 & 0 \\ 0 & +\tilde{\varepsilon}_B(i)\tilde{E}_B^\top(i)E_B(i) & 0 \\ 0 & 0 & 0 \end{bmatrix},$$

$$\begin{bmatrix} 0 & 0 & \tilde{E}_{C_z}^\top(i)F_{C_z}^\top(i,t)\tilde{D}_{C_z}^\top(i) \\ 0 & 0 & 0 \\ 0 & 0 & 0 \end{bmatrix} + \begin{bmatrix} 0 & 0 & 0 \\ 0 & 0 & 0 \\ \tilde{D}_{C_z}(i)F_{C_z}(i,t)\tilde{E}_{C_z}(i) & 0 & 0 \end{bmatrix}$$

$$\leq \tilde{\varepsilon}_{C_z}^{-1}(i)\begin{bmatrix} 0 & 0 & \tilde{E}_{C_z}^\top(i) \\ 0 & 0 & 0 \\ 0 & 0 & 0 \end{bmatrix}\begin{bmatrix} 0 & 0 & 0 \\ 0 & 0 & 0 \\ \tilde{E}_{C_z}(i) & 0 & 0 \end{bmatrix}$$

$$+\tilde{\varepsilon}_{C_z}(i)\begin{bmatrix} 0 & 0 & 0 \\ 0 & 0 & 0 \\ 0 & 0 & \tilde{D}_{C_z}(i) \end{bmatrix}\begin{bmatrix} 0 & 0 & 0 \\ 0 & 0 & 0 \\ 0 & 0 & \tilde{D}_{C_z}^\top(i) \end{bmatrix}$$

$$=\begin{bmatrix} \tilde{\varepsilon}_{C_z}^{-1}(i)\tilde{E}_{C_z}^\top(i)\tilde{E}_{C_z}(i) & 0 & 0 \\ 0 & 0 & 0 \\ 0 & 0 & \tilde{\varepsilon}_{C_z}(i)\tilde{D}_{C_z}(i)\tilde{D}_{C_z}^\top(i) \end{bmatrix}.$$

Based on these transformations and after using the Schur complement we get the following:

$$\begin{bmatrix} \begin{bmatrix} \tilde{J}(i) \\ +\tilde{\varepsilon}_A(i)\tilde{E}_A^\top(i)\tilde{E}_A(i) \\ +\tilde{\varepsilon}_{C_y}(i)\tilde{E}_{C_y}^\top(i)\tilde{E}_{C_y}(i) \end{bmatrix} & P(i)\tilde{B}(i) & \tilde{C}^\top(i) & \mathcal{T}(i) \\ \tilde{B}^\top(i)P(i) & -\gamma^2\mathbb{I}+\tilde{\varepsilon}_B(i)\tilde{E}_B^\top(i)\tilde{E}_B(i) & \tilde{D}^\top(i) & 0 \\ \tilde{C}(i) & \tilde{D}(i) & -\mathbb{I}+\tilde{\varepsilon}_{C_z}(i)\tilde{D}_{C_z}(i)\tilde{D}_C^\top(i) & 0 \\ \mathcal{T}^\top(i) & 0 & 0 & -\mathcal{W} \end{bmatrix} < 0,$$

with

$$\mathcal{T}(i) = \begin{bmatrix} P(i)\tilde{D}_A(i) & P(i)\tilde{D}_{C_y}(i) & P(i)\tilde{D}_B(i) & \tilde{E}_{C_z}^\top(i) \end{bmatrix},$$

$$\mathcal{W} = \begin{bmatrix} \tilde{\varepsilon}_A(i)\mathbb{I} & 0 & 0 & 0 \\ 0 & \tilde{\varepsilon}_{C_y}(i)\mathbb{I} & 0 & 0 \\ 0 & 0 & \tilde{\varepsilon}_B(i)\mathbb{I} & 0 \\ 0 & 0 & 0 & \tilde{\varepsilon}_{C_z}(i)\mathbb{I} \end{bmatrix}.$$

Let us now use $P(i) = \begin{bmatrix} X_1(i) & 0 \\ 0 & X_2(i) \end{bmatrix}$, with $X_1(i)$ and $X_2(i)$ as symmetric and positive-definite matrices, and try to write the parameters of the extended dynamics as a function of the original ones, that is,

$$
J_1(i) = \begin{bmatrix} \begin{bmatrix} A^\top(i)X_1(i) \\ +X_1(i)A(i) \\ +\sum_{j=1}^{N}\lambda_{ij}X_1(j) \end{bmatrix} & \begin{bmatrix} A^\top(i)X_2(i) \\ -C_y^\top(i)K_B^\top(i)X_2(i) \\ -K_A^\top(i)X_2(i) \end{bmatrix} \\ \begin{bmatrix} X_2(i)A(i) \\ -X_2(i)K_B(i)C_y(i) \\ -X_2(i)K_A(i) \end{bmatrix} & \begin{bmatrix} K_A^\top(i)X_2(i) \\ +X_2(i)K_A(i) \\ +\sum_{j=1}^{N}\lambda_{ij}X_2(j) \end{bmatrix} \end{bmatrix},
$$

$$
P(i)\tilde{B}(i) = \begin{bmatrix} X_1(i)B(i) \\ X_2(i)B(i) - X_2(i)K_B(i)D_y(i) \end{bmatrix},
$$

$$
\tilde{C}_z(i) = \begin{bmatrix} C_z(i) - K_C(i) & K_C(i) \end{bmatrix},
$$

$$
\tilde{D}_z(i) = D_z(i),
$$

$$
\tilde{\varepsilon}_A(i)\tilde{E}_A^\top(i)\tilde{E}_A(i) = \tilde{\varepsilon}_A(i)\begin{bmatrix} E_A^\top(i) \\ 0 \end{bmatrix}\begin{bmatrix} E_A(i) & 0 \end{bmatrix}
$$

$$
= \begin{bmatrix} \tilde{\varepsilon}_A(i)E_A^\top(i)E_A(i) & 0 \\ 0 & 0 \end{bmatrix},
$$

$$
\tilde{\varepsilon}_{C_y}(i)\tilde{E}_{C_y}^\top(i)\tilde{E}_{C_y}(i) = \tilde{\varepsilon}_{C_y}(i)\begin{bmatrix} E_{C_y}^\top(i) \\ 0 \end{bmatrix}\begin{bmatrix} E_{C_y}(i) & 0 \end{bmatrix}
$$

$$
= \begin{bmatrix} \tilde{\varepsilon}_{C_y}(i)E_{C_y}^\top(i)E_{C_y}(i) & 0 \\ 0 & 0 \end{bmatrix},
$$

$$
\tilde{\varepsilon}_B(i)\tilde{E}_B^\top(i)\tilde{E}_B(i) = \tilde{\varepsilon}_B(i)E_B^\top(i)E_B(i),
$$

$$
\tilde{\varepsilon}_{C_z}(i)\tilde{D}_{C_z}(i)\tilde{D}_{C_z}^\top(i) = \tilde{\varepsilon}_{C_z}(i)D_{C_z}(i)D_{C_z}^\top(i),
$$

$$
\mathcal{T}(i) = \begin{bmatrix} P(i)\tilde{D}_A(i) & P(i)\tilde{D}_{C_y}(i) & P(i)\tilde{D}_B(i) & \tilde{E}_C^\top(i) \end{bmatrix}
$$

$$
= \begin{bmatrix} \begin{bmatrix} X_1(i)D_A(i) \\ X_2(i)D_A(i) \end{bmatrix} & \begin{bmatrix} 0 \\ -X_2(i)K_B(i)D_{C_y}(i) \end{bmatrix} \\ \begin{bmatrix} X_1(i)D_B(i) \\ X_2(i)D_B(i) \end{bmatrix} & \begin{bmatrix} E_{C_z}^\top(i) \\ 0 \end{bmatrix} \end{bmatrix}
$$

$$
= \begin{bmatrix} \mathcal{T}_1(i) \\ \mathcal{T}_2(i) \end{bmatrix},
$$

with

$$
\mathcal{T}_1(i) = \begin{bmatrix} X_1(i)D_A(i) & 0 & X_1(i)D_B(i) & E_{C_z}^\top(i) \end{bmatrix},
$$

$$
\mathcal{T}_2(i) = \begin{bmatrix} X_2(i)D_A(i) & -X_2(i)K_B(i)D_{C_y}(i) & X_2(i)D_B(i) & 0 \end{bmatrix}.
$$

This gives us the following:

$$\begin{bmatrix} J_{X_1} & \mathcal{U}(i) & X_1(i)B(i) \\ \mathcal{U}^\top(i) & J_{X_2} & \begin{bmatrix} X_2(i)B(i) \\ -X_2(i)K_B(i)D_y(i) \end{bmatrix} \\ B^\top(i)X_1(i) & \begin{bmatrix} B^\top(i)X_2(i) \\ -D_y^\top(i)K_B^\top(i)X_2(i) \end{bmatrix} & -\gamma^2\mathbb{I}+\varepsilon_B(i)E_B^\top(i)E_B(i) \\ C_z(i)-K_C(i) & K_C(i) & D_z(i) \\ \mathcal{T}_1^\top(i) & \mathcal{T}_2^\top(i) & 0 \end{bmatrix}$$

$$\left.\begin{matrix} C_z^\top(i)-K_C^\top(i) & \mathcal{T}_1(i) \\ K_C^\top(i) & \mathcal{T}_2(i) \\ D_z^\top(i) & 0 \\ -\mathbb{I}+\varepsilon_{C_z}(i)D_{C_z}(i)D_{C_z}^\top(i) & 0 \\ 0 & -\mathcal{W} \end{matrix}\right] < 0,$$

where

$$\begin{aligned} J_{X_1} &= A^\top(i)X_1(i)+X_1(i)A(i)+\sum_{j=1}^N \lambda_{ij}X_1(j) \\ &\quad +\tilde\varepsilon_A(i)E_A^\top(i)E_A(i)+\tilde\varepsilon_{C_y}(i)E_{C_y}^\top(i)E_{C_y}(i), \\ J_{X_2} &= K_A^\top(i)X_2(i)+X_2(i)K_A(i)+\sum_{j=1}^N \lambda_{ij}X_2(j), \\ \mathcal{U}(i) &= A^\top(i)X_2(i)-C_y^\top(i)K_B^\top(i)X_2(i)-K_A^\top(i)X_2(i). \end{aligned}$$

Letting $Y(i)=X_2(i)K_A(i)$, $Z(i)=X_2(i)K_B(i)$, and $W(i)=K_C(i)$, we get

$$\begin{bmatrix} J_{X_1} & \mathcal{U}(i) & X_1(i)B(i) \\ \mathcal{U}^\top(i) & J_{X_2} & X_2(i)B(i)-Z(i)D_y(i) \\ B^\top(i)X_1(i) & B^\top(i)X_2(i)-D_y^\top(i)Z^\top(i) & -\gamma^2\mathbb{I}+\tilde\varepsilon_B(i)E_B^\top(i)E_B(i) \\ C_z(i)-W(i) & W(i) & D_z(i) \\ \mathscr{T}_1^\top(i) & \mathscr{T}_2^\top(i) & 0 \end{bmatrix}$$

$$\left.\begin{matrix} C_z^\top(i)-W^\top(i) & \mathscr{T}_1(i) \\ W^\top(i) & \mathscr{T}_2(i) \\ D_z^\top(i) & 0 \\ -\mathbb{I}+\varepsilon_{C_z}(i)D_{C_z}(i)D_{C_z}^\top(i) & 0 \\ 0 & -\mathcal{W} \end{matrix}\right] < 0,$$

where

$$\begin{aligned} J_{X_1} &= A^\top(i)X_1(i)+X_1(i)A(i)+\sum_{j=1}^N \lambda_{ij}X_1(j) \\ &\quad +\tilde\varepsilon_A(i)E_A^\top(i)E_A(i)+\tilde\varepsilon_{C_y}(i)E_{C_y}^\top(i)E_{C_y}(i), \\ J_{X_2} &= Y^\top(i)+Y(i)+\sum_{j=1}^N \lambda_{ij}X_2(j), \end{aligned}$$

$$\mathcal{U}(i) = A^\top(i)X_2(i) - C_y^\top(i)Z^\top(i) - Y^\top(i),$$
$$\mathcal{T}_1(i) = \left[\, X_1(i)D_A(i) \; 0 \; X_1(i)D_B(i) \; E_{C_z}^\top(i) \, \right],$$
$$\mathcal{T}_2(i) = \left[\, X_2(i)D_A(i) \; -Z(i)D_{C_y}(i) \; X_2(i)D_B(i) \; 0 \, \right].$$

The following theorem gives the results for the design of the gains of the $\mathcal{H}_\infty$ filter.

Theorem 77. *Let γ and R be, respectively, a given positive constant and a symmetric and positive-definite matrix representing the weighting of the initial conditions. If there exist sets of symmetric and positive matrices $X_1 = (X_1(1), \cdots, X_1(N)) > 0$, $X_2 = (X_2(1), \cdots, X_2(N)) > 0$, and matrices $Y = (Y(1), \cdots, Y(N))$, $Z = (Z(1), \cdots, Z(N))$, and $W = (W(1), \cdots, W(N))$, and sets of positive scalars $\tilde{\varepsilon}_A = (\tilde{\varepsilon}_A(1), \cdots, \tilde{\varepsilon}_A(N))$, $\tilde{\varepsilon}_B = (\tilde{\varepsilon}_B(1), \cdots, \tilde{\varepsilon}_B(N))$, $\tilde{\varepsilon}_C = (\tilde{\varepsilon}_C(1), \cdots, \tilde{\varepsilon}_C(N))$, and $\tilde{\varepsilon}_{C_y} = (\tilde{\varepsilon}_{C_y}(1), \cdots, \tilde{\varepsilon}_{C_y}(N))$ satisfying the following LMIs for every $i \in \mathscr{S}$ for all admissible uncertainties:*

$$\begin{bmatrix} J_{X_1} & \mathcal{U}(i) & X_1(i)B(i) \\ \mathcal{U}^\top(i) & J_{X_2} & X_2(i)B(i) - Z(i)D_y(i) \\ B^\top(i)X_1(i) & B^\top(i)X_2(i) - D_y^\top(i)Z^\top(i) & -\gamma^2\mathbb{I} + \tilde{\varepsilon}_B(i)E_B^\top(i)E_B(i) \\ C_z(i) - W(i) & W(i) & D_z(i) \\ \mathcal{T}_1^\top(i) & \mathcal{T}_2^\top(i) & 0 \end{bmatrix}$$
$$\begin{matrix} C_z^\top(i) - W^\top(i) & \mathcal{T}_1(i) \\ W^\top(i) & \mathcal{T}_2(i) \\ D_z^\top(i) & 0 \\ -\mathbb{I} + \tilde{\varepsilon}_{C_z}(i)D_{C_z}(i)D_{C_z}^\top(i) & 0 \\ 0 & -\mathcal{W} \end{matrix} \Bigg] < 0, \quad (5.36)$$

$$X_1(r_0) + X_2(r_0) < \gamma^2 R, \qquad (5.37)$$

then there exists a filter of the form (5.19) such that the estimation error is stochastically stable and bounded by

$$\|z - \hat{z}\|_2 \leq \gamma \left[\|w\|_2^2 + x_0^\top R x_0 \right]^{\frac{1}{2}}. \qquad (5.38)$$

The filter gains are given by

$$\begin{cases} K_A(i) = X_2^{-1}(i)Y(i), \\ K_B(i) = X_2^{-1}(i)Z(i), \\ K_C(i) = W(i). \end{cases} \qquad (5.39)$$

If the initial conditions are equal to zero, the previous results become easier and are given by the following corollary.

Corollary 17. *Let the initial conditions of system (5.1) be zero. Let γ and R be, respectively, a given positive constant and a symmetric and positive-definite matrix representing the weighting of the initial conditions. If there* *exist symmetric and positive matrices* $X_1 = (X_1(1), \cdots, X_1(N)) > 0$, $X_2 = (X_2(1), \cdots, X_2(N)) > 0$, *and matrices* $Y = (Y(1), \cdots, Y(N))$, $Z = (Z(1), \cdots, Z(N))$, *and* $W = (W(1), \cdots, W(N))$, *and sets of positive scalars* $\tilde{\varepsilon}_A = (\tilde{\varepsilon}_A(1), \cdots, \tilde{\varepsilon}_A(N))$, $\tilde{\varepsilon}_B = (\tilde{\varepsilon}_B(1), \cdots, \tilde{\varepsilon}_B(N))$, $\tilde{\varepsilon}_C = (\tilde{\varepsilon}_C(1), \cdots, \tilde{\varepsilon}_C(N))$, *and* $\tilde{\varepsilon}_{C_y} = (\tilde{\varepsilon}_{C_y}(1), \cdots, \tilde{\varepsilon}_{C_y}(N))$ *satisfying the LMIs* (5.36)–(5.37) *for every* $i \in \mathscr{S}$, *then there exists a filter of the form* (5.19) *such that the estimation error is stochastically stable and bounded by*

$$\|z - \hat{z}\|_2 \leq \gamma \|w\|_2. \tag{5.40}$$

The filter gains are given by

$$\begin{cases} K_A(i) = X_2^{-1}(i)Y(i), \\ K_B(i) = X_2^{-1}(i)Z(i), \\ K_C(i) = W(i). \end{cases} \tag{5.41}$$

The minimal noise attenuation level γ that can be verified by the filter of the form of (5.19) can be obtained by solving the following optimization problem:

$$\mathcal{P}_1 : \begin{cases} \min\limits_{\substack{v > 0, \\ X_1 = (X_1(1), \cdots, X_1(N)) > 0, \\ X_2 = (X_2(1), \cdots, X_2(N)) > 0, \\ Y = (Y(1), \cdots, Y(N)), \\ Z = (Z(1), \cdots, Z(N)), \\ W = (W(1), \cdots, W(N)),}} v, \\ \\ \text{s.t.} \\ \Psi_v(i) < 0, \\ X_1(r_0) + X_2(r_0) < vR, \end{cases}$$

where $\Psi_v(i)$ is obtained from (5.36) by replacing γ^2 by v. Thus, if the convex optimization problem $\mathcal{P}_1$ has a solution v, then by using Theorem 77, the corresponding error of the filter (5.19) is stable with noise attenuation level $\sqrt{v}$.

Example 67. To show the usefulness of the results on robust $\mathscr{H}_\infty$ filtering, let us consider the same system of Example 66 with the following extra data:

- mode #1:

$$D_A(1) = \begin{bmatrix} 0.1 \\ 0.01 \\ 0.0 \end{bmatrix}, \qquad E_A(1) = \begin{bmatrix} 0.1 & 0.2 & 0.0 \end{bmatrix},$$

$$D_B(1) = \begin{bmatrix} 0.1 \\ 0.01 \\ 0.0 \end{bmatrix}, \qquad E_B(1) = \begin{bmatrix} 0.1 \end{bmatrix},$$

$$D_{C_y}(1) = \begin{bmatrix} 0.2 \end{bmatrix}, \qquad E_{C_y}(1) = \begin{bmatrix} 0.1 \end{bmatrix},$$

$$E_{C_z}(1) = \begin{bmatrix} 0.01 & 0.2 & 0.02 \end{bmatrix}, \quad D_{C_z}(1) = \begin{bmatrix} 0.2 \end{bmatrix},$$

- mode #2:

$$D_A(2) = \begin{bmatrix} 0.2 \\ 0.1 \\ 0.0 \end{bmatrix}, \qquad E_A(2) = \begin{bmatrix} 0.1 \ 1 \ 0.02 \end{bmatrix},$$

$$D_B(2) = \begin{bmatrix} 0.1 \\ 1 \\ 0.0 \end{bmatrix}, \qquad E_B(2) = \begin{bmatrix} 0.02 \end{bmatrix},$$

$$D_{C_y}(2) = \begin{bmatrix} 0.1 \end{bmatrix}, \qquad E_{C_y}(2) = \begin{bmatrix} 0.01 \end{bmatrix},$$

$$E_{C_z}(2) = \begin{bmatrix} 0.0 \ 0.01 \ 0.1 \end{bmatrix}, \quad D_{C_z}(2) = \begin{bmatrix} 0.3 \end{bmatrix}.$$

The required positive scalars are

$$\varepsilon_A(1) = \varepsilon_A(2) = \varepsilon_B(1) = \varepsilon_B(2) = \varepsilon_{C_y}(1) = \varepsilon_{C_y}(2) = \varepsilon_{C_y}(1) = \varepsilon_{C_y}(2) = 0.1.$$

Letting $\gamma = 3.1$ and solving LMIs (5.36), we get

$$X_1(1) = \begin{bmatrix} 0.8091 & -0.0968 & -0.2500 \\ -0.0968 & 0.4640 & 0.0817 \\ -0.2500 & 0.0817 & 0.5950 \end{bmatrix},$$

$$X_1(2) = \begin{bmatrix} 0.8839 & -0.1779 & -0.3102 \\ -0.1779 & 0.2617 & 0.1025 \\ -0.3102 & 0.1025 & 0.4106 \end{bmatrix},$$

$$X_2(1) = \begin{bmatrix} 1.1596 & 0.0027 & -0.4042 \\ 0.0027 & 0.7079 & 0.1684 \\ -0.4042 & 0.1684 & 0.7864 \end{bmatrix},$$

$$X_2(2) = \begin{bmatrix} 1.1558 & -0.2274 & -0.5523 \\ -0.2274 & 0.2575 & 0.0930 \\ -0.5523 & 0.0930 & 0.5223 \end{bmatrix},$$

$$Y(1) = \begin{bmatrix} -1.9846 & 0.7822 & 0.6965 \\ -0.2953 & -1.3059 & 0.2741 \\ 0.2570 & -0.2890 & -1.4980 \end{bmatrix},$$

$$Y(2) = \begin{bmatrix} -4.2764 & -1.4911 & 0.3010 \\ -0.2820 & -1.9178 & -0.0669 \\ -2.4172 & -2.4012 & -3.2604 \end{bmatrix},$$

$$Z(1) = \begin{bmatrix} -0.0252 \\ 0.2978 \\ -0.1132 \end{bmatrix}, \qquad Z(2) = \begin{bmatrix} 4.4270 \\ 2.8044 \\ 4.9579 \end{bmatrix},$$

$$W(1) = \begin{bmatrix} 0.1212 \ 0.6145 \ 0.0194 \end{bmatrix},$$

$$W(2) = \begin{bmatrix} 0.4689 \ 0.2171 \ 0.1646 \end{bmatrix},$$

which gives the following gains:

$$K_A(1) = \begin{bmatrix} -1.9212 & 0.8501 & -0.1623 \\ -0.2661 & -1.9645 & 0.9071 \\ -0.6037 & 0.4902 & -2.1826 \end{bmatrix},$$

$$K_A(2) = \begin{bmatrix} -13.4326 & -9.6045 & -5.7118 \\ -6.5793 & -11.3307 & -0.9274 \\ -17.6598 & -12.7355 & -12.1169 \end{bmatrix},$$

$$K_B(1) = \begin{bmatrix} -0.1343 \\ 0.4972 \\ -0.3195 \end{bmatrix}, \qquad K_B(2) = \begin{bmatrix} 21.2821 \\ 19.3760 \\ 28.5458 \end{bmatrix},$$

$$K_C(1) = \begin{bmatrix} 0.1212 & 0.6145 & 0.0194 \end{bmatrix},$$

$$K_C(2) = \begin{bmatrix} 0.4689 & 0.2171 & 0.1646 \end{bmatrix}.$$

5.4 Notes

This chapter dealt with the filtering problem of the class of stochastic switching systems. Two types of filters have been studied and LMI-based results have been established to design these filters. The design of robust filters is also tackled and the LMI procedure to design filters established. Most of the results presented in this chapter are based on the work of the author and his coauthors.

6

Singular Stochastic Switching Systems

In the previous chapters we dealt with many classes of systems, including regular stochastic switching systems with and without Brownian disturbance. For these classes of systems we studied the stability and stabilization problems. Many stabilization techniques have been considered and most of the results we developed are in the LMI framework, which makes the results powerful and tractable.

In reality, not all the systems are regular and we may encounter physical systems that cannot be modeled by the previous class of systems. These systems may be modeled more adequately by the class of singular systems shown in Chapter 1 for electrical circuit. In the literature, these systems are also referred to as descriptor systems, implicit systems, generalized state-space systems, semi-state systems, or differential-algebraic systems. Singular systems arise in many practical systems such as electrical circuits, power systems, and networks (for more examples see [31] and the references therein). The goal of this chapter is to introduce the class of singular stochastic switching systems and consider some of the problems treated earlier to see how we can extend the previous results to this case.

The rest of this chapter is organized as follows. In Section 6.1, the different problems are stated and the necessary assumptions given. Section 6.2 deals with the stability problem of the class of singular stochastic switching systems, and LMI conditions are developed to check if a given system is stochastically stable. The robust stability problem is also considered. Section 6.3 treats the stabilization problem and its robustness. State feedback controller is considered and LMI design approaches are developed to synthesize the stabilizing and the robust stabilizing controllers.

6.1 Problem Statement

Let us consider a dynamical singular system defined in a probability space $(\Omega, \mathcal{F}, \mathbb{P})$ and assume that its dynamics are described by the following differential systems:

$$\begin{cases} E\dot{x}(t) = A(r(t),t)x(t) + B(r(t),t)u(t), \\ x(0) = x_0, \end{cases} \tag{6.1}$$

where $x(t) \in \mathbb{R}^n$ is the state vector; $x_0 \in \mathbb{R}^n$ is the initial state; $u(t) \in \mathbb{R}^m$ is the control input; $\{r(t), t \geq 0\}$ is a continuous-time Markov process taking values in a finite space $\mathscr{S} = \{1, 2, \cdots, N\}$ that describes the evolution of the mode at time t; E is a known singular matrix with rank $(E) = n_E < n$; and $A(r(t), t) \in \mathbb{R}^{n \times n}$ and $B(r(t), t) \in \mathbb{R}^{n \times m}$ are matrices with the following forms for every $i \in \mathscr{S}$:

$$\begin{aligned} A(i,t) &= A(i) + D_A(i)F_A(i,t)E_A(i), \\ B(i,t) &= B(i) + D_B(i)F_B(i,t)E_B(i), \end{aligned}$$

with $A(i)$, $D_A(i)$, $E_A(i)$, $B(i)$, $D_B(i)$ and $E_B(i)$ are real known matrices with appropriate dimensions; and $F_A(i,t)$ and $F_B(i,t)$ are unknown real matrices that satisfy

$$\begin{cases} F_A^\top(i,t)F_A(i,t) \leq \mathbb{I}, \\ F_B^\top(i,t)F_B(i,t) \leq \mathbb{I}. \end{cases} \tag{6.2}$$

The Markov process $\{r(t), t \geq 0\}$ that represents the switching between the different modes is described by the following probability transitions:

$$\mathbb{P}\left[r(t+h) = j | r(t) = i\right] = \begin{cases} \lambda_{ij}h + o(h), & \text{when } r(t) \text{ jumps from } i \text{ to } j , \\ 1 + \lambda_{ii}h + o(h), & \text{otherwise.} \end{cases} \tag{6.3}$$

where λ_{ij} is the transition rate from mode i to mode j with $\lambda_{ij} \geq 0$ when $i \neq j$ and $\lambda_{ii} = -\sum_{j=1, j \neq i}^N \lambda_{ij}$ and $o(h)$ is such that $\lim_{h \to 0} \frac{o(h)}{h} = 0$.

As we did in the previous chapters, we assume that the matrix Λ belongs to a polytope, that is,

$$\Lambda = \sum_{k=1}^\kappa \alpha_k \Lambda_k, \tag{6.4}$$

where κ is a positive given integer; $0 \leq \alpha_k \leq 1$, with $\sum_{k=1}^\kappa \alpha_k = 1$; and Λ_k is a known transition matrix and its expression is given by

$$\Lambda_k = \begin{bmatrix} \lambda_{11}^k & \cdots & \lambda_{1N}^k \\ \vdots & \ddots & \vdots \\ \lambda_{N1}^k & \cdots & \lambda_{NN}^k \end{bmatrix}, \tag{6.5}$$

where λ_{ij}^k keeps the same meaning as before.

Remark 23. The uncertainties that satisfy the conditions (6.2) and (6.4) are referred to as admissible. The uncertainty term in (6.2) is supposed to depend on the system mode $r(t)$ and on time t.

Throughout this chapter, we assume that the system state $x(t)$ and the system mode $r(t)$ are completely accessible for feedback when necessary.

Remark 24. The matrix E is supposed to be singular, which makes the dynamics (6.1) different from those usually used to describe the behavior of the time-invariant dynamical systems as considered in the previous chapters.

Remark 25. Notice that when E is not singular, (6.1) can be transformed easily to the class of Markov jump linear systems and the results developed earlier can be used to check the stochastic stability, design the appropriate controller that stochastically stabilizes this class of systems, and even design the appropriate filter.

The following definitions will be used in the rest of this chapter.

Definition 19. *(Dai [31])*

i.   *System (6.1) is said to be regular if the characteristic polynomial $\det(sE - A(i))$ is not identically zero for each mode $i \in \mathscr{S}$.*
ii. *System (6.1) is said to be impulse free, if $\deg(\det(sE - A(i))) = rank(E)$ for each mode $i \in \mathscr{S}$.*

Definition 20. *System (6.1) with $u(t) \equiv 0$ is said to be stochastically stable if there exists a constant $M(x_0, r_0) > 0$ such that the following holds for any initial conditions (x_0, r_0):*

$$\mathbb{E}\left[\int_0^\infty x^\top(t)x(t)|x_0, r_0 \right] \leq M(x_0, r_0). \tag{6.6}$$

Definition 21. *System (6.1) is said to be stochastically stabilizable if there exists a control*

$$u(t) = K(i)x(t), \tag{6.7}$$

with $K(i)$ a constant matrix such that the closed-loop system is stochastically stable.

The robust stochastic stability and the robust stochastic stabilizability are defined in a similar manner. For more details, we refer the reader to Chapter 2.

The goal of this chapter is to develop LMI-based stability conditions for system (6.1) with $u(t) \equiv 0$ and design a state feedback controller of the form (6.7) that stochastically stabilizes the class of systems under study. The robust stochastic stability and robust stabilization are also treated.

Before closing this section, let us give a transformation that facilitates the determination of the solution of system (6.1). For simplicity, let us assume that the uncertainties are all equal to zero either in the matrix $A(i)$ or the matrix $B(i)$ for all $i \in \mathscr{S}$. In fact, for singular systems, we know that we can find, using Jordan canonical form decomposition, nonsingular matrices $\widehat{M}(i)$ and $\widehat{N}(i)$ for this purpose (see Dai [31]). The searched transformation will divide the state vector into components named, respectively, slow and fast and denoted, respectively, by $\eta_1(t) \in \mathbb{R}^{n_1}$ and $\eta_2(t) \in \mathbb{R}^{n_2}$ with $n = n_1 + n_2$. This can be obtained by choosing

$$\widehat{N}(i)\eta(t) = x(t).$$

Using this transformation, the nominal dynamics become

$$E\widehat{N}(i)\dot{\eta}(t) = A(i)\widehat{N}(i)\eta(t) + B(i)u(t).$$

If we pre-multiply this equation by $\widehat{M}(i)$ and use the fact that

$$\widehat{M}(i)E\widehat{N}(i) = \begin{bmatrix} \mathbb{I} & \mathbf{0} \\ \mathbf{0} & N(i) \end{bmatrix},$$

$$\widehat{M}(i)A(i)\widehat{N}(i) = \begin{bmatrix} A_1(i) & \mathbf{0} \\ \mathbf{0} & \mathbb{I} \end{bmatrix},$$

$$\widehat{M}(i)B(i) = \begin{bmatrix} B_1(i) \\ B_2(i) \end{bmatrix},$$

where $N(i) \in \mathbb{R}^{n_2 \times n_2}$ a nilpotent matrix (i.e., a square matrix whose eigenvalues are all 0, it is also a square matrix, $N(i)$, such that $N^p(i) = 0$ for some positive integer power p) and $\mathbb{I} \in \mathbb{R}^{n_1 \times n_1}$ an identity matrix with appropriate dimension, with $n_1 + n_2 = n$, we get

$$\begin{bmatrix} \mathbb{I} & 0 \\ 0 & N(i) \end{bmatrix} \begin{bmatrix} \dot{\eta}_1(t) \\ \dot{\eta}_2(t) \end{bmatrix} = \begin{bmatrix} A_1(i) & 0 \\ 0 & \mathbb{I} \end{bmatrix} \begin{bmatrix} \eta_1(t) \\ \eta_2(t) \end{bmatrix} + \begin{bmatrix} B_1(i) \\ B_2(i) \end{bmatrix} u(t),$$

which can be rewritten as follows:

$$\dot{\eta}_1(t) = A_1(i)\eta_1(t) + B_1(i)u(t),$$
$$N(i)\dot{\eta}_2(t) = \eta_2(t) + B_2(i)u(t).$$

For the first differential equation, the solution is given by (Boukas [9])

$$\eta_1(t) = e^{A_1(i)t}\eta_1(0) + \int_0^t e^{A_1(i)(t-\tau)}B_1(i)u(\tau)d\tau,$$

which is completely determined by the initial conditions and the values of the control in the interval $[0, t]$.

For the second differential equation, let us assume that control $u(t)$ is p-times piecewise continuously differentiable. Therefore, by differentiating p-times and at each differentiation pre-multiplying by $N(i)$, we have

$$N(i)\eta_2^{(1)}(t) = \eta_2(t) + B_2(i)u(t),$$
$$N^2(i)\eta_2^{(2)}(t) = N^1(i)\eta_2^{(1)}(t) + N^1(i)B_2(i)u^{(1)}(t),$$
$$\vdots$$
$$N^p(i)\eta_2^{(p)}(t) = N^{p-1}(i)\eta_2^{(p-1)}(t) + N^{p-1}(i)B_2(i)u^{(p-1)}(t),$$

where $\eta_2^{(k)}(t)$ and $u^{(k)}(t)$ stand, respectively, for the k-times derivative of $\eta_2(t)$ and $u(t)$.

Using the fact that $N^p(i) = 0$ and by adding all these equations we get

$$0 = \eta_2(t) + \sum_{k=1}^{p-1} N^k(i)B_2(i)u^{(k)}(t),$$

which gives in turn

$$\eta_2(t) = -\sum_{k=1}^{p-1} N^k(i)B_2(i)u^{(k)}(t).$$

These equations are useful for stochastic stability. As when $u(t) = 0$ for all $t \geq 0$, the stochastic stability of the system (6.1) is brought to the one of the first differential equation related to the stochastic stability of the slow state variable linked to the matrix $A_1(i)$, $i \in \mathcal{S}$, since the second gives $\eta_2(t) = 0$ for all $t \geq 0$.

6.2 Stability Problem

This section addresses the stability problem of system (6.1) with $u(t) \equiv 0$.

Let us assume that $u(t) = 0$ for $t \geq 0$ and all the uncertainties on the state matrix are equal to zero, and study the stochastic stability of the nominal system. Our concern is to establish LMI conditions that can be used to check if a given dynamical system belonging to the class of systems we are considering in this chapter is stochastically stable. The following theorem states the first result of stability of such a class of systems.

Theorem 78. *System (6.1) is regular, impulse-free, and stochastically stable if there exists a set of nonsingular matrices $P = (P(1), \cdots, P(N))$ such that the following coupled LMIs hold for every $i \in \mathcal{S}$:*

$$\begin{cases} E^\top P(i) = P^\top(i)E \geq 0, \\ A^\top(i)P(i) + P^\top(i)A(i) + \sum_{j=1}^N \lambda_{ij} E^\top P(j) < 0. \end{cases} \tag{6.8}$$

Proof: Under the condition of Theorem 78, we first show the regularity and absence of impulses of system (6.1). Based on what was presented earlier, choose two nonsingular matrices $\widehat{M}$ and $\widehat{N}$ such that

$$\widehat{M}E\widehat{N} = \begin{bmatrix} \mathbb{I} & 0 \\ 0 & 0 \end{bmatrix}$$

and write

$$\widehat{M}A(i)\widehat{N} = \begin{bmatrix} \widehat{A}_1(i) & \widehat{A}_2(i) \\ \widehat{A}_3(i) & \widehat{A}_4(i) \end{bmatrix}, \quad \widehat{M}^{-\top}P(i)\widehat{N} = \begin{bmatrix} \widehat{P}_1(i) & \widehat{P}_2(i) \\ \widehat{P}_3(i) & \widehat{P}_4(i) \end{bmatrix}.$$

By the first relation of (6.8) it can be shown that $\widehat{P}_2(i) = 0$. Pre- and post-multiplying the second inequality of (6.8) by $\widehat{N}^\top$ and $\widehat{N}$, respectively, gives

$$\begin{bmatrix} * & * \\ * & \widehat{A}_4^\top(i)\widehat{P}_4(i) + \widehat{P}_4^\top(i)\widehat{A}_4(i) \end{bmatrix} < 0,$$

where $*$ will not be used in the following development. Then we have

$$\widehat{A}_4^\top(i)\widehat{P}_4(i) + \widehat{P}_4^\top(i)\widehat{A}_4(i) < 0,$$

which implies that $\widehat{A}_4(i)$ is nonsingular. System (6.1) is therefore regular and impulse-free.

Next we show the stochastic stability. Let $\mathscr{L}$ denote the weak infinitesimal generator of the Markov process $\{(x(t), r(t)), t \geq 0\}$. Let us consider the following Lyapunov function:

$$V(x(t), r(t)) = x^\top(t)E^\top P(r(t))x(t),$$

where $P(i)$, $i \in \mathscr{S}$, is a solution of (6.8).

The infinitesimal operator $\mathscr{L}$ of the Markov process $\{(x(t), r(t)), t \geq 0\}$ acting on $V(.)$ and emanating from the point (x, i) at time t, where $x(t) = x$ and $r(t) = i$ for $i \in \mathscr{S}$, is given by:

$$\mathscr{L}V(x(t), i) = \dot{x}^\top(t)E^\top P(i)x(t) + x^\top(t)E^\top P(i)\dot{x}(t)$$
$$+ \sum_{j=1}^{N} \lambda_{ij} x^\top(t)E^\top P(j)x(t).$$

Now if we use the fact that $E^\top P(i) = P^\top(i)E$ holds, we get

$$\mathscr{L}V(x(t), i) = \dot{x}^\top(t)E^\top P(i)x(t) + x^\top(t)P^\top(i)E\dot{x}(t)$$
$$+ \sum_{j=1}^{N} \lambda_{ij} x^\top(t)E^\top P(j)x(t)$$
$$= x^\top(t)\left[A^\top(i)P(i) + P^\top(i)A(i) + \sum_{j=1}^{N} \lambda_{ij}E^\top P(j) \right]x(t)$$
$$= x^\top(t)\Theta(i)x(t),$$

with $\Theta(i) = A^\top(i)P(i) + P^\top(i)A(i) + \sum_{j=1}^{N} \lambda_{ij} E^\top P(j)$.

Therefore, if the following holds:

$$A^\top(i)P(i) + P^\top(i)A(i) + \sum_{j=1}^{N} \lambda_{ij} E^\top P(j) < 0,$$

then

$$\mathscr{L}V(x(t), i) \le -\alpha \|x(t)\|^2,$$

where $\alpha = \min_{i \in \mathscr{S}}\{\lambda_{min}(-\Theta(i))\} > 0$.

Using Dynkin's formula, we obtain

$$\mathbb{E}[V(x(t), i)] - \mathbb{E}[V(x_0, r_0)] = \mathbb{E}\left[\int_0^t [\mathscr{L}V(x_s, r(s))] ds | x_0, r_0\right]$$

$$\le -\alpha \mathbb{E}\left[\int_0^t \|x_s\|^2 ds | x_0, r_0\right].$$

Since $\mathbb{E}[V(x(t), i)] \ge 0$, the last equation implies

$$\alpha \mathbb{E}\left[\int_0^t \|x_s\|^2 ds | x_0, r_0\right] \le \mathbb{E}[V(x(t), i)] + \alpha \mathbb{E}\left[\int_0^t \|x_s\|^2 ds | x_0, r_0\right]$$

$$\le \mathbb{E}[V(x_0, r_0)], \forall t > 0.$$

This proves that the system under study is stochastically stable and this completes the proof of Theorem 78.   □

Remark 26. Notice that when the matrix $E = \mathbb{I}$, the first condition of (6.8) becomes $P(i) = P^\top(i), i \in \mathscr{S}$, which means in this case that the matrix $P(i), i \in \mathscr{S}$ is symmetric and positive-definite and the results of the previous theorem become those of Chapter 2.

The above theorem provides LMI-based test conditions for system (6.1) to be stochastically stable. To illustrate the effectiveness of these results, let us give a numerical example.

Example 68. Consider a singular linear system with two modes, that is, $\mathscr{S} = \{1, 2\}$, and assume that its dynamics are described by (6.1) and its data are given by

- mode #1:

$$A(1) = \begin{bmatrix} -1.0 & 0.0 & -3.0 \\ 3.0 & 0.0 & -1.0 \\ 0.0 & -1.0 & 1.0 \end{bmatrix},$$

- mode #2:

$$A(2) = \begin{bmatrix} -1.0 & 0.0 & -2.0 \\ 2.0 & 0.0 & -1.0 \\ 0.0 & -1.0 & 1.0 \end{bmatrix}.$$

The switching between the two modes is described by the following transition rates:

$$\Lambda = \begin{bmatrix} -2 & 2 \\ 1 & -1 \end{bmatrix}.$$

The singular matrix E is given by

$$E = \begin{bmatrix} 1.0 & 0.0 & 0.0 \\ 0.0 & 0.0 & 1.0 \\ 0.0 & 0.0 & 0.0 \end{bmatrix}.$$

With the above data, solving LMIs (6.8) gives the following solution:

$$P(1) = \begin{bmatrix} 0.8210 & 0.0950 & 0.0000 \\ 0.0950 & 0.8762 & 0.0000 \\ 2.5594 & 0.5731 & -0.5290 \end{bmatrix},$$

$$P(2) = \begin{bmatrix} 0.8063 & 0.1086 & 0.0000 \\ 0.1086 & 0.9203 & 0.0000 \\ 1.7199 & 0.5175 & -0.5396 \end{bmatrix}.$$

Therefore, according to Theorem 78, the system under study is stochastically stable.

Let us now consider the effects of the uncertainties on the state matrix with $u(t) = 0$ for $t \geq 0$ and see how the stochastic stability conditions can be modified to guarantee that the system (6.1) is regular, impulse-free, and robustly stochastically stable for all admissible uncertainties. We are interested in establishing conditions that can help us to check if a given system with the dynamics (6.1) is regular, impulse-free, and robustly stochastically stable.

Based on Theorem 78, the free uncertain system (6.1) (with $u(t) \triangleq 0, \forall t \geq 0$) will be regular, impulse-free, and robustly stochastically stable if there exists a set of nonsingular matrices $P = (P(1), \cdots, P(N))$ such that the following holds for every $i \in \mathscr{S}$:

$$\begin{cases} E^{\top} P(i) = P^{\top}(i) E \geq 0, \\ P^{\top}(i) A(i, t) + A^{\top}(i, t) P(i) + \sum_{j=1}^{N} \lambda_{ij} E^{\top} P(j) < 0. \end{cases}$$

Notice that the second LMI can be rewritten as

$$P^{\top}(i) A(i) + A^{\top}(i) P(i) + \sum_{j=1}^{N} \lambda_{ij} E^{\top} P(j)$$

$$+P^\top(i)D_A(i)F_A(i)E_A(i) + E_A^\top(i)F_A^\top(i)D_A^\top(i)P(i) < 0.$$

Using Lemma 7 in Appendix A, we have

$$P^\top(i)D_A(i)F_A(i)E_A(i) + E_A^\top(i)F_A^\top(i)D_A^\top(i)P(i)$$
$$\le \varepsilon_A^{-1}(i)P^\top(i)D_A(i)D_A^\top(i)P(i) + \varepsilon_A(i)E_A^\top(i)E_A(i).$$

Using this inequality and the Schur complement, we get the sufficient conditions of the following theorem.

Theorem 79. *System (6.1) is robustly stochastically stable if there exist a set of nonsingular matrices $P = (P(1), \cdots, P(N))$ and a set of positive scalars $\varepsilon_A = (\varepsilon_A(1), \cdots, \varepsilon_A(N))$ such that the following set of LMIs holds for each $i \in \mathscr{S}$ and for all admissible uncertainties:*

$$\begin{cases} E^\top P(i) = P^\top(i)E \ge 0, \\ \begin{bmatrix} \tilde{J}(i) & P^\top(i)D_A(i) \\ D_A^\top(i)P(i) & -\varepsilon_A(i)\mathbb{I} \end{bmatrix} < 0, \end{cases} \tag{6.9}$$

with $\tilde{J}(i) = A^\top(i)P(i) + P^\top(i)A(i) + \sum_{j=1}^N \lambda_{ij}E^\top P(j) + \varepsilon_A(i)E_A^\top(i)E_A(i)$.

Example 69. To show the usefulness of the results of this theorem, let us consider the two-mode system of the previous example with the following data:

- mode #1:

$$D_A(1) = \begin{bmatrix} 0.1 \\ 0.2 \\ 0.0 \end{bmatrix},$$
$$E_A(1) = \begin{bmatrix} 0.2 & 0.1 & 0.1 \end{bmatrix},$$

- mode #2:

$$D_A(2) = \begin{bmatrix} 0.2 \\ 0.1 \\ 0.0 \end{bmatrix},$$
$$E_A(2) = \begin{bmatrix} 0.1 & 0.2 & 0.1 \end{bmatrix}.$$

Let $\varepsilon_A(1) = \varepsilon_A(2) = 0.1$ and solve the set of coupled LMIs (6.9) to give

$$P(1) = \begin{bmatrix} 0.2935 & 0.0189 & 0.0000 \\ 0.0189 & 0.2810 & 0.0000 \\ 0.9045 & 0.2037 & -0.1724 \end{bmatrix},$$

$$P(2) = \begin{bmatrix} 0.2694 & 0.0243 & 0.0000 \\ 0.0243 & 0.2953 & 0.0000 \\ 0.5620 & 0.1736 & -0.1808 \end{bmatrix},$$

which gives two nonsingular matrices. Therefore the system is robustly stochastically stable.

If we now consider uncertainties on system matrix and transition rates, we can easily establish the results in the following corollary.

Corollary 18. *System (6.1) is robustly stochastically stable if there exist a set of nonsingular matrices $P = (P(1), \cdots, P(N))$ and a set of positive scalars $\varepsilon_A = (\varepsilon_A(1), \cdots, \varepsilon_A(N))$ such that the following holds for each $i \in \mathscr{S}$ and for all admissible uncertainties:*

$$
\begin{cases}
E^\top P(i) = P^\top(i)E \\
\begin{bmatrix} \tilde{J}(i) & P(i)D_A(i) \\ D_A^\top(i)P(i) & -\varepsilon_A(i)\mathbb{I} \end{bmatrix} < 0,
\end{cases}
\tag{6.10}
$$

with

$$
\tilde{J}(i) = A^\top(i)P(i) + P^\top(i)A(i) + \sum_{k=1}^{\kappa}\sum_{j=1}^{N} \lambda_{ij}^k E^\top P(j) + \varepsilon_A(i)E_A^\top(i)E_A(i).
$$

Example 70. To show the usefulness of the results of this corollary, let us consider the two-mode system of the previous example with the following data:

$$
\Lambda_1 = \begin{bmatrix} -2 & 2 \\ 1 & -1 \end{bmatrix}, \alpha_1 = 0.6,
$$

$$
\Lambda_2 = \begin{bmatrix} -3 & 3 \\ 1 & -1 \end{bmatrix}, \alpha_2 = 0.4.
$$

Let $\varepsilon_A(1) = \varepsilon_A(2) = 0.1$ and solve the set of coupled LMIs (6.10) to give

$$
P(1) = \begin{bmatrix} 0.2928 & 0.0201 & 0.0000 \\ 0.0201 & 0.2820 & 0.0000 \\ 0.9039 & 0.2040 & -0.1738 \end{bmatrix}, \quad P(2) = \begin{bmatrix} 0.2685 & 0.0235 & 0.0000 \\ 0.0235 & 0.2929 & 0.0000 \\ 0.5597 & 0.1738 & -0.1793 \end{bmatrix},
$$

which gives two nonsingular matrices. Therefore the system is robustly stochastically stable.

Our goal in this section was to study the stochastic stability and robustness of the class of stochastic switching systems. Using a simple expression for the Lyapunov function and some algebraic calculations, we developed LMI conditions that can be used to check if a given system is stable. In the next section we establish design algorithms for stabilizing controllers.

6.3 Stabilization Problem

This section deals with the stabilization problem. We consider here only the design of the state feedback controller that achieves closed-loop dynamics that are regular, impulse-free, and stochastically stable. The block diagram of the closed-loop system under the state feedback controller is represented by Figure 6.1. As can be seen, complete access to the state vector and to the mode is crucial for this type of stabilization. Both the stabilization of the nominal system and the uncertain system are tackled in this chapter. For the stabilization problem of the nominal system, combining the system dynamics (6.1) and the controller expression (6.7) gives the following closed-loop dynamics:

$$\begin{cases} \dot{x}(t) = A_{cl}(r(t))x(t), \\ x(0) = x_0, \end{cases} \tag{6.11}$$

where $A_{cl}(r(t)) = A(r(t)) + B(r(t))K(r(t))$.

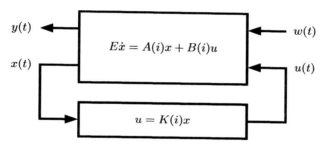

Fig. 6.1. State feedback stabilization block diagram (nominal system).

Based on the results of Theorem 78, these dynamics will be regular, impulse-free, and stochastically stable if there exists a set of nonsingular matrices $P = (P(1), \cdots, P(N))$ such that the following set of coupled LMIs holds for each $i \in \mathscr{S}$:

$$\begin{cases} E^\top P(i) = P^\top(i)E \geq 0, \\ P^\top(i)A_{cl}(i) + A_{cl}^\top(i)P(i) + \sum_{j=1}^{N} \lambda_{ij}E^\top P(j) < 0. \end{cases}$$

Pre- and post-multiply the second LMI, respectively, by $P^{-\top}(i)$ and $P^{-1}(i)$ to get

$$A_{cl}(i)P^{-1}(i) + P^{-\top}(i)A_{cl}^\top(i) + \sum_{\substack{j=1 \\ j \neq i}}^{N} \lambda_{ij}P^{-\top}(i)E^\top P(j)P^{-1}(i)$$

$$+ \lambda_{ii}P^{-\top}(i)E^\top < 0. \tag{6.12}$$

If the following holds for each $i \in \mathscr{S}$:

$$E^\top P(j) \le \epsilon_j P^\top(j) P(j), \epsilon_j > 0, \tag{6.13}$$

then a sufficient condition for (6.12) is

$$A_{cl}(i)P^{-1}(i) + P^{-\top}(i)A_{cl}^\top(i) + \lambda_{ii}P^{-\top}(i)E^\top$$
$$+ \sum_{\substack{j=1 \\ j\ne i}}^{N} \lambda_{ij}\epsilon_j P^{-\top}(i)P^\top(j)P(j)P^{-1}(i) < 0.$$

If we define

$$G_i = \left[\sqrt{\lambda_{i1}}P^{-\top}(i), \cdots, \sqrt{\lambda_{ii-1}}P^{-\top}(i), \sqrt{\lambda_{ii+1}}P^{-\top}(i), \cdots, \sqrt{\lambda_{iN}}P^{-\top}(i) \right],$$
$$J_i = \text{diag}\left[\epsilon_1^{-1}P^{-1}(1)P^{-\top}(1), \cdots, \epsilon_{i-1}^{-1}P^{-1}(i-1)P^{-\top}(i-1), \right.$$
$$\left. \epsilon_{i+1}^{-1}P^{-1}(i+1)P^{-\top}(i+1), \cdots, \epsilon_N^{-1}P^{-1}(N)P^{-\top}(N) \right],$$

then we obtain

$$\sum_{\substack{j=1 \\ j\ne i}}^{N} \lambda_{ij}\epsilon_j P^{-\top}(i)P^\top(j)P(j)P^{-1}(i) = G_i J_i^{-1} G_i^\top.$$

Using this we have

$$\begin{bmatrix} \tilde{J}_0(i) & G_i \\ G_i^\top & -J_i \end{bmatrix} < 0, \tag{6.14}$$

where $\tilde{J}_0(i) = A_{cl}(i)P^{-1}(i) + P^{-\top}(i)A_{cl}^\top(i) + \lambda_{ii}P^{-\top}(i)E^\top$.
Based on Lemma 9 in Appendix A, we get the following for each $i \in \mathscr{S}$:

$$\epsilon_i^{-1}P^{-1}(i)P^{-\top}(i) = P^{-1}(i)\left(\epsilon_i\mathbb{I}\right)^{-1}P^{-\top}(i) \ge P^{-1}(i) + P^{-\top}(i) - \epsilon_i\mathbb{I},$$

that is, $J_i \ge W_i$, which implies the following sufficient condition:

$$\begin{bmatrix} \tilde{J}_0(i) & G_i \\ G_i^\top & -W_i \end{bmatrix} < 0, \tag{6.15}$$

where

$$W_i = \text{diag}\left[P^{-1}(1) + P^{-\top}(1) - \epsilon_1\mathbb{I}, \cdots, P^{-1}(i-1) + P^{-\top}(i-1) - \epsilon_{i-1}\mathbb{I}, \right.$$
$$\left. P^{-1}(i+1) + P^{-\top}(i+1) - \epsilon_{i+1}\mathbb{I}, \cdots, P^{-1}(N) + P^{-\top}(N) - \epsilon_N\mathbb{I} \right].$$

Using the expression of $A_{cl}(i)$, letting $X(i) = P^{-1}(i)$, and $Y(i) = K(i)X(i)$, and noting that (6.13) can be rewritten as

$$X^\top(i)E^\top \le \epsilon_i\mathbb{I},$$

we get the following results for the stabilization.

Theorem 80. *There exists a state feedback controller of the form (6.7) such that the closed-loop system (6.1) is regular, impulse-free, and stochastically stable if there exist a set of nonsingular matrices $X = (X(1), \cdots, X(N))$, a set of matrices $Y = (Y(1), \cdots, Y(N))$, and a set of positive scalars $\epsilon = (\epsilon_1, \cdots, \epsilon_N)$ such that the following set of coupled LMIs holds for each $i \in \mathscr{S}$:*

$$\epsilon_i \mathbb{I} \geq X^\top(i)E^\top = EX(i) \geq 0, \tag{6.16}$$

$$\begin{bmatrix} \widehat{J}(i) & S_i(X) \\ S_i^\top(X) & -\mathcal{X}_i(X) \end{bmatrix} < 0, \tag{6.17}$$

where

$$\widehat{J}(i) = A(i)X(i) + X^\top(i)A^\top(i) + B(i)Y(i) + Y^\top(i)B^\top(i) + \lambda_{ii}X^\top(i)E^\top,$$
$$\mathcal{X}_i(X) = \text{diag}\left[X(1) + X^\top(1) - \epsilon_1\mathbb{I}, \cdots, X(i-1) + X^\top(i-1) - \epsilon_{i-1}\mathbb{I}, \right.$$
$$\left. X(i+1) + X^\top(i+1) - \epsilon_{i+1}\mathbb{I}, \cdots, X(N) + X^\top(N) - \epsilon_N\mathbb{I}\right],$$
$$S_i(X) = \left[\sqrt{\lambda_{i1}}X^\top(i), \cdots, \sqrt{\lambda_{ii-1}}X^\top(i), \sqrt{\lambda_{ii+1}}X^\top(i), \cdots, \sqrt{\lambda_{iN}}X^\top(i)\right].$$

The stabilizing controller gain is given by $K(i) = Y(i)X^{-1}(i)$, $i \in \mathscr{S}$.

Example 71. Let us consider a two-mode dynamical singular system with dynamics described by (6.1) and assume that the data are given by

- mode #1:

$$A(1) = \begin{bmatrix} 1.0 & 0.5 & 1.0 \\ -0.2 & 1.0 & 2.0 \\ 0.0 & 0.0 & 0.0 \end{bmatrix}, \quad B(1) = \begin{bmatrix} 0.0 & 0.2 \\ 1.0 & 0.0 \\ -0.1 & 1.0 \end{bmatrix},$$

- mode #2:

$$A(2) = \begin{bmatrix} -1.2 & 0.3 & 0.0 \\ 1.2 & 1.0 & 0.0 \\ 0.0 & 0.4 & 0.0 \end{bmatrix}, \quad B(2) = \begin{bmatrix} 0.0 & 0.2 \\ 1.2 & 0.0 \\ -0.1 & 1.2 \end{bmatrix}.$$

The switching between the two modes is described by the following transition matrix:

$$\Lambda = \begin{bmatrix} -2 & 2 \\ 1 & -1 \end{bmatrix}.$$

The singular matrix E is given by

$$E = \begin{bmatrix} 1 & 0 & 0 \\ 0 & 1 & 0 \\ 0 & 0 & 0 \end{bmatrix}.$$

Let us fix the parameters ϵ_1 and ϵ_2 to the following values:

$$\epsilon_1 = \epsilon_2 = 0.1.$$

With the above set of data, solve (6.16)–(6.17) to get the following solution:

$$X(1) = \begin{bmatrix} 0.0714 & -0.0015 & 0.0 \\ -0.0015 & 0.0743 & 0.0 \\ -0.0541 & 0.0050 & 0.1936 \end{bmatrix},$$

$$X(2) = \begin{bmatrix} 0.0855 & -0.0008 & 0.0 \\ -0.0008 & 0.0784 & 0.0 \\ -0.0556 & 0.0028 & 0.2090 \end{bmatrix},$$

$$Y(1) = \begin{bmatrix} 0.1640 & -0.1970 & -0.0000 \\ -0.1443 & -0.4094 & -0.2375 \end{bmatrix},$$

$$Y(2) = \begin{bmatrix} -0.1040 & -0.1620 & -0.0000 \\ -0.0159 & -0.0393 & -0.1702 \end{bmatrix}.$$

In view of Theorem 80, we conclude that the system under study is stochastically stabilizable and a set of stabilizing gains is given by

$$K(1) = \begin{bmatrix} 2.2430 & -2.6094 & -0.0000 \\ -3.0602 & -5.4891 & -1.2269 \end{bmatrix},$$

$$K(2) = \begin{bmatrix} -1.2355 & -2.0803 & -0.0000 \\ -0.7199 & -0.4804 & -0.8142 \end{bmatrix}.$$

Let us now consider robust stabilization using a state feedback controller of the form (6.7). As we did for the nominal case, let see how we can extend the previous results on the theorems on stabilization. Combining the expression of the controller and the dynamics of the system, we get the following closed-loop system:

$$\begin{aligned}
E\dot{x}(t) &= A(r(t),t)x(t) + B(r(t),t)K(r(t))x(t) \\
&= [A(r(t)) + B(r(t))K(r(t)) + D_A(r(t))F_A(r(t),t)E_A(r(t)) \\
&\quad + D_B(r(t))F_B(r(t),t)E_B(r(t))K(r(t))] x(t) \\
&= A_{cl}(r(t),t)x(t),
\end{aligned}$$

with

$$\begin{aligned}
A_{cl}(r(t),t) &= A(r(t)) + B(r(t))K(r(t)) + D_A(r(t))F_A(r(t),t)E_A(r(t)) \\
&\quad + D_B(r(t))F_B(r(t),t)E_B(r(t))K(r(t)).
\end{aligned}$$

Based on Theorem 80, the closed-loop uncertain system is regular, impulse-free, and stochastically stable if there exist a set of nonsingular matrices $X = (X(1), \cdots, X(N))$, a set of matrices $Y = (Y(1), \cdots, Y(N))$, and a set of positive scalars $\epsilon = (\epsilon_1, \cdots, \epsilon_N)$ such that the following set of coupled LMIs holds for each $i \in \mathscr{S}$:

$$\begin{cases} \varepsilon_i \mathbb{I} \geq X^\top(i)E^\top = EX(i) \geq 0, \\ \begin{bmatrix} \widehat{J}(i) & \mathcal{S}_i(X) \\ \mathcal{S}_i^\top(X) & -\mathcal{X}_i(X) \end{bmatrix} < 0, \end{cases}$$

where

$$\begin{aligned} \widehat{J}(i) &= [A(i) + D_A(i)F_A(i)E_A(i)]\, X(i) + X^\top(i)\, [A(i) + D_A(i)F_A(i)E_A(i)]^\top \\ &+ [B(i) + D_B(i)F_B(i)E_B(i)]\, Y(i) + Y^\top(i)\, [B(i) + D_B(i)F_B(i)E_B(i)]^\top \\ &+ \lambda_{ii} X^\top(i)E^\top, \\ \mathcal{X}_i(X) &= \mathrm{diag}\, \big[X(1) + X^\top(1) - \epsilon_1 \mathbb{I}, \cdots, X(i-1) + X^\top(i-1) - \epsilon_{i-1}\mathbb{I}, \\ & \qquad X(i+1) + X^\top(i+1) - \epsilon_{i+1}\mathbb{I}, \cdots, X(N) + X^\top(N) - \epsilon_N \mathbb{I} \big], \\ \mathcal{S}_i(X) &= \big[\sqrt{\lambda_{i1}}X^\top(i), \cdots, \sqrt{\lambda_{ii-1}}X^\top(i), \sqrt{\lambda_{ii+1}}X^\top(i), \cdots, \sqrt{\lambda_{iN}}X^\top(i)\big], \end{aligned}$$

and the controller gain in this case is given by $K(i) = Y(i)X^{-1}(i)$, $i \in \mathscr{S}$.

Using Lemma 7 in Appendix A, we get

$$\begin{aligned} & D_A(i)F_A(i)E_A(i)X(i) + X^\top(i)\, [D_A(i)F_A(i)E_A(i)]^\top \\ & \leq \varepsilon_A(i)D_A(i)D_A^\top(i) + \varepsilon_A^{-1}(i)X^\top(i)E_A^\top(i)E_A(i)X(i), \\ & D_B(i)F_B(i)E_B(i)Y(i) + Y^\top(i)\, [D_B(i)F_B(i)E_B(i)]^\top \\ & \leq \varepsilon_B(i)D_B(i)D_B^\top(i) + \varepsilon_B^{-1}(i)Y^\top(i)E_B^\top(i)E_B(i)Y(i). \end{aligned}$$

Using the Schur complement we have the following results.

Theorem 81. *There exists a state feedback controller of the form (6.7) such that the closed-loop system (6.1) is regular, impulse-free, and robustly sto-chastically stable if there exist a set of nonsingular matrices $X = (X(1), \cdots, X(N))$, a set of matrices $Y = (Y(1), \cdots, Y(N))$, and sets of positive scalars $\epsilon = (\epsilon_1, \cdots, \epsilon_N)$, $\varepsilon_A = (\varepsilon_A(1), \cdots, \varepsilon_A(N))$, and $\varepsilon_B = (\varepsilon_B(1), \cdots, \varepsilon_B(N))$ such that the following set of coupled LMIs holds for each $i \in \mathscr{S}$:*

$$\varepsilon_i \mathbb{I} \geq X^\top(i)E^\top = EX(i) \geq 0, \qquad (6.18)$$

$$\begin{bmatrix} \widehat{J}(i) & X^\top(i)E_A^\top(i) & Y^\top(i)E_B^\top(i) & \mathcal{S}_i(X) \\ E_A(i)X(i) & -\varepsilon_A(i)\mathbb{I} & 0 & 0 \\ E_B(i)Y(i) & 0 & -\varepsilon_B(i)\mathbb{I} & 0 \\ \mathcal{S}_i^\top(X) & 0 & 0 & -\mathcal{X}_i(X) \end{bmatrix} < 0, \qquad (6.19)$$

where

$$\begin{aligned} \widehat{J}(i) &= A(i)X(i) + X^\top(i)A^\top(i) + B(i)Y(i) + Y^\top(i)B^\top(i) \\ &+ \varepsilon_A(i)D_A(i)D_A^\top(i) + \varepsilon_B(i)D_B(i)D_B^\top(i) + \lambda_{ii}X^\top(i)E^\top, \\ \mathcal{X}_i(X) &= \mathrm{diag}\, \big[X(1) + X^\top(1) - \epsilon_1 \mathbb{I}, \cdots, X(i-1) + X^\top(i-1) - \epsilon_{i-1}\mathbb{I}, \\ & \qquad X(i+1) + X^\top(i+1) - \epsilon_{i+1}\mathbb{I}, \cdots, X(N) + X^\top(N) - \epsilon_N \mathbb{I} \big], \end{aligned}$$

$$\mathcal{S}_i(X) = \left[\sqrt{\lambda_{i1}}X^\top(i), \cdots, \sqrt{\lambda_{ii-1}}X^\top(i), \sqrt{\lambda_{ii+1}}X^\top(i), \cdots, \sqrt{\lambda_{iN}}X^\top(i)\right].$$

The stabilizing controller gain is given by $K(i) = Y(i)X^{-1}(i)$, $i \in \mathscr{S}$.

Example 72. To show the usefulness of this theorem, let us consider the two-mode system of Example 71 with the following extra data:

- mode #1:

$$D_A(1) = \begin{bmatrix} 0.1 \\ 0.2 \\ 0.1 \end{bmatrix}, \quad E_A(1) = \begin{bmatrix} 0.2 & 0.1 & 0.1 \end{bmatrix},$$

$$D_B(1) = \begin{bmatrix} 0.2 \\ 0.1 \\ 0.1 \end{bmatrix}, \quad E_B(1) = \begin{bmatrix} 0.1 & 0.2 \end{bmatrix},$$

- mode #2:

$$D_A(2) = \begin{bmatrix} 0.1 \\ 0.2 \\ 0.1 \end{bmatrix}, \quad E_A(2) = \begin{bmatrix} 0.2 & 0.1 & 0.1 \end{bmatrix},$$

$$D_B(2) = \begin{bmatrix} 0.2 \\ 0.1 \\ 0.1 \end{bmatrix}, \quad E_B(2) = \begin{bmatrix} 0.1 & 0.2 \end{bmatrix}.$$

The matrix E is given by the following expression:

$$E = \begin{bmatrix} 1.0 & 0.0 & 0.0 \\ 0.0 & 0.0 & 1.0 \\ 1.0 & 0.0 & 0.0 \end{bmatrix}.$$

Fix the parameters ϵ_1, ϵ_2, $\varepsilon_A(1)$, $\varepsilon_A(2)$, $\varepsilon_B(1)$, and $\varepsilon_B(2)$ to the following values:

$$\epsilon_1 = \epsilon_2 = 0.1,$$
$$\varepsilon_A(1) = \varepsilon_A(2) = 0.1,$$
$$\varepsilon_B(1) = \varepsilon_B(2) = 0.1.$$

Solving LMIs (6.18)–(6.19) gives

$$X(1) = \begin{bmatrix} 0.0697 & -0.0010 & 0.0 \\ -0.0010 & 0.0730 & 0.0 \\ -0.0517 & -0.0014 & 0.1733 \end{bmatrix},$$

$$X(2) = \begin{bmatrix} 0.0856 & -0.0008 & 0.0 \\ -0.0008 & 0.0772 & 0.0 \\ -0.0552 & 0.0022 & 0.1985 \end{bmatrix},$$

$$Y(1) = \begin{bmatrix} 0.1028 & -0.1850 & -0.2393 \\ -0.1330 & -0.1397 & -0.2571 \end{bmatrix},$$

$$Y(2) = \begin{bmatrix} -0.1033 & -0.1601 & 0.0104 \\ -0.0121 & -0.0609 & -0.1789 \end{bmatrix},$$

which gives the following gains:

$$K(1) = \begin{bmatrix} 0.4145 & -2.5563 & -1.3807 \\ -3.0365 & -1.9826 & -1.4835 \end{bmatrix},$$

$$K(2) = \begin{bmatrix} -1.1926 & -2.0862 & 0.0522 \\ -0.7296 & -0.7696 & -0.9015 \end{bmatrix}.$$

As we did for the stochastic stability case, let us also consider the case where we have uncertainties on the system matrices and on the transition rates. In this case, the corresponding results can be established following the same steps as before. These results are given by the following corollary.

Corollary 19. *There exists a state feedback controller of the form (6.7) such that the closed-loop system (6.1) is regular, impulse-free, and robustly stochastically stable if there exist a set of matrices* $X = (X(1), \cdots, X(N))$, $Y = (Y(1), \cdots, Y(N))$, *and sets of positive scalars* $\epsilon = (\epsilon_1, \cdots, \epsilon_N)$, $\varepsilon_A = (\varepsilon_A(1), \cdots, \varepsilon_A(N))$, *and* $\varepsilon_B = (\varepsilon_B(1), \cdots, \varepsilon_B(N))$ *such that the following set of coupled LMIs holds for each* $i \in \mathscr{S}$:

$$\epsilon_i \mathbb{I} \geq X^\top(i)E^\top = EX(i) \geq 0, \tag{6.20}$$

$$\begin{bmatrix} \widehat{J}(i) & X^\top(i)E_A^\top(i) & Y^\top(i)E_B^\top(i) & S_i(X) \\ E_A(i)X(i) & -\varepsilon_A(i)\mathbb{I} & 0 & 0 \\ E_B(i)Y(i) & 0 & -\varepsilon_B(i)\mathbb{I} & 0 \\ S_i^\top(X) & 0 & 0 & -\mathcal{X}_i(X) \end{bmatrix} < 0, \tag{6.21}$$

where

$$\widehat{J}(i) = A(i)X(i) + X^\top(i)A^\top(i) + B(i)Y(i) + Y^\top(i)B^\top(i)$$

$$+\varepsilon_A(i)D_A(i)D_A^\top(i) + \varepsilon_B(i)D_B(i)D_B^\top(i) + \sum_{k=1}^{\kappa} \alpha_k \lambda_{ii}^k X^\top(i)E^\top,$$

$$\mathcal{X}_i(X) = \text{diag}\left[X(1) + X^\top(1) - \epsilon_1 \mathbb{I}, \cdots, X(i-1) + X^\top(i-1) - \epsilon_{i-1}\mathbb{I}, \right.$$

$$\left. X(i+1) + X^\top(i+1) - \epsilon_{i+1}\mathbb{I}, \cdots, X(N) + X^\top(N) - \epsilon_N\mathbb{I} \right],$$

$$S_i(X) = \left[\sqrt{\sum_{k=1}^{\kappa} \alpha_k \lambda_{i1}^k} X^\top(i), \cdots, \sqrt{\sum_{k=1}^{\kappa} \alpha_k \lambda_{ii-1}^k} X^\top(i), \sqrt{\sum_{k=1}^{\kappa} \alpha_k \lambda_{ii+1}^k} X^\top(i), \right.$$

$$\left. \cdots, \sqrt{\sum_{k=1}^{\kappa} \alpha_k \lambda_{iN}^k} X^\top(i) \right].$$

The stabilizing controller gain is given by $K(i) = Y(i)X^{-1}(i)$, $i \in \mathscr{S}$.

Example 73. To show the usefulness of this theorem, let us consider the two-mode system of Example 72 with the following extra data:

$$\Lambda_1 = \begin{bmatrix} -1 & 1 \\ 1 & -1 \end{bmatrix}, \alpha_1 = 0.6,$$

$$\Lambda_2 = \begin{bmatrix} -2 & 2 \\ 1 & -1 \end{bmatrix}, \alpha_2 = 0.4.$$

Solving LMIs (6.20)–(6.21) gives

$$X(1) = \begin{bmatrix} 0.0714 & -0.0010 & 0.0 \\ -0.0010 & 0.0735 & 0.0 \\ -0.0592 & -0.0013 & 0.1853 \end{bmatrix},$$

$$X(2) = \begin{bmatrix} 0.0819 & -0.0008 & 0.0 \\ -0.0008 & 0.0754 & 0.0 \\ -0.0526 & 0.0017 & 0.1927 \end{bmatrix},$$

$$Y(1) = \begin{bmatrix} 0.1168 & -0.1818 & -0.2684 \\ -0.1417 & -0.1378 & -0.2367 \end{bmatrix},$$

$$Y(2) = \begin{bmatrix} -0.1002 & -0.1521 & 0.0058 \\ -0.0177 & -0.0545 & -0.1657 \end{bmatrix},$$

which gives the following gains:

$$K(1) = \begin{bmatrix} 0.4024 & -2.4926 & -1.4486 \\ -3.0709 & -1.9355 & -1.2778 \end{bmatrix},$$

$$K(2) = \begin{bmatrix} 1.2220 & -2.0290 & 0.0303 \\ -0.7749 & -0.7108 & -0.8601 \end{bmatrix}.$$

6.4 Constant Gain Stabilization

Previously we assumed complete access to the system mode $r(t)$ at each time to compute the state feedback controller of the form (6.7). This assumption could for some physical reasons be violated making the developed results unusable. To overcome this, we can design a constant gain state feedback controller that does not require access to the mode. The structure of such controller is given by the following form:

$$u(t) = \mathcal{K}x(t), \tag{6.22}$$

where $\mathcal{K}$ is a common gain for all the modes to be determined.

Remark 27. Since the gain $\mathcal{K}$ is common to all the modes, we do not need to switch the gain as done previously when the system mode switches from one mode to another. Notice that this may be restrictive in some cases.

Before designing such a controller, let us modify our previous stochastic stability conditions.

Corollary 20. *System (6.1) is regular, impulse-free, and stochastically stable if there exists a nonsingular matrix P such that the following coupled LMIs hold for every $i \in \mathscr{S}$:*

$$\begin{cases} E^\top P = P^\top E, \\ A^\top(i)P + P^\top A(i) < 0. \end{cases} \tag{6.23}$$

Proof: The proof of this corollary is similar to the one of Theorem 78. In fact, if we choose a Lyapunov candidate function of the following form:

$$V(x(t), r(t) = i) = x^\top(t)Px(t), \quad \text{when} \quad r(t) = i,$$

we get from Theorem 78:

$$\begin{cases} E^\top P = P^\top E, \\ A^\top(i)P + P^\top A(i) + \sum_{j=1}^{N} \lambda_{ij} E^\top P < 0. \end{cases}$$

And since $\sum_{j=1}^{N} \lambda_{ij} = 0$, we get the results of the corollary. $\square$
For the robust stochastic stability, we get the following results.

Corollary 21. *System (6.1) is robustly stochastically stable if there exist a nonsingular matrix P and a set of positive scalars $\varepsilon_A = (\varepsilon_A(1), \cdots, \varepsilon_A(N))$ such that the following set of LMIs holds for each $i \in \mathscr{S}$ and for all admissible uncertainties:*

$$\begin{cases} E^\top P = P^\top E \geq 0, \\ \begin{bmatrix} \tilde{J}(i) & P^\top D_A(i) \\ D_A^\top(i)P & -\varepsilon_A(i)\mathbb{I} \end{bmatrix} < 0, \end{cases} \tag{6.24}$$

with $\tilde{J}(i) = A^\top(i)P + P^\top A(i) + \varepsilon_A(i)E_A^\top(i)E_A(i)$.

Proof: The proof of this corollary is similar to the one of Theorem 79 and the details are omitted. $\square$
With these results, let us now focus on the design of a constant gain stabilizing controller for the nominal system and the uncertain system. Let us first handle the nominal case. Plugging the controller expression in the system dynamics gives

$$\begin{aligned} E\dot{x}(t) &= A(i)x(t) + B(i)\mathcal{K}x(t), \\ &= [A(i) + B(i)\mathcal{K}]x(t), \\ &= A_{cl}(i)x(t). \end{aligned}$$

Based on Corollary 20, the closed-loop dynamics will be regular, impulse-free, and stochastically stable if there exists a nonsingular matrix P such that the following set of LMIs holds for each $i \in \mathscr{S}$:

$$\begin{cases} E^{\top} P = P^{\top} E, \\ A_{cl}^{\top}(i)P + P^{\top} A_{cl}(i) < 0, \end{cases}$$

which gives in turn:

$$\begin{cases} E^{\top} P = P^{\top} E, \\ A^{\top}(i)P + P^{\top} A(i) + P^{\top} B(i)\mathcal{K} + \mathcal{K}^{\top} B^{\top}(i)P < 0. \end{cases}$$

Notice that the second matrix inequality of these conditions is nonlinear in the design parameters $\mathcal{K}$ and P. To put it in the LMI setting, pre- and post-multiply this matrix inequality, respectively, by $P^{-\top}$ and P^{-1} to get

$$P^{-\top} A^{\top}(i) + A(i)P^{-1} + P^{-\top}\mathcal{K}^{\top} B^{\top}(i) + B(i)\mathcal{K}P^{-1} < 0.$$

Letting $X = P^{-1}$ and $Y = \mathcal{K}X$ gives

$$X^{\top} A^{\top}(i) + A(i)X + Y^{\top} B^{\top}(i) + B(i)Y < 0.$$

The results of this development are summarized by the following corollary.

Corollary 22. *There exists a state feedback controller of the form (6.22) such that the closed-loop system (6.1) is regular, impulse-free, and stochastically stable if there exist a nonsingular matrix X and a matrix Y such that the following set of coupled LMIs holds for each $i \in \mathscr{S}$:*

$$X^{\top} E^{\top} = EX \geq 0, \tag{6.25}$$
$$A(i)X + X^{\top} A^{\top}(i) + B(i)Y + Y^{\top} B^{\top}(i) < 0. \tag{6.26}$$

The stabilizing controller gain is given by $\mathcal{K} = YX^{-1}$.

For the robust constant gain stabilization, the closed-loop state equation will be regular, impulse-free, and robustly stochastically stable for all the admissible uncertainties if there exists a nonsingular matrix P such that the following set of LMIs holds for every $i \in \mathscr{S}$ and for all admissible uncertainties:

$$\begin{cases} E^{\top} P = P^{\top} E \geq 0, \\ \bar{A}^{\top} P + P^{\top} \bar{A}(i) < 0, \end{cases}$$

with $\bar{A}(i) = A(i) + D_A(i)F_A(i)E_A(i) + B\mathcal{K} + D_B(i)F_B(i)E_B(i)\mathcal{K}$.

The second matrix inequality can be rewritten as follows:

$$A^{\top}(i)P + P^{\top} A(i) + P^{\top} B\mathcal{K} + \mathcal{K}^{\top} B^{\top}(i)P + P^{\top} D_A(i)F_A(i)E_A(i)$$

$$+E_A^\top(i)F_A^\top(i)D_A^\top(i)P + P^\top D_B(i)F_B(i)E_B(i)\mathcal{K}$$
$$+\mathcal{K}^\top E_B^\top(i)F_B^\top(i)D_B^\top(i)P < 0.$$

Pre- and post-multiply this matrix inequality, respectively, by $P^{-\top}$ and P^{-1} to give

$$P^{-\top}A^\top(i) + A(i)P^{-1} + B\mathcal{K}P^{-1} + P^{-\top}\mathcal{K}^\top B^\top(i) + D_A(i)F_A(i)E_A(i)P^{-1}$$
$$+P^{-\top}E_A^\top(i)F_A^\top(i)D_A^\top(i) + D_B(i)F_B(i)E_B(i)\mathcal{K}P^{-1}$$
$$+P^{-\top}\mathcal{K}^\top E_B^\top(i)F_B^\top(i)D_B^\top(i) < 0.$$

Letting $X = P^{-1}$, $Y = \mathcal{K}X$, and using Lemma 7 in Appendix A, we get

$$D_A(i)F_A(i)E_A(i)X + X^\top E_A^\top(i)F_A^\top(i)D_A^\top(i)$$
$$\leq \varepsilon_A(i)D_A(i)D_A^\top(i) + \varepsilon_A^{-1}(i)X^\top E_A^\top(i)E_A(i)X,$$
$$D_B(i)F_B(i)E_B(i)Y + Y^\top E_B^\top(i)F_B^\top(i)D_B^\top(i)$$
$$\leq \varepsilon_B(i)D_B(i)D_B^\top(i) + \varepsilon_B^{-1}(i)Y^\top E_B^\top(i)E_B(i)Y.$$

Using this and the Schur complement we get the following results.

Corollary 23. *There exists a state feedback controller of the form (6.22) such that the closed-loop system (6.1) is regular, impulse-free, and robustly stochastically stable if there exist a nonsingular matrix X, a matrix Y, and sets of positive scalars $\varepsilon_A = (\varepsilon_A(1), \cdots, \varepsilon_A(N))$ and $\varepsilon_B = (\varepsilon_B(1), \cdots, \varepsilon_B(N))$ such that the following set of coupled LMIs holds for each $i \in \mathscr{S}$ and for all admissible uncertainties:*

$$X^\top E^\top = EX \geq 0, \tag{6.27}$$

$$\begin{bmatrix} \tilde{J}(i) & X^\top E_A^\top(i) & Y^\top E_B^\top(i) \\ E_A(i)X & -\varepsilon_A(i)\mathbb{I} & 0 \\ E_B(i)Y & 0 & -\varepsilon_B(i)\mathbb{I} \end{bmatrix} < 0, \tag{6.28}$$

with

$$J(i) = A(i)X + X^\top A^\top(i) + B(i)Y + Y^\top B^\top(i) + \varepsilon_A(i)D_A(i)D_A^\top(i)$$
$$+\varepsilon_B(i)D_B(i)D_B^\top(i).$$

The stabilizing controller gain is given by $K = YX^{-1}$.

Remark 28. When the uncertainties are on the transition rate matrix, the results of the previous corollaries will remain valid and can be used either to check the robust stochastic stability or to design the constant gain state feedback controller.

6.5 Notes

This chapter dealt with the singular class of stochastic switching systems. The stability problem and the stabilization problem have been considered and LMI conditions were developed. The conditions we developed in this chapter are tractable using commercial optimization tools. The content of this chapter is mainly based on the work of the author and his coauthors.

Appendix A

Mathematical Review

Our goal is to make this book self-contained and easy to understand. This appendix recalls some concepts used in this book and is organized as follows. In Section A.1, basic concepts on linear algebra are given. In Section A.2, matrix theory is reviewed and some important results are given to facilitate the understanding of the results developed in this book. In Section A.3 we present the stochastic processes and their links to what we are treating in this volume. In Section A.4, some important lemmas that are useful for our analysis and design are presented and some proofs are given.

A.1 Linear Algebra

In this section some important concepts in real analysis are reviewed. The material presented here is only an introduction to some concepts and for more details we refer the reader to the appropriate books.

Let $\mathcal{K}$ be the field of real numbers and $\mathcal{B}$ a class of objects. $\mathcal{B}$ is a linear space over the field $\mathcal{K}$ if the following hold:

- sum: for any x and y belonging to $\mathcal{B}$, their sum is also an element of $\mathcal{B}$, i.e.,

$$\forall x, y \in \mathcal{B}, x + y \in \mathcal{B},$$

- product: for any $x \in \mathcal{B}$, its product by any scalar $\alpha \in \mathcal{K}$ is also an element of $\mathcal{B}$, i.e.,

$$\forall x \in \mathcal{B}, \forall \alpha \in \mathcal{K}, \alpha x \in \mathcal{B},$$

- commutative:

$$\forall x, y \in \mathcal{B}, x + y = y + x,$$

- associative:

$$\forall x, y, z \in \mathscr{B}, [x + y] + z = x + [y + z],$$

$$\forall \alpha, \beta \in \mathscr{K}, \forall x \in \mathscr{X}, [\alpha\beta] x = \alpha [\beta x],$$

- zero element:

$$\forall x \in \mathscr{B}, \exists v \in \mathscr{B}, x + v = x,$$

- distributivity of sum/product:

$$\forall x, y \in \mathscr{B}, \forall \alpha \in \mathscr{K}, \alpha [x + y] = \alpha x + \alpha y,$$

- distributivity of product/sum:

$$\forall \alpha, \beta \in \mathscr{K}, \forall x \in \mathscr{X}, [\alpha + \beta] x = \alpha x + \beta x.$$

The elements of $\mathscr{B}$ are called vectors and any subset of $\mathscr{B}$ that is a linear space is called a subspace. Therefore, for a given subspace $\mathscr{B}_0$, if we have n vectors, $x_1, \cdots, x_n$ we have the following:

- a vector $x \in \mathscr{B}_0$ is a linear combination of vectors $x_1, \cdots, x_n$ that belong to $\mathscr{B}_0$ if the following holds:

$$x = \alpha_1 x_1 + \cdots + \alpha_n x_n = \sum_{j=1}^{n} \alpha_j x_j,$$

- vectors $x_1, \cdots, x_n$ are linearly dependent if and only if

$$0 = \alpha_1 x_1 + \cdots + \alpha_n x_n = \sum_{j=1}^{n} \alpha_j x_j,$$

and there is some $\alpha_j \neq 0$,
- vectors $x_1, \cdots, x_n$ are linearly independent if

$$0 = \alpha_1 x_1 + \cdots + \alpha_n x_n = \sum_{j=1}^{n} \alpha_j x_j,$$

which implies $\alpha_j = 0$ for $j = 1, \cdots, n$. In this case, the set of vectors $x_1, \cdots, x_n$ is called a basis of the subspace $\mathscr{B}_0$. Also, the maximum number of linearly independent vectors is called the dimension of the considered linear space.

For two vectors x and y with n components, respectively, the scalar product, which is also called the inner product, is defined as

$$\langle x, y \rangle \doteq \sum_{j=1}^{n} x_j y_j = x^\top y = y^\top x.$$

The scalar product satisfies the following properties:

- $\langle x, x \rangle > 0$ for all $x \neq 0$,
- $\langle x, y \rangle = \prec y, x \succ$,
- $\langle \alpha x + \beta y, z \rangle = \alpha \langle x, z \rangle + \beta \langle y, z \rangle$ for all α and β in $\mathbb{R}$.

When two vectors x and y are orthogonal, we have $\prec x, y \succ = \prec y, x \succ = 0$.

Let x be a vector that belongs to the subspace $\mathscr{B}_0$. The norm of x denoted by $\|x\|$ is a nonnegative-valued function of x satisfying the following properties:

- $\|x\| \geq 0$ for all $x \in \mathscr{B}_0$ and $\|x\| = 0$ if and only if $x = 0$,
- $\|\alpha x\| = |\alpha| \|x\|$ for all α elements of $\mathbb{R}$ and $x \in \mathscr{B}_0$,
- $\|x + y\| \leq \|x\| + \|y\|$ for all x and y belonging to the subspace $\mathscr{B}_0$.

The subspace is called normed linear space if for every x in this subspace the norm of x is finite.

The usual norms are

- $\|x\|_1 = \sum_{j=1}^{n} |x_j|$, 1-norm,
- $\|x\|_2 = \sqrt{x^\top x} = \left[\sum_{j=1}^{n} x_j^2 \right]^{\frac{1}{2}}$, 2-norm or Euclidean norm,
- $\|x\|_p = \left[\sum_{j=1}^{n} |x_j|^p \right]^{\frac{1}{p}}$, p-norm $(1 \leq p < \infty)$,
- $\|x\|_\infty = \sup_j |x_j|$, ∞-norm.

Two norms a and b are equivalent if there exist two positive constants M and N such that the following holds:

$$N\|x\|_a \leq \|x\|_b \leq M\|x\|_a.$$

For vector functions, we can also talk about norms. Let $\mathscr{B}$ be a linear space of Lebesgue-measurable functions defined on $\mathbb{R}$. The L_p-norm of a function $f(t) \in \mathscr{B}$ is defined by

$$\|f\|_{L_p} = \left[\int_{-\infty}^{\infty} |f(t)|^p dt \right]^{\frac{1}{p}}, \text{ when } 1 \leq p < \infty,$$

$$\|f\|_{L_\infty} = \sup_{-\infty < t < \infty} |f(t)|, \text{ when } p = \infty.$$

A.2 Matrix Theory

In this section we recall certain concepts on matrix theory. The material here is only an introduction to the matrix theory. For more details on the different concepts, we refer the reader to the appropriate books.

A rectangular (real or complex) matrix with n rows and m columns is defined as follows:

$$A = \begin{bmatrix} a_{11} & a_{12} & \cdots & a_{1m} \\ a_{21} & a_{22} & \cdots & a_{2m} \\ \vdots & \vdots & \ddots & \vdots \\ a_{n1} & a_{n2} & \cdots & a_{nm} \end{bmatrix}.$$

When the number of rows is equal to the number of columns, that is, $n = m$, the real matrix is called a square matrix. The transpose of the matrix A is obtained by interchanging the rows and columns, that is,

$$A^{\top} = \begin{bmatrix} a_{11} & a_{21} & \cdots & a_{n1} \\ a_{12} & a_{22} & \cdots & a_{n2} \\ \vdots & \vdots & \ddots & \vdots \\ a_{1m} & a_{2m} & \cdots & a_{nm} \end{bmatrix}.$$

The conjugate matrix $\bar{A}$ is obtained from the matrix A (when A is a complex matrix) by replacing every element of the matrix A by its complex conjugate, that is,

$$\bar{A} = \begin{bmatrix} \bar{a}_{11} & \bar{a}_{12} & \cdots & \bar{a}_{1m} \\ \bar{a}_{21} & \bar{a}_{22} & \cdots & \bar{a}_{2m} \\ \vdots & \vdots & \ddots & \vdots \\ \bar{a}_{n1} & \bar{a}_{n2} & \cdots & \bar{a}_{nm} \end{bmatrix}.$$

A square matrix A is

- symmetric if $A = A^{\top}$,
- Hermitian if $A = \bar{A}^{\top}$,
- skew-symmetric if $A = -A^{\top}$,
- skew-Hermitian if $A = -\bar{A}^{\top}$.

The minor $M(i, j)$ of a square matrix A of size n is the determinant of $(n-1) \times (n-1)$ matrix formed from A by crossing out the ith row and the jth column. Notice also that each element a_{ij} has a cofactor denoted by $C(i, j)$ that differs from $M(i, j)$ at most by a sign change, that is,

$$C(i, j) = (-1)^{i+j} M(i, j).$$

The determinant of a square matrix A of size n is defined by

$$\det(A) = \sum_{j=i}^{n} a_{jk} C(j, k).$$

A singular matrix A has a determinant equal to zero, $\det(A) = 0$, and a nonsingular matrix A has a nonzero determinant, $\det(A) \neq 0$.

The inverse of a square matrix, A (when it exists), denoted by A^{-1} is defined as follows:

$$A^{-1} = \frac{A^+}{\det(A)},$$

where A^+ is the adjoint matrix of A defined as $A^+ = [C(i,j)]^\top = C(j,i)$.
A matrix A is said to be

- normal if $A\bar{A}^{-1} = \bar{A}^\top A$,
- involutory if $A^{-1} = A$,
- orthogonal if $A^{-1} = A^\top$,
- unitary if $A^{-1} = \bar{A}^\top$.

The trace of a square matrix A is defined as

$$\mathrm{tr}(A) \doteq \sum_{i=1}^{n} a_{ii}.$$

For a square matrix of size n, we have $\mathrm{tr}(A) = \sum_{i=1}^{n} \lambda_i$, and $\det(A) = \Pi_{i=1}^{n}\lambda_i$ with λ_i is the eigenvalue of the matrix A. The eigenvalues of a symmetric matrix are all real.

A symmetric square matrix of size n is

- positive-definite if all the eigenvalues of A are positive,
- positive-semi-definite if all the eigenvalues of A are zero or positive,
- negative-definite if all the eigenvalues of $-A$ are positive,
- negative-semi-definite if all the eigenvalues of $-A$ are zero or positive.

To end this section on matrix algebra, let us give some useful identities:

- for any matrix A we have

$$\left[A^\top\right]^\top = A,$$

- for any matrices A and B, we have

$$[A + B + C]^\top = A^\top + B^\top + C^\top,$$

- for any matrices A, B, and C, we have

$$[ABC]^\top = C^\top B^\top A^\top,$$

$$\begin{bmatrix} A & B \\ C & ABC \end{bmatrix}^\top = \begin{bmatrix} A^\top & C^\top \\ B^\top & C^\top B^\top A^\top \end{bmatrix},$$

- for any matrix A that has an inverse, we have

$$\left[A^{-1}\right]^{-1} = A,$$

- for any matrix A that has an inverse, we have

$$\left[A^{\top}\right]^{-1} = \left[A^{-1}\right]^{\top},$$

- for any matrices A and B with appropriate dimensions that the product AB has an inverse, we have

$$[AB]^{-1} = B^{-1}A^{-1},$$

- for any matrix A, we have

$$\operatorname{tr}\left[A^{\top}\right] = \operatorname{tr}\left[A\right],$$

- for any matrices A and B with appropriate dimensions, we have

$$\operatorname{tr}\left[A + B\right] = \operatorname{tr}\left[A\right] + \operatorname{tr}\left[B\right],$$

- for any matrices A and B with appropriate dimensions, we have:

$$\operatorname{tr}\left[AB\right] = \operatorname{tr}\left[BA\right]$$

- for any matrices A and P (nonsingular) with appropriate dimensions, we have

$$\operatorname{tr}\left[P^{-1}AP\right] = \operatorname{tr}\left[A\right],$$

- for any square matrix A, we can always express it as a sum of a symmetric matrix A_1 and a skew-symmetric matrix A_2, that is,

$$A = A_1 + A_2,$$

with

$$A_1 = \frac{A + A^{\top}}{2}, A_2 = \frac{A - A^{\top}}{2}.$$

Let us now give some useful relations for integration and differentiation of vectors and matrices. If we assume, for instance, that the state vector $x(t) \in \mathbb{R}^n$ is given by

$$x(t) = \begin{bmatrix} x_1(t) \\ \vdots \\ x_n(t) \end{bmatrix},$$

then we have

$$\dot{x}(t) = \begin{bmatrix} \dot{x}_1(t) \\ \vdots \\ \dot{x}_n(t) \end{bmatrix},$$

$$
\int_0^t x(s)ds = \begin{bmatrix} \int_0^t x_1(s)ds \\ \vdots \\ \int_0^t x_n(s)ds \end{bmatrix}.
$$

For a matrix $P(t) \in \mathbb{R}^{n \times m}$ defined as follows:

$$
P(t) = \begin{bmatrix} p_{11}(t) & p_{12}(t) & \cdots & p_{1m}(t) \\ p_{21}(t) & p_{22}(t) & \cdots & p_{2m}(t) \\ \vdots & \vdots & \ddots & \vdots \\ p_{n1}(t) & p_{n2}(t) & \cdots & p_{nm}(t) \end{bmatrix},
$$

we have

$$
\dot{P}(t) = \begin{bmatrix} \dot{p}_{11}(t) & \dot{p}_{12}(t) & \cdots & \dot{p}_{1m}(t) \\ \dot{p}_{21}(t) & \dot{p}_{22}(t) & \cdots & \dot{p}_{2m}(t) \\ \vdots & \vdots & \ddots & \vdots \\ \dot{p}_{n1}(t) & \dot{p}_{n2}(t) & \cdots & \dot{p}_{nm}(t) \end{bmatrix},
$$

$$
\int_0^t P(s)ds = \begin{bmatrix} \int_0^t p_{11}(s)ds & \int_0^t p_{12}(s)ds & \cdots & \int_0^t p_{1m}(s)ds \\ \int_0^t p_{21}(s)ds & \int_0^t p_{22}(s)ds & \cdots & \int_0^t p_{2m}(s)ds \\ \vdots & \vdots & \ddots & \vdots \\ \int_0^t p_{n1}(s)ds & \int_0^t p_{n2}(s)ds & \cdots & \int_0^t p_{nm}(s)ds \end{bmatrix}.
$$

The following useful relations are also of interest in this book:

- for any two matrices with appropriate dimensions $A(t)$ and $B(t)$, we have

$$
\frac{d}{dt}[A(t) + B(t)] = \dot{A}(t) + \dot{B}(t),
$$

$$
\frac{d}{dt}[A(t)B(t)] = \dot{A}(t)B(t) + A(t)\dot{B}(t),
$$

$$
\frac{d}{dt}[A^{-1}(t)] = -A^{-1}\dot{A}(t)A^{-1},
$$

- for any function $v(x(t), t)$ that represents a Lyapunov function (a scalar function), we have

$$
\frac{d}{dt}v(x(t), t) = \frac{\partial v}{\partial t} + \left[\frac{\partial v}{\partial x}\right]^{\mathsf{T}} \left[\frac{\partial x}{\partial t}\right],
$$

- for a vector x that may represent the state vector and a matrix B with appropriate dimension, we have

$$
\frac{\partial (B^{\mathsf{T}} x)}{\partial x} = B = \frac{\partial (x^{\mathsf{T}} B)}{\partial x},
$$

- for a vector x that may represent the state vector and a matrix A with appropriate dimension, we have

$$\frac{\partial (Ax)}{\partial x} = A^{\top},$$

$$\frac{\partial \left(x^{\top} A^{\top}\right)}{\partial x} = A,$$

- for a vector x that may represent the state vector and a matrix P with appropriate dimension that is assumed to be symmetric and positive-definite, we have

$$\frac{\partial \left(x^{\top} P x\right)}{\partial x} = \left[P + P^{\top}\right] x = 2 P x.$$

As we did for vectors, we can also introduce norms for matrices. Therefore, for a vector $x \in \mathbb{R}^n$, the norm of a matrix A is defined by

$$\|A\| = \sup_{x \neq 0} \frac{\|Ax\|}{\|x\|} = \sup_{\|x\|=1} \|Ax\|,$$

where sup stand for supremum or the least upper bound. This norm is defined through the norm of the vector x and it is referred to as the induced norm. Notice that for different $\|x\|$ we have different $\|A\|$. We have the following for the useful norms. For any matrix $A \in \mathbb{R}^{n \times n}$ with the components a_{ij}, we have

- for the 1-norm of x, i.e.: $\|x\|_1$, we have $\|A\|_1 = \max_j \left(\sum_{i=1}^{n} |a_{ij}|\right) =$ largest column absolute sum;
- for the 2-norm of x, i.e.: $\|x\|_2$ (Euclidean norm), we have $\|A\|_2 = \sqrt{\lambda_{max}}$, with λ_{max} is the largest eigenvalue of the matrix $A^{\top} A$;
- for the ∞-norm of x, i.e.: $\|x\|_\infty$, we have $\|A\|_\infty = \max_i \left(\sum_{j=1}^{n} |a_{ij}|\right) =$ largest row absolute sum.

The following properties hold for the norms of matrices:

- $\|A\| \geq 0$ and $\|A\| = 0$ if and only if the matrix A is equal to the null matrix,
- $\|aA\| = |a|\|A\|$ for all $a \in \mathbb{R}$,
- $\|A + B\| \leq \|A\| + \|B\|$,
- $\|AB\| \leq \|A\|\|B\|$.

A.3 Markov Process

In this section we deal with the definitions of certain terms used in probability theory. The material presented here is only an introduction of probability concepts and for more details about these concepts, we refer the reader to the appropriate books.

A probability space consists of

- a set Ω;
- a collection $\mathscr{F}$ of subsets of Ω, called events, including the set Ω and verifying the following properties:
  1. if A is an event, then the complement $\bar{A} = \{\omega \in \Omega|\ \omega \notin A\}$ is also an event; the complement of the set Ω is the empty set, denoted by $\emptyset$, which is also an event;
  2. if A_1, A_2 are two events, then their intersection $A_1 \cap A_2$ and their union $A_1 \cup A_2$ are also events;
  3. if $A_1, A_2, \cdots, A_k$ are events, then $\sqcup_{k=1}^{\infty} A_k$ and $\sqcap_{k=1}^{\infty} A_k$ are also events;
- a function $\mathbb{P}(.)$ that assigns to each event A a real number, called the probability of the event, satisfying the following properties:
  1. $\mathbb{P}(A) \geq 0$ for every event A;
  2. $\mathbb{P}(\Omega) = 1$;
  3. $\mathbb{P}(A_1 \cup A_2) = \mathbb{P}(A_1) + \mathbb{P}(A_2)$ for every pair of disjoint events A_1 and A_2;
  4. $\mathbb{P}(\sqcup_{k=1}^{\infty} A_k) = \sum_{k=1}^{\infty} \mathbb{P}(A_k)$ for every sequence of mutually disjoint events $A_1, A_2, \cdots, A_k$.

The function $\mathbb{P}(.)$, referred to as a probability measure, assigns to each event a real positive number between 0 and 1.

Let us consider probability space $(\Omega, \mathscr{F}, \mathbb{P})$. A one-dimensional random variable defined on this probability space is a function from Ω to $\mathbb{R}$, i.e.: $x : \omega \to \mathbb{R}$, such that for every scalar b the set $\{\omega \in \Omega | x(\omega) \leq b\}$ is an event, i.e., it belongs to the collection $\mathscr{F}$.

An n-dimensional random vector $x = (x_1, x_2, \cdots, x_n)$ is an n-tuple of random variables $x_1, x_2, \cdots, x_n$; each one is defined on the same probability space $(\Omega, \mathscr{F}, P)$.

The distribution function (or cumulative distribution function) $F : \mathbb{R} \to \mathbb{R}$ of a random variable x is defined by

$$F(z) = \mathbb{P}(\{\omega \in \Omega | x(\omega) \leq z\}).$$

It is defined as the probability that random variable takes a value less than or equal to z.

The distribution function $F : \mathbb{R}^n \to \mathbb{R}$ of a random vector $\mathbf{x} = \{x_1, x_2, \cdots, x_n\}$ is defined by

$$F(z_1, z_2, \cdots, z_n) = P(\{\omega \in \Omega | x_1(\omega) \leq z_1, x_2(\omega) \leq z_2, \cdots, x_n(\omega) \leq z_n, \}).$$

Given the distribution function of a random vector $\mathbf{x}$, the (marginal) distribution function of each random variable x_i is obtained by

$$F_i(z_i) = \lim_{\substack{z_j \to \infty \\ j \neq i}} F(z_1, z_2, \cdots, z_n).$$

When the random variables $x_1, x_2, \cdots, x_n$ are independent, we have

$$F(z_1, z_2, \cdots, z_n) = F(z_1)F(z_2) \cdots F(z_n).$$

When the random variable is continuous, we say that the random variable vector x is characterized by a piecewise probability density function $f : \mathbb{R}^n \to \mathbb{R}$ if $f(.)$ is piecewise continuous and

$$F(z_1, z_2, \cdots, z_n) = \int_{-\infty}^{z_1} \int_{-\infty}^{z_2} \cdots \int_{-\infty}^{z_n} f(y_1, y_2, \cdots, y_n) dy_1 dy_2 \cdots dy_n.$$

The expected value of a random variable x is defined by

$$\mathbb{E}\{x\} = \int_{-\infty}^{\infty} z dF(z),$$

provided that the integral above is defined.

The expected value of a random vector is defined as

$$\mathbb{E}\{x\} = (\mathbb{E}\{x_1\}, \mathbb{E}\{x_2\}, \cdots, \mathbb{E}\{x_n\}).$$

The conditional probability of two events A and B is defined by

$$\mathbb{P}(B|A) = \begin{cases} \frac{\mathbb{P}(A \cap B)}{\mathbb{P}(A)} & \text{if } \mathbb{P}(A) > 0, \\ 0 & \text{if } \mathbb{P}(A) = 0. \end{cases}$$

If $B_1, B_2, \cdots$ are a countable (possibly finite) collection of mutually exclusive and exhaustive events (i.e., the sets B_i are disjoint and their union is Ω) and A is an event, then we have

$$\mathbb{P}(A) = \sum_i \mathbb{P}(A, B_i)$$

From this we can prove that

$$\mathbb{P}(A) = \sum_i \mathbb{P}(B_i)\mathbb{P}(A|B_i).$$

This is called the theorem of total probability. The conditional expectation of a vector x is defined by

$$\mathbb{E}\{x|w\} = \int_{R^n} z dF(z|w),$$

provided that the integral is well defined.

Let the state of a physical system be observed at the discrete moments of time $n = 0, 1, 2, \cdots$ and let X_n denote this state at time n. Let X_n take values in the state space $\mathscr{S}$ with finite or countably infinite number of states. The conditional probability or transition probability is described by the following relation:

$$\mathbb{P}\left[X_{n+1} = x_{n+1} | X_n = x_n, \cdots, X_0 = x_0\right].$$

In practice many systems have the property that, given the present state, the past states have no influence on the future. This property is called the *Markov property*, and systems having this property are called *Markov chains*. The Markov property is defined by the requirement that

$$\mathbb{P}\left[X_{n+1} = x_{n+1} | X_n = x_n, \cdots, X_0 = x_0\right] = \mathbb{P}\left[X_{n+1} = x_{n+1} | X_n = x_n\right],$$

for every choice of n and the numbers $x_n, \cdots, x_0$.

Example 74. To show the concept of the Markov chain, let us assume that the state of a given machine can be described by a Markov chain with two states denoted, respectively, as operational and under repair. Let X_n be the state of the machine at time n. When the machine is operational at time n, we write $X_n = 1$ and when it is under repair we write $X_n = 0$. Therefore the state space in this case is $\mathscr{S} = \{1, 0\}$. The transition probabilities in this case are

$$\mathbb{P}\left[X_{n+1} = 1 | X_n = 0\right] = p,$$
$$\mathbb{P}\left[X_{n+1} = 0 | X_n = 1\right] = q.$$

Let the initial state be given by the following probability:

$$\mathbb{P}\left[X_0 = 0\right] = \pi(0).$$

Based on the probability theory we have the following probability transition matrix:

$$\mathbb{P} = \begin{bmatrix} 1 - p & p \\ q & 1 - q \end{bmatrix}.$$

Now try to estimate the state of the machine at time $n + 1$ when the initial probability of the initial state at time 0 is given. We need to compute $\mathbb{P}\left[X_{n+1} = 0\right]$ and $\mathbb{P}\left[X_{n+1} = 1\right]$. Using the fact that the machine is described by a Markov chain we get

$$\mathbb{P}\left[X_{n+1} = 0\right] = \mathbb{P}\left[X_{n+1} = 0 \text{ and } X_n = 0\right] + \mathbb{P}\left[X_{n+1} = 0 \text{ and } X_n = 1\right],$$

which gives

$$
\begin{aligned}
\mathbb{P}\left[X_{n+1} = 0\right] &= \mathbb{P}\left[X_{n+1} = 0 | X_n = 0\right]\mathbb{P}\left[X_n = 0\right] \\
&\quad + \mathbb{P}\left[X_{n+1} = 0 | X_n = 1\right]\mathbb{P}\left[X_n = 1\right] \\
&= (1 - p)\mathbb{P}\left[X_n = 0\right] + q\mathbb{P}\left[X_n = 1\right] \\
&= (1 - p)\mathbb{P}\left[X_n = 0\right] + q\left(1 - \mathbb{P}\left[X_n = 0\right]\right) \\
&= (1 - p - q)\mathbb{P}\left[X_n = 0\right] + q.
\end{aligned}
$$

This relation is valid for each n, therefore:

- for $n = 1$, we have

$$\mathbb{P}[X_1 = 0] = (1 - p - q)\pi(0) + q,$$

- for $n = 2$, we have

$$\begin{aligned}
\mathbb{P}[X_2 = 0] &= (1 - p - q)\mathbb{P}[X_1 = 0] + q \\
&= (1 - p - q)^2 \pi(0) + q[1 + (1 - p - q)],
\end{aligned}$$

- for n, it can be proven that the probability at repair state at time n is given by

$$\mathbb{P}[X_n = 0] = (1 - p - q)^n \pi(0) + q \sum_{j=0}^{n-1} [1 - p - q]^j .$$

If we assume that $p + q > 0$, in this case we get

$$\sum_{j=0}^{n-1} [1 - p - q]^j = \frac{1 - (1 - p - q)^n}{p + q}.$$

Using this we have

$$\mathbb{P}[X_n = 0] = \frac{q}{p + q} + (1 - p - q)^n \left[\pi(0) - \frac{q}{p + q}\right],$$

$$\mathbb{P}[X_n = 1] = \frac{p}{p + q} + (1 - p - q)^n \left[\pi(1) - \frac{1}{p + q}\right].$$

If we assume that p and q are neither equal to zero nor equal to one, then $0 < p + q < 2$, which implies that $|1 - p - q| < 1$. In this case we can let $n \to \infty$, which gives

$$\lim_{n \to \infty} \mathbb{P}[X_n = 0] = \frac{q}{p + q},$$

$$\lim_{n \to \infty} \mathbb{P}[X_n = 1] = \frac{p}{p + q}.$$

This example shows how to perform the computation of the steady state of the Markov chain when it exists. Now consider the general case in which the Markov chain X_n has more than two states in the state space $\mathscr{S}$. Let us restrict ourselves to the computation of the m-step transition function $P^m(x, y)$, which gives the probability of going from x to y in m steps. Let $\mathbb{P}[y, x]$ denote the $\mathbb{P}[X_1 = y | X_0 = x]$, i.e.:

$$\mathbb{P}[y, x] = \mathbb{P}[X_1 = y | X_0 = x], \forall x, y \in \mathscr{S}.$$

It is such that

$$\begin{cases} \mathbb{P}\left[y,x\right] \geq 0, \\ \sum_{y \in \mathscr{S}} \mathbb{P}\left[y,x\right] = 1, \forall x \in \mathscr{S}. \end{cases}$$

If the Markov chain has stationary probabilities, we see that:

$$\mathbb{P}\left[X_{n+1} = y | X_n = x\right] = \mathbb{P}\left[y,x\right].$$

It follows from the Markov property that

$$\mathbb{P}\left[X_{n+1} = y | X_0 = x_0, \cdots, X_n = x\right] = \mathbb{P}\left[y,x\right].$$

The initial distribution of the Markov chain is

$$\pi(x) = \mathbb{P}\left[X_0 = x\right], \ x \in \mathscr{S}.$$

This initial distribution satisfies the following:

$$\begin{cases} \pi(x) \geq 0, \\ \sum_{x \in \mathscr{S}} \pi(x) = 1. \end{cases}$$

The joint distribution of $X_0, \cdots, X_n$ can easily be expressed in terms of the transition function and the initial distribution. In fact, we have

$$\begin{aligned} \mathbb{P}\left[X_1 = x_1, X_0 = x_0\right] &= \mathbb{P}\left[X_0 = x_0\right] \mathbb{P}\left[X_1 = x_1 | X_0 = x_0\right] \\ &= P\left[x_1, x_0\right] \pi(x_0). \end{aligned}$$

We can use the Markov property to get

$$\begin{aligned} \mathbb{P}&\left[X_2 = x_2, X_1 = x_1, X_0 = x_0\right] \\ &= \mathbb{P}\left[X_1 = x_1, X_0 = x_0\right] \mathbb{P}\left[X_2 = x_2 | X_1 = x_1, X_0 = x_0\right] \\ &= \mathbb{P}\left[x_2, x_1\right] \mathbb{P}\left[x_1, x_0\right] \pi(x_0). \end{aligned}$$

For the more general case, we get

$$\begin{aligned} \mathbb{P}&\left[X_n = x_n, \cdots, X_1 = x_1, X_0 = x_0\right] \\ &= \mathbb{P}\left[x_n, x_{n-1}\right] \mathbb{P}\left[x_{n-1} = x_{n-2}\right] \cdots \mathbb{P}\left[x_2, x_1\right] P\left[x_1, x_0\right] \pi(x_0). \end{aligned}$$

For the m-step, we have

$$\begin{aligned} \mathbb{P}&\left[X_{n+m} = x_{n+m}, \cdots X_n = x_n, X_{n-1} = x_{n-1}, \cdots, X_0 = x_0\right] \\ &= \mathbb{P}\left[X_{n+m} = x_{n+m}, \cdots, X_{n+1} = x_{n+1} | X_n = x_n, \cdots, X_0 = x_0\right] \\ &\quad \times \mathbb{P}\left[X_n = x_n, \cdots, X_0 = x_0\right], \end{aligned}$$

which gives

$$\mathbb{P}\left[X_{n+m} = x_{n+m}, \cdots, X_{n+1} = x_{n+1} | X_n = x_n, \cdots, X_0 = x_0\right]$$

$$= \frac{\mathbb{P}\left[X_{n+m} = x_{n+m}, \cdots, X_n = x_n, X_{n-1} = x_{n-1}, \cdots, X_0 = x_0\right]}{\mathbb{P}\left[X_n = x_n, \cdots, X_0 = x_0\right]}$$

$$= \frac{\mathbb{P}\left[x_{n+m}, x_{n+m-1}\right] P\left[x_{n+m-1}, x_{n+m-2}\right] \cdots P\left[x_1, x_0\right] \pi(x_0)}{\mathbb{P}\left[x_n, x_{n-1}\right] \cdots \mathbb{P}\left[x_1, x_0\right] \pi(x_0)}$$

$$= \mathbb{P}\left[x_{n+m}, x_{n+m-1}\right] \cdots \mathbb{P}\left[x_{n+1}, x_n\right].$$

Let $A_0, \cdots, A_{n-1}$ be subsets of $\mathscr{S}$. It follows that

$$\mathbb{P}\left[X_{n+m} = y_m, \cdots, X_{n+1} = y_1 | X_n = x, \cdots, X_{n-1} \in A_{n-1}, X_0 \in A_0\right]$$
$$= \mathbb{P}\left[y_m, y_{m-1}\right] \cdots P\left[y_1, x\right].$$

Let $B_1, \cdots, B_m$ be subsets of $\mathscr{S}$. It follows that

$$\mathbb{P}\left[X_{n+m} \in B_m, \cdots, X_{n+1} \in B_1 | X_n = x, \cdots, X_{n-1} \in A_{n-1}, X_0 \in A_0\right]$$
$$= \sum_{y_1 \in \mathscr{S}} \cdots \sum_{y_m \in \mathscr{S}} \mathbb{P}\left[y_m, y_{m-1}\right] \cdots \mathbb{P}\left[y_1, x\right].$$

The m-step transition function $P^m\left[y, x\right]$, which gives the probability of going from x to y, is defined by

$$\mathbb{P}^m\left[y, x\right] = \sum_{y_1} \cdots \sum_{y_{m-1}} \mathbb{P}\left[y, y_{m-1}\right] \cdots P\left[y_2, y_1\right] \mathbb{P}\left[y_1, x\right].$$

It can be shown that

$$P^{n+m}\left[y, x\right] = \sum_z P^n\left[z, x\right] P^m\left[y, z\right].$$

It can also be shown that

$$\mathbb{P}\left[X_n = y\right] = \sum_z P^n\left[X_n = y, X_0 = x\right]$$
$$= \sum_x \mathbb{P}\left[X_n = y | X_0 = x\right] \mathbb{P}\left[X_0 = 0\right].$$

We see that

$$\mathbb{P}\left[X_n = y\right] = \sum_x \mathbb{P}^n\left[y, x\right] \pi(x).$$

Let us now consider the case of continuous-time Markov process with finite state space $\mathscr{S}$. In this case, the stochastic process is denoted by $X(t)$ and we have

$$P\left[X(t) \in A | X(s) = x(s), s \leq \tau\right] = P\left[X(t) \in A | X(\tau) = x(\tau)\right],$$

where A is a subset of $\mathscr{S}$ and t any time larger than τ.

Based on what we developed for the Markov chain, we can write

$$\mathbb{P}\left[X(t)=i\right]=\sum_{j}\mathbb{P}\left[X(t)=i|X(\tau)=j\right]\mathbb{P}\left[X(\tau)=j\right].$$

If we now replace t by $t+h$ and τ by t in the previous relation, we get

$$\mathbb{P}\left[X(t+h)=i\right]=\sum_{j}\mathbb{P}\left[X(t+h)=i|X(t)=j\right]\mathbb{P}\left[X(t)=j\right].$$

Let us define λ_{ij} as follows if it exists:

$$\lambda_{ij}=\lim_{h\to 0}\frac{\mathbb{P}\left[X(t+h)=j|X(t)=i\right]}{h} \tag{A.1}$$

or

$$\mathbb{P}\left[X(t+h)=j|X(t)=i\right]=\lambda_{ij}h+o(h) \tag{A.2}$$

with $\lim_{h\to 0}\frac{o(h)}{h}$.

Let us compute the $\mathbb{P}\left[X(t+h)=i\right]$. This can be given by

$$\mathbb{P}\left[X(t+h)=i\right]=\sum_{j\in\mathscr{S}}\mathbb{P}\left[X(t+h)=i|X(t)=j\right]\mathbb{P}\left[X(t)=j\right]. \tag{A.3}$$

Using now the definition of λ_{ij} we get:

$$\mathbb{P}\left[X(t+h)=i\right]=\sum_{j\neq i\in\mathscr{S}}\lambda_{ij}h\mathbb{P}\left[X(t)=j\right]$$
$$+\mathbb{P}\left[X(t+h)=i|X(t)=i\right]\mathbb{P}\left[X(t)=i\right]+o(h).$$

To shorten our notation let us define

$$p_i\left[t\right]=\mathbb{P}\left[X(t)=i\right].$$

Then the previous expression becomes

$$p_i\left[t+h\right]=\sum_{j\neq i\in\mathscr{S}}\lambda_{ij}hp_j\left[t\right]+\mathbb{P}\left[X(t+h)=i|X(t)=i\right]\mathbb{P}\left[X(t)=i\right]+o(h).$$

By using the conditional expectation that satisfies

$$\sum_{j\in\mathscr{S}}\mathbb{P}\left[X(t+h)=j|X(t)=i\right]=1,$$

we get

$$\mathbb{P}\left[X(t+h)=i|X(t)=i\right]=1-\sum_{j\neq i}\mathbb{P}\left[X(t+h)=j|X(t)=i\right]$$
$$=1-\sum_{j\neq i}\lambda_{ij}h+o(h).$$

Plugging this in the previous equation we get

$$p_i\,[t+h] = \sum_{j\neq i \in \mathscr{S}} \lambda_{ij} h p_j\,[t] + \left[1 - \sum_{j\neq i} \lambda_{ij} h\right] p_i\,[t] + o(h).$$

Since $\lambda_{ii} = -\sum_{j\neq i} \lambda_{ij}$ we have

$$p_i\,[t+h] = \sum_{j\neq i \in \mathscr{S}} \lambda_{ij} h p_j\,[t] + \lambda_{ii} h p_i\,[t] + p_i\,[t] + o(h),$$

which becomes

$$p_i\,[t+h] = \sum_{j\in\mathscr{S}} \lambda_{ij} h p_j\,[t] + p_i\,[t] + o(h).$$

Now if we use the Taylor expansion we get

$$p_i\,[t] + \frac{dp_i\,[t]}{dt} = \sum_{j\in\mathscr{S}} \lambda_{ij} h p_j\,[t] + p_i\,[t] + o(h),$$

which gives finally

$$\frac{dp_i\,[t]}{dt} = \sum_{j\in\mathscr{S}} \lambda_{ij} h p_j\,[t].$$

Remark 29. The steady-state regime corresponds to the case where we have

$$\frac{dp_i\,[t]}{dt} = 0,$$

which gives in turn

$$\sum_{j\in\mathscr{S}} \lambda_{ij} p_j = 0.$$

Notice that the amount of time the Markov process $\{r(t), t \geq 0\}$ spent in mode i before making a transition into a different mode follows an exponential distribution with rate ν_i. Also, when the Markov process leaves the mode i, it will reach the mode j that is different from mode i with a probability $\mathbb{P}_{ij}$.

Let $\{r(t), t \geq 0\}$ be a stochastic Markov process defined on $(\Omega, \mathscr{F}, \mathbb{P})$ and taking values in a finite set $\mathscr{S}$. For each $t, \omega \to r(t)(\omega)$ is a measurable map from $(\Omega \times \mathscr{F}) \to \mathscr{S}$.

Let a dynamical system of the class of systems we are studying in this book subject to an additive noise be described by the following differential equation:

$$dx(t) = A(r(t))x(t)dt + B(r(t))u(t)dt + f(x(t), r(t))dw(t), \qquad \text{(A.4)}$$

where $x(t)$ is the state of the system; $r(t)$ is the mode and $w(t)$ is Wiener process acting on the system; and $A(r(t))$, $B(r(t))$, and $f(x(t), r(t))$ are known with some appropriate dimensions and characteristics.

Let $x(t)$ be the solution of the system (A.4) when the initial conditions are $x(t) = x$ and $r(t) = i$ at time t when the control $u(t) = 0$ for all $t \geq 0$.

Definition 22. *The operator $\mathscr{L}(.)$, also called the averaged derivative at point $(t, x(t) = x, r(t) = i)$, is defined by the following expression:*

$$\mathscr{L}V(x(t), r(t))$$
$$= \lim_{h \to 0} \frac{1}{h} \left[\mathbb{E} \left[V(x(t+h), r(t+h)) | x(t) = x, r(t) = i \right] - V(x, i) \right],$$

where $\mathbb{E}[.|.]$ is the conditional mathematical expectation.

The value of $\mathscr{L}V(.)$ can be interpreted as an averaged value of the derivative of the function $V(x, i)$ along all realizations of the Markov process $\{(x(t), r(t)), t \geq 0\}$ emanating from the point (x, i) at time t.

Remark 30. The operator $\mathscr{L}V(.)$ is also referred to as the weak infinitesimal operator of the process $\{(x(t), r(t)), t \geq 0\}$.

Let the initial conditions of the process $\{(x(t), r(t)), t \geq 0\}$ be fixed to (x, i). Let us denote by $\mathscr{L}_{ij}V(x, i) = V(x(t+h), r(t+h)) - V(x, i)$. At any time $\tau \in (t, t+h]$ we have two possible events:

- the process $\{r(t), t \geq 0\}$ does not jump, i.e.: $r(\tau) = i$ for $\tau \in (t, t+h]$ and the probability of this event is $1 + \lambda_{ii}h + o(h)$; then $\mathscr{L}_{ii}V(x, i)$ is given by

$$\mathscr{L}_{ii}V(.) = \left[\frac{\partial V(x, i)}{\partial s} + \left[\frac{\partial V(x, i)}{\partial x} \right]^\top [A(i)x(t)] \right.$$
$$\left. + \frac{1}{2} [x(t+h) - x]^\top \frac{\partial V^2(x, i)}{\partial x^2} [x(t+h) - x] \right] h,$$

where

$$V_x^\top(x, i) = \left[\frac{\partial V(x(t), i)}{\partial x} \right]^\top = \left[\frac{\partial V(x(t), i)}{\partial x_1}, \cdots, \frac{\partial V(x(t), i)}{\partial x_n} \right],$$
$$V_{xx}(x, i) = \frac{\partial V^2(x(t), i)}{\partial x^2} = \left[\frac{\partial V^2(x(t), i)}{\partial x_k \partial x_m} \right]_{k=1, m=1}^n,$$

- the process $\{r(t), t \geq 0\}$ jumps from mode i to $j \neq i$, i.e.: $r(\tau) = j$ for $\tau \in (t, t+h]$ and the probability of this event is $\lambda_{ij}h + o(h)$; then $\mathscr{L}_{ij}V(x, i)$ is given by

$$\mathscr{L}_{ij}V(.) = V(x(t+h), j) - V(x, i).$$

Remark 31. Notice that all the partial derivatives are computed at the point (t, x, i).

Using these relations, the possible events, and the fact that $x^\top P x = tr\left[Pxx^\top\right]$, and after dividing by h and letting h go to zero, we get

$$\mathscr{L}V(x(t), i) = \frac{\partial V(x(t), i)}{\partial s} + \left[\frac{\partial V(x(t), i)}{\partial x}\right]^\top [A(i)x(t)]$$

$$+ \frac{1}{2}\frac{\partial V^2(x(t), i)}{\partial x^2} f(x(t), i) f^\top(x(t), i)$$

$$+ \sum_{j=1}^{N} \lambda_{ij} \left[V(x(t), j) - V(x(t), i)\right].$$

When the control is not equal to zero and when the corresponding infinitesimal operator is denoted by $\mathscr{L}_u V(.)$ we get the following results.

Theorem 82. *Let $V(x(t), r(t))$ be a function from $\mathbb{R}^n \times \mathscr{S}$ into $\mathbb{R}$ such that $V(x(t), r(t))$ and $V_x(x(t), r(t))$ are continuous in x for any $r(t) \in \mathscr{S}$ and such that $|V(x(t), r(t))| < \gamma(1 + \|x\|)$ for a constant γ, the generator $\mathscr{L}_u$ of $(x(t), r(t))$ under an admissible control law u, for $x(t)$ solution of (A.4) and $\{r(t), t \geq 0\}$ a continuous-time Markov process taking values in a finite state space $\mathscr{S}$ with transition rate matrix Λ, is given by*

$$\mathscr{L}_u V(x(t), r(t)) = [A(r(t))x(t) + B(r(t))u(t)]^\top V_x(x(t), r(t))$$

$$+ \frac{1}{2}tr\left(f^\top(x(t), r(t))V_{xx}(x(t), r(t))f(x(t), r(t))\right)$$

$$+ \sum_{j=1}^{N} V(x(t), j). \tag{A.5}$$

A.4 Lemmas

This section presents a number of results that are extensively used in different proofs of the proposed theorems in this book. These results are given in the form of lemmas. The first lemma is critical in our development since it is used to cast a nonlinear problem into the framework of an LMI.

Lemma 2. *Let X and Y be two real constant matrices of compatible dimensions; then the following equation:*

$$\pm X^\top Y \pm Y^\top X \leq X^\top X + Y^\top Y,$$

holds.

Proof: The proof follows from the following inequality:

$$0 \leq [X \mp Y]^\top [X \mp Y].$$

Expanding the right-hand side of this inequality implies that

$$0 \leq X^\top X \mp X^\top Y \mp Y^\top X + Y^\top Y,$$

which gives the result

$$\pm X^\top Y \pm Y^\top X \leq X^\top X + Y^\top Y.$$

This ends the proof of the lemma.   □

Lemma 3. *Let X and Y be real constant matrices of compatible dimensions, then the following equation:*

$$X^\top Y + Y^\top X \leq \varepsilon X^\top X + \varepsilon^{-1} Y^\top Y,$$

holds for any $\varepsilon > 0$.

Proof: The proof follows from the following inequality:

$$0 \leq \left[\sqrt{\varepsilon} X^\top - \frac{1}{\sqrt{\varepsilon}} Y^\top \right] \left[\sqrt{\varepsilon} X - \frac{1}{\sqrt{\varepsilon}} Y \right].$$

Expanding the right-hand side of this inequality implies that

$$0 \leq \varepsilon X^\top X - X^\top Y - Y^\top X + \varepsilon^{-1} Y^\top Y,$$

which gives in turn the desired result. This ends the proof of the lemma.   □

Lemma 4. *(Schur Complement) Let the symmetric matrix M be partitioned as*

$$M = \begin{bmatrix} X & Y \\ Y^\top & Z \end{bmatrix},$$

with X and Z being symmetric matrices. We have

1. *M is nonnegative-definite if and only if either*

$$\begin{cases} Z \geq 0, \\ Y = L_1 Z, \\ X - L_1 Z L_1^\top \geq 0, \end{cases} \tag{A.6}$$

or

$$\begin{cases} X \geq 0, \\ Y = X L_2, \\ Z - L_2^\top X L_2 \geq 0, \end{cases} \tag{A.7}$$

holds, where L_1, L_2 are some (nonunique) matrices of compatible dimensions;

2. *M is positive-definite if and only if either*

$$\begin{cases} Z > 0, \\ X - YZ^{-1}Y^\top > 0, \end{cases} \tag{A.8}$$

or

$$\begin{cases} X > 0, \\ Z - Y^\top X^{-1}Y > 0. \end{cases} \tag{A.9}$$

The matrix $X - YZ^{-1}Y^\top$ is called the Schur complement $X(Z)$ in M.

Proof: We begin with proving (A.6).

Necessity: Clearly, $Z \geq 0$ is necessary. We can prove the necessity of $X - YZ^{-1}Y^\top$. Let x be a vector and partition it as $x = \begin{bmatrix} x_1 \\ x_2 \end{bmatrix}$. According to the partitioning of M, we have

$$x^\top M x = x_1^\top X x_1 + 2x_1^\top Y x_2 + x_2^\top Z x_2. \tag{A.10}$$

Let x_2 be such that $Zx_2 = 0$. If $Yx_2 \neq 0$, let $x_1 = -\alpha Y x_2, \alpha > 0$. Then

$$x^\top M x = \alpha^2 x_2^\top Y^\top X Y x_2 - 2\alpha x_2^\top Y^\top Y x_2,$$

which is negative for a sufficiently small $\alpha > 0$. Therefore

$$X x_2 = 0 \Longrightarrow Y x_2 = 0, \forall x_2,$$

which implies

$$Y = L_1 Z \tag{A.11}$$

for some (nonunique) L_1.

Since $M \geq 0$, the quadratic form (A.10) has for any x_1 a minimum over x_2. Thus differentiating (A.10) with respect to $x_2^\top$ we have

$$0 = \frac{\partial(x^\top M x)}{\partial x_2^\top} = 2Y^\top x_1 + 2Z x_2 = 2ZL_1^\top x_1 + 2Z x_2,$$

where

$$ZL^\top x_1 = -Z x_2. \tag{A.12}$$

Using (A.11) and (A.12) in (A.10), we find that the minimum of $x^\top M x$ over x_2 and for any x_1 is

$$\min_{x_2} x^\top M x = x_1^\top \left[X - L_1 Z L_1^\top \right] x_1,$$

which proves the necessity of $X - L_1 Z L_1^\top \geq 0$.

Sufficiency: The conditions (A.6) are therefore necessary for $M \geq 0$, and since together they imply that the minimum of $x^\top M x$ over x_2 for any x_1 is nonnegative, they are also sufficient.

By the same argument, conditions (A.7) can be derived as those of (A.6) by starting with X. (2.) is a direct corollary of (1). This ends the proof of the Lemma 4. □

Let $U \in \mathbb{R}^{n \times k}$. $U_\perp$ is said to be the orthogonal complement of U if $U^\top U_\perp = 0$ and $[U U_\perp]$ is of maximum rank (which means that $[U U_\perp]$ is nonsingular).

Lemma 5. *Let G, U, and V be given matrices with G being symmetric.*

1. *Then there exists a matrix X such that the following inequality holds:*

$$G + U X V^\top + V X^\top U^\top > 0, \tag{A.13}$$

if and only if the following ones hold:

$$\begin{cases} U_\perp^\top G U_\perp > 0, \\ V_\perp^\top G V_\perp > 0, \end{cases} \tag{A.14}$$

where $U_\perp, V_\perp$ are the orthogonal complements of U and V respectively;

2. *$U_\perp^\top G U_\perp > 0$ holds if and only if there exists a scalar σ such that*

$$G - \sigma U U^\top > 0.$$

Proof: (See Boyd et al. [21] pp. 32–33).

Using Lemma 5, we can eliminate some matrix variables in a matrix inequality; therefore a nonlinear problem can be cast into the LMI framework. This reduces the computation burden significantly for the problem under consideration.

Lemma 6. *Let $X \in \mathbb{R}^{n \times n}$ and $Y \in \mathbb{R}^{n \times n}$ be symmetric and positive-definite matrices. Then there exists a symmetric and positive-definite matrix $P > 0$ satisfying $P = \begin{bmatrix} Y & \# \\ \# & \# \end{bmatrix}, P^{-1} = \begin{bmatrix} X & \# \\ \# & \# \end{bmatrix}$ if and only if $X - Y^{-1} \geq 0$.*

Lemma 7. *(See [64]) Let Y be a symmetric matrix and H, E be given matrices with appropriate dimensions and F satisfying $F^\top F \leq \mathbb{I}$. Then we have*

1. *for any $\varepsilon > 0$, $HFE + E^\top F^\top H^\top \leq \varepsilon H H^\top + \frac{1}{\varepsilon} E^\top E$;*
2. *$Y + HFE + E^\top F^\top H^\top < 0$ holds if and only if there exists a scalar $\varepsilon > 0$ such that the following holds: $Y + \varepsilon H H^\top + \varepsilon^{-1} E^\top E < 0$.*

Lemma 8. *(see [61]) Let A, D, F, E be real matrices of appropriate dimensions with $\|F\| \leq 1$. Then we have*

1. *for any matrix $P > 0$ and scalar $\varepsilon > 0$ satisfying $\varepsilon \mathbb{I} - EPE^\top > 0$,*

$$(A + DFE)P(A + DFE)^\top$$
$$\leq APA^\top + APE^\top(\varepsilon \mathbb{I} - EPE^\top)^{-1}EPA^\top + \varepsilon DD^\top; \quad \text{(A.15)}$$

2. *for any matrix $P > 0$ and scalar $\varepsilon > 0$ satisfying $P - \varepsilon DD^\top > 0$,*

$$(A + DFE)^\top P^{-1}(A + DFE) \leq A^\top(P - \varepsilon DD^\top)^{-1}A + \varepsilon^{-1}E^\top E.$$

Lemma 9. *(see [12])*

1. *For any $x, y \in \mathbb{R}^n$,*

$$\pm 2x^\top y \leq x^\top X x + y^\top X^{-1} y \quad \text{(A.16)}$$

holds for any $X > 0$.

2. *For any matrices U and $V \in \mathbb{R}^{n \times n}$ with $V > 0$, we have*

$$UV^{-1}U^\top \geq U + U^\top - V^\top. \quad \text{(A.17)}$$

Proof: The proof of (i) is trivial and can be found in the Appendix of [12]. For the proof of (ii), note that since $V > 0$, we have the following:

$$(U - V)V^{-1}(U - V)^\top > 0,$$

which yields

$$UV^{-1}U^\top - UV^{-1}V^\top - VV^{-1}U^\top + V \geq 0.$$

This gives the desired results and ends the proof of the lemma. □

References

1. M. Ait Rami and L. El Ghaoui, "Robust state-feedback stabilization of jump linear systems via LMIs," *International Journal of Robust and Nonlinear Control*, vol. 6, pp. 1015–1022, 1996.
2. M. D. S. Aliyu and E. K. Boukas, "Finite and infinite horizon $\mathcal{H}_\infty$ control for stochastic nonlinear systems," *IMA Journal of Mathematical Control and Information*, vol. 17, no. 3, pp. 265–279, 2000.
3. ——, "Robust $\mathcal{H}_\infty$ control for Markovian jump nonlinear systems," *IMA Journal of Mathematical Control and Information*, vol. 17, no. 3, pp. 295–308, 2000.
4. W. J. Anderson, *Contionus-time Markov chains*. Springer-Verlag, Berlin, 1991.
5. J. Bao, F. Deng, and Q. Luo, "Exponential stabilizability of Markovian jumping systems with distributed time delay and uncertain nonlinear exogenous disturbances," in *International Conference on Machine Learning and Cybernetics*, vol. 2, November 2003, pp. 841–846.
6. K. Benjelloun and E. K. Boukas, "Mean square stochastic stability of linear time-delay system with Markov jumping parameters," *IEEE Transactions on Automatic Control*, vol. 43, no. 10, pp. 1456–1459, 1998.
7. K. Benjelloun, E. K. Boukas, and O. L. V. Costa, "$\mathcal{H}_\infty$-control for linear time-delay systems with Markovian jumping parameters," *Journal of Optimization Theory and Applications*, vol. 105, no. 1, pp. 73–95, 1998.
8. K. Benjelloun, E. K. Boukas, and H. Yang, "Robust stabilizability of uncertain linear time-delay systems with Markovian jumping parameters," *Journal of Dynamic Systems, Measurement, and Control*, vol. 118, pp. 776–783, 1996.
9. E. K. Boukas, *Systèmes Asservis*. Éditions de l'École Polytechnique de Montréal, Montréal, 1995.
10. E. K. Boukas and Z. K. Liu, "Robust stability and $\mathcal{H}_\infty$ control of discrete-time jump linear systems with time delay: An LMI approach," *ASME, Journal of Dynamic Systems, Measurement, and Control*.
11. ——, "Delay-dependent robust stability and stabilization of uncertain linear systems with discrete and distributed time delays," in *IEEE American Control Conference*, Arlingtion, Virginia, June 25–27, 2001.
12. ——, *Deterministic and stochastic time-delay systems*. Birkhauser, Boston, Cambridge, MA, 2002.

13. ——, *Systems with time delay, stability, stabilization, $\mathcal{H}_\infty$ and their robustness.* In *Decison and control in management science Essays in Honor of A. Haurie,* G. Zaccour, Ed. Kluwer Academic Publisher, Boston, MA, 2002.

14. E. K. Boukas, Z. K. Liu, and G. X. Liu, "Delay-dependent robust stability and $\mathcal{H}_\infty$ control of jump linear systems with time delay," *International Journal of Control,* vol. 74, no. 4, pp. 329–331, 2001.

15. E. K. Boukas, Z. K. Liu, and P. Shi, "Delay-dependent stability and output feedback stabilization of Markovian jump systems with time delay," *IEE, Proceedings Control Theory and Applications,* vol. 149, no. 5, pp. 379–387, 2002.

16. E. K. Boukas and P. Shi, "Stochastic stability and guaranteed cost control of discrete-time uncertain systems with Markovian jumping parameters," *International Journal of Robust and Nonlinear Control,* vol. 8, no. 13, pp. 1155–1167, 1998.

17. ——, "$\mathcal{H}_\infty$ control for discrete-time linear systems with Frobenius norm-bounded uncertainties," *Automatica,* vol. 35, no. 9, pp. 1625–1631, 1999.

18. E. K. Boukas, P. Shi, and K. Benjelloun, "On stabilization of uncertain linear systems with Markovian jumping parameters," *International Journal of Control,* vol. 72, pp. 842–850, 1999.

19. E. K. Boukas and H. Yang, "Stability of discrete-time linear systems with Markovian jumping parameters," *Mathematics Control, Signals and Systems,* vol. 8, pp. 390–402, 1995.

20. ——, "Exponential stability of stochastic Markovian jumping parameters," *Automatica,* vol. 35, no. 8, pp. 1437–1441, 1999.

21. S. Boyd, L. El-Ghaoui, E. Feron, and V. Balakrishnan, *Linear Matrix Inequalities in System and Control Theory.* SIAM, Philadelphia, 1994.

22. Y. Y. Cao and J. Lam, "Stochastic stabilizability and $\mathcal{H}_\infty$ control for discrete-time jump linear systems with time delay," in *Proceedings of the 14th IFAC World Congress,* Beijing, China, 1999.

23. ——, "Robust $\mathcal{H}_\infty$ control of uncertain Markovian jump systems with time delay," *IEEE Transactions on Automatic Control,* vol. 45, no. 1, pp. 77–83, 2000.

24. Y. Y. Cao, Y. X. Sun, and J. Lam, "Delay-dependent robust $\mathcal{H}_\infty$ control for uncertain systems with time-varying delays," *IEE Proceedings D, Control Theory and Applications,* vol. 145, no. 3, 1998.

25. O. L. V. Costa and E. K. Boukas, "Necessary and sufficient conditions for robust stability and stabilizability of continuous-time linear systems with Markovian jumps," *Journal of Optimization Theory and Applications,* 1998.

26. O. L. V. Costa, J. B. R. do Val, and J. C. Geromel, "Continuous-time state-feedback $\mathcal{H}_2$ control of Markovian jump linear systems via convex analysis," *Automatica,* vol. 35, no. 2, pp. 259–268, 1999.

27. O. L. V. Costa and M. D. Fragoso, "Necessary and sufficient conditions for mean square stability of discrete-time linear systems subject to Markovian jumps," in *Proceedings of the 9th International Symposium on Mathematics Theory of Networks and Systems,* Kobe, Japan, 1991, pp. 85–86.

28. ——, "Stability results for discrete-time linear systems with Markovian jumping parameters," *Journal of Mathematical Analysis and Applications,* vol. 179, no. 2, pp. 154–178, 1993.

29. O. L. V. Costa and S. Guerra, "Stationary filter for linear minimum mean square error estimator of discrete-time Markovian jump systems," *IEEE Transactions on Automatic Control,* vol. 47, no. 8, pp. 1351–1356, August 2002.

30. O. L. V. Costa and R. P. Marques, "Mixed $\mathcal{H}_2/\mathcal{H}_\infty$ control of discrete-time Markovian jump linear systems," *IEEE Transactions on Automatic Control*, vol. 43, pp. 95–100, 1998.

31. L. Dai, *Singular Control Systems, Lecture Notes in Control and Information Sciences*, Springer-Verlag, New York, NY, 1989.

32. D. P. de Farias, J. C. Geromel, J. B. R. do Val, and O. L. V. Costa, "Output feedback control of Markov jump linear systems in continuous time," *IEEE Transactions on Automatic Control*, vol. 45, no. 5, pp. 944–949, 2000.

33. C. E. de Souza and M. D. Fragoso, "$\mathcal{H}_\infty$ control for linear systems with Markovian jumping parameters," *Control Theory and Advanced Technology*, vol. 9, no. 2, pp. 457–466, 1993.

34. V. Dragan and T. Morozan, "Stability and robust stabilization of linear stochastic systems described by differential equations with Markovian jumping and multiplicative white noise," Institute of Mathematics of Romanian Academy Science, Technical report 17/1999, 1999.

35. ——, "Stability and robust stabilization to linear stochastic systems described by differential equations with Markovian jumping and multiplicative white noise," *Stochastic Analysis and Applications*, vol. 20, no. 1, pp. 33–92, 2000.

36. V. Dragan, P. Shi, and E.-K. Boukas, "Control of singularly perturbed systems with Markovian jump parameters: an $\mathcal{H}_\infty$ approach," *Automatica*, vol. 35, no. 8, pp. 1369–1378, August 1999.

37. F. Dufour and P. Bertrand, "The filtering problem for continuous-time linear systems with Markovian switching coefficients," *Systems and Control Letters*, vol. 23, no. 5, pp. 453–461, 1994.

38. ——, "An image-based filter for discrete-time Markovian jump linear systems," *Automatica*, vol. 32, no. 2, pp. 241–247, 1996.

39. J. Ezzine and D. Karvanoglu, "On almost-sure stabilization of discrete-time parameter systems, an LMI approach," *International Journal of Control*, vol. 68, no. 5, pp. 1129–1146, 1997.

40. X. Feng, K. A. Loparo, Y. Ji, and H. J. Chizeck, "Stochastic stability properties of jump linear systems," *IEEE Transactions on Automatic Control*, vol. 37, no. 1, pp. 38–53, 1992.

41. B. A. Francis, *A Course in $\mathcal{H}_\infty$ Control Theory*. Springer-Verlag, New York, 1987.

42. S. B. Gershwin, *Manufacturing Systems Engineering*. Prentice-Hall, Englewood Cliffs, NJ, 1994.

43. I. I. Gihman and A. V. Skorohod, *Stochastic Differential Equations*. Springer-Verlag, New York, NY, 1965.

44. Y. Ji and H. J. Chizeck, "Controllability, stabilizability, and continuous-time Markovian jump linear quadratic control," *IEEE Transactions on Automatic Control*, vol. 35, no. 7, pp. 777–788, 1990.

45. I. Y. Kats and A. A. Martynyuk, *Stability and Stabilization on Nonlinear Systems with Random Structure*. Taylor and Francis, London, New York, NY, 2002.

46. L. H. Keel and L. Bhattacharyya, "Robust, fragile, or optimal," *IEEE Transactions on Automatic Control*, vol. 42, pp. 1098–1105, 1997.

47. I. E. Kose, F. Jabbari, and W. E. Schmitendorf, "A direct characterization of $\mathcal{L}_2$-gain controllers for LPV systems," *IEEE Transactions on Automatic Control*, vol. 43, no. 9, pp. 1302–1407, 1998.

48. N. N. Krasovskii and E. A. Lidskii, "Analysis design of controller in systems with random attributes–Part 1," *Automatic Remote Control*, vol. 22, pp. 1021–1025, 1961.

49. H. J. Kushner, *Stochastic Stability and Control*. Academic Press, New York, NY, 1967.

50. H. Liu, F. Sun, K. He, and Z. Sun, "Design of reduced-order $\mathscr{H}_\infty$ filter for Markovian jumping systems with time delay," *IEEE Transactions on Circuits and Systems II*, vol. 51, pp. 607–612, November 2004.

51. M. S. Mahmoud and P. Shi, *Methodologies for control of jump time-delay systems*. Kluwer, Boston, MA, 2003.

52. M. Mariton, *Jump Linear Systems in Automatic Control*. Marcel Dekker, New York, NY, 1990.

53. K. S. Narendra and S. S. Tripathi, "Identification and optimization of aircraft dynamics," *Journal of Aircraft*, vol. 10, no. 4, pp. 193–199, 1973.

54. G. L. Olsder and R. Suri, "Time optimal control of parts-routing in a manufacturing system with failure prone machines," in *Proceedings of 19th IEEE Conference on Decision and Control*, 1980, pp. 722–727.

55. S. Sethi and Q. Zhang, *Hierarchical Decision Making in Stochastic Manufacturing Systems*. Birkhauser, Boston, 1994.

56. P. Shi and E. K. Boukas, "$\mathscr{H}_\infty$ control for Markovian jumping linear systems with parametric uncertainty," *Journal of Optimization Theory and Applications*, vol. 95, no. 1, pp. 75–99, 1997.

57. P. Shi, E. K. Boukas, and R. K. Agarwal, "Control of Markovian jump discrete-time systems with norm bounded uncertainty and unknown delays," *IEEE Transactions on Automatic Control*, vol. 44, no. 11, pp. 2139–2144, 1999.

58. ——, "Kalman filtering for continuous-time uncertain systems with Markovian jumping parameters," *IEEE Transactions on Automatic Control*, vol. 44, no. 8, pp. 1592–1597, 1999.

59. A. Stoorvogel, *The $\mathscr{H}_\infty$ Control Problem*. Prentice-Hall, Englewood Cliffs, NJ, 1992.

60. D. D. Sworder, "Feedback control of a class of linear systems with jump parameters," *IEEE Transactions on Automatic Control*, vol. 14, no. 1, pp. 9–14, 1969.

61. Y. Wang, L. Xie, and C. E. de Souza, "Robust control of a class of uncertain systems," *Systems and Control Letters*, vol. 19, pp. 139–149, 1992.

62. Z. Wang, J. Lam, and X. Liu, "Exponential filtering for uncertain Markovian jump time-delay systems with nonlinear disturbances," *IEEE Transactions on Circuits and Systems II*, vol. 51, no. 5, pp. 262–268, May.

63. W. M. Wonham, "Random differential equations in control theory." In *Probabilistic Methods in Applied Mathematics*, A. T. Bharucha-Reid, Ed. Academic Press, New York, NY, 1971.

64. L. Xie, "Output feedback $\mathscr{H}_\infty$ control of systems with parameter uncertainty," *International Journal of Control*, vol. 63, no. 4, pp. 741–750, 1996.

65. S. Xu, T. Chen, and J. Lam, "Robust $\mathscr{H}_\infty$ filtering for uncertain Markovian jump systems with mode-dependent time delays," *IEEE Transactions on Automatic Control*, vol. 48, no. 5, pp. 900–907, May 2003.

66. K. Zhou, J. C. Doyle, and K. Glover, *Robust Optimal Problem*. Prentice-Hall, Upper Saddle River, NJ, 1996.

Index